Fifth Edition

Qualitative Research Methods in Human Geography

EDITED BY

Iain Hay

Meghan Cope

OXFORD
UNIVERSITY PRESS

OXFORD
UNIVERSITY PRESS

Oxford University Press is a department of the University of Oxford.
It furthers the University's objective of excellence in research, scholarship,
and education by publishing worldwide. Oxford is a registered trade mark of
Oxford University Press in the UK and in certain other countries.

Published in Canada by
Oxford University Press
8 Sampson Mews, Suite 204,
Don Mills, Ontario M3C 0H5 Canada

www.oupcanada.com

First and second editions published by Oxford University Press, 253 Normanby Road,
South Melbourne, Victoria, 3205 Australia. Copyright © 2000, 2005 Iain Hay.
Third Edition published by Oxford University Press Canada in 2010.

Fourth Edition published by Oxford University Press Canada in 2016.

Library and Archives Canada Cataloguing in Publication

Title: Qualitative research methods in human geography / edited by Iain Hay and Meghan Cope.
Names: Hay, Iain, editor. | Cope, Meghan, editor.

Description: Fifth edition. | Includes bibliographical references and index.
Identifiers: Canadiana (print) 20200327240 | Canadiana (ebook) 20200327259 | ISBN 9780199034215
(softcover) | ISBN 9780199034222 (ebook)

Subjects: LCSH: Human geography—Methodology—Textbooks. | LCSH: Qualitative research—
Textbooks. | LCGFT: Textbooks.

Classification: LCC GF21 .Q35 2021 | DDC 304.2072—dc23

Cover image: © Plasteed/Shutterstock
Cover design: Laurie McGregor
Interior design: Sherill Chapman

Oxford University Press is committed to our environment.
Wherever possible, our books are printed on paper which comes from
responsible sources.

Printed by Sheridan Books, Inc., United States of America

1 2 3 4 — 21 22 23 2

Contents

Boxes and Figures

Boxes

Figures

Contributors

Jamie Baxter, BA (Hons), MA, PhD, is Professor of Geography and Associate Dean of Graduate Affairs at Western University. He has also worked as an associate professor of geography at the University of Calgary. Jamie's research interests involve environmental risks from hazards, community responses to technological hazards, geography of health, noxious facility siting, and methodology. His recent projects have involved case studies of risk perception in communities living with toxic waste and urban pesticides. He served as principal investigator for a study entitled Environmental Inequity in Canada. Jamie has published in a wide array of journals, including *Transactions of the Institute of British Geographers*, *Journal of Risk Research*, *Risk Analysis*, *Environmental Planning and Management*, *Social Science and Medicine*, and *The Canadian Geographer*.

Lawrence D. Berg, BA (Dist.), MA, DPhil, began his career as a lecturer in the School of Global Studies at Massey University, Aotearoa New Zealand. Lawrence now works in Canada at the University of British Columbia where he has been a Canada Research Chair and is now Professor of Critical Geography and Leader of the Social, Spatial and Economic Justice Research Cluster at the UBC Institute for Community Engaged Research. He has a diverse range of academic interests in radical geography, focusing especially on issues relating to analyses of identity politics and place and the emplaced cultural politics of knowledge production. Lawrence has been editor of both *SITES* and *The Canadian Geographer*, and he was one of the founding editors of *ACME: An International E-Journal for Critical Geographies*. Lawrence's current research focuses on the neo-liberalization of academic knowledge production at a range of geographical scales.

Matt Bradshaw, BA (Hons), MEnvStudies, PhD, is Principal Fisheries Management Officer (Dive Fisheries), Department of Primary Industries, Parks, Water and Environment, Tasmanian Government, Australia. His current academic interests include economic geography, the social impact assessment of fisheries, and community involvement in local government planning. Matt has published in *Antipode*, *Applied Geography*, *Area*, *Australian Geographical Studies*, *Environment and Planning A*, *Fisheries Research*, and *New Zealand Geographer*, among others.

Jenny Cameron, DipTeach, BAppSc, MA, PhD, is Conjoint Associate Professor with the Discipline of Geography and Environmental Studies at the University of Newcastle, Australia. Jenny's research is in the area of diverse and community economies, and she is a member of the Board of Directors of the Community Economies Institute. She has conducted focus groups and facilitated workshops as part of action research projects in a range of settings, including working with economically marginalized groups to develop community economic projects and co-researching with firms that are manufacturing in just and sustainable ways. Her research features in her co-authored book with J.K. Gibson-Graham and Stephen Healy, *Take Back the Economy: An Ethical Guide for Transforming our Communities* (University of Minnesota Press, 2013), and in journals that include *Antipode*, *Economic Geography*, and *Local Environment*. She is also committed to communicating research outcomes to a wide audience, and has produced a documentary on asset-based community economic development, resource kits, and research reports based on qualitative data.

John Paul (JP) Catungal, BA (Hons), MA, PhD, is Assistant Professor at the Institute for Gender, Race, Sexuality and Social Justice at the University of British Columbia, Canada. An urban social geographer by training, JP's research to date has concerned how queer, racialized, and migrant community organizing—especially in education, health promotion, and social services—articulate critiques of intersectional structural violence and visions for more socially just futures. His current research on Filipino Canadian community organizing in Vancouver, British Columbia, examines the emergence and evolution of student groups, mentorship programs, and hometown associations as sites in and through which Filipino Canadians negotiate their place and belonging as racialized subjects in a settler colonial city. He was co-editor of *Filipinos in Canada: Disturbing Invisibility* (2012) as well as a 2017 special issue of the journal *TOPIA* on "Feeling queer, feeling Asian, feeling Canadian." He has also published articles and book chapters on the queer-people-of-colour politics of HIV/AIDS organizing; urban geographies of Filipino Canadian youth; the social justice implications of creative city policymaking; and the politics of teaching, learning, and knowledge production.

Meghan Cope, AB, MA, PhD, is Professor in the Department of Geography at the University of Vermont, US. Her work has always blended qualitative research with a sustained interest in the everyday, particularly for marginalized groups. Meghan's current research project, *Mapping American Childhoods*, is a historical geography of childhood in early twentieth-century United States. Meghan's work on blending qualitative research with geographic information systems (GIS) resulted in a co-edited book (with Sarah Elwood) on this topic in 2009 (*Qualitative GIS*, SAGE). These interests have expanded into digital geographies and the mix of qualitative research with GeoHumanities. Meghan has published more than 25 articles and chapters on methods, children's geographies, urban post-industrial decline, feminist geography, and participatory research and pedagogy.

Dydia DeLyser, BA, MA, PhD, is Associate Professor of Geography at California State University, Fullerton, US. Her qualitative research—largely on landscape, tourism, and social memory in the American West—has been both ethnographic and archival. Dydia teaches cultural geography, qualitative methods, research design, and writing at the graduate and undergraduate levels. She strives to make her teaching, research, writing, and community service overlap, and she regularly combines those with her personal interests in historic sites, and historic things. Dydia's work has appeared in such journals as the *Annals of the Association of American Geographers*, *Area*, *cultural geographies*, *Social & Cultural Geography*, *Progress in Human Geography*, and *Transactions of the Institute of British Geographers*; and her book, *Ramona Memories: Tourism and the Shaping of Southern California* (University of Minnesota Press, 2005). She serves as Editor-in-Chief of *cultural geographies*.

Robyn Dowling, BEc (Hons), MA, PhD, is an urban geographer who is currently Dean of the School of Architecture, Design and Planning at the University of Sydney, Australia. Her research has focused on the character of everyday lives in cities, especially suburbs, as summarized in her co-authored book *Home* (with Alison Blunt). Her more recent interests are in the changing nature of urban governance in response to technological disruption and climate change. Robyn has published widely across urban and cultural geography, including a series of progress reports (2016–18) on qualitative methods for *Progress in Human Geography*. She is a Fellow of the Academy of Social Sciences in Australia, and was Editor of *Transactions of the Institute of British Geographers* (2016–20).

Kevin Dunn, BA (Hons), PhD, is Professor in Human Geography and Urban Studies, as well as Pro Vice-Chancellor of Research, and Penrith campus Provost at Western Sydney University, Australia. Like other contributors to this volume, Kevin has a broad range of academic interests, including geographies of racism, transnationalism and migrant settlement, identity and place, media representations of place and people, and the politics of heritage and memorial landscapes. His PhD focused on opposition to mosques in Sydney. He co-authored *Introducing Human Geography* (Longman-Pearson Education Australia, 2000), which won the Australian Award for Excellence in Publishing, and *Landscapes* (Pearson, 2003) and has published more than 50 chapters and articles in various books and in journals including *Australian Geographer*, *Australian Geographical Studies*, *Environment and Planning A* and *D*, *Social & Cultural Geography*, and *Urban Studies*.

Mabel Denzin Gergan, BA, MA, PhD, is Assistant Professor of Asian Environmental Studies, Department of Asian Studies, Vanderbilt University, US. Mabel's research in South Asia combines political ecology, tribal/Indigenous studies, anti-colonial and materialist theory. So far her research has focused on the Indian Himalayan borderlands and the oppositional trends shaping the relationship between the margins and the centre, characterized on the one hand by state-led infrastructural interventions in this region and on the other, through the movement of racialized bodies from the borderland to India's urban heartland.

Iain Hay, BSc (Hons), MA, PhD, GradCertTertEd, MEdMgmt, LittD, is Matthew Flinders Distinguished Emeritus Professor of Geography in the College of Humanities, Arts and Social Sciences at Flinders University. He is former foundation Dean (Education) for the College. Iain completed his PhD at the University of Washington as a Fulbright Scholar and received a LittD from the University of Canterbury for work on geographies of domination and oppression. He is author or editor of many books including *Handbook on Wealth and the Super-Rich* (Elgar, 2016) and *How to be an Academic Superhero: Establishing and Sustaining a Career in the Social Sciences, Arts and Humanities* (Elgar 2017). He has had editorial roles with an array of journals, most recently as Editor-in-Chief of *Geographical Research*. Iain is a recipient of the Prime Minister's Award for Australian University Teacher of the Year and Fellow of the Academy of Social Sciences (UK). He is also a Principal Fellow of the Higher Education Academy. Iain has a long history of service to the Institute of Australian Geographers, including as its Secretary and President. He is currently First Vice-President of the International Geographical Union.

Jay T. Johnson, BA, MSW, PhD, is Professor in Geography and Indigenous Studies as well as Director of the Center for Indigenous Research, Science, and Technology at the University of Kansas, US. He is the co-author of *Being Together in Place: Indigenous Coexistence in a More Than Human World* (University of Minnesota Press, 2017) with Soren Larsen. His research interests concern the broad area of Indigenous peoples' cultural survival with specific regard to the areas of resource management, political activism at the local, national, and international levels, and the philosophies and politics of place, which underpin the drive for cultural survival. Much of his work is comparative in nature but focuses predominantly on Aotearoa New Zealand, the Pacific, and North America. Jay received the Association of American Geographers Enhancing Diversity Award in 2014 and an American Council of Learned Societies Collaborative Fellowship along with Soren Larsen.

Sara Kindon, BA (Hons), MA, DPhil, was recently promoted to become the first female professor of Geography at Victoria University of Wellington, Aotearoa New Zealand where she has worked since 1994. As a social geographer, she is best known for her scholarship on participatory action research, participatory geographies, and visual and creative methods; being a pioneer in the use of participatory video in geographic research with Indigenous Māori co-researchers; and *arpilleras* (Chilean tapestries) with colleague Marcela Palomino-Schalscha and Latin American women living in Wellington, Aotearoa New Zealand. Sara's research is cross-cultural, long-term and mostly community-based, but she has worked nationally and with large organizations like Te Papa Tongarewa the National Museum of New Zealand and the NZ Families Commission. Sara's work has been published in books and journals in the fields of human geography, development studies, and cross-cultural psychology, and in reports, exhibitions, and poetry. This work informs her leadership of a national tertiary network to support refugee-background students, and her interest in fostering more diverse and equitable forms of knowledge production and tertiary institutions. She has received prestigious awards for her teaching, research, and community engagement.

Hilda Kurtz, BFA, MA, PhD, is Professor and Chair of the Department of Geography at the University of Georgia, US. Hilda's research, grounded in attention to discourse, focuses on how marginalized social actors defend their spaces and norms of life and livelihood while asserting a place of belonging in a broader polity and economy. Empirically, her research has focused on how environmental justice and alternative food activists draw on a geographical imagination to strategize for social change. Hilda has published more than 30 articles and chapters on environmental justice, alternative food politics, biopolitics, racialization, and methods.

Clare Madge, BSc (Hons), PhD, is Emeritus Professor in the Department of Geography at the University of Leicester, UK. Her research interests are eclectic, ranging through post-colonial and feminist sensibilities, social media, and internet-mediated research and creative forms of world-writing, particularly poetry. Clare has published widely in international journals, within and beyond geography.

Juliana Mansvelt, BA (Hons), PhD 1994, is Associate Professor in Geography in the School of People, Environment and Planning, Massey University, Aotearoa New Zealand. A social geographer, her teaching and research interests lie in the geographies of everyday life with emphasis on the socialities, spatialities, and subjectivities surrounding ageing and consumption. She is the author of *Geographies of Consumption* (SAGE, 2005) and editor of *Green Consumerism: An A-Z Guide* (SAGE, 2011) and has contributed to a number of books and journals on topics related to learning and teaching, qualitative research, ageing, and consumption. In 2006, Juliana was awarded a New Zealand National Tertiary Teaching Award for sustained excellence and in 2013 she received the New Zealand Geographical Society Award for distinguished service to geography.

Pauline M. McGuirk, BA (Hons), Dip Ed, PhD, is Senior Professor of Human Geography at the University of Wollongong, Australia, where she is Director of the Australian Centre for Culture, Environment, Society and Space. Pauline's research focuses on urban political geography, particularly through critical studies of urban governance, its changing geographies, practices, and politics. She has published widely in leading international journals, with more than 100 refereed publications.

Pauline was co-editor of *Progress in Human Geography* (2015–19) and is on the editorial boards of *Territory, Politics, Governance, Irish Geography, Geographical Research*, and *Geographical Compass*. She was elected a Fellow of the Academy of Social Sciences in Australia in 2017, awarded the Geographical Society of New South Wales's MacDonald Holmes Medal in 2019, and elected as a Fellow of the Institute of Australian Geographers in 2009.

Phillip O'Neill, BA (Hons), Dip Ed, MA (Hons), PhD, is Professorial Research Fellow at Western Sydney University. Phillip is also Director of his university's Centre for Western Sydney. Phillip is a member of a number of international editorial boards, is a regular newspaper columnist and a frequent national media commentator and public speaker on economic change and regional development issues. Phillip is widely published in the fields of industrial, urban, and regional economic change and has a keen interest in the financing and operation of infrastructure in urban areas.

Eric Pawson, MA, DPhil, is Emeritus Professor of Geography at the School of Earth and Environment, University of Canterbury, Aotearoa New Zealand. He has developed a particular interest in community-based learning and undergraduate research, partly in response to the experience of the Christchurch earthquakes. Although now retired, Eric still teaches a graduate class, and is actively involved with the Australasian Council for Undergraduate Research. He is engaged in a range of post-earthquake projects in the city, particularly to do with the future of its red-zoned lands. Eric is the lead author of *The New Biological Economy: How New Zealanders Are Creating Value from the Land* (Auckland, 2018), and is a past President of the Ako Aotearoa Academy of national tertiary teaching award winners.

Michael Roche, MA, PhD, DLitt, is Professor of Geography in the School of People, Environment and Planning at Massey University at Palmerston North in Aotearoa New Zealand. His publications based on archival research on historical and contemporary aspects of forestry and agriculture have appeared in *Historical Geography, Journal of Historical Geography, Environment and History, International Review of Environmental History, New Zealand Geographer, South Australian Geographical Journal, Australian Journal of Forestry, Historical Atlas of New Zealand*, and "Te Ara the online encyclopedia of New Zealand." Michael is a Life Fellow of the New Zealand Geographical Society, was a member of the New Zealand Geographic Board Ngā Pou Taunaha o Aotearoa, and is a former editor of *New Zealand Geographer*. He is a holder of the Distinguished New Zealand Geographer Medal (2010).

Darius Scott, PhD, is Assistant Professor of Geography at Dartmouth College, Hanover, New Hampshire, US. Darius recently completed a postdoctoral fellowship at the Center for Health Equity Research at the University of North

Carolina School of Medicine. His research uses personal narratives to understand the impacts of longstanding inequity on communities marginalized by race and sexuality, along with how collectivity forms in the face of such inequity. Past research was supported by the Ford Foundation and involved recording the oral histories of Black long-time residents of rural Orange County, North Carolina, US, communities. Darius's postdoctoral research was supported by the National Science Foundation and uses oral history to weigh HIV's impact on Black gay men's relationships with rural communities in the American South.

Sara Smith, BA, MA, PhD, is Associate Professor of Geography at the University of North Carolina, Chapel Hill, US. She is a feminist political geographer interested in the relationship between territory, bodies, and the everyday. Sara's research seeks to understand how politics and geopolitics are constituted or disrupted through intimate acts of love, friendship, and birth. She has worked on these questions in the Ladakh region of India's Jammu and Kashmir State in relation to marriage and family planning, and is now engaged in a project on marginalized Himalayan youth. She also pursues these issues as they emerge in the national (US) and global context, through developing work on race, biopolitics, and the future.

Elaine Stratford, BA (Hons), MTertMgt, PhD, is Professor of Geography, University of Tasmania, Australia. Her research centres on questions related to how we flourish over the lifecourse in place and on the move. Elaine has published widely in leading journals such as *Annals of the American Association of Geographers*, *Transactions of the Institute of British Geographers*, *Geoforum*, *Environment and Planning A* and *B*, *Health and Place*, and *Political Geography*, as well as in numerous edited collections. Her most recent monographs are *Geographies, Mobilities, and Rhythms over the Lifecourse* (Routledge, 2014) and *Home, Nature, and the Feminine Ideal* (Rowman & Littlefield International, 2019).

Sarah Turner, BSocSci (Hons), MSocSci, PhD, is Professor in the Department of Geography, McGill University, Quebec, Canada. She published *Indonesia's Small Entrepreneurs: Trading on the Margins* (Routledge Curzon, 2003), and co-authored with Christine Bonnin and Jean Michaud *Frontier Livelihoods: Hmong in the Sino-Vietnamese Borderlands* (University of Washington Press, 2015). She also edited *Red Stamps and Gold Stars: Fieldwork Dilemmas in Upland Socialist Asia* (University of British Columbia Press, 2013). Sarah's current research interests include the livelihoods of upland ethnic minority communities in the Sino-Vietnamese borderlands, and the resistance strategies of informal workers including street vendors and motorbike taxi drivers in Hanoi, Vietnam. She strives to better understand and reduce the ethical dilemmas that come with conducting research with marginalized groups in socialist states, and extend appropriate ways to give back to the communities with whom she works. She also appreciates the rewards that come from working with and teaching about a range of innovative qualitative methods. Sarah has published a number of refereed articles and book chapters across these themes and is a co-editor of the journal *Geoforum*.

Gordon Waitt, MA, PhD, is Senior Professor, School of Geography and Sustainable Communities, University of Wollongong, Australia. Gordon currently teaches qualitative research methods. His research interest focuses on vulnerable bodies and employs qualitative methods to understand the behaviour of cyclists, gamblers, senior drivers, and low-income households. He is co-author of *Tourism and Australian Beach Cultures: Revealing Bodies* (Channel View Publications, 2012) and *Household Sustainability: Challenges and Dilemmas in Everyday Life* (Edward Elgar, 2013). Gordon has published widely in journals including *Gender, Place & Culture*, *Social & Cultural Geography*, *Australian Geographer*, *Annals of the Association of American Geographers*, and *Transactions of the Institute of British Geographers.*

Annette Watson, BA, MA, PhD, is Associate Professor in the Department of Political Science, and Director for the Master of Science in Environmental and Sustainability Studies at the College of Charleston, South Carolina, US. Annette's research focuses on the politics of natural resource management using geographic field methods and mixed-method research design, including participatory and co-productive research strategies with Indigenous communities. Her publications appear in journals including *Social & Cultural Geography*, *Environmental Management*, *Journal of Environmental Management*, and *GeoJournal*, and, in partnership with tribal/First Nations, she has facilitated numerous technical reports for the National Park Service and the Arctic Council.

Jamie Winders, BA, MA, PhD, is Professor of Geography and Director of the Autonomous Systems Policy Institute at Syracuse University, US. In addition to her new work on autonomous systems, artificial intelligence, and emerging technologies, she is best known for her research on immigration, racial politics, and social belonging. Jamie is the author of *Nashville in the New Millennium: Immigrant Settlement, Urban Transformation, and Social Belonging* (Russell Sage, 2013) and co-editor of *The Wiley-Blackwell Companion to Cultural Geography* (Wiley-Blackwell, 2013). Jamie teaches human, cultural, and urban geography, as well as courses on race, migration, qualitative methods, feminist theories, and public policy. She is Editor-in-Chief of *International Migration Review* and an associate editor at *cultural geographies*. She has published in outlets including *Annals of the Association of American Geographers*, *Transactions of the Institute of British Geographers*, and *Antipode* and has an ongoing collaboration with Barbara Ellen Smith on social reproduction.

Preface

Qualitative Research Methods in Human Geography provides a succinctly written, comprehensive, and accessible guide to thinking about and practising qualitative research methods. This exciting fifth edition has been revised to respond to reviewer and publisher suggestions as well as to reflect recent changes in the field. Maintaining its twenty-chapter length but featuring subtle organizational reform, the book continues its extensive but concise coverage of the field. All revised chapters have been carefully updated with fresh references and a look at new issues and technologies in the field that have arisen in the past five years. Several chapters have been revised significantly by a new, invigorated group of authors; we are also excited to include an entirely new chapter on Solicited Journals and Narrative Mapping. With this edition we wanted to make a concerted effort to include voices that had been missing or not adequately represented in previous volumes. Our intentional recruitment of diverse authors was wonderfully successful as all seven of our new authors in this edition are women and/or scholars of colour, and there is rich topical diversity in their work, particularly an emphasis on social justice, Indigenous issues, and matters of race/racism.

Qualitative Research Methods in Human Geography is still aimed primarily at an audience of upper-level undergraduate students, but feedback on ways in which the previous editions have been used indicates that the volume is of considerable interest and value to people commencing theses and post-graduate study, and indeed, some professional geographers with many decades of research experience have said—if only quietly!—that they have learned a great deal from the previous editions. Chapters have been written with the dual intent of providing novice researchers with clear ideas on how they might go about conducting their own qualitative research thoroughly and successfully, and offering university academics a teaching-and-learning framework around which additional materials and exercises on research methods can be developed. The text maintains its longstanding dedication to the provision of practical guidance on methods of qualitative research in geography.

Without realizing it at the time, I (Iain) started work on this book in 1992 when I was asked to teach Research Methods in Geography at Flinders University in South Australia. I developed lectures and extensive sets of notes for my classes. I also referred students to helpful texts of the time, such as Bernard (1988), Kellehear (1993), Patton (1990), Sarantakos (1993), and Sayer (1992). All of these books came from disciplines other than geography or referred to the social sciences in general. It disturbed me that despite the renewed emphasis on qualitative research and the teaching of qualitative methods in the discipline, no geographer had at that time produced an accessible text on the day-to-day practice of qualitative research.

In the meantime, during their visits to my office, publishers' representatives asked what sorts of books might be useful in my teaching. I often mentioned the need for a good text that dealt with qualitative research methods in geography. That message came back to haunt me when I was asked to write a book on qualitative methods for Oxford University Press. I agreed, but rather than writing a book offered instead to try to draw together the great expertise and energies of a group of active and exciting geographers from Australia and Aotearoa New Zealand in the shape of an edited collection.

I was particularly keen to prepare a book that might contribute in some small way to unsettling the Anglo-American dominance of scholarly publishing. At the time it was difficult to find good scholarly texts in Australia and Aotearoa New Zealand on almost any aspect of geography that did not take for granted a northern perspective. Despite their fine geographical work Antipodean scholars with their Antipodean examples appeared, for the most part, to be largely absent from international book markets. Moreover, Australian and Aotearoa New Zealand students were often expected to make the "translations" demanded of books acclaiming places like London, Los Angeles, New York, and Newcastle upon Tyne as paradigmatic examples of "global" processes. I wanted to produce a book that took up these challenges and spoke back to the "centre."

To some extent, my own ambition for the first edition matched that of the editors of the Meridian Series, a collaboration of the Institute of Australian Geographers and Oxford University Press, who had encouraged me to prepare the book. *Meridian: Australian Geographical Perspectives* was intended to build on the maturity of geographical research in Australia. And, as a New Zealander by birth and upbringing, I was keen to ensure the inclusion of high quality scholarship from that eastern side of the Tasman Sea too. So it was that the first edition of *Qualitative Research Methods in Human Geography* emphasized Australian and Aotearoa New Zealand contributions. Given the changing contributor profile and publishing location of this fifth edition, which alter the book's hemispherical balance and focus significantly, there is considerable irony and success marked by this trajectory.

While a number of other very useful volumes on research methods in geography have emerged over the past two decades (e.g., Clifford, Cope, Gillespie, & French, 2016; DeLyser et al., 2009; Gomez & Jones, 2010; Limb & Dwyer, 2001; Moss, 2002), *Qualitative Research Methods in Human Geography* has been received very well, not only in Australia and Aotearoa New Zealand but also much farther afield. Indeed, it was the book's growing popularity in Canada in particular that led to the third edition's shift from Oxford University Press, Australia, to Oxford University Press, Canada. Canadian students, as it turned out, welcomed a book that originated from neither Cambridge, UK, nor Cambridge, Massachusetts. With the book's move to Ontario came a request from the publishers for more Canadian examples and perhaps some Canadian contributors to join those from Australia and Aotearoa New Zealand. Excellent new authors from Canada and farther afield for the third and fourth editions helped satisfy these requirements and also began to give the book a different international complexion. At the same time the book found growing traction with students and professors on the southern side of North America's 49th parallel and for this edition there has emerged a need to reshape aspects of the book to better serve the needs of this extended audience. The outcome is a fifth edition that, through editors, authors, contents, and readership, maintains its antipodean genealogy while gradually becoming integrated with the academic axis to which it was originally intended to speak. This, I like to think, is symptomatic of success rather than sell-out.

To help ease the transformation of the book into slightly more cosmopolitan form and to introduce a bright new perspective on recent developments in the field I asked Meghan Cope to join me as editor for the fifth edition. Meghan is long-known to me as an active and highly credible scholar of qualitative methods; as a chapter contributor to an earlier edition of the book; and through her significant leadership in qualitative research methods in the United States. Meghan has been able to draw on her extensive connections and abilities very successfully to help extend the scope and scale of this new edition of *Qualitative Research Methods in Human Geography*

This fifth edition is revised to respond to constructive comments made about its predecessors. For instance, the well-received emphases on reflexivity, diversity, and ethical practice across the book have been deepened. Several chapters have been modified substantially in recognition of the ways in which electronic technologies are affecting the conduct of qualitative research. And to support the book's learning-and-teaching roles each chapter offers one or two in-class exercises to accompany the earlier edition's review questions and suggestions for supplementary reading.

In overview, *Qualitative Research Methods in Human Geography* is subdivided loosely into three main sections: introducing qualitative research in human geography, the scope and practice of qualitative research, and making sense of your data. Some might regard it ill-considered to impose such separations. We acknowledge, for instance, the ways in which writing—which might be seen superficially as a part of communication—is embedded in the process of practising qualitative research in human geography. Writing is actually part of "making sense." Similarly, practising qualitative research typically implies ongoing interpretation and reinterpretation. The ordering of the book's contents is the product of the medium in which the contents are communicated (a book of linear structure). It is a device to help make more quickly comprehensible a book of 20 chapters. And it follows the organizing sequence used in many research methods classes—such as the one that originally gave rise to this book—for which this might serve as a textbook.

The first section of the book deals with introducing qualitative research methods in human geography. In Chapter 1, we (Meghan and Iain) situate qualitative research methods within the context of geographic inquiry, and against a backdrop of broader trends regarding the growth of artistic methods, a "rhizomatic" approach to the production of knowledge, and the shifting place of science in global political discourse. We outline the range of qualitative techniques commonly used in human geography and explore the relationship between those methods and the recent history of geographic thought. On this foundation, in Chapter 2 John Paul Catungal and Robyn Dowling build a review of some critical issues associated with qualitative research. These issues include power relationships between researchers and their co-researchers, questions of subjectivity and intersubjectivity, and points of ethical regulation and concern. Similar issues are taken up in further detail by Mabel Gergan and Sara Smith who present in Chapter 3 an illuminating examination of cross-cultural qualitative research. Chapter 4 by Jay T. Johnson and Clare Madge is a tremendous contribution, exploring empowering methodologies through an examination of feminist and Indigenous research approaches. In Chapter 5, Hilda E. Kurtz discusses compelling ways of introducing or proposing qualitative research projects to others. Chapter 6 by Elaine Stratford and Matt Bradshaw examines the difficult but vitally important matters of design and rigour in qualitative research.

The second section of the book focuses on the scope and practices of qualitative research methods in human geography. Each of the chapters offers concise yet comprehensive, similarly structured, outlines of good practice in some of the main forms of qualitative research practice employed in geography. The section begins with a chapter by Jamie Baxter exploring research design issues associated with case studies. Jamie sets out the value of case study methodology for understanding specific situations as well as for developing theory. In Chapter 8 Annette Watson discusses participant observation, drawing on her rich experience doing ethnography through the National Park Service (US) in collaboration with eight Alaska Native communities. In Chapter 9 Kevin Dunn draws on his extensive experience of interviewing in cross-cultural settings to provide a complete and valuable outline of interviewing practice. Acknowledging some of the effects of the internet on qualitative research practice, this rich chapter includes substantial material on electronic means of interviewing. Darius Scott follows this up with a chapter setting out the basic scope of oral history, discussing the ways in which its practice differs from interviewing and describing how oral history can enhance our understandings of space, place, region, landscape, and environment. In Chapter 11, Jenny Cameron considers the research potential of focus groups in geography, outlines key issues to take into account when planning and conducting successful focus groups, and provides an overview of strategies for analyzing and presenting the results. This chapter includes material on online focus groups.

Continuing its consideration of the scope and practices of qualitative research methods, the book then turns to two chapters on texts. These deal, respectively, with archival research and the use of questionnaires in qualitative human geography. Historical geographer Michael Roche provides a helpful and substantially revised chapter on the use of archival resources, a sometimes neglected yet enormously productive and rewarding activity for human geographers, and an area being transformed by digital technologies. In Chapter 13, Pauline M. McGuirk and Phillip O'Neill explore the ways in which questionnaire surveys—including those offered online—can be used in qualitative research in human geography. It may be helpful to read this chapter in conjunction with the chapter on research design and rigour by Stratford and Bradshaw.

In a wholly new chapter, Sarah Turner dives into exploring solicited journals and narrative mapping, both of which have been increasingly—and creatively—activated in human geography. This chapter is a particularly nice combination of a set of data-gathering (or data-generating) techniques with analytical and presentation strategies. Jamie Winders then guides us through the rapidly-changing terrain of emerging digital geographies in Chapter 15. Recognizing that the specific digital programs (apps, platforms, and media) change very quickly, Jamie demonstrates that fundamental issues of ethical, dependable, and transformative qualitative research practice remain at the core of all digital geographies. Then, taking up some of the issues about participant status for researchers that Annette Watson introduced in Chapter 8, as well as matters of empowering research raised by Jay T. Johnson and Clare Madge, Sara Kindon offers an excellent chapter on participatory action research. There is also a strong complementarity between Sara's chapter and that by Mabel Gergan and Sara Smith: it is helpful to consider them in conjunction with each other.

Four chapters in a section on making sense of your data follow. In the first of these, Gordon Waitt helps uncover the taken-for-granted, through a discussion of Foucauldian discourse analysis. In the second chapter in this section Meghan Cope examines the organization, coding, and analysis of qualitative data, identifying key strategies for practising trustworthy analysis in qualitative geographic research. The book then turns from explicit discussions of interpretation to matters of communication as a means of making sense. Juliana Mansvelt and Lawrence D. Berg provide the penultimate chapter of the book, discussing some of the practical and conceptual issues surrounding representation of qualitative research findings. They make the important point that writing not only reflects our findings but constitutes how and what we know about that work. And finally, in their chapter "Small Stories, Big Impact," Dydia DeLyser and Eric Pawson offer assistance to qualitative researchers beginning with the vital task of communicating results for public consumption.

This book has chapters of varying lengths, densities, and complexity. While some of the chapters are challenging to read, we trust that the challenges lie in comprehending the ideas presented, not in wading through obfuscatory text. In every case, authors have striven to present material that is as clear and well-illustrated as possible. As editors with a long-standing interest in effective communication, we have devoted considerable attention to the matter of clarity.

While the chapters of this volume offer a comprehensive overview of qualitative research methods in human geography, the book is not intended as a one-stop resource or prescriptive outline for qualitative researchers and students. Instead, it is meant as a starting point and framework. As noted earlier, each chapter directs readers to a number of useful resources that may be consulted to follow up the material introduced here. There is, of course, a consolidated list of all references cited for readers who might wish to inquire even further. Chapters also include review questions and review exercises intended to support individual reflection as well as classroom engagement. These questions might also serve as prompts for ideas for quite different exercises.

As with earlier editions this one continues to feature an extensive glossary, which has been thoroughly updated. While authors have worked to ensure that chapters are written in language accessible to undergraduate readers, there are—without doubt—terms that may be unfamiliar to many readers. The glossary should help resolve that sort of difficulty. Many of the terms in the glossary are drawn from the lists of key words associated with each chapter.

Finally, and importantly, we wish to acknowledge the contributors without whom this volume would not have been possible. We admire their tolerance, persistence, and fortitude and we thank each of them for their efforts, their patience, and their continuing dedication to this collection. And we hope that all of our efforts for this edition continue to support the undertaking of thoughtful, sensitive, and high quality qualitative research in human geography, wherever it is practised.

Iain Hay
Adelaide, South Australia

Meghan Cope
Burlington, Vermont, United States

Acknowledgements

My great thanks go to Meghan Cope for accepting the challenge of co-editing this fifth edition and for the tremendous skill, care, and intellect she has brought to the project. It is so much better for these wonderful contributions. I would also like to express my appreciation to the many scholars who, as authors and reviewers, have contributed to the enduring success of this text's earlier editions as well as to the newest generation who have so generously reshaped this edition. Thanks to Alaric Maude for sparking the idea for this book way back in the 1990s and to Amy Hick for outstanding editing.

And a vast thank you to Tania for her enduring love and support.

Iain Hay

I want to thank Iain Hay for inviting me to join him as co-editor on this fifth edition of what has been an essential and enriching text for nearly two decades. Many thanks also to two students who helped me review the first drafts of chapters in the summer of 2019—Sofie Schmidt (University of Vermont) and Gemayel Gaxcon (California State–Fullerton). I also thank all the authors—I have learned so much from them and I am continually in awe of geographers' innovative, energizing, and compassionate research practices.

Finally, thank you to my family. Alan, Geneva, and Celia cheered me on and kept my tea mug full.

Meghan Cope

PART

Introducing Qualitative Research

Where Are We Now? Qualitative Research in Human Geography

Meghan Cope and Iain Hay

Chapter Overview

In this chapter we start by asking what is qualitative research and, after identifying common characteristics, goals, and approaches, we ask further questions about its use, why it is trustworthy, and illustrate our answers with examples. We then frame contemporary qualitative work within critical perspectives and the context of present global discourses regarding knowledge and science. We finish the chapter with illustrations of creative new methods.

Introduction

Qualitative research is deeply familiar to all of us, even if we have never formally been taught its practices, because we use its core strategies throughout our lives. That is, as social beings, we learn about culture and environments, interpret others' actions and statements, figure out our place, try to make sense of the world, and tell stories to each other that build on previous experience. The distinction between our everyday social interactions and doing research is partly a matter of intentionality and purpose, but both are still very much rooted in our sociality. As researchers, we formulate questions driven by curiosity and concern; we design a plan of inquiry that is systematic and transparent in order to be trustworthy and convincing; we critically ponder our own biases and assumptions; and we attempt to communicate our findings in compelling ways to our audiences. But, although we need to embrace our social selves, we also need the discipline and structure of a research plan in order to do qualitative research, including thinking critically about what we want to know, how we engage with people, common pitfalls, and best practices of qualitative inquiry, analysis, and representation. This book attempts to help you build on your existing social knowledge to channel and refine your research in stimulating, productive, and creative ways.

So, what exactly is qualitative research? We can begin to approach this by looking at three elements of all research: the types and sources of *data*, the kinds of *questions* researchers ask, and, at a deeper level, the ways that *knowledge* is produced.

Looking at these in turn, qualitative research engages primarily with *non-numerical* data, which includes a wide range of forms such as texts, photos and video, oral recordings, drawings and sketches, maps, observations of human behaviour, historical documents, material artifacts, and social media posts (Hay, 2020). Further, data sources are interpreted in many ways, including such techniques as narrative analysis, discourse analysis, and coding practices (see Chapters 14, 17, and 18 for discussion). Interpretation allows researchers to make the bridge between real-world, empirical data and deeper insights based on relations, connections, processes, and theories.

Second, in terms of the types of *questions* being asked, qualitative research in human geography is typically concerned with understanding social–spatial processes and people's everyday lives in past or present contexts. A more geographical way to frame this is to consider how places are produced through social action, but social action is also constrained and shaped by place. These can be represented by two dimensions of inquiry proposed decades ago by Sayer (1992), but still very relevant:

- What are the *structures* of society and how are they created, maintained, legitimized, and/or resisted?
- What are individuals' *experiences* of places and events?

Lastly, in considering **knowledge production**, qualitative research in human geography rests on **epistemologies** (ways of knowing) that value and seek to understand human experiences, emotions, relationships, differences, injustices, cultural practices, and the production of place. Many epistemologies in contemporary human geography are rooted in critical perspectives, such as feminism and critical race theory, that directly confront oppression and inequality. In more abstract terms, qualitative research helps shed light on the human condition, ways of being in the world, philosophical concerns about meaning and "truth," and designing better futures.

To guide the flow of our discussion, we have organized this chapter along the lines of several common questions that are raised about qualitative research, such as its use, strengths and limitations, and role in the broader project of understanding our world.

What *Use* Is Qualitative Research in Human Geography?

Qualitative research is useful for the overarching academic goal of knowledge production in several dimensions: providing contextual and personal *explanations* for trends identified in quantitative studies; generating new insights into people's *experiences*, lives, emotions, and communities; and mobilizing those insights to *explore* and build new theories of the human condition, the production of meaning, and human–environment relations. Here we first examine the value of qualitative approaches as compared to, or complementary to, quantitative research. We then demonstrate the ways that qualitative research does not merely serve as a counterpoint to quantitative work but is, in fact, a robust and reliable means of knowledge production in its own right.

Numerically based, statistically analyzed *quantitative* data have enormous relevance in all areas of geography, from climate change to pandemics, and from watersheds to digitally connected smart cities. Embracing qualitative research does not dispute the power of insights generated by quantitative operations—indeed, much of the trend toward **mixed methods** recognizes the complementary benefits of interlocking approaches. However, while quantitative data is very good at showing the extent of a phenomenon (how common it is, how widespread, and where it is), and demonstrating important *correlations* between multiple forces, there are always gaps, outliers, and spaces for potential understanding that are less well-served by quantitative work. In this sense, qualitative research plays an important role as a complementary set of inquiries in a given project, providing process-based insights, understandings of the lived experience of different conditions, and generating possible causal explanations for phenomena. (See Sayer [2010] for an extended discussion.)

Consider this statement from an article in the field of health geographies, which is a research area with long-standing practices of both quantitative and qualitative inquiry. The authors had noted quantitative trends of high levels of smoking among bisexuals in California and wanted to understand more:

> *In-depth understanding* of how and why high rates of smoking persist among bisexual people is needed to develop interventions to reduce associated health disparities . . . This article works toward *explaining* why bisexuals' smoking rates are even higher than their gay and lesbian counterparts' by *exploring the everyday smoking contexts and practices* of bisexual young adults. (McQuoid et al., 2019, p. 2, emphasis added)

This short excerpt exemplifies some of the motivations for researchers to engage in qualitative approaches: to explain a phenomenon, understand people's experiences, gain a sense of context, improve people's lives, and address injustices.

But beyond merely filling the gaps left by powerful numerical studies, qualitative research also allows—even compels—us to ask fundamentally different types of questions from the start. Qualitative research projects typically lie along a spectrum between more concrete work on one end and more abstract work on the other. The empirical/concrete end of the spectrum informs and substantiates the abstract/theoretical end, and the reverse is also true—the different dimensions are woven together through critical analysis and reflection. An excellent example of this is shown in the questions McQuoid and her co-authors (2019, p. 2, emphasis added) used to frame their research on bisexuals' smoking:

1. *Where* are participants' smoking and craving episodes concentrated in everyday life?
2. What *contextual* factors and place-based practices drive **spatial** and temporal patterns of smoking?
3. *How*, if at all, does bisexual identity *interplay* with these contextual factors and place-based practices of smoking?

Note that McQuoid et al.'s first question is a concrete one: where do people smoke? Importantly, the research team did not just take a GPS point to answer this (though spatial data were included), but asked about surroundings, with some participants stating that they smoked in their parked cars or avoided certain spaces. The second question blends concrete and abstract, looking at context such as social pressures, work schedules, and accessibility of public smoking spaces. The final question is more abstract, working toward a theory of identity and place.

Indeed, it is not uncommon to begin a project by noticing an interesting pattern on a map or in trend data, digging into its constitutive processes, engaging with local communities who are actively producing their lives *in place* while they are affected by broader processes of political and economic power, and then arriving at deep questions of the meaning of a place's humanity. Further, the concrete–abstract continuum does not imply a lock-step process toward an end-point, but rather helps us articulate various facets of a project that all contribute to each other. It may be, for example, that one day a new philosophical revelation (abstract side of the continuum) occurs to you and you decide to investigate more empirical, descriptive data (concrete side) to see if it holds up to further scrutiny. In that sense, all aspects of research are iterative—they build on each other, inform new lines of inquiry, spark subsequent rounds of data-gathering, and shape our analytical approaches in an ongoing way.

In many instances, questions about the relationships between people and place do not lend themselves well to large statistical studies and yet they have profound implications for community well-being, local policy-making, and for better scholarly understanding of different social groups. The methods needed to generate data to answer these questions require centring the human experience, concerted trust-building work with local participants, lots of patient listening, creative engagement with diverse groups, and sensitive **reflexivity** on the part of researchers to recognize and work from their own **positionalities** (in terms of gender identity and sexuality, "expert" status, class privilege, and risks of cultural exploitation).

For these reasons, qualitative research is often called **intensive**, in comparison to the **extensive** strategies of quantitative work. Sayer has been influential in this vein, helping many of us articulate what we are searching for in our research, when he writes of data that are "qualitative and concern processes, activities, relations, and episodes of events rather than statistics on particular characteristics" (1992, p. 242). He goes on to argue:

> By looking at the actual relations entered into by identifiable agents, the interdependencies between activities and between characteristics can be revealed; for example, how waged work and domestic work commitments are integrated in time and space. The results are more vivid because they describe individuals and their activities concretely rather than in the bloodless categories of statistical indicators such as "socio-economic group." (Sayer, 1992, p. 242)

Thus, qualitative research holds the power to provide explanations, bring insights into people's experiences, and foster greater exploration in order to generate better understanding of places, geography, and relations between nature and society.

What Are Some Examples of Qualitative Methods? What's New in the Field?

There is no master list of all the qualitative methods one might use (though the chapters in this volume cover many of the most common ones), in part because researchers are highly adaptive and creative in tailoring techniques to the goals of the project, the participant population, and the limits of time and funds. It is also important to recognize that the boundary between data collection and analysis is often blurry in qualitative research—for example, if you find a set of nineteenth-century diaries in the archives of your local historical society and start transcribing the entries into categories in a database, is that data collection or analysis, or a bit of both woven together? We would argue that it is both. Further, in qualitative research the operations of data collection are often more accurately called data *creation* because the researcher is actively involved in generating the data, whether by conducting interviews with community elders or asking children to draw their routes from home to school or scraping geo-located tweets in a social media project. While in some senses having a flexible, seemingly boundless set of methods to choose from may be disorienting, in another sense it allows researchers a great deal of discretion and creativity (and responsibility!) in designing their projects.

That said, however, we can identify broad categories of methods that help us consider the scope of options. One category is **oral methods**, which includes any type of talking with and listening to people: interviews, focus groups, oral histories, audio-recorded stories, participant observation, and many more. A second category is **text**-based methods, which includes investigating archives and other historical documents, solicited journals, written stories, textual answers to open-ended survey questions, travel diaries, etc. We could perhaps add a third category of observational/participatory methods that are rooted in long traditions of **ethnography** and have developed many offshoots.

Of course, much of qualitative research in human geography engages with methods that do not fall neatly into these groupings, such as image interpretation, landscape analysis, participatory mapping, and **PhotoVoice**. Even beyond these, a recent and exciting expansion of creative engagements in geography has emerged through artistic, performative, poetic, and visual means, often collectively referred to as **GeoHumanities**. At a moment in disciplinary history when originality, experimentation, and pushing the boundaries of tradition are highly valued, we are seeing the rise of diverse and ground-breaking approaches to making meaning in geographic terms. Indeed, one of the leading figures identifying the "creative (re)turn" in geography, Harriet Hawkins, has noted that "even a cursory cataloguing brings to light geographical practices of cabaret, drawing, film-making, photography, theatre, dance, art, poetry and exhibition curation, to list but a few" (Hawkins, 2019, p. 963). Significantly, the move toward online academic journals has opened up new possibilities for sharing creative works that are either impossible or poorly represented on the printed page; Hawkins notes several geography publications (including *ACME, cultural geographies, Gender, Place & Culture, and GeoHumanities*) that have allocated digital space and developed publication policies for multi-media projects, animations, video essays, and other documentation of creative practices. Importantly, as de Leeuw and Hawkins (2017) point out, creative geographic practice flows through numerous channels: in some cases the creative product serves as the *object of study* (such as when a researcher examines artwork for themes of place attachment); in other cases the creative process itself is a *means of inquiry* (such as a PhotoVoice project that explores Indigenous youths' experiences of mobility (Goodman, Snyder, and Wilson, 2018), and in some instances the creative practice serves as the method of *making geographical meaning* (such as the **geopoetics** work of Clare Madge, see below). These developments are explored further, with examples from some of the contributors to this book, at the end of this chapter.

The creative turn is contemporaneous and often intersects with the expansion of **digital geographies** (see also Chapter 15). This has been aided by the growth in user-friendly tools for data collection, analysis, and representation. As new digital tools and capacities have been developed, the potential for examining "geographies produced *through*, produced *by*, and *of* the digital" (Ash, Kitchin, and Leszczynski, 2018, p. 27, emphasis added) have been equally captivating to qualitative researchers as to quantitative modelers. Qualitative data collection techniques have been rapidly digitized, including digital voice recorders and cameras for interviews and focus groups, scanners for digitizing hand-drawn materials, online survey tools for open-ended responses, smartphone apps with GPS tracking for participant journals (see Chapter 14), and optical character recognition

(OCR) for digitizing the riches of historical archives and making them searchable (Chapter 12). This digitization has been incremental, unevenly adopted, and not without a certain amount of handwringing by practitioners who worry about losing something along the way—perhaps a sense of intimacy with one's data?

However, this is not just the digitization of traditional methods of inquiry; in fact, we are seeing the emergence of qualitative data that are born digital, indicative of geographies produced *by* the digital. These include exploring the content and geo-location of tweets (Jung, 2015; Poorthuis et al., 2016), the gendered nature of *OpenStreetMap* entries (Stephens, 2013) and of "new spatial media" (Leszczynski & Elwood, 2015), as well as the entrance of **volunteered geographic information** (VGI) into analyses of the digital minutiae of mundane daily life (Sui, Elwood, & Goodchild, 2012). Although many of these practices involve the quantification of data, there are also many qualitative operations here, such as evaluating gender bias, investigating community perceptions of "risk," and exploring the diverse experiences, emotions, and motivations of social actors. Similarly, methods of interpretation are diverse and constantly evolving. In addition to changing the tools of data collection, the digital turn has also had significant impact on analysis, such as in **computer-aided qualitative data analysis software** (CAQDAS) (see Watson & Till, 2010; and Chapter 18).

The intersecting shifts in the discipline of geography toward artistic and digital engagements are both the result of and provide support for a continued exploration of critical perspectives focused on diverse social groups, processes of marginalization and colonialism, and **ontologies** of difference. That is, these shifts call us to move beyond thinking of social groups as mere categories (for example, race and gender) toward a deeper realm of thinking about the very meanings and intersections of identity, place, and knowledge. Indeed, this volume reflects broader moves in the discipline toward **decolonizing research**, defined by Catungal and Dowling in Chapter 2 as "research whose goals, methodology, and use of research findings contest imperialism and other oppression of peoples, groups, and classes by challenging the cross-cultural discourses, **asymmetrical power relationships**, and institutions on which they are based," a charge taken up in virtually every chapter of this new edition. Decolonizing research involves not just the adoption of particular methods, but of epistemologies and ontologies based on the dismantling of oppression globally. For example, LaToya Eaves (2017, p. 84) points out:

> Black geographies is an intervention into the discipline by presenting knowledge of racialized spaces, bodies, and landscapes, undergirded by and perpetuated through colonial legacies, pushing the boundaries of critical geographies research. Scholars of Black geographies rely on the corporeal, the aesthetic, the creative, the spiritual, and the elemental (earth, air, water, and fire) as texts with which to read into the meaning of Blackness, its accompanying implications of oppression(s), and its futuristic possibilities.

Thus, the expansion of research methods into more creative practices, as well as the exploratory work in the digital realm, demonstrate a vibrant experimentation, diligent groundedness in experiences, and sometimes playful and sometimes deadly serious approaches to new ways to make meaning in geography.

Some authors (see Johnson & Madge, Chapter 4, also Wilson, 2017) have identified these new approaches as **rhizomatic**, keying off the term used by Deleuze and Guattari (1987). This builds on the qualities of rhizomes (such as potatoes and grass), which develop new versions of themselves by sending out shoots laterally, vertically, and sometimes in unexpected places (such as the compost bin or cracks in the pavement). These are framed in contrast to the more predictable, vertically/hierarchically oriented, and stable qualities of something like a tree, which are referred to as arborescent. As Adkins (2015) points out, trees (and tree-like networks) frequently underlie our notions of things and related metaphors—they have roots that interact with conditions underground/out of sight, they have sturdy trunks that bend without breaking, they grow in predictable ways outward with new rings looking much like the previous ones, and they branch out to the sky, reaching for light and air, much as our theories and practices of social research do. But rhizomes are opportunistic, surprising, self-replicating but hybridized, and even born out of their own detritus as volunteers sprouting out-of-place. Rhizomes are connected underground in a complex system of dependence and community yet individuals can survive and start new colonies, adapting to—but also themselves changing—the local conditions. Taking these metaphors to heart, we can see the potential for envisioning creative practices, experimental methods, and playful new ways of making meaning that simultaneously draw from past existence and begin anew, often in unexpected ways.

Why Should We Trust Qualitative Research? Or, Isn't It Just Anecdotal Storytelling?

In order to be **trustworthy** all types of research should be constantly concerned with **ethics**, establishing and monitoring rigorous practices, acknowledging biases, and maintaining transparency (see Chapter 2). Most quantitative studies use statistical methods to ensure **rigour**, such as drawing from **representative samples** to make **generalizations** about a population, and/or using **replicability** to demonstrate a lack of bias (i.e., if a result can be reproduced by a different researcher, it can be understood to be more robust). However, these same mechanisms to ensure trustworthy research do not apply in most qualitative work due to the very personal, context-specific, and ephemeral nature of qualitative social data, as noted above by Sayer (1992).

Qualitative geographies have their own standards and guidelines for practice and representation. The chapters in this volume explore many techniques for ensuring that one's research is rigorous and trustworthy, and we highlight just three dimensions here, using a hypothetical study of low-income women's difficulties in finding and keeping full-time employment in diverse locations.

First, in the processes of data collection, adopting a process of **triangulation** using diverse information sources and methods can result in better **confirmability** and **dependability** of one's data (Baxter & Eyles, 1997). So, researchers could gather multiple, varied accounts through interviews with rural and urban women, conduct a PhotoVoice project with a sub-sample of those women, and examine large-scale quantitative labour statistics; these all offer different types of data but they can confirm each other's insights about barriers to women's employment.

Next, considering one goal of the production of knowledge is to contribute to broad understandings of the world, qualitative researchers might use the property of **transferability** to identify particular trends, practices, or social phenomena in one context that likely happen in other, similar contexts. For example, the project introduced here may find that low-income women experience difficulties in finding full-time employment due in part to transportation and child-care issues. Although the specific barriers for women in rural Saskatchewan might differ from those that women face in New York City or Sydney, there are structural commonalities (such as uneven gender divisions of labour at home and in the labour market) that can be transferred across settings, perhaps sparking new questions and subsequent data gathering. Further, making theoretical connections between uneven gender divisions of labour at home and work—the ways they are reinforced through cultural expectations, discrimination, and place-based support networks—might allow the researcher to use **abstraction**. Abstraction is a way of enabling us to make broader statements about how the world works based on connections between processes and contexts. So, in this case, we begin to be able to identify which collation of cultural assumptions, social supports, and (lack of) economic opportunities combine to generate varied experiences for low-income women seeking employment.

Finally, trustworthy qualitative research depends greatly on the **transparency** of the research process (show your work!) and the reflexivity of the researcher(s), as explored in multiple chapters of this volume. For this, rather than attempting the impossible task of eliminating bias, we need to recognize that bias is *always* present in the social world of research because researchers are themselves social beings with distinct positionalities (gender, race, age, dis/ability, sexuality, and other dimensions of identity) and worldviews. Qualitative researchers take the position that it is better to acknowledge and confront those perspectives through critical self-reflection than to ignore them or try to sweep them under the carpet. This practice is part of what Donna Haraway (1988) called **situated knowledge**—the recognition that what we know depends on our social, historical, and cultural locations.

Being in the World/Researching the World—New Dilemmas of Critical Inquiry

The most recent revival of qualitative research in geography, occurring over the past 25 years, has intersected with and supported a critical turn in the discipline in which post-modern, post-colonial, feminist, queer, and anti-racist epistemologies have gained significant traction. However, the rallying cries of these critical perspectives to scrutinize the production of mainstream knowledge, the dominance of a "scientific" objectivity, and the power of elites has become more complicated amidst recent global political shifts toward populist movements. One of the goals of critical geography has been to recognize multiple truths, such as when the children of a society have a different experience of public space from adults—both experiences are "true" but we often have only been made aware of only one because adults are assumed to be the "standard" users of public space (Valentine, 1996). In this case, arguing for the recognition of multiple truths is a political and scholarly act to challenge existing power relations on behalf of an oppressed group. However, the rise of populist, nationalist, and anti-immigrant political movements globally puts the notion of multiple truths on precarious, even dangerous, footing. It is important to distinguish carefully between advocating for the inclusion of oppressed social groups' experiences and world views in research for the purpose of greater social justice (as championed by critical geographies), and the populist/nativist claims of "alternative facts" or the framing of white people as being categorically oppressed or at risk of demographic replacement by the "invasion" of immigrants and people of colour, discourses that are for the purpose of exclusion, not justice.

Similarly, and relatedly, critical scholars have long scrutinized the ways that "science" has been used to oppress and violate marginalized groups, including women, people of colour, Indigenous communities, and incarcerated people (Haraway, 1988). In a very different critique of science, populist/nativist movements have capitalized on a growing anti-intellectualism in which long-standing practices of rigorous scientific inquiry have been challenged and disbelieved, most notably through climate change denial, but also in online anti-vaxxer sites and other conspiracy-driven narratives. How, then, can critical scholars simultaneously advocate for unheard voices and critique science as a knowledge-producing practice without fanning the flames of white supremacists and climate science–deniers? It is a tricky balance that I (Meghan) have had to confront as a scholar and as an instructor as I implore my students to be critical consumers of information and producers of knowledge, but also to trust rigorous findings.

In short, we need to carve a path forward by focusing on two important dimensions of ethical research: staying critical of grand scientific narratives but also staying open to *being convinced by reliable evidence generated by trustworthy research* (see, for an illuminating discussion, Stephens, Poon, and Tang, 2021). Critical scrutiny of data and how they were produced, anonymous peer review, researcher reflexivity, and transparency are more essential now than ever to fend off claims driven by ideologues. Ethical, rigorous, and critical qualitative research in geography plays an important role in this endeavour.

Exploring the Edges: Trends to Watch in Qualitative Human Geography

Some of the authors in this volume have been exploring the edges of qualitative research methods. For example, Clare Madge provides an explanation of geopoetics (see Box 1.1); Sara Kindon has made excursions into textile methods (Box 1.2); and I (Meghan) have been experimenting with story maps based on archival records from an early twentieth-century orphanage in Vermont (Box 1.3). These are included here as a demonstration of possibility, fresh takes, and a quickly changing landscape.

Lastly, in exploring the emerging trends to watch in this field, we return to those critical perspectives identified above that have had significant impact on the practice of qualitative research methods and the production of knowledge in human geography. Indeed, engaging with these shifts more deeply has been one

Clare Madge's Geopoetics

There are three main ways in which poetry has been used as a method by geographers: condensing research data to produce a research poem; the researcher producing an interpretive poem to express or interpret findings; or the participants producing a participant poem. For a research poem interviews are conducted and transcripts are condensed and rearranged by the researcher to produce a poem that crystalizes research data but no new substantive text is added. Research poems are thus created from nuggets of research prose and are based solely on primary data. Interpretive poems are more free-standing and here the researcher constructs poems to represent/interpret/express their research insights. They may use an amalgam of fieldwork data to do this, including interviews, photos, and secondary data. Participant poems are less common but here poems are elicited from participants as part of a research project. This usually requires support of a professional poet. This foregrounds participants' voices. For more on poetic methods see Magrane (2015) and Eshun and Madge (forthcoming).

BOX 1.2 Sara Kindon's Textile Methods

Geographers working with qualitative approaches have recently begun to interrogate creative practices (Carr & Gibson, 2017; Patchett, 2014, 2016; Price & Hawkins, 2017). In the UK, for example, some geographers have become interested in enduring practices such as crafting, knitting, and quilting and their positive impacts for health and well-being in an era of austerity (for knitting see Riley et al., 2013; quilting, Burt & Atkinson, 2012).

Elsewhere geographers have begun to experiment with creative practices as qualitative research methods in a range of contexts and projects. Some examples here include the use of embroidery and creative arts to understand how students engage and develop conceptual thinking in geography classrooms (Hofverberg & Kronid, 2018; Scoffham, 2016). As well, textile methods—specifically the making of Chilean *arpillera* tapestries—are being used to explore women's experiences of violence, migration, and resettlement. In this long-term project, *arpilleras* have been able to provide insights into women's intimate worlds and political agency when often their words fail them (Palomino-Schalscha et al., 2017).

Such experiments are logical given the widespread shift to co-production and participatory approaches across the social sciences. They also reflect the influence of relational and **more-than-human** approaches, which have called for methods that better attend to, and engage with, complex systems and interactions. With increasing interest in the materialities, spatialities, temporalities, embodiments, and affects of creative practices, it is likely that more new methods involving crafts and textiles will evolve to help geographers continue to make more sense of knowledge generated through experience and doing.

of the goals of this new edition. Geographers continue to expand the discipline by approaching research through feminism (see Moss [2002] and Chapters 4, 14, and 16), critical race theory and Black geographies (see McKittrick & Woods [2007] and Chapters 3 and 10), Indigenous and decolonial perspectives (see Johnson & Larsen [2013], Tuck & McKenzie [2015], Linda Tuhiwai Smith [2013], and Chapter 4), and queer positionalities (see Browne & Nash [2010] and Chapters 2 and 4). These are just some of the foundational texts, as well as chapters in this book, that touch on these issues; we cite them here to point scholars toward research methods and modes of practice that are more sensitive and responsive to the politics of difference and the lived realities of injustice.

Meghan Cope's Story Maps

Story maps are in some ways a very old technique of incorporating place-based data with people's lived experiences, migrations, and traumas—one need only think of ancient maps with labels and fantastical illustrations for *terra incognita* (unknown or undiscovered land) and places where "here be dragons" to recognize this. In the digital era, however, there are many new tools for telling stories with maps that blend spatial information with qualitative accounts, which range from low-bar, free, open-source programs (e.g., *StoryMap JS* from Knight Labs) to more complex but powerful proprietary systems such as ArcGIS's StoryMap extension. At the beginning of my project, *Mapping American Childhoods*, I was lucky enough to be given access to the handwritten logbooks of admission and placement records from the Home for Destitute Children, a state-wide agency located in Burlington, Vermont. My students and I transcribed the records for every year from 1899 to1941 into a database and then compiled sample stories of individual children or sibling groups. These children were often brought to the Home for brief periods of time while their families were in crisis, but many were brought to be surrendered for adoption, only to be "placed out" with one family after another in an unstable system that was the precursor to foster care. The image here shows a sample of a story map tracing the Campbell family (all names are pseudonyms), in which five brothers, aged four to 13, were brought to the Home in 1905 and sent out to multiple different families as indentured farm workers over subsequent years. By compiling data from the logbook as well as public records and period photographs, we were able to use the story map medium to try to understand the children's experiences as well as portray them to a wider public in engaging but sensitive ways. For more, see https://blog.uvm.edu/mcope-childhoods/.

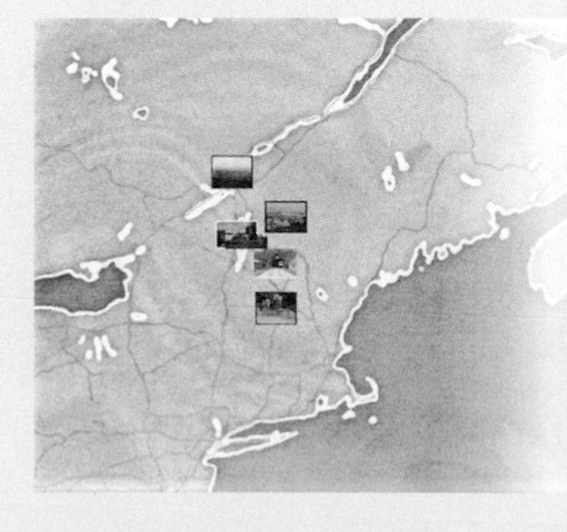

Conclusion

Our hope for this volume is that readers are both empowered to rely on the tried-and-true methods, such as interviews and participant observation, *and* inspired to attempt greater creativity in the range and mix of techniques from diverse standpoints and critical perspectives. The ethical, transparent, and trustworthy production of geographical knowledge that can help us understand—and honour—the human condition is worthy of our effort. We think it is worth emphasizing that, as we write this in mid-2020, when the coronavirus pandemic has upended everyday geographies, mass protests for racial and economic justice fill the streets of cities and towns across the world, and climate change sparks fires and floods, these skills and perspectives are needed more than ever to interpret, comprehend, and guide future actions.

Key Terms

epistemology
ethics
GeoHumanities
production of knowledge
rhizomatic
transferability
trustworthy

Review Questions

1. What are some examples of geographic research questions that may be answered using qualitative research methods?
2. What are some of the shared elements of both qualitative and quantitative research?
3. What are some practices that help ensure trustworthy and rigorous qualitative human geography?
4. How can drawing from multiple data sources strengthen a research project?

Review Exercise

In a classroom or other group setting, ask participants each of the following questions, one at a time.

1. Where are you from? (Closed survey questions on the board or screen, using four categories of place, for example: your university's city; the state, province or region you are in [excluding your university's city]; the rest of the nation [that is, excluding your university's state, province or region]; and outside-of-nation, then count how many identified themselves from each category).

2. What's the best thing about where you grew up? (Class members take two to three minutes to jot down answers. Then write responses on the board and create thematic groupings [e.g., natural elements, people/culture, sports, food] to demonstrate coding and making sense of qualitative data in a rigorous way).
3. Reflect on how each of the previous questions and their answers provide important, but different, knowledge about who is in your class? What else might be helpful to know?

Useful Resources

Clifford, N., Cope, M., Gillespie, S., & French, S. (Eds) (2016). *Key methods in human geography*. Thousand Oaks, London, New Delhi: SAGE Publications

DeLyser, D., Herbert, S., Aitken, S., Crang, M., & McDowell, L. (Eds) (2010). *The SAGE handbook of qualitative geography*. London: SAGE.

Haraway, D. (1988). Situated knowledges: The science question in feminism and the privilege of partial perspective. *Feminist Studies, 14*(3), 575–99.

Tuck, E., & McKenzie, M. (2015). *Place in research: Theory, methodology, and methods*. New York: Routledge.

Power, Subjectivity, and Ethics in Qualitative Research

John Paul Catungal and Robyn Dowling

Chapter Overview

This chapter introduces the importance of recognizing how relations of power and difference shape the conduct of research at all stages of the knowledge production process. It discusses the methodological implications of understanding research as situated in political, social, and institutional contexts, as well as of negotiating the material and ethical consequences of the researcher's own experiences and perspectives and how these affect relationships with research participants. We discuss three general themes that researchers need to be aware of: (1) research as embedded in particular cultural and political contexts, (2) critical reflexivity as a mode of personal reflection on how power shapes one's conduct as a researcher, and (3) ethical guidelines as tools for ensuring responsible research relationships and behaviours. Rather than offering prescriptions for dealing with power in research, we propose that researchers engage in critically reflexive practice to recognize and respond to the contextually specific ways that power manifests in their research.

Introduction

In Chapter 1, Meghan Cope and Iain Hay outlined the types of questions that motivate qualitative research, including human geographers' quest to understand how individuals and groups experience and negotiate spaces of community, work, leisure, family, and government, as well as the social, cultural, and political structures, institutions, and **discourses** that shape them. This chapter is concerned with how the conduct of qualitative research is shaped by **power** relations of various kinds: among researchers and participants; among researchers and institutions; among **participants** themselves; and among researchers and other researchers. It considers how and why researchers' and participants' **subjectivities** (relative to people's background, positionality, locational context, and life experiences) matter for qualitative research, and some important **ethical** considerations that human geographers must account for as part of the process of doing qualitative research.

Power affects qualitative research in several ways. First, at its foundation, qualitative research is a social process that involves interactions between individuals, institutions, and/or communities. The conduct of research itself is shaped by power. Personal and group identities influence and shape the research context and relationships between researchers and participants. For example, **interviews** might be thought of as conversations between researchers and participants that take place in contexts of broader power relations and histories. Researchers who want to interview members of marginalized communities may find that participants' socio-economic, racial, gendered, and sexual positionings in society may affect their willingness or capacity to participate in research. Similarly, researchers who **study up** and conduct research with people in positions of power (e.g., corporate executives) may experience barriers to access. **Focus group** research might yield results that are themselves shaped by broader power relations (see Chapter 11). You might, for example, find that men speak more in focus groups than women, which is illustrative of how broader gender norms manifest in the research process. Qualitative researchers may attempt to mitigate the influence of such power relations by conducting a focus group or two specifically for women, but then may find that racial or class differences among women and the very definition of who identifies with the category "woman" may affect participation and results. These examples suggest that neither researchers nor the research process can be treated as separate from broader societal structures, norms, and discourses. Indeed, they illustrate that because the conduct of research is embedded in society, it is, by necessity, shaped by the very processes, structures, and inequalities that shape society more broadly. In turn, research has the capacity to shape and reshape society itself. Qualitative research often investigates issues that are the subject of intense political and social debate, such as the root causes of homelessness or the regulation of environmental issues. Research can influence how the public and decision makers think about and respond to such issues. Researchers may mobilize their institutionally backed and publicly recognized positions as "experts" to enter into media, planning, and policy spheres and thus influence legal, governmental, and public responses to, say, health care, immigration, crime, or urban development.

As qualitative researchers, we must be attentive to the various ways that power relations shape our conduct of research. Power relations matter across all qualitative methods and all phases of research. They cannot be ignored in part because, at a practical level, they have the capacity to affect the success of the research project itself. Issues related to power have a bearing on recruitment and participation, the collection and interpretation of data, and the circulation and influence of results. This chapter places these issues squarely at the heart of qualitative research practice. To do this, we outline three issues that human geographers must consider as part of their conduct of research and make some recommendations for how to deal with them.

The chapter begins with the importance of power and subjectivity in the conduct of research. It also pays particular attention to how a researcher's social positions and **intersubjective** relations might influence research at multiple phases. It then moves into a discussion of some ethical considerations in qualitative research and the role of institutional review processes within universities and funding agencies in responding to these considerations. It also discusses the genesis of institutional review processes and their limitations.

A word of caution to begin: in general, most of the chapters that follow offer practical guides to different qualitative methods. They explain how to conduct **participant observation,** how to set up and lead a focus group, and so forth. This chapter is different in at least two important respects. First, while it uses examples of specific methods to give flesh to its discussion of power in research, the issues that it raises concern qualitative research more broadly. This chapter should therefore be used in combination with those on particular methods to help you think about the specific challenges you are likely to encounter in your research. Second, this chapter does not—and cannot—offer hard and fast rules on conducting ethical research that is responsive to matters of power and intersubjectivity. As we discuss below, this is partly because we occupy different positions in broader grids and geographies of power. It matters who we are, who we are in relationship with in the research process, what kinds of questions and issues we are concerned with, where we are conducting our research, and what we want to do with our research. Our ethical responses to issues of power in research must account for this combination of considerations.

Placing Power in Research

Decisions about which research topics to pursue, what are appropriate and worthwhile methods of investigation, the "right" ways to relate to sponsors of and participants in research, and the appropriate modes of writing and communication of results all involve ethical questions. These questions include how researchers ought to behave, the role of research in the pursuit of social change, and whether and how research methods are just. Research ethics are broadly defined as being about "the conduct of researchers and their responsibilities and obligations to those involved in the research, including sponsors, the general public, and most importantly, the subjects of the research" (O'Connell-Davidson & Layder, 1994, p. 55). How might we understand what these responsibilities and obligations are, to whom we owe them, and on what principles they are based? How do our multiple subjectivities influence our understandings and pursuit of these responsibilities and obligations?

Whether or not we recognize it, who we are as researchers and our **positionalities** in society (including our biographical characteristics, social locations, and formative experiences) matter greatly for our capacity to produce knowledge through research. For those of us who are formally employed by or studying at universities, our capacity to be in the university is itself shaped by the fact that universities are spaces of both privilege and inequality. For a long time, and still to a large degree, geographical **knowledge production** has been the privileged preserve of white men with social and economic standing (Pulido, 2002). This continues to influence how members of historically marginalized communities, including Black, Indigenous, and racialized people; women; LGBTQ+ people; people living in poverty; and people with disabilities experience research as a field (see also Chapter 4). Indeed, barriers to full participation of these marginalized communities continue to shape not only the demographic composition of human geography as a discipline, but also the kinds of topics, approaches, and problems that are considered proper for geographical research (Oswin, 2019).

Race, gender, sexuality, class, and ability continue to be important **axes of difference** that shape which bodies of knowledge are in place and out of place within the discipline. This is significant for the conduct of research in several ways. Juanita Sundberg (2003, p. 180) argues, for example, that "the researcher's geographic location, social status, race, and gender fundamentally shape the questions asked, the data collected, and the interpretation of the data." In short, positionalities matter in research, including in terms of shaping the relationship between researchers and research participants. Sundberg (2003) writes about the ways that masculinist methods in Latin Americanist geographies shape not only the ways that gender relations are taken up (or not) as a research problem, but also how geographers conduct research, including who they consider important subjects to be engaged in research. The researcher's positionality also matters in terms of shaping how they relate to and experience field sites. For example, participant observation research on clubs and other nightlife spaces might be experienced differently by differently gendered and sexualized researchers, owing to the ways that such spaces are shaped by socio-cultural practices and logics such as **cis-heteronormativity** and **patriarchy**. Similarly, a researcher who might be interested in conducting ethnographic research in religious sites would likely be expected to follow behavioural norms and dress codes, in part to ensure that their presence does not disrupt other people's participation in those spaces.

Insider research is one mode of research where particular kinds of similarities in identities and experiences, rather than differences, enable a researcher to relate to participants in particular ways. In this mode of research, scholars are able to mobilize their shared identities and experiences as community insiders to negotiate research relationships, establish rapport, and gain access to certain participants where marked dissimilarity might pose a barrier. A refugee studies researcher who is a former refugee might relate well to prospective participants who are current refugees based on common personal histories of asylum. In this case, prospective participants might read into the researcher's insider status certain assumptions of trustworthiness and a certain level of understanding that might influence their decision to participate. Examples in geography include research by Filipinx Canadian geographer May Farrales (2019) on pageant politics among Filipinx migrant communities in Vancouver, British Columbia, and that of lesbian geographer Gill Valentine (1993) on lesbian negotiations of everyday heteronormative spaces.

It is worth clarifying that insider status does not, in and of itself, remove power relations or ethical dilemmas from research. Indeed, insider research can pose its own challenges, as taken-for-granted norms or loyalties might erect further barriers to research. For example, work on the geographies of small communities by a researcher from such a community might find some people unwilling to participate for fear of public exposure that might affect their status in the community. This issue is even more acute in, say, focus group research, where confidentiality among participants is more difficult to ensure. Conversely, in such a setting, participants might participate but be quite measured about the stories or views of their communities that they choose to share with the researcher and other participants.

Another challenge that insider research could pose involves how the terms of insider status itself are defined between the researcher and the research subjects. There is a politics to claiming belonging to a community when the definition of who belongs in that community is itself in question. The simple employment of a singular criterion for insider status (e.g., nationality) could possibly mask crucial differences and power relations within communities. Research on migrant communities, for example, might employ **methodological nationalism** by taking countries for granted as "natural" units for recruiting participants thus reproducing status quo understandings of national community. Such methodological nationalism might be reproduced by researchers who are themselves migrants from the same nation-state as their participants (Nowicka & Cieslik, 2014). Failure to check your assumptions of loyalty to nation, or of shared understandings of national boundaries and histories, could affect not only your data collection process (e.g., recruitment), but also your interpretation of results.

Insider status is not always guaranteed. Indeed, the research process itself can make visible the ways that claims to such status are precarious. Ju Hui Judy Han's (2010) ethnographic research with Korean/American evangelical religious missionaries in East Africa is instructive of the limitations of single-axis (e.g, nation-based) definitions of insider status. She notes that her research was enabled in part by her insider status as a Korean American scholar with fluency in Korean, but her identities as a lesbian scholar and social justice activist who is quite critical of religious missions (including what she perceived to be homophobic, racist, and geopolitical goals) affected her experience of her ethnographic embeddedness. She shares an experience of feeling unsafe in the presence of an older male research participant who demanded her pious commitment to Christianity, a moment where gender, age, and religious differences undercut assumptions of ethnic solidarity in the research relationship. To enable her research to proceed and for her own safety, Han had to make use of strategic forms of distancing and discretion in her interactions with her participants.

By contrast, outsider research is where scholars conduct research in communities or contexts in which they do not have personal involvement. The history of knowledge production in geography and anthropology is arguably characterized by such an approach, since research in the mode of exploring worlds and cultures unknown to European researchers was the dominant paradigm in these disciplines for a long time (Driver, 2001; Smith, 2012). Such scholarship is premised on the racial and developmental difference between the researching self and the researched **Other**. **Voyeurism** often characterizes this mode of research, as research subjects are reduced to objects to be known, that is, gazed at, documented, classified, and at times collected (e.g., as artifacts). Through this mode of scholarship, researchers come to perceive themselves not only as experts on Others, but also as those with the means, power, and capital to shape how the world is known, that is, to arrange the world into discrete regions, cultures, and societies.

Power differentials and social differentiations are especially acute in such research, and particularly so where research is conducted in contexts of **asymmetrical relationships** between the researcher and research participants. Research in the service of imperial conquest or humanitarian rescue are characterized by such power differentials, where the researched Other is framed as needing to be helped by the expert knowledge and institutional pull of outsider researchers. Such a benevolent posture is premised on the (supposedly) insurmountable difference between the researcher and the research subject.

This research approach continues to bedevil scholarship today, not only in far flung places of Other cultures, but in close proximity to researchers' own contexts. For example, scholarship on marginalized urban communities has often been conducted by scholars privileged in racial, class, and institutional status and who are themselves outsiders to these spaces. Indeed, in cases where over-research of specific areas conducted by outsider researchers has taken place but with no perceptible change to participants' quality of life, scholars have documented research fatigue among residents, who resist further participation in research that they deem **extractive**, useless and even dangerous in terms of exacerbating the stigmatization of these areas as problem spaces that require outsider interventions (Clark, 2008). In the case of Vancouver's Downtown Eastside (DTES) neighbourhood, for example, Neufeld et al. (2019) argue that "[t]here are too many stories of communities and individuals in the DTES who have felt disrespected by research. Research can increase inequality, contribute to stigma, exploit peoples' pain, exhaust community members and typically benefits researchers much more than it benefits the DTES" (p. 7). Neufeld et al. (2019) make clear that the existence of histories of exploitation and the continuation of power differentiation between researcher and research participants does not necessarily mean that all research needs to stop. Instead, research needs to be approached in responsible ways. Their work offers important pointers for how to pursue forms of more ethical research in such contexts.

Another mode of research that is characterized by differences between researchers and participants is studying up, where participants themselves occupy higher positions of power relative to researchers by virtue of the former's status as, say, political or economic elites. David Ley's (2010) work on millionaire migrants and Iain Hay's (2013) edited collection on the geographies of the super-rich offer examples of such an approach (though it is important to also note how racial difference characterizes the former's research relationship). Linda McDowell's (1998) work on highly positioned workers in merchant banking in London is another. In this mode of research, scholars might come into their research critical of the agendas of the elites, but are interested in their participation as a means of gaining more in-depth information about their lives, goals, practices, and strategies and to acquire knowledge that might not be readily available in the public domain. Power and difference crop up in specific ways in studying up. Elite agendas might exert themselves onto the research process, influencing the quality of interactions between researchers and participants. Powerful participants could mobilize, for example, classed and gendered positions to dictate the tenor of their research relationships, to make demands of a researcher's time, and to shape the researcher's approach. Margaret Desmond (2004) reflects on how this issue came up during her research on elites in the biotechnology sector. She documents how corporate participants sought to curry her favour by providing transportation to and from the train station to her research site, the company's headquarters. She interprets this as partly an attempt by her participants to gain her trust and possibly to influence her findings, to co-opt her research towards their own corporate ends.

The ability of researchers to conduct research with elite participants can itself be shaped by who a researcher is and where the researcher is from. For instance, in his research on World Bank–funded land policy implementation in Guatemala, Kevin Gould (2010) notes the ways that being a white American man enabled him to tap into the sustained histories of US and white presence in international development policymaking, which allowed him to gain access to these privileged spaces of power with relative ease.

It is not that research cannot proceed when political, social, and cultural differences are present. Simply put, it is impossible to scrub the research process of power and difference. But as researchers, we cannot proceed under the illusion that we are immune from long histories, powerful institutions, and structural systems that produce and sustain norms, inequalities, differentiations, and other relations of power. We also cannot completely shield our research practices from the presence of these power relations. What we are compelled to do is ensure that we account for and minimize the negative influence of these power relations in our research, particularly so as to minimize, as much as possible, the harms we might expose our participants and ourselves to during the research process. In order to do this, we must be able to name and recognize the presence and influence of power and difference in our research. We do this not to work towards **objectivity**, an impossible task that is premised on the idea that knowledge can be purified of power, but precisely to recognize that knowledge production itself is situated in networks of power and in processes of social differentiation (Haraway, 1988).

Human geographers have responded in multiple ways to the recognition of power and difference in the research process. One way is through involving participants themselves in shaping the conduct of research at various points of the process, from the identification of problems to be investigated to the interpretation of data and the communication of results (see also Chapter 16 and 19). This is an especially important tactic in research characterized by the presence of asymmetrical power relations. One example is Geraldine Pratt's (2004, 2012) research on the gendered, racial, class, and familial geographies of Filipina caregiver migrations to Canada, which she conducted in the context of a sustained long-term research relationship and collaboration with the Philippine Women Centre (PWC) of BC, a migrant workers' advocacy organization by and for Filipina migrants based in Vancouver. Pratt worked in partnership with PWC leadership and members to identify research directions and methods and to create deliberate space for iterative feedback and reworking of interpretations. Several of Pratt's publications are formally co-authored with PWC and other research partners, which illustrates how scholarly communication can itself involve the collaboration of research participants (see, for example, Pratt in collaboration with the Philippine Women Centre, 2009; Pratt in collaboration with the Philippines–Canada Task Force on Human Rights, 2008).

Critical Reflexivity and the Place of Positionality

Such participatory forms of research might not always be the most appropriate approach in all research contexts. A second response that recognizes and negotiates relations of power is **critical reflexivity**. Critical reflexivity begins with the recognition that the practice of research, as well as researchers and our participants, do not operate in a vacuum, but in fields of power and ongoing histories of social differentiation (see also Chapters 4 and 8). Kim England (1994) understands reflexivity as a process of constant, self-conscious scrutiny of the self as researcher and of the research process. Being reflexive requires that researchers understand that they are active participants in the research process and that their identities, commitments, histories, and approaches are part of the production of knowledge. The key to critical reflexivity is not necessarily to minimize the impact of the researcher as an actor, but to recognize how the researcher's situatedness affects the research. Being reflexive means analyzing your own situation as if it were something you were studying. It requires asking throughout the research process: what is happening? What social relations are being enacted during the research? How, if at all, are these influencing the kinds of questions guiding the research, the act of data collection, the role of participants, and the analysis of results?

Reflexivity on one's positionalities and their impact on research is an important part of responsible research practice. It requires accounting for how one's racial, sexual, gender, class, and other identifications, in their **intersectionalities**, influence the research process. For example, a male-identifying labour geographer investigating workplace cultures in the creative industries might find that male workers are willing to let him witness their displays of masculinist interactions (e.g., misogynistic or homophobic jokes) and even ask him to participate (e.g., to join them for after-work drinks). However, this access might be more restrictive if his masculinity does not conform to their normative conceptions, such as if the researcher is effeminate in presentation or does not drink for religious or health reasons. Another illustrative example might involve a geographer of disability researching the schooling experiences of people with spinal cord injuries, who might be able to find nuances in participants' narratives if they themselves have had to navigate institutions and spaces in a similar way, compared to another geographer who has not.

Gillian Rose (1997) cautions against an individualistic approach to reflexivity, which can take the form of a researcher announcing their identities as their act of being reflexive. She finds that this confessional mode does not adequately account for the powerful force of positionalities on research—that is, the ways that they bear materially on knowledge production. In other words, merely saying that one was raised in an upper-class family does not get adequately at how this socio-economic standing might affect the conduct of research. The power relations involved in a geographer tapping into familial networks to access elite subjects (e.g., CEOs) differs from the power relations involved if the same geographer does research with poor communities. The recognition of your positions is a mere first step in understanding their impact in research practice.

Critical reflexivity can be rewarding in that it can initiate new research directions or different analytical insights. However, it also poses possible difficulties. For one, as researchers, we are accustomed to training our analytical gaze outward, not to ourselves, but to our research participants. It might therefore be useful to think of reflexivity not as an inward gaze, but as the task of positioning oneself in relation to people with whom we are in a research relationship. Another difficulty with critical reflexivity is that the impact of our identities and experiences might not always be visible to us, especially if we have been conditioned to take them for granted as normal. The often unmarked nature of cisgender identity, for example, might mean that cisgender researchers often won't have to worry about being able to access restroom facilities safely during a visit to an archival facility where only binary coded restrooms might exist. Finally, while it is more common now for geographers to be forthcoming about their practices of reflexivity vis-à-vis how their positionalities shape their research, such practices were frowned upon as unnecessary, even improper, research practice in the past. This was especially so as objectivity was idealized in geographic knowledge production based on the idea that the researcher should take themselves out of the research process as much as possible. As a result, many geographers have not always considered, much less written about and published on, how their positionalities affect the research process.

Three particular tools might be useful in developing more reflexive practice. The first is the Power Flower exercise, which is commonly used in the fields of education, social work, and social services. This exercise helps catalogue one's positionalities in relation to various axes of social difference and encourages critical reflection on how these positionalities translate materially to power, access, and safety in different contexts. The Power Flower exercise can be a first step in encouraging researchers to reflect on how their various identities might affect their research process (see Box 2.1).

BOX 2.1

Power Flower Exercise

Discussion of Power Flower

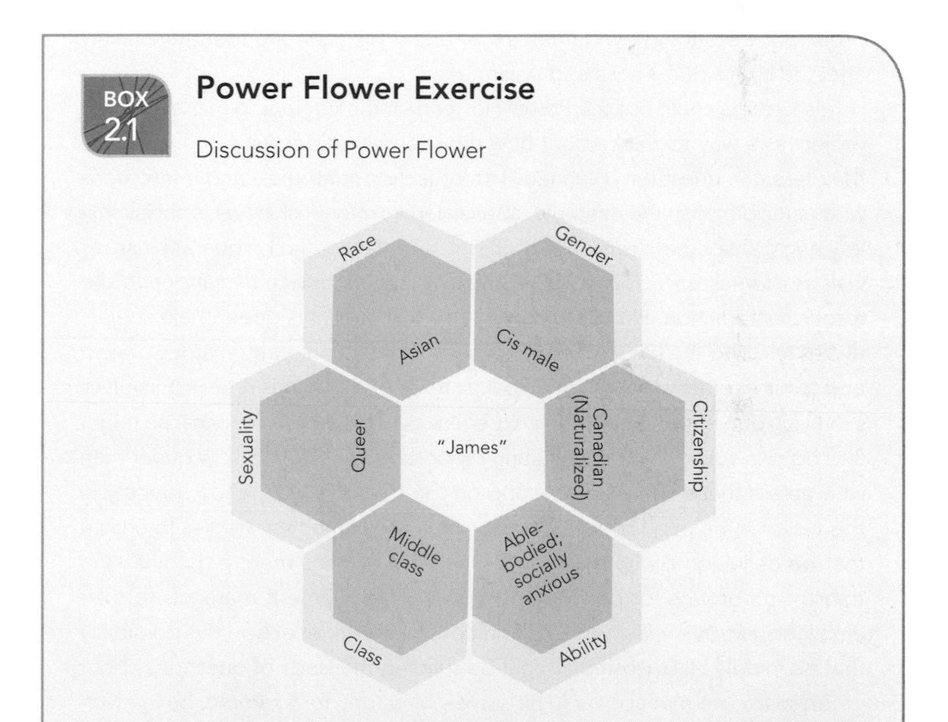

One of the earliest iterations of Power Flower appears in "Educating for a Change," a resource guide developed by Arnold et al. (1991). They describe it as "a tool . . . for looking at who we are in relation to those who wield power in society" (p. 13). The Power Flower activity is a way of thinking about our various subject positions and of considering who we are in

(continued)

relation to broader systems of power. It can be a personal reflection tool for us to consider our proximity or distance to normative social positionings in society and what benefits or disadvantages might arise from our subject positions. As researchers, we can also use this tool as a starting point for thinking personally about how individual subject positions might shape our research. Educators might also use this exercise to enliven collective discussions of how the personal and the political converge in research practice, though an abundance of caution must be exercised in terms of compelling the public revelation of personal reflections prompted by this exercise.

In the graphic above, a fictional researcher named James has identified, in the inner darker petals, his subject positions in relation to various relations of power as identified in the outer lighter petals. While the different subject positions seem visually separate from each other, in everyday life, James is likely to experience them simultaneously and in relation to each other. For instance, how James experiences his cis male identity is likely affected by his queerness and Asianness.

Researchers could use the Power Flower near the beginning of the research project as a way to think about how their various subjectivities could shape their research questions, planned data collection strategies, and interactions with participants. In the example, above, James might ask how his social anxiety might shape his decisions on what data collection technique/s to use, as well as how his participants might relate to him (and vice versa) during the research interaction. He might consider how, if at all, his queer male subjectivities relate to the kinds of research questions that are central to his project and to his interactions with participants. If, for example, his research involves surveying queer men about their experiences of local green spaces, he might find his own subjectivities facilitating rapport building with fellow queer men who access these spaces. Reflecting on the role of his cisgender identity in this research, he might shape his participant recruitment strategies to ensure that his definition of "queer men" is inclusive of trans men, e.g., by asking trans men's groups for permission to post his recruitment materials to their email listservs or social media. In addition, he might reflect on the possibility that his middle class position might be shaping the kinds of questions about green space use that appear in his survey. He might, for example, be inadvertently focusing on leisure activities (e.g., hiking) as opposed to ones related to survival (e.g., foraging, rough sleeping).

Construct your own Power Flower, identifying relations of power that you think might be pertinent to your research. Reflect on how your subjectivities might be affecting your research questions and data collection. In your own personal reflection on your subjectivities, you might add additional petals in order to identify more relations of power. Having used the Power Flower exercise to begin your critical self-reflection of your intersectional subject positions, think about how these subject positions affect different phases of your research process. Pay attention to your emotions during the research process as well, as these emotions often offer a window into our embeddedness, as researchers, in systems of power (Catungal, 2017).

Second, a **research diary** is a kind of autobiography of the research process, a space for a researcher's methodical account of their reflections, observations, and experiences of the research itself. It can act as a repository for cataloguing the unfolding of the research process, including the emotions—joys, excitements, frustrations, disappointments, anger, etc.—that one experiences during the research. These emotional accounts could themselves be windows into the impact of power relations in the research process itself (Catungal, 2017; Laliberté & Schurr, 2016). (See Box 2.2.)

The Research Diary as a Tool for the Reflexive Researcher

Your efforts to be reflexive will be enhanced if you keep a research diary. The contents of a research diary are slightly different from those of a fieldwork diary. While a **fieldwork diary**, or field notes, contains your qualitative data—including observations, conversations, and maps—a research diary is a place for recording your reflexive observations. It contains your thoughts and ideas about the research process, its social context, and your role in it. It is the place where you can record your own experiences and feelings throughout the research, including any challenges and difficulties that you face in the process. You could start your research diary by writing answers to the questions posed in the checklist below.

On beginning:

- Why am I interested in my research topic? How do my positionalities shape my interests and my planned approach to studying them?
- How does my research topic make me feel? Are there particular emotions (e.g., anger, excitement) that motivate my interest in my research topic? What might these emotions reveal about my investments in my research and in my own sense of self as a researcher?

(continued)

- What are some of the power dynamics at play in the general social situation that I am researching? What sort of power dynamics do I expect between myself and my informants? What feelings do these power dynamics elicit in me as a researcher?
- In what ways am I an insider and/or outsider with respect to this research topic? How might my position affect my capacity to do this research?
- What challenges might my positionalities cause? Will any of them be insurmountable? How might my research participants feel about my positioning in relation to the research topic?
- What ethical issues might impinge upon my research (for instance, privacy, informed consent, harm, coercion, deception)? How might I, as a researcher, respond to these possible challenges and issues?

During data collection:

- What role do the different subject positions present during data collection play in shaping the research interaction? How did I respond to the impact of these subject positions?
- How do my different subject positions affect how I feel during the conduct of research? In relation to research participants, interactions, and settings?
- How does the conduct of research make me feel? Are there particular moments in the research that made me feel specific emotions?
- Remember to take notes throughout the data collection and keep them in a research diary. Note down emotional responses that research interactions might elicit at different points during the data collection process. Pay attention to your embodied and affective reactions to research participants and the narratives they offer.

After data collection:

- Did my perspective and opinions change during the research?
- Were there issues or power relations that I did not anticipate during data collection? How did I respond to them? How did the emergence of these issues or power relations make me feel?

During writing and interpretation:

- What voices or perspectives are absent in the corpus of data that I am analyzing and writing about? How does this impact the usefulness of the data?
- What social and conceptual assumptions underlie my interpretations?
- How is my writing reproducing and/or challenging power relations?
- What kinds of emotions motivate my framing and interpretation of data?

Finally, in the spirit of extending Rose's (1997) caution against treating positionality in an individualistic way, geographers could develop methods for sharing their reflections on their positionalities, dilemmas, and experiences with other researchers. Many of us already share our experiences and seek and provide feedback with each other in the informal spaces of conversations during breaks, meals, or hangouts. We might formalize check-ins with each other not as incidental to, but as a part of, our research practice. Indeed, in team-based forms of qualitative research, collective sharing and analysis of positionalities and experiences can be a key part of negotiating group dynamics, parsing out divisions of labour, and providing support to each other. Serin Houston, Jennifer Hyndman, James McLean, and Arif Jamal (2010a, 2010b) have written, for example, about the analytical benefits and methodological challenges of collaborative research, including how their differentiated positionalities (including institutional roles) came to shape their dynamics as a research team, their data collection strategies, and their interpretation of results. Rather than treating reflexivity as a private exercise, it can thus become a means to tap into each other for wisdom and resources on how to analyze specific experiences and to strategize to ensure a more responsible and ethical research practice, whether individually or in teams.

Additional Modes of Ethical Accountability

How we as researchers ought to conduct ourselves in relation to our participants, our peers, our sponsors, and others who are involved in some way in research is an issue that we deal with not only personally and interpersonally, but also in relation to the institutions and communities in and with which we do our work. In addition to reflexive research practice, institutional and community-based ethics guidelines and processes also exist to ensure that research is done in responsible ways and in accordance with broadly accepted principles.

Those of us who do research in a professional context, whether as faculty members or as students, are governed by institutional guidelines. For example, the American Association of Geographers' "Statement on Professional Ethics," most recently revised in 2009, is meant to "encourage consideration of the relationship between professional practice and the well-being of the peoples, places, and environments that make up our world" (AAG, 2009). Moreover, in recognition of the history of exploitative colonial research relations in the field of geography, the AAG Indigenous People's Specialty Group provides more specific ethical guidelines for geographers who work with Indigenous people. Their "Declaration of Key Questions about Research Ethics with Indigenous Communities" draws on broader protocols to support ethical research with Indigenous communities, encouraging geographers to "begin building mutually beneficial relationships with Indigenous nations—to bring more integrity to our field, and to enrich geographic inquiry" ("AAG Indigenous Peoples Specialty Group's Declaration of Key

Questions about Research Ethics with Indigenous Communities," n.d., 2). Their declaration works in tandem with broader guidelines and principles on research with Indigenous communities (see also Chapters 4 and 8). The First Nations Information Governance Committee (FNIGC), for example, identifies Ownership, Control, Access, and Possession (OCAP) as key principles in research with First Nations communities. According to FNIGC, OCAP "recognizes community rights and interests in their information" and "ensures that First Nations own their information and respects the fact that they are stewards of their information" (First Nations Information Governance Centre, 2019). The FNIGC is clear that each First Nations community may interpret and implement OCAP principles on their own terms. The emergence of OCAP principles is one example of marginalized communities responding proactively to histories of exploitative research. Other such examples exist, such as the Denver Principles, which insist on Greater Involvement of People living with HIV/AIDS (GIPA) and the Meaningful Involvement of People living with HIV/AIDS (MIPA), including in research. One of us (Catungal) is currently involved in a community-based project that embeds GIPA and MIPA in decision-making across all aspects of research, from the composition of the research team and the recruitment of participants to the interpretation and communication of results to the broader public.

There has been an increasing formalization of ethical reviews (Israel & Hay, 2006), including as part of the work of research funding agencies. In the Canadian context, for example, federally funded research involving human subjects is governed by the "Tri-Council Policy Statement: Ethical Conduct for Research Involving Humans" (2018) (aka TCPS 2). Researchers who work in Canadian universities and/or receive public funding for their research are required to complete the TCPS 2 Course on Research Ethics, which consists of eight online modules. Ethics committees within institutions implement these directives and guidelines. These are called institutional review boards (USA), research ethics boards (REB) (Canada), or human research ethics committees (Australia, UK), and usually exist at the level of the university, though some departments might require formal reviews internally before ethics submissions are forwarded to university level boards. Some boards are differentiated by areas of research or by institutional affiliation. For instance, at the University of Toronto, ethics review might be done through the Social Sciences, Humanities and Education REB, or the Health Sciences Review, depending on one's institutional home at the university. A third REB, focused specifically on HIV/AIDS research, is in charge of all reviews on HIV/AIDS research regardless of the researcher's institutional affiliation. Knowledge of the schedules, procedures, and expectations of our institutions is key to our capacity to conduct research on time and in full compliance with institutional requirements. Ethics applications usually require researchers, including students, to articulate their detailed consideration of methods, including possible ethical dilemmas, participant selection processes, privacy procedures, and storage of data. These documents take considerable time to prepare, particularly when research supervisors and ethics boards require revisions from the researcher.

Research might also need to go through formal review under the auspices of boards external to the university. For example, a Research Committee exists at the Vancouver School Board (VSB), in British Columbia, Canada, which has oversight over research to be conducted with VSB students and on VSB schools. Qualitative research in hospitals (e.g., interviews on family members' experiences of waiting rooms) also usually requires approval from ethics committees specific to each hospital. Community organizations might also have their own approval processes for researchers seeking to work with them or to access the people or communities that they serve. For example, the Nuu-chah-nulth Research Ethics Committee (2008) is tasked with approving research projects being conducted with more than one Nuu-chah-nulth First Nations community in British Columbia. In addition to process, the Committee also identifies respect for persons, beneficence, and justice as key principles for the conduct of research in Nuu-chah-nulth communities.

Institutional review processes in research institutions are concerned with the researcher's responsibility to research participants, though they are also increasingly concerned, including for legal and liability reasons, with the researcher's safety. Privacy, informed consent, and potential risk or harm are matters that review boards often attend to when assessing ethics proposal submissions. These are addressed, in turn, below.

Privacy and Confidentiality

Qualitative research often involves asking personal questions or observing people, and thus involves knowledge that is usually not publicly available. Ethics committees are typically concerned that private details about individuals are not released into the public without participants' consent. For example, they might ask about how you intend on storing research materials (e.g., recordings, field notes, transcripts, consent forms) securely and who will have access to them. You might also have to be clear about how you intend to use research data in scholarly communication, ensuring where needed that your informants are not identifiable to others. There are various techniques to ensuring the **anonymity** of participants, including using pseudonyms and masking identifying characteristics (for example, occupation, location) in written or oral presentations of research findings. It is, however, not always possible or even ideal for researchers to ensure anonymity. This is especially the case for public figures, as in O'Neill's (2001) research on Broken Hill Proprietary Company Ltd., in which he identified the firm and the executives with whom he spoke. Pratt's research, described above, also names the organizations that she works with. Indeed, in her case, it would not be ideal to anonymize the organization given their desire to ensure their full participation with all aspects of the research, including writing up results.

Informed Consent

For most geographical research, participants must consent to being part of your research. In other words, researchers have to acquire people's permission before they can involve them. Permission requires **informed consent**, a decision that a participant must arrive at with sufficient knowledge of what participation entails (e.g., being part of a focus group), what sorts of issues will be explored and what the researcher's expectations are (e.g., the time commitment). Informed consent also usually involves letting prospective participants know how the knowledge they share will be handled (e.g., who will have access to audio recordings) and what provisions there are for ensuring their safety (e.g., anonymity or **confidentiality**). It is important for researchers to honestly disclose to participants some of the projected benefits and possible harms involved in the research, as this information might be part of how someone decides whether or not to participate in the research.

There are some cases where informed consent is not always possible or ideal, but the lack of informed consent is not always a barrier to acquiring approval from ethics committees. Researchers seeking to make use of **deception** must be able to justify why deception is necessary. In cases where deception has been employed by geographers, this has sometimes been done to safeguard researchers and/or their participants. For example, Maguire et al. (2019) discuss the emotionally difficult work of conducting covert research on "heteroactivist" organizations, a case where publicly identifying oneself as lesbian, gay, or an ally to LGBTQ+ communities could result not only in foreclosed research on "antagonistic organizations," but also in possible harm to researchers themselves. In other cases, deception in geographical research has been done in the service of mobilizing research for social justice, sometimes at the behest of community research partners. Paul Routledge (2002) has written about adopting the persona of Walter Kurtz, a tour operator, in his research in collaboration with two Goan organizations who were seeking to resist tourist development. His piece discusses, in part, how his research participants utilized his positionality as a white man in their activisms, a form of collaboration that he welcomed not just to enable his research, but also because he aligned himself and his research with the political goals of these organizations. Similarly, Rob Wilton (2000, p. 95) writes about withholding his own political stances in order to conduct research with opponents of group homes for disabled people. He writes: ". . . my own commitment to greater social justice for people with disabilities, and a belief that community opposition is motivated by negative stereotypes of disability and difference, justified this approach."

The use of deception clearly brings up difficult ethical issues. You should think carefully and seek advice before contemplating a research project that involves deception. Pamela Moss's (1995) discussion of why and how she used deception in her study of domestic workers may be particularly useful.

Potential Risk or Harm

Your research should not expose yourself or your informants to harm—physical or psychological/social. As social scientists, it is unlikely that you will be subjecting participants to physical harm, though this is possible in certain circumstances. The emergence of walking interviews as a method, where researchers and participants walk together as a means of discussing participants' life-worlds, could expose both to physical risks associated with being in public spaces, though it might be that such a risk is not any worse than what they might encounter in their everyday lives. For social scientists, the risk of harm is often psycho-social, involving the possibility of making participants upset or angry or recalling past trauma, for example.

The presence of this risk does not mean that research cannot proceed. Rather, it requires that you build into your research plans specific strategies for responding to these risks should they arise. How might a human geographer respond, for example, when an interviewee starts crying as a result of recounting experiences of grief? It would be prudent for the interviewer to pause to give the interviewee an opportunity not only to collect themselves, but also to decide whether they still would like to continue their participation in the research. Prior to the interview itself, the researcher might already have primed the participant of the possibility of being upset due to the topic of the research as well as of their options as a participant, e.g., to discontinue their involvement with the research or to redirect the interview towards other topics. Many ethics committees also recommend that researchers have the contact information of a counselor ready to provide to participants, should they require professional psychological assistance.

You should also avoid putting yourself at risk during the research as much as possible, and where there is possibility of risk, to come up with strategies to minimize harm to yourself. For example, a researcher planning a participant observation project on backcountry hiking cultures could be asked by an institutional ethics committee and/or a research supervisor to consider potential dangers while conducting fieldwork. They might ask for more thorough information on the precautions that the researcher intends on taking in the event, for example, of injury or getting lost in the wilderness. They might also be asked to reconsider their methods for studying the topic, for instance, switching to interviewing avid hikers instead.

Although important, ethical rules and ethics committees are not unproblematic for qualitative researchers in human geography. At a general level, the universalist ethical stance and biomedical model of research embedded in these guidelines involve rigid codes that cannot always deal with "the variability and unpredictability of geographic research" (Hay, 1998, p. 65). There is a danger in employing universalistic understandings of what is ethical, for example, especially in cross-cultural contexts where differences in conceptualizing ethics between researchers and participants might exist (Tauri, 2014). Debate also exists about the primacy of written consent over, say, oral consent, where the latter might be more appropriate in situations where participants are unable to sign documents due to, for example, illiteracy or disability (Dawson & Kass, 2005; Gordon, 2000), or where participants come from a cultural context where one's word is as valid, if not more so, than a written document (Coram, 2011). The individualized nature of informed consent processes might also not work in contexts where community or collective consent may be more appropriate (Butz, 2008; and see Chapter 8). Finally, in relation to harm, it is not always possible to predict the impact of research on participants, especially over a longer term (Bailey, 2001). However, this does not mean that it is not important to anticipate possible issues or outcomes.

Institutional ethics processes are crucial, but they are only one component of ethical research practice. Acquiring formal approval does not absolve us as researchers from paying attention to the workings of power and social difference in our research. Critical reflexivity throughout the research process, and not just during the data collection stage, is crucial to ensuring ethical, responsible, and responsive practice.

Conclusion

Many researchers have been trained to think that the best research is the distillation of universal truths unencumbered by the workings of power. At best, this approach misunderstands the relationship between research and power. As we have noted in this chapter, power suffuses all aspects of research because research involves relationships and takes place in worlds shaped by power. One of our tasks, as researchers, is thus to linger with and consider how best to approach our research relationships, behaviours, and practices in ways that are ethical and responsible.

The qualitative research methods described in subsequent chapters are knowledge production practices that will involve you in different kinds of social relations and responsibilities. Awareness of and responsiveness to these relations and responsibilities are crucial to ethical research practice. Critical reflexivity requires self-conscious scrutiny of yourself and the social nature of research and close attention to how your research interactions and the information you collect are socially conditioned. How, in other words, are your social roles and the nature of your research interactions inhibiting and enhancing the information you are gathering? This is not an easy task, since it is not always possible to anticipate or assess accurately the ways in which our personal characteristics and situatedness in systems of power affect how we conduct research or the information we collect. While challenging, reflexive attention to power, situatedness, and subjectivity is crucial to more ethically and socially grounded research practice.

Given the contextual and multiple ways that power relations relate to research practice, this chapter has only begun to identify some ways that power, subjectivity, and difference matter for the conduct of research. We are unable to offer universal rules, in part because different research questions and methods and differently positioned researchers and participants will produce different ethical dilemmas and relations of power. Instead, we invite you, as researchers, to consider how the issues we discuss might matter in your own research and institutional context.

Key Terms

asymmetrical relationships
axes of difference
collaborative research
critical reflexivity
informed consent
insider (research)
intersectionality
intersubjectivity
methodological nationalism
outsider (research)
positionality
research diary
research ethics
situated knowledge
studying up
subjectivity
voyeurism

Review Questions

1. What forms of power relations may be part of a qualitative research project?
2. How are qualitative methods intersubjective? How might social relations like gender or race influence the collection of data?
3. How can you identify critical reflexivity in others' work? Read a study in human geography in your field of interest that uses qualitative methods. Is the researcher reflexive about the research process? If so, how? If not, what sort of questions would you like to ask the researcher about the research process?
4. Outline and explain the kinds of issues of concern to university ethics committees.
5. Discuss why the acquisition of ethics clearance from institutions does not guarantee that a research will unfold ethically.

Review Exercises

1. Each university has its own research ethics review processes. What does your university's process entail? Identify the ways that your university asks you to deal with questions related to informed consent, confidentiality, privacy, and risk. If you were conducting focus groups involving students and discussing their perceptions and experience of the university campus, which of the ethics approval questions would be particularly important? What kinds of power relations might be important to pay attention to in this research project?
2. Alison Bain and Catherine Nash (2006), reflecting on their research on the Toronto Women's Bathhouse Committee, write about "the researcher's body as a tool for data collection in the process of ethnographic fieldwork." Robyn Longhurst, Elsie Ho and Lynda Johnston (2008) similarly argue that the body is "a primary tool through which all interactions and emotions filter in accessing research subjects and their geographies" (p. 208). Read these articles and discuss how power and difference manifested during the researchers' process of conducting their studies. Compare and contrast the ways that they negotiated these issues in and through their bodies. Discuss the role their positionalities played in their fieldwork.

Useful Resources

AAG Indigenous Peoples Specialty Group's Declaration of Key Questions about Research Ethics with Indigenous Communities. (n.d.). Available online: www.indigenousgeography.net/IPSG/pdf/IPSGResearchEthics-Final.pdf.

Tri-Council Policy Statement: Ethical Conduct for Research Involving Humans. (2018). www.pre.ethics.gc.ca/eng/policy-politique_tcps2-eptc2_2018.html

American Association of Geographers. (2009). Statement on Professional Ethics. www.aag.org/cs/about_aag/governance/statement_of_professional_ethics

Butz, D. (2008). Sidelined by the guidelines: Reflections on the limitations of standard informed consent procedures for the conduct of ethical research. *ACME: An International E-Journal for Critical Geographies*, *7*(2), 239–59.

England, K.V. (1994). Getting personal: Reflexivity, positionality, and feminist research. *The Professional Geographer*, *46*(1), 80–9.

First Nations Information Governance Committee. (2019). The First Nations Principles of OCAP. https://fnigc.ca/ocap

Indigenous Peoples Specialty Group. (n.d.). Declaration of Key Questions about Research Ethics with Indigenous Communities. www.indigenousgeography.net/IPSG/pdf/IPSGResearchEthicsFinal.pdf

Israel, M., & Hay, I. (2006). *Research ethics for social scientists. Between ethical conduct and regulatory compliance.* London: SAGE.

Neufeld, S.D., Chapman, J., Crier, N., Marsh, S., McLeod, J., & Deane, L.A. (2019). Research 101: A process for developing local guidelines for ethical research in heavily researched communities. *Harm Reduction Journal, 16*, 41. https://harmreductionjournal.biomedcentral.com/articles/10.1186/s12954-019-0315-5

Nuu-chah-nulth Tribal Council Research Ethics Committee. (2008). Protocols and principles for conducting research in a Nuu-chah-nulth context. https://icwrn.uvic.ca/wp-content/uploads/2013/08/NTC-Protocols-and-Principles.pdf

Wilton, R. (2002). Sometimes it's OK to be a spy: Ethics and politics in geographies of disability. *Philosophy and Geography, 1*, 91–7.

Reaching Out: Cross-cultural Research

Mabel Gergan and Sara Smith

Chapter Overview

Most, if not all, research projects will involve cross-cultural research—that is, doing research across forms of difference that require cultural translation, whether that is differences of class or racialization, ethnicity, gender, region, or intersections of these. In this chapter, through an extended discussion of our research with Himalayan youth, we put forward questions and strategies to guide cross-cultural research so that it can be meaningful, ethical, and academically rigorous. The chapter begins with a discussion of how we came to conduct art-based research in a small mountain town in the Ladakh region of India. After this, we engage with literature on decolonizing research methodologies, to evoke the big questions undergirding research engagement across difference. In this chapter you will encounter guiding questions to consider, as well as practical strategies and examples of cross-cultural research gone awry. We make an argument to keep your attention not only on cultural difference, but how this difference is often complicated by global and local power structures that must be addressed if you hope to do research *alongside* those you work with rather than *on* a community.

Introduction

We met before the workshop in the café on the second floor of Lehling Bookshop, while down below in the main bazaar young people were checking out each other's outfits and women sell apricots and cabbage. We were there to hash out plans for a youth workshop in the mountain town of Leh, Ladakh, in Jammu and Kashmir, India's northernmost state. How would 12 young people from different backgrounds talk openly about their experiences? They had a few things in common: all the participants are from Ladakh, in the Indian Himalayas, but had significant experience outside the region—they had been sent outside as teenagers to pursue higher education, either in the lowland cities of India or even farther afield, in the United States. At the same time, within the group, there was a mix of gender, class backgrounds, and life experience (though the group is more religiously homogeneous than we had hoped). Our goal was for them to share what studying far from home has meant to their lives. But how could we make this interesting and productive for the students?

Mabel pointed out that Sara had recently worked on a 'zine (a do-it-yourself style booklet) with a feminist collective (FLOCK). What if we tried the same thing here in Leh? Would they find it too childish, to be wielding scissors and markers as adults? Or would this playfulness be freeing, and enable new ways of speaking to one another? We pitched the idea to Tashi Morup, projects director of the organization we collaborated with (LAMO, the Ladakh Arts and Media Organisation), and to Rinchen Dolma, who had been working on the project at LAMO. To our delight, they were both enthusiastic. We planned a two-day workshop, in which students would use photographic images that they had taken but also have access to local magazines, copies of family pictures, and blank paper. To broaden the conversation and include topics that some might not feel comfortable addressing openly, we asked students to write anonymous short letters: to their parents, to future Ladakhi students, to "the rest of India," or to others they wished to address. As researchers working across cultural difference, this open and creative process allowed us to learn more about the worlds of these students in a collaborative format driven by the students themselves, and not solely guided by questions that we had curated. By partnering with a local organization, we had developed relationships that allowed us to check in on our ideas and make sure they would be feasible.

During the days of the workshop we moved between small clustered conversations and group discussion in which we pulled together themes we saw emerging from student descriptions of their experiences. A few young women made a page about gender in Ladakh. A student trained as an artist made creative and funny collages about student life and about changing Ladakh. The letters were put into collages of messages for students' parents, for future students, and for the rest of India. Heartfelt and poignant, they were a promise to care for mothers and fathers who sacrificed to send their children to school, and a plea for a world free of discrimination and corruption. With the help of LAMO and the students, we created an exhibition of students' photographs with captions and strung the 'zine pages across one of the rooms. We also included a map with different colour strings that showed where the students were from and their travels, as well as the travels of their parents and grandparents. At the opening of the exhibition, Ladakhi youth and elders wandered through expressing moments of surprise and of familiarity with the sentiments and images.

In this chapter, we use our collaborative research as a starting place to engage with the promise and challenges of undertaking **cross-cultural research**. After defining cross-cultural research, we begin with background on our project and a quick overview of foundational literature on this topic. We also encourage the reader to ask: what are your motivations for engaging in cross-cultural research? Are you the right person for this work? We then turn to a set of questions to begin from when conducting cross-cultural research, and examine our guiding principles. We conclude by encouraging careful engagement with cross-cultural work and highlighting inspiring examples.

What Does It Mean to Do Cross-cultural Research?

Fiona Smith (2016, p. 157) defines cross-cultural research as "researching 'other' cultures using other languages. This may often involve working in distant places but can also include working with 'other' communities closer to home." By putting **Other** in quotation marks, Smith reminds us to question who is "other" to us. As Smith (2016, p. 157) observes, the idea of the researcher as intrepid adventurer traveling to distant lands has been central to traditional understandings of geographic research: "To the popular imagination, travel to 'other' places and cultures defines 'Geography' as a subject, particularly as it is presented in publications such as the *National Geographic* or in television programmes about grand adventures." This imaginary is inflected with gendered and racialized expectations about who the geographer is and what constitutes an *exotic* destination. Geographers have implored us to interrogate this imaginary carefully (even if that is what drew you to the discipline), and to question the effects of such thinking (Rose 1996; Sharp and Dowler 2011; Sparke 1996). We define cross-cultural research as research requiring some degree of *translation*—not necessarily translation between languages, which might be shared, but rather the translation of values and concepts, which have meanings and associations that are contextual. This is an important starting point, as the idea of research as necessitating engagement with a defined Other, has roots in the colonial sciences, implies that people cannot work in their own communities, and undergirds the colonial idea that researchers have a right to know (Simpson 2007; Smith 1999).

We might begin then by thinking carefully about how we understand the idea of cross-cultural. We are all **insiders** and **outsiders** in different ways, even if we consider ourselves to be from a given place (see Chapter 2). You might be doing research closer to home, or farther away, but it is important not to create binaries of insider/outsider that obscure other kinds of difference: sharing someone's ethnic or racialized identity may give you confidence that you understand someone's experiences—however, if they have a different class background from your own, have a different sexual orientation or gender identity, you might miss or not be attuned to how this has shaped their perspectives and experiences. Howitt (Howitt and Stevens, 2005, p. 41) writes that "most human geographic research is cross-cultural, because we are drawn into thinking about other people's constructions of place, other people's ways of reading their cultural landscapes—even when they are the landscapes that we live in ourselves in our everyday lives!"

Engaging with this question, Robina Mohammad (2001), has written compellingly about the nuances of insider research that reveals a complex terrain. In her early research career within the South Asian **diaspora** in the UK, she found that she was racialized as someone who "'belonged' to the local Pakistani 'community,'" in ways that supposedly gave her "a superior, almost organic knowledge of the 'community' not accessible to 'outsiders'—for example white people," (Mohammad, 2001, p. 101) but at the same time, her Pakistani identity was often questioned within diasporic circles. At other times, her position as somewhat of an insider made her "too close for comfort" for some research participants. Doing research in this liminal state led her to complicate the idea of insider/outsider research (Mohammad 2001; see also Sharp and Dowler, 2011). Similarly, Farhana Sultana (2007) writes about her experience as an urban Bangladeshi woman with education and class privilege, doing research in rural Bangladesh on the gendered dimensions of water resources management. Being from Bangladesh she was accepted as *deshi* (local) but her class difference (her hairstyle, sartorial choices, ability to speak English) "othered" her and made it difficult to forge friendships, especially with the women in her field site.

Even if you come from a different part of the world you might connect to people in your research site based on shared political interests, shared class background, gendered experiences, life stage, or other lived experiences. Conversely, even if you are doing research with your neighbour, the different ways you are racialized, your differing gendered experiences, your age, or your cultural expectations as an academic might make understanding one another a process of cultural translation (Whitson, 2017). It is also important to consider that historical and **archival research** can also be a kind of cross-cultural work in which issues of translation of ideas across time are just as contentious and require just as much care as communication between living people (Arondekar, 2005; Mawani, 2012). To connect across cultural difference, it can be effective to establish shared political commitments or have conversations about global processes—in our case, for instance, we have had conversations with students on the recent rise of right-wing authoritarianism in the US (see Chapter 1), and on gender norms that occur across sites.

Why do cross-cultural research? As we discuss below, it is important to concede that not all research is necessary or ethical. But what is lost if we give up research across cultural contexts? In our research with young Ladakhis, we wish to do creative, **ethical**, responsible research that unravels rather than strengthens the colonial structures that we have encountered. A beginning point is to make your research ethical and respectful of the place you are in or community you are working with—but this is only a starting point. It's not enough simply to not be insensitive in interpersonal relations or to make friends—the stakes are higher than this and being "nice" is not enough (Parker, et al., p. 2017). We do not want our own ideas to overwhelm those with whom we are working, and we want our analysis to be recognizable and legible to those we are accountable to in our field sites. Our hope is to address cultural differences in a clear-headed and politically accountable manner that sees difference not as a problem, but as a generative site.

Working with Young People in the Indian Himalayas

The workshop we describe above is part of an ongoing project on young people's experience of moving between the margins of the Indian state and its major urban centres for education. Across the world, young people are leaving their small towns to travel for higher education. Our project seeks to understand how that movement, from places that are linguistic, racial, cultural, and religious margins, to large Indian cities, changes young people's sense of self and their political orientation. The Himalayan borderlands have a complex relationship with "mainland" India. While both of us have strong personal ties and political commitments to Ladakh (we discuss this in more detail later), we had to be careful in framing our research not to reiterate existing stereotypes of remoteness and exoticism associated with the region. These stereotypes are a product of the colonial process of **knowledge production**, whereby European and even Indian explorers and researchers diagnosed the region as marginal and remote in need of "upliftment" and "integration" with the rest of India, with Ladakhis seen as "incapable of participating in decisions about their own future" (Van Beek, 1999, p. 436). It is likely that in your region, there are similar tropes about folks from "rural backwaters," agricultural regions, or difficult-to-access spaces. Perhaps you have had to work to overcome stereotypes about the place you are from. These misrepresentations have far-reaching political and social effects especially for young Ladakhis in Indian cities. Our hope with this project is to challenge such misrepresentations by amplifying the creative and powerful voices of Ladakhi youth.

This work began with a summer of interviews with young people studying far from their home in Ladakh, in Indian cities like Jammu, Delhi, and Chandigarh. The idea for the research had emerged from Sara's casual conversations with family members who had undergone this movement between the Himalayas and urban India, and Mabel's own personal experience of college life. You may not have the same kinds of connections to the place where you will do research, or your first research project might lead to the kind of engagements that enable these kinds of intuitive connections in the future. With limited funding, Sara ran a small pilot study in 2014–15. The larger project was then designed in consultation with LAMO leadership. At this stage, the planning and budgeting was done with attention to how the research could be useful and interesting to the organization, and how resources could be directed toward their goals and not only serve those of the US-based researchers (see also Chapters 4 and 16). Securing **participation** through partnership with a local organization is a step toward a more ethical research, but this is still researcher-driven.

Methods are always contextual—for instance some communities may be delighted to engage in mapping work that they can use for advocacy (Elwood,

2008), while for other communities, even the suggestion of a mapping project may be met with suspicion and derail a project due to histories of surveillance and dispossession. Working cross-culturally, it is crucial for you to do the necessary research to find out how the methods you choose will be received in context. **Interviews** have their limits as staged encounters between two people who may not know each other well—perhaps particularly in a cross-cultural context. In the interview setting, the narrative may be driven by the questions the interviewer asks, and often the interviewer simply does not know enough to ask good questions (see Chapter 9 for best practices in interviewing). Combining interviews with group discussions can shift this dynamic a little bit, as organic discussions can emerge between people in the discussion, and the role of the researcher is downplayed. This project has added the use of visual material to this mix.

BOX 3.1

Visual Methods in Cross-cultural Research

Our collaborative media project is based on the principles of **PhotoVoice** and photo-elicitation (Shah, 2015; Wang & Burris, 1994) and builds upon the creative use of photography and film to work collaboratively with research participants (see details in data analysis section below). It is also designed as a creative way to respond to the challenges and ethical demands of working with young people (Cahill, 2007; Quijada Cerecer, Cahill, & Bradley, 2013; Valentine, Butler, & Skelton, 2001). Sara had used PhotoVoice previously and found it revealed a rich depth of data not elicited in interviews (S. Smith, 2013; Wang & Burris, 1994). PhotoVoice enables a productive disruption of the interviewer–interviewee dynamic. Students are on their own or with friends when they take photos, and are thus able to reveal surprising or unexpected interests and ideas, or capture experiences that are more difficult to put into words: a simple breakfast consumed each day in a hurry, a dingy hostel room made welcoming through food and friendship, a flashy outfit that could not be worn back home. In explaining the photo to the researcher, the student further reveals details not captured by the photo (e.g., "I am smiling with my classmate, but actually we hate each other.") and is able to unsettle the roles of expert and collaborator. Photographs and interviews together flesh out the experience of daily life and photographs also provide a way to quickly interact in the field, in ways that are accessible to audiences beyond the academic sphere, with its esoteric language and long delays involved in writing for publication.

In addition to traditional PhotoVoice, our research design incorporated a number of methods, including longitudinal and go-along interviews. **Longitudinal interviews** required conducting interviews with the same set of students over the three-year period of our research project. Since students were in different stages of their educational journey, at the end of the second year of the project many had graduated and were considering their future options. Longitudinal interviews allowed us to track how their experience studying outside Ladakh might have shifted their opinions and perspectives over time. **Go-along interviews** are walking tours led by the research participant and are a relatively new methodological innovation. Mabel conducted go-along interviews with a few students in Delhi, where they walked through their college campuses, the neighbourhoods they lived in, and favourite hangout spots. Go-along interviews can be particularly effective at eliciting biographical stories, social realms, and **spatial** practices (Carpiano, 2009; Kusenbach, 2003). In addition, they have the potential to destabilize research structures of **power** and introduce elements of sensory reactions to the weather, sights, and smells, and accompanying discoveries that occur while walking together. Needless to say, a lot of trial and error was involved in developing these methods and some were more successful than others. Methods we chose allowed us to engage with our young research participants in creative and meaningful ways. But more than this, these methods worked well not because they were foolproof strategies for eliciting information; rather, it was the generosity of our research participants and the relationship of trust and respect between us that allowed for meaningful conversations

When you are designing a research project, choosing which qualitative research method works best for your context ultimately depends on the relationships you are able to cultivate with the people and/or organizations in your field site. Below we will discuss strategies for making these connections. This relationship is fundamental to the process of knowledge production and, as we discuss below, is also a central concern driving feminist, and **decolonizing**, and postcolonial scholarship on qualitative research.

Cross-cultural Research as Problem and Promise

In her foundational book, *Decolonizing Methodologies*, Linda Tuhiwai Smith (1999, p. 1) writes that research "is probably one of the dirtiest words in the indigenous world's vocabulary." A wealth of scholarship demonstrates the ways that research has been **extractive**—that is, knowledge has been extracted from one place or community and used in another, without deep engagement or even compensation. Research and the creation of knowledge have historically been part of global power structures that served empires (Said, 1979) by creating the idea of racial distinction in order to justify colonization and colonial violence (Lowe, 2015; Sharpe, 2009; Stoler, 1997, 2006). Through imposing these **epistemologies**—these structures of knowing—on a global scale, the colonized were taught that their own knowledge systems were inferior and adopted European forms of knowledge that furthered the penetration of these ways of knowing into their minds and bodies (Fanon 2008). This legacy is not easily overcome and, in fact, if we look at universities today we can see the ways it continues to shape who occupies positions of power, who is understood to be "research **subject**," what constitutes "data," who is understood to be capable of producing knowledge, and even who must fight to be perceived as a legitimate scholar (see, e.g., Cottom, 2018; Matthew, 2016; Puwar, 2004).

As a number of scholars have argued, even **participatory research** "can amount to nothing more than enlisting local cooperation in a research project that continues to be driven by outsider researchers' definitions of its purposes, methods, and use" (Howitt & Stevens, 2005, p. 57). Others have cautioned about academic scholarship that claims to be able to "decolonize" or work against colonialism, asserting strongly that our priorities should be to repatriate the land and life of Indigenous people rather than to use decolonization as a metaphor (Tuck & Yang, 2012). Richa Nagar (2017; see also Nagar & Shirazi, 2019), drawing on her long-term research with feminist collectives and social movements in India, offers the concept of **radical vulnerability** as an approach to questions of difference. Radical vulnerability, she writes, is about "reminding ourselves and one another of the violent histories and geographies that we inherit and embody despite our desires to disown them;" our willingness to be vulnerable with each other stems from our commitments to particular sites and struggles—a **situated solidarity** (Nagar & Shiraz, 2019, p. 239). Nagar and other feminist scholars remind us that qualitative research is the process of translating what we see and hear in our field site into narratives that we hope will transform or draw attention to unjust social relationships. Fundamental to such translation work are relationships of trust, vulnerability, and openness between the researcher and the communities with which they work. For new researchers this might seem like a daunting task and *it is,* but one worth undertaking since Nagar also reminds us that "the joys and lessons of moving and creating together in a radically vulnerable mode are often deeper than the sacrifices made by individual travellers" (p. 242). Juanita Sundberg (Sundberg, 2014; see also TallBear, 2014) draws inspiration from the Zapatistas who practise ***preguntando caminamos*** or "asking we walk"—a commitment to relationships of reciprocity and mutuality where one learns by walking, listening, talking, and doing. A first step in challenging the extractive model of knowledge production, therefore, is to start viewing ourselves not as experts but as co-travellers, walking and learning alongside communities.

Considering the history of research, contemporary power structures, and the constraints of working within academic structures (for instance, the necessity of condensed research periods and the need to publish relatively quickly), it is crucial to remember that sometimes the most ethical thing to do is to step away from a research idea, even a good one. You might not be the right person to do this research at this time—maybe the research you have embarked upon turns out to be something that the community cannot risk having go public. Maybe there is not an adequate way to write about the topic without reproducing stereotypes. Or perhaps people simply are not interested in sharing this information with you (see also Chapters 5 and 6). Smith (2016) points to researchers who have decided to keep some of their research out of their published work, or even changed topics or field sites after realizing the cross-cultural research they had begun could not be sustained in an ethical manner (Chapter 2). If we acknowledge that knowledge production is a deeply political process, then how is this understanding reflected in our own research methodologies?

Guiding Questions

Cross-cultural research is both rewarding and risky. Here we provide some guiding questions to consider as you pursue your research. We provide examples from our own research project to illustrate how we sought to address these questions.

What Is My Position in Relation to Power Structures?

Many researchers begin their careers because they passionately love the topic, the place, or the people with whom they wish to work. They may also desperately desire to make a difference in the world—to help mitigate the effects of climate change on the world's most vulnerable, to find ways that migrants can live in dignity and safety, to right historical wrongs for racialized groups in their hometowns, or to study why and how powerful people exploit their workers. And yet, when we enter (or return to) communities as experts, we become part of a web of power relations ourselves and may do more harm than good. If we are relatively new to a community, we may not even realize how things like our educational credentials, gender, regional background, or racialized appearance shape how people interact with us, or transform relations in a place where we grew up. You might cringe at the word "expert," and avidly disavow this title for yourself and still be perceived as an expert when you show up at a community organization or a doorstep. Or you might be perceived as another ill-informed, extractive, and bumbling researcher in a community that has seen many. In some cases, you will also have to confront the legacies of another researcher who came for a summer, asked invasive questions, caused community arguments, and was never heard from again.

Situating Yourself in Your Research

It is important to assess clear-headedly where you are in relation to existing power structures. This may require a frank accounting of how your self-narrative may not be reflective of your good intentions. Does your linguistic, ethnic, caste, or class background align you with those with whom you will be speaking? Or is it something you will need to work through, possibly with each person you speak with? How do other factors intersect with this one? These questions, as feminist theorists argue, are ultimately about **positionality, reflexivity**, and **identity** (Nagar & Geiger, 2007; Sultana, 2007). The researcher's positioning with regards to the social and political context of their study (positionality), cannot be disentangled from their race, class, gender, age, sexuality, and ability (identity) among other things, and requires a self-awareness of their biases and blind spots (reflexivity) (Mills, Durepos, & Wiebe, 2010). For many, this line of questioning can uncover personal histories and unexamined privileges. This can be an unpleasant and perhaps even painful experience but even seasoned researchers will tell you that these questions are never fully resolved and can come up even after years of working with the same community.

In our research with Ladakhi youth, we are confronted with questions of positionality and identity at every stage of this still-unfolding project. For this project, both of us have complex and contingent ties to the region. Mabel's father is from the minority Christian community in Ladakh and she has close relatives living in Leh. Like the students, she has the experience of studying in Delhi and Mumbai but, having never lived in Ladakh, she does not speak or understand Ladakhi. Moreover, having lived in the US since 2010, where she received her doctorate and is now working, heightens her outsider status. Sara is an outsider, a white woman from the US. Being married into a Ladakh family for nearly 20 years has shifted some aspects of that outsiderness, but does not mitigate the degree to which economics, race, and colonial history shape her experience in Ladakh. For us it was therefore important to collaborate with a local organization like LAMO that has a long-standing relationship with the community, especially young Ladakhis.

If you are grappling with these questions, then chances are you might also be concerned with how to make academic research meaningful to the people you interview and/or collaborate with in the field.

How Will I Make My Research Meaningful across Cultural Realms?

As academics, in some ways our research is *always* cross-cultural, unless our research is on academia itself. Alongside any public-facing work that we do, our jobs or studies require us to engage in the world of academic literature, which is itself a culture, and a rather specific and insular one at that! This means that the research that we do is translated into the world of theory, peer-review, and advancing arguments. In some cases, this means that those we do research with might find our articles confusing, jargon-filled, or, worse, meaningless and boring.

If we take this situation seriously it might become a site of potential, rather than *only* a problem. The crux of this is: how can your research be meaningful both to the people you work with "in the field," (however you define this), *and* to others in geography? All cross-cultural research requires translation of various kinds, not limited to language itself. At the same time, working across language and cultural difference can add a layer of complication—even to the extent that it may not be easy to translate certain words or ideas and this can lead to significant misunderstandings (for examples see Howitt & Stevens, 2005, p. 55). In Ladakh, for instance, there are several local words for *Buddhist* and *Muslim*—however, each comes with a certain set of associations implying a political stance, such that in this case, it may be safest to use the English words rather than local terms.

To address these issues, we began from a topic that is deeply of interest to folks in Ladakh, and worked to make the research questions and methods appropriate and accessible to those with whom we would be speaking. We shared our proposal early on with LAMO staff to make it align with their existing goals, and incorporated features that would be compelling and useful for those involved in the project. For example, the directors of LAMO suggested that our first workshop (in summer 2016) include a short component teaching basics of film photography, led by Otsal, a local photographer, and an overview of the research process so that by participating in the project the students would also gain practical skills and knowledge.

In the workshop itself, we tried (though often struggled) to analyze and organize student comments into themes while we were in the room, so we could gauge whether students could see themselves in our analysis. Subsequent to the workshop, we held a public exhibition and advertised it to the general public, and encouraged student participants to bring friends and family. Reflections from the older generation and non-Ladakhis in the audience brought up important continuities and differences in the experience of this generation of Ladakhis. After this, the exhibition remained up in the centre, where people could read the 'zine pages as well as view the photographs and captions written by the young people themselves. As we have been writing up our findings, we have been sharing them with young people for their feedback. In considering how you can make your research meaningful across cultural realms, consider sharing results early and often in a range of accessible forms, and seeking reflections and feedback from those you are working with about what is and is not useful for them.

How Will I Be Sensitive to Cultural Differences?

Cross-cultural research is rewarding but also complex. If you are not from the research context in which you are working, or even if you consider it home but have spent a great deal of time elsewhere you may face challenges in how your actions are interpreted or in understanding why people behave the way they do. Read as much as you can about context. Do not take big steps until you have the lay of the land, and check things like your research process and your interview questions with collaborators for things that might offend or cause confusion.

Cultivate relationships with local experts, academics, or activists, as appropriate to your project, and check in with them when you find yourself confused or unsure of next steps or your analysis. It is important that you appreciate the labour this will take on their end and reciprocate! Can you proofread or provide feedback on grants or reports for them? Do they want you to set up a website to sell their local products? Is that feasible on your timeline? Can they co-author papers with you? Be aware of how much space you are taking up when people help you, and try to make the relationship one of reciprocity, not one in which you, the researcher, extract generosity and knowledge without reciprocation. This means saying yes to some things that may (at least at first glance) *have no benefit to you or your project.* If you live in the place where your research will take place, show up for your organization's events, and see if you can help out even when it is not pertinent to your own research.

In cross-cultural research, you may be working across two or more languages. If you intend to develop an extended research project in a place, it is crucial to work for fluency in the language. It is also important to attend to how language matters in the place where you work. For instance, in Latin America, if you choose to learn Spanish or Portuguese but do not learn the specific Indigenous language or languages in the place where you work, you must question who is excluded and what is missed. Even as you gain fluency, it is important to check and double check your own interpretation work, with attention to how nuances of class, gender, and ethnicity are infused into language and may alienate or put people off (see Smith (2016) for more discussion).

Depending on context, language may also become a challenge for you in academic scholarship and publishing: will you be able to read the academic literature and engage with scholars writing about the same topic as you? When you write up your results, will your translations be able to capture the intention of those who spoke with you (see also Bujra, 2006; Müller, 2007; Smith, 1996)?

It is difficult for most geographers to spend more than a year, at a stretch, in the field. If you are still a student, your time and financial constraints may be even more challenging. If you are in the field for a shorter period, the sense of urgency you feel might not be shared by the people you work with in the field who will have their own rhythms of life and prior commitments. It is crucial to be considerate of people's generosity, and also to keep in mind the potential of radical vulnerability.

You may need to expose more of your own personal life, economic position, and relation to power structures than you feel comfortable doing. When you are challenged in this way, consider what you are asking of those in your field site. For our long-term collaborative work, our local research assistants, Rinchen Dolma and Rigzin Chodon, remain our points of contact, and underscore the potential of working with local assistants if you have funding. For students on a short-term project away from home, you may find working with a local organization greatly increases your ability to work ethically and productively with a local community. If you are working locally, be frank and realistic about your limitations, and make sure that the organization you wish to work with will feel their efforts on your behalf are reciprocated.

What Is My Timeframe?

Careful cross-cultural research requires meaningful conversations and engagement before, during, and after the project. Does your academic schedule allow for this? Or can you scale down your project so that you can prioritize this engagement? Whenever possible, find an organization or group of individuals with whom your own research goals align. If there are no such organizations, do consider the ethics and accuracy of a research agenda that is not relevant to local folks. Try to make your work useful to this group, and accept their feedback, even when you cannot take it all on board or even when it is difficult to hear. The earlier you can do this, the better! Ideally, if you have prior ties to your research site or have the opportunity to do a summer pilot study before designing your major research project, you can share an early draft of your proposal with people in the place and receive feedback on it from them. Consider criticism of your project as a sign of trust and cultivate it. Many Indigenous groups are now developing their own research protocols and internal research review boards (Chapter 4). Seek these out and respect their integrity. In politically contentious areas, take care that the gatekeepers and leaders in a place do not steer your research in such a way that you do not end up encountering only the majority perspective.

Seeking funding for research often makes the process of aligning your goals with those of your collaborators tricky. Funding agencies ask that you establish the larger importance of the project beyond the field site, but in working with a local organization, it is necessary to prioritize their goals rather than to helicopter in with abstract ideas. Working this out requires conversation and negotiation. For an undergraduate or graduate student with a limited time to try out a project, you might be more constrained; however, you could consider contacting an organization as early as you can and seeing how your research might contribute to its existing goals.

Our project involved starting small with an experimental workshop before the longer-term project was developed. This established that some methods like PhotoVoice would be effective, and that we could work well with LAMO staff. A student project usually does not allow for such long-term engagement, but could still allow for collaboration with a local organization. Can your advisor or other contacts that you have put you in touch with someone to begin a conversation? Can you make phone or video calls before arrival to make sure you are on the same page with collaborators or, better yet, make a small trip in advance of your research or choose to work with a community in which you have prior connections?

How Will I Know I'm Getting Things Right?

If you are doing research in a context not your own, there are many ways things can go wrong. Often researchers begin by speaking to those in leadership roles in a community. This can be a great start, but sometimes it means that you end up obtaining only the perspectives of local elites and those from elsewhere in the social hierarchy who share their views. This might be fine, if that is your intent, but if your aim is to understand how a diverse range of people see the world, this can be a problem. Be aware of how these dynamics may shape your research. Additionally, make sure to pilot your question (test them out on a few folks who will give you frank feedback) before beginning your research in earnest. Furthermore, as you begin to interpret your findings and write about them, how will you know if your own interpretations make sense to those with whom you have spoken? You may consider **member checking**, or sharing your results with those you worked with in the field to see how their interpretation aligns with or differs from your own.

Ideally, by the time you are wrapping up your research, you will have developed relationships of trust and reciprocity and be able to have frank and open conversations with people in your research site about these questions. In some contexts, it may be difficult to establish these relationships, but you can at minimum discuss your preliminary findings with the people you engage with and get their reactions. For us, knowing that we are getting things right has been the feeling of excitement we see when young Ladakhis and other folks from India's mountain margins encounter our work and immediately respond with their own examples, with connection, and with delight at seeing a story they recognize. This manifests as invitations to attend events or participate in organizations, positive feedback on drafts of articles, and having young people reach out to us—for instance to share art related to their experience or to ask us to come and interview them to get their story.

As you develop relationships with those you are researching alongside, your priorities may also change. Howitt and Stevens (2005) recount their realizations in the field that their preconceptions about local communities' needs had not been accurate, leading them to shift focus—for instance, in Howitt's case, to shift from studying aboriginal land claims (research that could later have been used against the communities) to studying mining. Researchers' impulses to study communities that are facing structural oppression sometimes lead them to research those communities—when in fact members of those communities would prefer that people **study up** (though this is a fraught term), that is, to study those in positions of power and the mechanisms of power that create oppression. As you design and carry out your research, particularly if you are new, make sure to get feedback on each stage of your research project in order to make sure you are accountable to your collaborators.

What If Things Go Wrong?

As you embark on your research, it is important not to expect to get things right all the time. Such an expectation may lead you to conceal mistakes and weaknesses, rather than to embrace failure as part of the process. That is not an excuse to be sloppy, but rather an encouragement to pay attention and learn from points at which you know you could have done better. Each next step in research will hopefully be one in which you learn from those with whom you are engaged.

For example, the project discussed in this chapter emerged in part as a response to frustrations that Sara felt with her dissertation research. While that research (on how intimate life was connected to decisions about family life and marriage) was pertinent and meaningful in the context of individual conversations and dinnertime talk, many of the people with whom Sara spoke were not particularly interested in her analysis or in the outcomes of the research. They were happy to help and to get things off their chest, but after that, it was difficult to continue the conversation. The project with young people was designed to respond to this failure, by creating more interactive venues in which we could learn along the way what made sense or rang hollow for our young participants, and so that they might be interested in some of the research outcomes.

We each enter the field with different situated positions, meaning that the structures of privilege and oppression that we are embedded within follow us into research. In anthropology, Maya Berry et al. (2017) have discussed the complexity of "fugitive anthropology," of navigating fieldwork while positioned variously as women of colour, as Black women, as queer women. This means that the stakes of fieldwork can be very high, and as, for example, a Black woman researcher, and/or a researcher who is queer or transgender, you may simply face different kinds of obstacles for which you have not been prepared (perhaps particularly if your training is in a majority white department in which expertise at handling a range of researcher positionalities has not been fostered). You may find it necessary to develop a support network of peers or develop a buddy system for research in order to figure out how to navigate experiences that your mentors may not have encountered themselves—such as racist remarks in the field or threat of sexual assault. Fortunately, geographers and anthropologists now are grappling with these experiences and decentring expectations of stoic research performed by a universal unmarked subject often assumed to be a white, heterosexual, able-bodied man

(Berry et al., 2017; Faria & Mollett, 2016; Kohl & McCutcheon, 2015; Lopez & Gillespie, 2016; Proudfoot 2015).

Of course, you will also encounter more mundane problems in the field: interpersonal awkwardness, anxiety due to the need to perform the role of extrovert in your research even if that is not your natural state, keeping participants interested in your project, managing **field notes** in a timely manner. These logistical issues are also things you will want to plan for (for instance, taking careful field notes about things you did not understand may become even more useful in a cross-cultural context, in which you slowly gain cultural fluency over time). For many of the day-to-day problems you might encounter, the answer is often to remember the bigger picture and to go back to your root commitments. Let your ethical commitments guide how you respond to them and seek advice from trusted peers and people in the field.

What Will the Afterlives of this Research Be?

Cross-cultural research begins from the challenge of translation: will the things you write be meaningful to the people you are writing about? Are you revealing information that could sow conflict? Or, worse, can your research be used by state power structures to further marginalize people? You may need to keep some compelling "data," to yourself. This could mean masking identifying information, but in small communities this may be next to impossible. You may also keep in mind the possibility of "ethnographic refusal," in the terms of Mohawk anthropologist Audra Simpson (2007). That is, those you had been interested in working with may wish to refuse any kind of participation. If you are a member of the community that you are doing research with, the cross-cultural element may require your own careful curating of what is shared for and about your community. Sometimes refusal is the most ethical choice: refusal to be translated into academic language, refusal to trade on interpersonal connections, refusal to be made legible to the academy (Simpson, 2007, 2014; see also Lugones (2010) on the politics of translation). Being "from" a community or ethnic group, or sharing one or more aspects of your identity with the group does not give you free rein—you need to think about the limits of any one voice to speak for a group of people.

In our project, interviews and workshops with students that culminated in the exhibition continue to have a strong afterlife as presentations at conferences, research publications (such as this one), and also as forms of community engagement. The publications from this research are in different stages of the process. And in reading for and writing these articles we have been introduced to the exciting scholarship of other researchers working on young people in other parts of the Himalayas. The 'zine pages remain on display in the LAMO library, and we have heard from their staff that parents of young people in particular stop to look at the 'zine and comment that it provides them a glimpse into their children's lives, who may not directly share their experience with their parents. These 'zine pages were uploaded to a blog by Chris Neubert, a doctoral student working with Sara, and you can find them here: https://lamocards.wordpress.com/.

Another important afterlife of this research is the household survey, conducted by Rigzin Choden and Rinchen Dolma from LAMO, which focuses on the opinions and desires of parents, elders, and community leaders regarding education-driven migration. While prior qualitative research and media reports suggest a growing population of Himalayan students studying outside the region (McDuie-Ra, 2012), the extent of student migration and the experiences of their original families and communities are not yet widely known or understood, and no data exist on this population in the Ladakh region. We hope this will be an important resource for both community members and future researchers.

Conclusion

Reaching out across cultural difference is both fraught and full of potential. It is not an endeavour to be embarked on lightly, but carefully, with purpose, and collaboratively. As we note above, for most of modern history, when research was conducted across geographic spaces and cultural differences, it fed the purposes of colonization, empire, and obscured rather than revealed our human reality. As you embark on cross-cultural research, it is important to consider that if you find yourself in a position of power in relation to those positioned as Other in relation to you, sometimes it may make more sense to study the power structures and your place within them, rather than those who are subject to those structures. We conclude with examples of researchers who are cognizant of these questions and this history and seek to work against these still-existing power structures.

For Richa Nagar, even at home, working across difference must be nuanced, complicated, and "muddy." In *Playing with Fire* (2006), the Sangtin Writers Collective, a group of eight women, Anupamlata, Ramsheela, Reshma Ansari, Richa Singh, Shashi Vaishya, Shashibala, Surbala, and Vibha Bajpayee, from across class, caste, and education differences, worked together with Nagar to understand their shared political goals and divergent life experiences in order to develop a collective ethos for societal change. The book, published both in Hindi and in English, demonstrates how even among those sharing cultural knowledge, collective purpose is something that is built.

R. Aida Hernandez Castillo, a legal and activist anthropologist, has devoted her career to working alongside Indigenous women, mainly from Mexico and Guatemala. Her cross-cultural work has involved learning Maya and other Indigenous concepts and theorizations to work alongside Indigenous feminists protecting their land (for an overview of her work see Castillo (2016)). This has involved long-term trust building and translation—for instance translating theorizations of oppression such as Crenshaw's (1991) intersectionality—into the context of Indigenous/state relations, but also translating Indigenous ideas into anthropological theorizing.

All scholars start with a first project. As graduate students, for example, Michelle Daigle and Margaret Ramírez began thinking through the limits of current scholarship and possibilities for decolonial praxis in geography (see Daigle & Ramírez, 2019; Naylor, et al., 2018). Daigle is Mushkegowuk (Cree), a member of Constance Lake First Nation in Treaty 9. Her dissertation on Indigenous resurgence involved cross-cultural work to understand Cree belonging in the face of settler colonial capitalism (Daigle, 2016; 2018). Ramírez's dissertation approached Oakland, California, as a borderland, thus cross-cultural analysis, research, and translation, was at the heart of her understanding of racial capitalism and settler colonialism in Oakland as well as the Black and Latinx theorizations fighting against it and the settler colonialism (Ramírez, 2020). In her MA work, Pavithra Vasudevan co-produced a short film on water and environmental racism with Reverend William Kearney of Warren County, North Carolina, and went on to co-author an article with him (Vasudevan & Kearney, 2016). This led to her dissertation in Badin, North Carolina, writing a play based on **oral histories**, and having it performed in Badin as well as elsewhere, and also writing about the experience of being, in her words, a "Brown scholar in Black studies" (Vasudevan, 2019; 2021).

As you begin your journey as a researcher it may help to reflect on early career projects in which young researchers begin what might become a lifelong trajectory of deep engagement and "walking alongside."

Key Terms

cross-cultural research
diasporic
go-along interviews
identity
longitudinal interviews
PhotoVoice
positionality
preguntando caminamos
radical vulnerability
reflexivity
situated solidarity

Review Questions

1. What is positionality? How might your positionality shape your research?
2. How would you define cross-cultural research in your own words, in relation to your region of interest or sub-disciplinary interests?

Review Exercises

1. Everyone is an "insider," to one or several groups. Think about a group that you feel you belong to (this could be anything from an ethnic or religious group to a club or affinity group—like your soccer team or book club or an activist organization). What are some of the norms and values in the group? What are some of the power dynamics that affect how people interact within the group? If a researcher intended to study a particular topic within your group, what are the "deal breakers" that might close the doors to them? What are the steps that one might take to become a helpful and welcomed part of the group? Once you have taken notes on the above, pair up with another student in the class. Tell them about the group and then have them guess the answers to your questions. After they have done so, compare notes. What did the "insider" know that the outsider did not? Did the outsider have questions, insight, or reactions that you, the insider did not expect? What does this teach you about the research process?
2. Peruse a popular magazine or website that regularly showcases cross-cultural representations (e.g., *National Geographic*, *Travel + Leisure*). Choose one example of a representation of a community or group of people. Pick one example of what you consider to be a problematic representation of a community or people group—what troubles you? Can you find voices of people from those communities on social media, and see how their way of representing themselves is different? In the locale of your university, is there a neighbourhood, a region in the area, or another place that has a certain set of representations that the residents embrace or reject? How is this representation manifest?
3. Find a recent article in an academic journal matching your research interests, or one you have been assigned in class lately and analyze it based on the principles outlined in this chapter. How does the author explain their research process? Can you see how they have engaged with the community, or only the results of the research? Who is named? Who is cited? How are concepts that cross cultures explained? What stage of their career is the author in, and how might that have affected their possibilities (e.g., was it based on MA or PhD research, thus constrained in some ways)? After reading the article, what questions are you left with?
4. Creative and arts-based approaches can elicit very different responses from standard interview questions. In small groups, come up with a research question, and then two ways of addressing it: a short set of three interview questions and a small creative prompt (to draw a picture or make a collage, for instance). Now do a practice run: have your small group break up and mingle with the other groups, with each member

trying *either* the interview or the creative prompt. Reconvene and discuss how your responses differ and what kinds of information you gathered from the different approaches.

Useful Resources

Faria, Caroline, & Mollett, Sharlene. (2016). Critical feminist reflexivity and the politics of whiteness in the "field." *Gender, Place & Culture, 23*(1), 79–93.

Jazeel, T. (2014). Subaltern geographies: Geographical knowledge and postcolonial strategy. *Singapore Journal of Tropical Geography*, *35*(1), 88–103.

Kohl, Ellen, & McCutcheon, Priscilla. (2015). Kitchen table reflexivity: Negotiating positionality through everyday talk. *Gender, Place & Culture, 22*(6), 747–63.

Nagar, R. (2002). Footloose researchers, "traveling" theories, and the politics of transnational feminist praxis. *Gender, Place & Culture*, *9*(2), 179–86.

Nagar, Richa. (2019). Hungry translations: The world through radical vulnerability, The 2017 Antipode RGS-IBG Lecture. *Antipode, 51*(1), 3–24.

Nagar, R., & Geiger, S. (2007). Reflexivity, positionality and identity in feminist fieldwork revisited. In *Politics and practice in economic geography* (pp. 267–78). SAGE.

Sharp, Joanne, & Dowler, Lorraine. (2011). Framing the field. In Vincent J. Del Casino Jr, Mary E. Thomas, Paul Cloke, & Ruth Panelli (Eds), *A companion to social geography* (pp. 146–60). Malden, MA: Blackwell Publishing Ltd.

Smith, Linda Tuhiwai. (2013). *Decolonizing methodologies: Research and indigenous peoples* (2nd edition). Zed Books Ltd.

Sultana, F. (2007). Reflexivity, positionality and participatory ethics: Negotiating fieldwork dilemmas in international research. *ACME: An international E-journal for Critical Geographies*, *6*(3), 374–85.

TallBear, Kim. (2014). Standing with and speaking as faith: A feminist-Indigenous approach to inquiry. *Journal of Research Practice*, *10*(2), 17.

Empowering Methodologies: Feminist and Indigenous Approaches

Jay T. Johnson and Clare Madge

Chapter Overview

This chapter explores empowering methodologies through an examination of feminist and Indigenous research approaches. First, we consider what might be involved in employing empowering methodologies and suggest that such methodologies have deep roots in feminist research practice, which is discussed further in this chapter. Through time, features of feminist research practice have been incorporated into, and been changed by, many other research approaches. We then examine one of these approaches—that of Indigenous geographies—and outline some of the specific empowerment strategies currently being employed by Indigenous researchers. Having briefly outlined some of the potentials and challenges of feminist and Indigenous approaches for qualitative researchers, we outline some key issues to consider when using empowering methodologies.

Introduction

In this chapter we are concerned with the empowering and transformative potential and mechanisms of qualitative research. We consider first what is meant by *empowering research*. Such research is not simply about studying "something;" through its objectives and day-to-day research practices, rather, empowering research holds significant transformative potential for those involved. Thus, according to Raju (2005, p. 194), **empowerment** is "a process of undoing internalized oppression" and, in the case of women's empowerment, "it is also about changing social and cultural forms of **patriarchy** that remain the sites of women's domination and oppression." Women are recognized as proactive agents who can exercise **power** to alter the process of empowerment and participate in social change. However, for Louis (2007, p. 131), empowering research with Indigenous communities must "be conducted respectfully, from an Indigenous point of view" and should have "meaning that contributes to the community. If research does not benefit the community by extending the quality of life for those in the community then it should not be done."

At its simplest, empowerment therefore refers to the process of increasing the social, political, spiritual, economic, and/or psychological potential of individuals and communities. It often concerns groups that have been marginalized from decision-making processes through discrimination based on historically constructed unequal relations of power (on the basis of gender, race, ethnicity, sexuality, age, class, religion, nation, dis/ability). The process of empowerment aims to undo or overcome oppression and increase opportunities, knowledge, skills, collective action, and choices for those groups routinely pushed to the margins of society. It can also disrupt further attempts to deny improvement to marginalized groups' opportunities. Empowering research aims to support these groups to (re)shape organizations, policies, institutions, and everyday encounters affecting their lives. It can also challenge normative assumptions and negative stereotypes about these groups to promote greater justice or equity. Such research can be instigated and developed by marginalized groups, or through collaboration with researchers as allies who have access to useful resources, knowledge, and skills (see Chapters 2, 3, 8, and 16). In sum, according to Scheyvens (2009, p. 464) "empowerment means activation of the confidence and capabilities of previously disadvantaged or disenfranchised individuals or groups so that they can exert greater control over their lives, mobilize resources to meet their needs, and work to achieve social justice."

However, doing empowering research is not easy. Research that aims to transform the lives of the people it is working with, or to challenge hegemonic power relations and promote social justice, cannot be simply and quickly achieved. It can involve hard work, frustrations, contradictions, uncomfortable **reflexivity**, reinforced power relations (as well as successes, progressive change, and satisfaction). It should not be assumed, for example, that such communities wish to be transformed, nor that there are shared meanings about what empowerment might entail or how it might be attained. For example, many Indigenous communities face the false dichotomy between development and tradition, leading to conflicts between elders, youth, and community leaders. Rather, empowerment unfolds, is resisted, and is transformed through the process of research and there are no straightforward guarantees of liberatory research (de Leeuw et al., 2012). Indeed, in a critical appraisal of empowerment, Ansell (2014) argues for the need to adopt a relational approach that recognizes the need to transform power relations at multiple levels, while Wijnendaele (2014) suggests that emotions and embodied knowledges are crucial elements in bringing about such social transformation. An understanding of these types of complexities involved in the process of doing empowering research has been deeply influenced by feminist research practice.

Feminist Research Practice

There is no single route to conducting empowering feminist research because there are many different approaches to feminist geography. As Moss and Al Hindi (2010, p. 1) note, there are "myriad ways of being feminist, engaging in feminist praxis and producing feminist geographies." Moreover, feminist research practice has changed over time and has been conducted in different ways in different places. There is therefore no single story or totalizing account about empowerment and feminist research practice: it is diverse, sometimes contradictory, and overlaps with, draws on, and influences other bodies of geographical work, as we will discuss.

Early debates about feminist research practice were concerned with many questions. What makes geography research feminist? Are there any distinctly feminist methods? To what use should such methods be put? Answers circulated in a range of special journal issues and edited collections around the political goals that might be afforded through attention to the design, analysis, and dissemination of a research project and to the diverse methods that might be used to achieve these feminist aims (see *Antipode*, 1995; *Canadian Geographer*, 1993; Jones et al., 1997; *Professional Geographer*, 1994, 1995; WGSG, 1997). There was a (contested) understanding that no particular research methods were distinctly feminist. Rather, it was more important to consider the work to which the methods were put and to choose methods appropriate for answering the research questions and addressing the aims of the research. In other words, it was the **epistemological** stance taken towards the methods that was important in achieving feminist goals. However, what many studies in feminist geography did have in common was their political and intellectual goal of changing the world they sought to research by engaging in social and political change. Such transformative feminist research often initially foregrounded women and/or gender as the primary social relation (see Box 4.1).

BOX 4.1

Example of Research Empowering Women

Based on the Dangme West district of Ghana, a paper by Charlotte Wrigley-Asante explores how poverty reduction programs (PRPs) with credit components can reduce women's vulnerability to poverty and significantly improve their socio-economic status through access to financial and non-financial resources. This has, in some cases, improved gender relations at the household level, with some women being recognized as earners of income and contributors to household budget. Other women, however, still regard their spouses as "heads" and require their consent in decisions even related to their own personal lives, and their improved economic status has created confrontation between spouses. The paper recommends that assisting organizations must address "power relations" at the household level, otherwise socio-cultural norms and practices, underpinned by **patriarchal** structures, will remain cages for rural women.

Source: Wrigley-Asante, Charlotte. (2012). Out of the dark but not out of the cage: Women's empowerment and gender relations in the Dangme West district of Ghana. *Gender, Place & Culture* 19(3), 344–63.

So feminist geography research involved politicizing a methodology through feminism to conduct research that was often pro-women, anti-oppression, or based on social justice (Moss, 2002a), thereby challenging male dominance, making women's lives visible, and exposing gender inequalities (England, 2006). It was recognized that this involved the whole project, from the initial decision to undertake research on a specific topic, to the members of the research team and the specific methods selected, to presenting the final outcomes.

As feminist geography asserted itself (*ACME*, 2003; Bondi et al., 2002; *Gender, Place & Culture*, 2002; Moss, 2002b; Moss & Al Hindi, 2008), there was increasing recognition that there was no singular feminist geography political project, precisely because understandings of feminism were grounded in specific histories, cultures, places, and biographies. As feminism became reconstituted into diverse *feminisms* that challenged the idea of a "universal female identity" (often based on the unstated assumptions of white, heterosexual, able-bodied, Western norms), so too came increasing awareness of the multiple oppressions affecting women's lives, which demanded that feminist geography go beyond gender as the central construct, to recognize its intersection with lifecourse/age, class, race, ethnicity, sexuality, dis/ability, place, nation, and religion (see, for example, Blanch (2013)). Drawing on influences from queer, **post-colonial**, and ecofeminist theory, and anti-racist and transgender politics, there was increased recognition of the multiplicity of social relations of difference and the myriad hierarchies of power that were involved in research with diverse social groups. This recognition of the differences between women (and men) is sometimes referred to as "third-wave feminism." It led to some demanding questions for feminist geographers through the deconstruction of the category of gender that had initially formed such an important foundation of feminist geography enquiry. Thus, Jenkins, Jones, and Dixon. (2003) asked, if gender inequality were no longer privileged, what made a feminist project distinct from other critical human geography projects? In response, Raghuram and Madge (2007, p. 221) suggested that diverse feminist geographers still shared a (polyvocal) interest "in challenging the varied forms and effects of gendered power differentials as they intersect with a host of other factors, such as race, class and nation, and in a commitment to dialogic, pedagogic, research and political practices." Sharp (2005, p. 305) concurred that it was "not just the processes through which data is collected . . . that makes it feminist, but also the way in which projects are conceptualized and how we as researchers act as people (ethically, politically, emotionally) while engaged in the process."

As Sharp (2005) was intimating, three key themes were emerging out of constantly evolving feminist geography research practice. First, feminist geography research critiqued *ways of knowing* (epistemology) by contesting **objectivity** and validating **subjective** experience, acknowledging the situatedness and non-universality of knowledge creation and demanding awareness of the importance of context in producing what could only ever be a partial understanding of any research situation. This created new understandings of what counted as "knowledge." But it also presented challenges to *ways of researching* (**methodology**, the second theme): accounts were replete with discussions of the complexity of power relations and **ethical** issues involved throughout the research process, the multiple and shifting identities of all those involved, and how these influenced both the knowledge created and contested the boundaries of the field of research. Feminist geographers were at the forefront of discussions surrounding reflexivity, **positionality**, politics, and accountability. Emerging out of these debates were reflections about the ambivalent yet **embodied** nature of the field (Parr, 2001; Sundberg, 2003); the emotional entanglements involved in feminist research, in which research subjects were viewed as knowledge agents (Bondi, 2003; Chacko, 2004); and the complexities of establishing collaborative ways of presenting findings through alternative writing strategies (Sharp Browne, and Thien, 2004). And thirdly, feminist geography research was also significant in stimulating debate surrounding *the politics of research*, promoting ideas about conducting research that allowed "silenced" voices to be heard; recognizing the multiplicity of viewpoints, voices, and locations that geographical investigations might entail; and retaining a sharp focus on the social and political empowering potentials of research.

These features of feminist geography research practice have been highly influential beyond feminist geography: they have formed important elements of debates and developments in qualitative research in cultural, queer, sexual, emotional, children's, participatory, post-colonial, and Indigenous geographies, for example. Perhaps this should come as no surprise, for feminist geography has always been in iterative dialogue and contestation, both changing and being changed by, other sub-disciplinary ideas, languages, and political visions. Through this process Sharp (2005) argues that although an awareness of, and sensitivity to, gender has been mainstreamed in geography, the feminist political project still operates on the discipline's margins (Sharp, 2009, p. 77). Despite this, many feminist research practices have become commonly accepted aspects of qualitative research (for example, consideration of reflexivity, positionality, research power relations, **situated knowledges**, emotions), although feminist geographers are also still at the forefront of troubling over and unsettling those very practices (see Billo & Hiemstra, 2013; Chattopadhyay, 2013; Faria & Mollett, 2016). Thus feminist geographers have highlighted the tricky and messy nature of empowering research in practice (see Attanapola Brun, and Lund, 2013; Nagar, 2013).

So perhaps feminist geography research practice can best be envisaged as **rhizomatic** (Moss & Al Hindi, 2010), constantly changing and being changed. The political visions of feminist geography are not sedimented into simple stages of historical progression nor do feminist research practices seeking empowerment remain unaltered. Rather, there is a situation of both continuity and change, with some tried and tested research methods and political intentions coexisting alongside newer ideas and practices. Thus today feminist geography research projects employing empowering methodologies are greatly varied. Some projects focus on empowering women (Buang & Momsen, 2013); others on exposing "naturalized" gender power relations or highlighting patriarchal assumptions, practices, and male bias (Bee, 2013; Zanotti, 2013); others on considering boys/men, manhood, and on challenging hegemonic masculinity (Faria, 2013; Lahiri-Dutt, 2013). Other projects interrogate heteronormativity through a focus on sexualities routed through queer theory or transgender politics (Dominey-Howes, Gorman-Murray, and McKinnon, 2013; Rosenberg, 2017; Selen, 2012), while others are more concerned with geographies of Blackness, questioning the universalism of Western gender theories and feminist readings of gender and sexuality (Bailey & Shabazz, 2013; He, 2013; Sullivan, 2018). Others explore how feminist politics might be revitalized through the dynamic networks of new media (McLean & Maalsen, 2013), through reinvigorated discussions surrounding patriarchy and everyday sexism (Valentine Jackson and Mayblin, 2014) or via engagement with anti-colonial scholarship (Coddington, 2017). However, as Moss (2005, p. 42) summarizes, what still distinguishes research as feminist is that "it deals with power in some way—whether conceived as something to be held, exerted, deployed, mobilized, sought after, or refused, or as something structural and inevitable, despotic and concentrated, or dispersed and everywhere." Next we examine one research approach that has drawn on, and influenced, feminist research practices: that of Indigenous geographies.

Indigenous Research

Geography has a long history of supporting the colonial expansion of Europeans into Africa, Asia, the Americas, and the Pacific. Wherever explorers ventured, the grid of the cartographer was soon to follow. Geography's colonial history has been documented by a number of geographers since the mid-1970s. Godlewska and Smith's (1994) *Geography and Empire*, for example, detailed the history of geography's complicity in European colonialism and laid a foundation for geographers interested in shifting the focus, or **decolonizing**, contemporary research within the discipline. One part of decolonizing the discipline has been focused on examining *how* research is undertaken.

At the same time that geographers, and other social scientists, began to reflect on their role in constructing and perpetuating colonialism, Indigenous peoples around the globe were coming together to redefine their tribal or regional conflicts with colonial and settler-state powers within a new, global anti-colonial narrative. This articulation of a global indigenism, or **indigeneity** (see Niezen, 2003) has resulted in relocating Indigenous–state conflicts from the national to international scale. The establishment of the United Nations Permanent Forum on Indigenous Issues (2000) and the Declaration on the Rights of Indigenous Peoples (2007) are two milestones within this global narrative around indigeneity.

As Indigenous peoples have been articulating their struggles within international forums, they have also been rejecting their long history as the "natural"

subjects of Western research. This push back by Indigenous peoples has led to a significant shift in Indigenous-focused research by geographers since 2000. While Indigenous peoples have been the focus of research by geographers since the rise of geography as a modern discipline in the early nineteenth century (see von Humboldt, 1811), much of this work has been exploitative in nature. It was research "on" and "about" Indigenous peoples and geared toward controlling and dominating populations who had only recently come under the jurisdiction of European crowns.

This method of doing research "on" Indigenous peoples has remained the dominant paradigm for the past two centuries and only began to shift once Indigenous communities began to articulate their own research methodologies and agendas. The crucial moment of paradigm shift within the academy is marked by many as the publication of Linda Tuhiwai Smith's (1999) influential book *Decolonizing Methodologies* (second edition 2013). Smith's book provided an opening for dialogue not only on how research methodologies have harmed Indigenous communities around the globe by aiding the EuroAmerican colonial enterprise, but also by asserting that research could and should be conducted "for" and "with," instead of "on" and "about" Indigenous communities. While it may seem like a straightforward and common-sense notion that Indigenous communities should be active collaborators and participants, deciding what research is done in and on their communities and homelands, this is a relatively new idea within the academy (see Box 4.2 for an example).

BOX 4.2 Example of Research Empowering Indigenous Peoples

Based on a research partnership involving the James Bay Cree community of Wemindji, northern Quebec, and academic researchers at Canadian universities, this research for a conservation project that used protected areas documents the process of applying community-based participatory research principles as a political strategy to redefine relations with governments in terms of a shared responsibility to care for land and sea. The authors describe how empowering methods, including collaborative, equitable partnerships in all phases of the research; promotion of co-learning and capacity building among all partners; emphasis on local relevance; and commitment to long-term engagement, can provide the basis for a revamped community-based conservation that supports environmental protection while strengthening local institutions and contributing to cultural survival.

Source: Mulrennan, Monica E., Mark, Rodney, & Scott, Colin H. (2012). Revamping community-based conservation through participatory research. *The Canadian Geographer, 56*(2), 243–59.

The other crucial shift Smith's (1999) work has brought to the fore is even more significant. For Smith, the logical progression of decolonized research methodologies is an inevitable assertion of Indigenous methods—methods conceived and articulated from non-Western world views. Smith describes a set of research projects within a framework based on her experience as a Māori woman. However, Smith did not simply imply that all Indigenous research need originate within a Māori perspective but instead she opened the door for different Indigenous communities to articulate their own specific Indigenous research methodologies. Hence, Indigenous communities need no longer be constrained by the Western research paradigm that has laboured to colonize them for centuries. Today, Indigenous communities frequently control not only the research agenda in their own communities, but that research can also take place using methodologies conceived within their own **ontology**.

Smith's ground-breaking work has opened space, albeit at the margins of the academy, for a deeper dialogue on how **Indigenous methodologies** might be conceived and articulated. Books by Shawn Wilson (2008) and Margaret Kovach (2009) pushed these boundaries forward for a broader Indigenous studies audience. Journal special editions brought this discussion within the sphere of the geographic community (*American Indian Culture and Research Journal*, 2008; *Canadian Geographer*, 2012; *Cultural Geographies, 2009, Geografiska Annaler B*, 2006; *Geographical Research*, 2007), which has provided a platform for expansion. Although the specific articulations concerning how Indigenous research methodologies should operate are as varied as the individuals and communities voicing their opinions, a few key concepts seem to have reached consensus status.

These key concepts, first identified by Harris and Wasilewski (2004) through their work with *Americans for Indian Opportunity*, articulate an indigeneity that cuts across the obvious differences between Indigenous groups throughout the Americas and beyond. These key concepts, commonly referred to as the 4 Rs—relationship, responsibility, reciprocity, and redistribution—were determined to be core shared values articulated through two decades worth of meetings and discussion among diverse Indigenous groups in the 1980s and 1990s. They have also been adopted by Indigenous academics and are now commonly referenced as the core ethical values that should govern an Indigenous research methodology. This drive to articulate Indigenous research methodologies is, in part, as Louis (2007) has identified, because there is a significant difference between research done with Indigenous communities using Western methodologies and using Indigenous methodologies.

The first stage in any research project is the establishment of a *relationship* between the researcher and those with whom they intend to work (see also Chapters 2 and 8). This is the same for those intent on working with Indigenous communities, although the establishment of a relationship in this context implies a deeper sense of *responsibility* than might be expected in many research relationships. Russell Bishop (2005, p. 118) describes this establishment of an ongoing relationship as "the process of establishing an (extended family) relationship, literally by means of identifying, through culturally appropriate means, your bodily linkage, your engagement, your connectedness, and, therefore, an unspoken but implicit commitment to other people." Building such a research relationship by creating an extended family commitment around your shared interests requires showing your face to the community. This form of relationship cannot be negotiated through emails and phone calls. Many a non-Indigenous researcher has been confounded by a lack of response from Indigenous communities to their research queries. For many Indigenous communities, no response is a way of saying "no" without creating disharmony. This raises the fundamental question of whether a "no" response should be conceived as meaning "no" to the research being proposed or whether it merely means "no, not now." Sometimes developing a research relationship with an Indigenous community or individual may require many hours sitting in an office or home, perhaps drinking tea or coffee, and talking. This dialogue is not only about establishing the research and agreeing to its parameters, but it is also about developing trust across the complex power inequalities inherent in any relationship between academics and non-academics, particularly where research has played or continues to play a role in the colonial relationship (see also Chapters 3 and 8).

The myriad responsibilities one takes on in establishing such a relationship are founded within acts of *reciprocity*. This reciprocity, while predicated on both acts of giving and receiving, is motivated by giving: not giving as charity, but giving as honouring. As Harris and Wasilewski (2004, p. 493) describe, "at any given moment the exchanges going on in a relationship may be uneven. The Indigenous idea of reciprocity is based on very long relational dynamics in which we are all seen as 'kin' to each other." Building a research relationship, then, with an Indigenous community cannot be based on a "helicopter" approach where you drop into their lives for a short stay and then disappear with the information you need, never to return. These extended familial, research relationships require lasting and durable commitments; they require not only being hosted in the community but they also require a reciprocal hosting in your own home or institution. It is a cyclical reciprocity predicated on a continual renewal and sharing (see also Chapter 16).

Redistribution is inherent in sharing and serves to balance or rebalance relationships. As researchers, we are disciplined to view research as our possession: knowledge we have created and own. The truth, though, is that research never takes place in a vacuum free from the influence of those with whom we "do" research—in fact, they are aiding us in creating new knowledge. Frequently, it is just a translation of knowledge already commonly held by Indigenous communities to a non-Indigenous audience. Sharing gifts, whether they be material wealth, information, time, talent, or knowledge, is all a part of this obligation of redistribution. Within many Indigenous communities, this is referred to as a "give away" or potlatch. Central to this obligation is the maxim "to whom much is given, much is expected." It is through this redistribution that everyone in the community is valued.

While increasingly influenced by Indigenous world views in their formation, Indigenous research methodologies remain in dialogue with critical, participatory, and feminist methodologies. In recent years this dialogue has coalesced around discussions of decoloniality (Jazeel, 2017; Noxolo, 2017a; Radcliffe, 2017a, 2017b). This has involved contesting the unmarked white privileges and continued exclusions, experiences, and resistances of Indigenous and Black, Asian, and minority ethnic (BAME) geographers within the geography academy (Desai, 2017; Faria et al., 2019; Mullings & Mukherjee, 2019; Tolia-Kelly, 2017); decolonizing geography curricula to produce multi-epistemic learning and pedagogy (Daigle & Sundberg, 2017; Esson, 2019); challenging the "whiteness" of Anglophone concepts, assumptions, and epistemologies to produce critical accounts of the racialized formations of geographical knowledge production (Legg, 2017; Naylor et al., 2018; Noxolo, 2017b); and articulating and operationalizing decolonized methodologies (Tuck & McKenzie, 2015). This transformation of geographical concepts, institutions, curricula, and methodologies from colonial to decolonial sensibilities requires vigilance and determination. This is perhaps best articulated by Richie Howitt (2019, p. 11), in his recent address to the American Association of Geographers when he challenged geographers to adopt a new methodology focused on listening, asserting that: "Listening is somehow less strident, humbler, and more respectful than the imperial methodologies of possessing, occupying, imposing, and silencing."

Having discussed some of the potentials and challenges feminist and Indigenous research practices raise for qualitative researchers, in the next section we outline some key issues to consider in making the move towards using more empowering methodologies. This discussion is only a starting point—there are numerous other issues we have not had space to mention—and it is structured around approaching the research, doing the research, and the politics of the research, although of course these three processes are constantly interacting.

Using Empowering Methodologies

Approaching the Research

Early Beginnings: Creating a Long-Term Dialogic Relationship

According to Raghuram and Madge (2006, p. 275), the initial framing of research might be in terms of why the research is being conducted in the first place, an approach that forefronts the ethical issues of who gains from the research and why. The researcher must start to think about and work through in dialogue the power relations, inequalities, and injustices that enable and allow the research to occur and must be committed to working towards challenging these at different scales—the personal, the institutional, and the global–political. This might be attempted through a process of engaged pluralism (DeLyser & Sui, 2013, p. 10) where various "views are engaged, divergences openly tolerated, and differences dialogically embraced. Differences may not be resolved, but genuine engagement can lead to enhanced . . . creativity on all sides, stimulating new thought." This might, for example, be undertaken through the use of "talking circles" (see Evans et al., 2009, p. 903), in which the opportunity to speak is distributed sequentially around the circle and confrontational style argument is discouraged.

However, from the outset, it is also important to recognize that the research project may not always be instigated by the researcher. It is imperative to make space to listen for and respond to the self-determination of women's or Indigenous groups that might articulate a need for a specific research project or might initiate the research in the first instance using their conceptual notions, research designs, political intentions, and ethical review practices, which may differ from those of the researcher. Here the researcher might become an academic ally, acting as a conduit between the research community and academic institutions and public funding organizations through all stages of the research process. This process of "walking with" (Sundberg, 2014, p. 39) the research community might take the form of supporting and fostering the group's capabilities to undertake research themselves and advocating for the institutional and structural changes necessary to make this possible. Thus the researcher might actively work with (or in response to) the community or group from the inception of the project, including articulating initial research questions, writing grant proposals, agreeing on shared responsibilities in the implementation of the research, discussing redistribution of the resources for carrying out the research, and considering how results might be analyzed, written, and reported to produce different types of research products that may be differentially beneficial to the various groups involved in the research. This process is likely to involve lasting and sustained relationships, commitments, and obligations. It also involves recognition of the active political subjectivity of women's and Indigenous groups, which may have their own structures of power that shape research agendas, designs, and relations, as well as potential harmony and/or dissent within their groups. All of this can be daunting for younger researchers, so we recommend working alongside a mentor who has spent time developing the necessary relationships with communities as they develop their own connections and learn more about the needs of the community.

Moving the Centre: Making Space for Multiple Ontologies and Polycentric Epistemologies

A second key issue in approaching empowering research is the value of developing a research sensibility that is open and hesitant, that refuses "to allow the taken-for-granted to be granted" (Ahmed, 2004, p. 182, quoted by Sharp, 2009, p. 78).

An example of this refusal is given by Mishuana Goeman (2013). She argues that it is vital to support the efforts of Indigenous nations that are working to refocus research beyond replicating settler models of territory, jurisdiction, and race toward the remapping of settler geographies and the centring of Indigenous knowledges. This appreciation that different societies (or groups and individuals in society) might have distinct views of the world, or have diverse ways of being in the world, involves being receptive to the idea of **multiple ontologies** (Hunt, 2014). Furthermore, different groups may have diverse ways of knowing, asking different types of questions about the world and transmitting them in varied ways, signifying the need to validate **polycentric epistemologies** (Harding, 2011, p. 154). In other words, if we can start to understand and value that there are multiple world views and many different ways of conceptualizing knowledge (although these might initially be unfamiliar and difficult to comprehend), we can start to appreciate that the world is made up of manifold, heterogeneous, dynamic ways of being and knowing. This is a vision of a **pluriversal world**, in which many worlds belong (see Sundberg, 2014, p. 34).

Thinking about the world as pluriversal involves advocating and making space for multiple knowledge systems and life worlds that are legitimated *on their own terms* (see Eshun & Madge, 2016). In this process, "Western" knowledge loses its central and universal position and becomes one of a range of competing and contested knowledge systems. This suggests that Western knowledge might start to be regarded as a local or provincial knowledge (Chakrabarty, 2000)—knowledge that is locally produced but has gained its apparent universality through being projected outwards throughout the world through colonial and neo-colonial power relations. Thus, according to Escobar (1995), the domination of Western knowledge is explained not through a privileged proximity to the truth, but as a set of historical and geographical conditions tied up with the geopolitics of power. This move forces recognition that so-called "powerful" Western discourses are also partial and fragmentary, often involving knowledges and practices emanating from "Indigenous informants." In turn, this challenges the idea of a precise dichotomy between Indigenous and Western knowledge formations, moving us toward a position of multi-epistemic literacy (see Sundberg, 2014, p. 34). In making this conceptual relocation that unsettles the hegemony of Western knowledge and challenges the strict Indigenous–Western binary, space is cleared for Indigenous knowledge to be relocated as one of many legitimate and valid (albeit sometimes competing) knowledge formations. This enables moving beyond "anthropological particularism"—in which Indigenous knowledge is seen as unchanging, pristine, traditional, or local, in opposition to modern, universal, global, Western knowledge—toward a position in which *all* knowledge might include mysticism, spiritual ontologies, and ritualistic methodologies and in which *all* knowledge is considered partial and emerging, but at the same time also place-specific or situated.

Research with No Guarantees: Troubling over the Research Process

However, this creation of a long-term dialogic relationship, which identifies and validates multiple ontologies and polycentric epistemologies, is not easy to achieve. It will include a commitment to respond to issues raised by research communities, a willingness to engage in continuing dialogue that takes into account the conceptual landscape of all those involved in the research process, and an awareness that, despite a shared desire to participate in empowerment politics, there may be contested meanings about what empowerment might entail or how it might be achieved. In other words, from the outset of the research it is important that all parties involved acknowledge that there are *no guarantees* of successful emancipatory outcomes (de Leeuw, Cameron, and Greenwood, 2012; Noxolo et al., 2012). Rather, these outcomes must be carefully worked toward through everyday research practices and intimate research relations.

This process of conducting empowering research is not likely to be straightforward; indeed, it can create a range of complicated practical issues in institutions (e.g., universities and grant agencies) where a more limited model of research is espoused and reinforced. It also may be fraught with contradictory and potentially refuted relations and complex emotional investments because the creation of knowledge is never "innocent"—it is always entwined with differentiated relations of power. This constitutes what Smith (2005) has termed "Tricky Ground." This tricky ground concerns the troubling methodological, ethical, and political issues and inter-subjective relations that require continual communal reflexivity in the process of developing workable research relationships. For example, as researchers we should be acutely aware of the limits to our understanding, ceaselessly grappling with the production of academic work, to acknowledge the limitations of speaking for others. This involves being mindful of the risks of appropriation of knowledge creation while always being open to new ways of thinking about and understanding the world. (For examples of the complexities of cross-cultural dialogue see Chapter 3 in this volume as well as Desbiens and Rivard, 2014; Hunt, 2014; and Windchief and San Pedro, 2019). This open and hesitant approach is also important during the process of doing the research, as explored next.

Doing the Research

Employing a Multi-layered Reflexivity

As we outlined above, the development of a relationship is the first and primary component of any methodology that aims at empowerment within collaborative anti-colonial research. The relationship-building component of the research process can take many forms. Bishop (2005) has described the process as one of developing an extended family the common interests of which are the agreed-upon research goals and objectives. De Leeuw, Cameron, and Greenwood (2012) have described a process centred on friendships that extend beyond the research framework, allowing for a more profound critique of the research process and greater reflexivity. Reflexivity has, through feminist and Indigenous research approaches, become key to any collaborative, empowering research methodology.

The reflexivity we outline here, adapted from Ruth Nicholls' (2010) work, encourages a multi-layered approach. This first layer, or *self-reflexivity*, asks the researcher to explore the hidden assumptions about the research that originate within disciplinary structures or funding streams that enable the work to proceed. It also involves being self-reflexive about the epistemological and ontological assumptions that the researcher brings to the research project, a process that may well involve unlearning what one has already learned (see Sundberg, 2014, p, 39). This might be in terms of rethinking the questions asked, or delving deep into analytical and interpretative understanding in the field through ongoing dialogue, or making room for redefining terms of representations or conceptual framings. The researcher should also attempt to become cognizant of the complex and changing power relations inherent throughout the research process, particularly the (almost inevitably privileged) position researchers bring to the relationship. Thorough consideration of the often unmarked privileges of race, nation, religion, and class is especially important for non-Indigenous researchers in establishing a critical and dynamic relationship with Indigenous collaborators; such **axes of difference** are also significant for women from the global north working with women from other parts of the globe.

The development of a research relationship requires the researcher to carry this self-awareness into dialogue with others. This second layer of reflexivity, termed *interpersonal reflexivity* by Nicholls (2010), implies a relationality that necessitates evaluation of interpersonal encounters within particular institutional, geopolitical, and material situations. Recognizing one's role within the (changing) research relationship necessitates that researchers reflect on their ability to collaborate as opposed to lead, control, or delegate. The researcher is commonly placed in between the expectations of academic institutions and the community, navigating the intersection of ethical demands. As de Leeuw, Cameron, and Greenwood (2012, p. 188) observe, "researchers who carry out participatory projects quickly confront the mismatch between demands of the institutions within which they operate and their own commitment to build meaningful relationships with the people and places about which they care."

The third layer of reflexivity, termed *collective reflexivity* by Nicholls (2010), requires all participants to engage in a dialogue about the process of doing research together. What are the terms of participation? Who initiated the research project, and why? Who involved themselves, and why? Whose voices have been heard and what form has this taken? How was the research conceived and carried out, and how did this affect social change and practical knowing? Has the research process been transformative, affirming, cathartic, empowering, and if so, for whom, and if not, why not? This third layer of reflexivity entails a shift in the researcher's positionality, "a ceding of research control beyond the initial phase of negotiation, and extending participation into data collection, analysis, and distribution" (Nicholls, 2010, p. 25). This approach, founded within a radical pedagogy, pushes beyond mere information transfer towards a **critical consciousness** that Freire (2000) argued provides the foundation of empowerment. Conducting empowering research may also be promoted through the employment of dialogic research tools.

Dialogic Research Tools: Reworking the Field as a Methodological Site of Agency

Raghuram and Madge (2006, p. 276) argue for the need to explore methods that will make research questions more **dialogic**. But in advocating dialogue they do not presume that difference can be simply "dissolved" to attain complete understanding, "for there will always be degrees of incomprehensibility and continuing spaces 'in between.'" However, they do suggest that working through these in-between spaces can "bring moments of enlightenment; glimpses of the world through someone else's reality and a sense of the losses associated with privilege. From such moments of deep personal and political change more relevant research questions can arise, questions that can potentially challenge the 'master narrative' of northern-centred research."

One research method that might enable such dialogue is **storytelling**—a research tool "wherein personal, experiential geographies are conveyed in narrative form" (Cameron, 2012, p. 575). This approach to narrating research experiences presents expressive and affective methods that can uncover new understandings and perspectives, or be a means to express different world views or expressions of being in the world. There is also a political potential to storytelling to construct counter-narratives that test dominant discourses and produce social change (Gibson-Graham, 2006). A storytelling approach is particularly well-suited to collaborative research with Indigenous communities as there are many synergies with Indigenous forms of knowledge creation and sharing (Christensen, 2012). However, such an approach does raise questions concerning authorship, first- or third-person narration, and issues arising from translating oral stories into text. Storytelling also requires considerable attention to concerns of power and representation (Eshun & Madge, 2012; Garvin & Wilson, 1999). However, despite the complexities of employing more dialogic methods, they can enable inclusion of community research agendas, thus having potential to rework the field as a methodological site of agency (see also Chapter 10 on oral histories).

The importance of thinking critically about the field—the place and the specific context in which research occurs—has been stressed for some time by feminist geographers (see Moss (2005) for example). As research is place-specific and all knowledge production is situated, the critical consciousness advocated by Freire (2000) develops through everyday lived experiences (Johnson, 2012).

To understand the place-based struggles of different communities necessitates engagement with the experiences, conflicts, languages, and histories they rely on to construct their collective identity. Here the *place* of research (or the field) has potential to become an active location of empowerment. How such emancipatory change might be achieved in a particular place involves not only thinking about and doing the research, but also includes consideration of the political outcomes of the research process, as we explore next.

The Politics of the Research

Changing Ourselves: Breaking Out of the "Hall of Mirrors"

As is probably clear from the preceding discussion, the political outcomes of the research process can take many forms. Initially this might be considered in terms of changing ourselves by interrogating, destabilizing, and reconfiguring our underlying epistemological and ontological assumptions. This is an important process in recognizing but also challenging the links between geography and (neo-)colonialism and moving toward more decolonized versions and visions. Rose (1999, p. 177) observes that Western science "sets itself within a hall of mirrors . . . mistakes its reflection for the world, sees its own reflections endlessly, talks endlessly to itself, and, not surprisingly, finds continual verification of itself and its world view." So how might researchers break free of this "hall of mirrors" in their research to promote the process of empowerment and participate in social change?

For research outcomes to be empowering, we need to embrace the uncanny realization that multiple ontologies are not only possible but are also the lived reality of most of our fellow humans. Putting ourselves into a space within which, as researchers, we can begin to glimpse these alternate but equally valuable ontologies requires us to challenge commonly accepted frameworks and hidden assumptions in a shift towards breaking out of the hall of mirrors. This will involve the relentless need for a rigorous interrogation of the politics of speaking and writing, and being open to opportunities to criticize and challenge dominant world views and propose alternative agendas rather than adding to existing ways of thinking about the world and conducting research. This process is likely to be demanding, particularly in an era when colonial notions of gender, race, class, etc. are constantly being re-inscribed by populist movements globally through new inscriptions of anti-immigrant, racist, and misogynistic discourse. Avoiding co-option of groups on the margins of society and appropriation of their knowledges and world views will involve constant vigilance and an active political agency on behalf of all those involved in the research process. It will take courage and determination to avoid propagating underlying neo-colonial "business as usual with the odd tweak," and to instead move beyond existing knowledge formations, experts, and institutional structures of power.

Changing Institutional Structures and Processes: From Research "for" and "with" to Research "by"

It is clear that there will be no easy and definitive answers, but a careful working towards dismantling, or at minimum acknowledging the complexity of, the historically produced power geometries (of colonial, white supremacist, capitalist, heteronormative, ableist patriarchy) upon which geographical research is based is crucial. It is only by stepping outside of hegemonic systems of knowledge production that a shift in the paradigms of research can begin. As Kuhn's (1962) work has demonstrated, these shifts are frequently concurrent with social, political, and technological upheavals that not only upset the rationalized frameworks of science and research, but also question fundamental social structures that perpetuate colonialism, homophobia, racism, ableism, and sexism. Following the manner in which feminists have occupied and changed the academy and its knowledges, everyday practices, and politics, indigeneity as a social movement operating both inside and outside of the academy is now also placing pressure on these hegemonic societal structures of control. Unfortunately, though, these hegemonic social structures continue to be reinforced, just as the counter-hegemonic pressures grow louder.

By placing pressure on the academy, Indigenous geographers are beginning to uncover "the spaces between intellectual and lived expressions" of indigeneity, prising open gaps in regimes of knowledge production and providing "sites where ontological shifts are possible" (Hunt, 2014, p. 30). The trick to identifying these gaps to facilitate ontological shifts requires that researchers respect the autonomy and independence of Indigenous organizations (Sundberg, 2014). It also requires that institutions and researchers uncover and publicly acknowledge how they historically, and in the present, benefit from the dispossession of Indigenous communities (Daigle, 2019). By serving as allies in support of the self-determination of Indigenous nations, and as collaborative partners focused on the research agendas of those communities, we also serve to broaden the ontological foundations of our discipline and the academy. Serving as allies in research and struggle, though, is only one step in the process of aiding Indigenous research agendas. As Coombes (2012, p. 290) identifies, a Freirian approach to research identifies collaboration as a "mere intermediary step towards the democratization *and* dissemination of knowledge production itself." The final step is the fostering of "communities' capacities to complete research for themselves" (Coombes, 2012, p. 291; see also Chapter 16). This can be seen as the final prepositional shift, from research "for" and "with" to research "by" Indigenous peoples, leading toward Indigenous research sovereignty.

Conclusion

Qualitative research methodologies in geography have been significantly influenced by feminist and Indigenous research practices in the past few decades. Both feminist and Indigenous approaches have moved geographical research towards more empowering methodologies. However, as this chapter has illustrated, employing empowering methodologies is not easy and involves careful reflection regarding approaching the research (for example, the development of long-term dialogic relationships and the validation of multiple world views), doing the research (for example, employing a troubling multi-layered reflexivity and using dialogic research tools), and the politics of the research (for example, stepping outside hegemonic systems of knowledge production and challenging academic institutional structures). Nevertheless, if employed thoughtfully and compassionately, empowering methodologies can move qualitative research towards more inspiring, meaningful, and potentially transformational and equitable outcomes.

Key Terms

critical consciousness
decolonization
empowerment
indigeneity
Indigenous methods
multiple ontologies
pluriversal world
polycentric epistemologies

Review Questions

1. How would you define empowerment? In what circumstances might you use it as a strategy for qualitative geographical research, and why?
2. Critically discuss the potentials and limitations of empowering methodologies. Consider the ways in which these limitations might be ameliorated.
3. Find an example of geographic research that has used an empowering methodology. How was the project initiated? What were the outcomes of the research for the different individuals/groups involved? How would you evaluate the "success" of the project?
4. Do you need to be a woman to do feminist research? Can an Indigenous researcher only do effective research with Indigenous communities? What skills and attributes do you need to have or develop to be able to undertake research with such (diverse) groups using empowering methodologies?
5. Outline some of the ethical issues involved in using empowering methodologies in geography. How might you negotiate these issues through the research process?
6. Does place matter when using empowering methodologies? Justify your viewpoint.

Review Exercise

Read the following paper:

Eshun, G., & Madge, C. (2012). Now let me share this with you: Exploring poetry for postcolonial geography research. *Antipode, 44*(4), 1395–428.

In small groups, debate the following issues:

1. How was poetry used as an empowering methodology?
2. Consider some of the potentials and problems of using poetry as a research method. Do you consider this approach was successful? Justify your viewpoint.
3. How and why did the use of poetry allow the researcher to be "(de)centred" in the research process? How and why did it (re)inscribe marginality?
4. In what ways were issues of relationship, responsibility, reciprocity, and redistribution raised in this project?

Useful Resources

Daigle, M. (2019). The spectacle of reconciliation: On (the) unsettling responsibilities to Indigenous peoples in the academy. *Environment and Planning D: Society and Space*, https://doi.org/10.1177/0263775818824342.

Johnson, J., & Larsen, S. (Eds) (2013). *A deeper sense of place: Stories and journeys of collaboration in Indigenous research*. Corvallis, OR: Oregon State University Press.

Louis, R. (2007). Can you hear us now? Voices from the margin: Using Indigenous methodologies in geographic research. *Geographical Research, 45*(2), 130–9.

Moss, P. (Ed.) (2002). *Feminist geography in practice: Research and methods.* Oxford: Blackwell.

Parpart, J.L., Rai, S.M., & Staudt, K.A. (Eds) (2013). *Rethinking empowerment: Gender and development in a global/local world*. London: Routledge.

Writing a Compelling Research Proposal

Hilda E. Kurtz

Chapter Overview

Writing a compelling research proposal can be one of the most rewarding tasks you undertake as a student or researcher. It calls for you to distill many elements of what you know about your topic and your field into a plan for producing new knowledge in that topic area. Producing new knowledge is exciting and important. This is good work to be able to do.

This chapter offers some starting points for writing a persuasive research proposal by first considering the qualities of good research questions, then demonstrating how to identify the conceptual building blocks of a proposal, and, finally, explaining the purpose of each section of a research proposal and how they fit together.

What Is a Research Proposal and What Makes It Compelling?

A research proposal is both an *argument* about why it is important and exciting to ask and answer a set of questions about the world, and also a *plan* for how you will go about executing the project. The world is complex, so there are endless questions that can be asked about it. A compelling research proposal makes *your* questions seem especially important and exciting. It also makes the reader feel eager for you to go answer those questions, and confident in your ability to do so.

To elaborate a bit, a successful research proposal is engaging to read, hooking the reader in the first paragraph and building a sense of momentum and unfolding logically from there forward. The proposal is organized in support of one or several clear research questions and should aim to accomplish two objectives in support of those research questions. First, it should signal a debt to existing scholarship on a topic and identify how your project will make a contribution back into the literature; that is, it says something about what other researchers have made known about a topic, what is still unknown, and how your work will help fill a gap. Second, a research proposal must offer a well-defined plan for gathering and analyzing data in ways that will actually answer the research questions.

While it sounds obvious, answerability is a key part of what makes research questions compelling and is overlooked surprisingly often, even by experienced researchers!

Breaking it down a little further, a good research proposal is organized around **conceptual building blocks**, which should be neither too big nor too small, but just right in scope. A strong proposal traces a logic—your logic—about why a question is important enough to try to answer, how you will answer it, and what the implications of the project might be. Further, as an argument, the proposal should be well-grounded in previous scholarship, while also composed with clear, concise, and persuasive writing. Research proposals should be written in well-constructed sentences, with powerful language. Following social science writing conventions, the main point of each paragraph should be easily found in the first or second sentence. As a research *plan*, the proposal should be logically coherent, with a fairly fine level of detail of the research design, as well as distinct identification of how the data and methodologies selected will enable you to answer your research questions in rigorous and ethical ways (see Chapters 2 and 6).

Building Strong Research Questions

Good research questions do not spring fully-formed onto the page; they usually result from several rounds of refinement. As Monk and Bedford (2016) note in an earlier edition of this volume, there are many different paths toward identifying and refining your research questions, but they usually start with your own curiosity. Your research questions might have started forming in your mind based on personal experience, social observations, current events, or from scholarly literature you have read. They might concern an issue being discussed or debated in a larger forum, or might start taking shape as a result of conversations with or research presentations by other scholars. Any and all of these can lead you to engaging and answerable research questions.

Of the many questions you *might* ask in a research project, the ones you choose to work with should have certain qualities:

1. Your research questions should really hold *your* attention.

   You will be working with these questions in many ways, over some period of time: they should keep you interested. Intellectually, geography is a big tent, and there are many ways that a set of research questions can maintain our interest. Sometimes, they are closely related to our own life stories, and so they remain interesting in quite a personal way. Other times, research questions concern persistent and pressing issues in society, and the durability of the issue or problem keeps interest high. Sometimes, research questions are grounded in conceptual paradoxes or conundrums that excite the puzzle-solving part of your brain. There are many more ways, and no one right way, to become and stay interested in a set of research questions.

2. Your research questions should be answerable, based on data availability and your own methodological potential.

   An important question for all researchers to start with is "what would constitute *evidence* for answering my research question?" By identifying the evidentiary needs for the project first, moving toward finding or generating the data becomes simplified and then points you toward appropriate methods (this volume can help with that process). For instance, if your research question requires data on public opinions or people's experiences of an event, you may need to use oral methods to engage with participants directly (see Chapters 9, 10, and 11 on interviews, oral histories, and focus groups). If you need data on cultural practices or social behaviours, you might consider surveys or ethnographic approaches (see Chapters 7 and 8 on case studies and participant observation, Chapter 13 on questionnaires, Chapter 15 on digital geographies). Are you trying to understand the historical geography of a phenomenon? Consider archival investigations (see Chapter 12). Or, perhaps you want to work directly with a community or social group in a partnership, in which case participatory and empowering methods should take precedence (see Chapter 4 on empowering methods, Chapter 14 on solicited journals, Chapter 16 on participatory research). By tracing logically from examining what kind of evidence is needed, to which kinds of data can *supply* that evidence, and then to which types of methods generate those data, your project is strongly empirically grounded and you are less likely to encounter difficulties later.

   In turn, developing familiarity with and competency in various research methods is essential because it helps you think through how you will answer the questions you have posed for yourself. More specifically, this helps you sort out what kinds of questions you can and cannot answer with a given method. If you do not already have the needed methodological competency, ask yourself if you can reasonably gain it before or while doing the research project itself—this might be the perfect time and reason to learn new techniques. There are specific criteria for conducting good research that produces reliable data and best practices guidelines that help you do so ethically and with **critical reflexivity** (see Chapter 2 of this volume for a strong start). So, rather than just throwing together a research plan using the most common qualitative methods, you should consider that learning how to construct and analyze data gathering and analysis is a valuable skill and worthy of your time and effort.

3. Your research questions should be answerable within the timeframe you have available.

This and the previous points are really a matter of scope and it can take some trial and error to arrive at a workable research question that fits your capacities and timeframe. It is all too easy to ask a question that is too broad to answer within the timeframe available, but is also easy to start with a question that is too narrow and/or too descriptive. As a general guideline, to be answerable within the timeframe of a student project or even a thesis, research questions should be open-ended but they should initially seem a little too modest in scope—this is because no matter how simple they seem, they will often become surprisingly complex during the process of qualitative research. Take for example a seemingly simple question that has served as the basis for many student projects and theses: how do different stakeholders view/respond to an existing or prospective land use ordinance (such as one regulating backyard chickens, direct sales of farm produce, the location of windfarms, or the construction of hazardous waste facilities)? The question may seem straightforward, but layers of history, political activism, land-use practices, perceptions of land-owners' rights, environmental justice issues, municipal governance policies, and cultural ideals of property ownership quickly cause the project to become richer and more complex as you go, in part because the data are more textured and complicated than you might anticipate.

4. Your research questions should be expressed as something other people can care about.

   A research project is composed of empirical data, research methods, and a conceptual framework based on a theoretical argument, and each of these has the capacity to excite your readers and make them care about the findings. For instance, if you were doing a collaborative project with a community group, they might be most invested in your research design and empirical findings, while future editors of academic journals might find your conceptual argument and contribution to the scholarly literature the most compelling. Your project could expand on existing scholarship by asking research questions in a new social setting, or adding a new tool or concept to an existing body of research on the topic. Alternatively, your project could critique existing scholarship, explaining why the findings in that research area are incomplete or even wrong. There are many possible variations on these stances and many projects are a combination of contributing to *and* critiquing existing scholarship.

5. Your research questions should help fill a gap by adding something to a body of knowledge.

   The scholarly endeavour of **knowledge production** is ongoing, meaning there will always be something *missing* in the existing body of academic knowledge; this may be called a gap or more formally, a lacuna.

Identifying a meaningful gap can feel intimidating to researchers just starting out, but here are several strategies to help you. The gap might be the social setting, a place that researchers doing similar work have never studied; you can fill that gap by conducting your research in a new setting. The gap might be found in the analytical or methodological toolkit. For example, you might want to ask questions that have been addressed using mostly archival materials, and you might know of people whose lives were affected by the events recorded in those materials. If you conducted oral histories with some or all of those people, you would be addressing a methodological gap in the existing body of research, and thus offering new insights into how these events were experienced.

Sometimes the new approach that a researcher can add to fill a gap in the existing scholarship is not methodological, but more conceptual. Adding a concept from another body of research, outside of (your part of) geography can produce effective new insights into the issue being studied. For instance, environmental justice (EJ) research in geography started as a problem of spatial analysis in which spatial distributions of point-source pollution were mapped against demographic data about race and income. These were important projects, and demonstrated uneven burdens of pollution, but they generally focused on spatial patterns and not the social processes contributing to those patterns. Then critical scholars from geography and sociology took interest and starting adding more conceptual perspectives. Laura Pulido (1996) famously argued that the problem to focus on was not race as a census category (the US Census was and is the source of the demographic data being used for US-based EJ analyses). She argued that EJ researchers should be focusing on racism as a set of lived relations and institutional arrangements. Her paper opened the floodgates to a whole generation of EJ scholars using concepts grounded in critical social science to gain more insight into how and why pollution tends to most impact the socially and economically vulnerable, and the potential ways to address that inequity. Schematically, Pulido brought a conceptual toolkit to bear on a research topic that had been studied using quite different tools, generating considerable impact.

To sum up, the process of arriving at a compelling and answerable research question often draws on your own lived experience and observations, exposure to topical scholarly literature, and consideration of data and methodology; these can take significant time. Further, research questions typically get refined over several versions by (re)reading and assessing scholarly work, gaining competency in research methods to decide how best to answer question you have posed and determine what kinds of evidence are needed, and sharing your ideas with peers and mentors for feedback. Although much of this labor might be unseen in your final product, it makes the research and writing processes much smoother.

Set Your Building Blocks

Compelling research questions, and the proposals that offer a road map for answering them, are put together using analytical building blocks. A more formal term for these building blocks would be *operational concepts*, that is, concepts that can be operationalized or acted upon within the context of a research project. Confirming that you are working with the right analytical and/or conceptual building blocks is an essential step in developing an effective research proposal. If you have ever spent weeks or months or more on a research project that somehow refused to come into focus, you may have been working with concepts that were too big, and thus too difficult to operationalize. (This is remarkably common.) Big ideas and significant phenomena will help you hook a broad audience, but they will not give clear direction to a research project, because they are not the actual building blocks of your research project. In the course of a research project, big ideas can and must be examined, prodded, problematized, and interrogated in relation to more narrowly defined *operational concepts*, and also in relation to bounded social settings.

Take, for example, the big idea of climate justice. The climate justice movement names a goal of political and social struggle that is defined at the intersection of social justice and climate change, two very big ideas. As these two ideas intersect, they have the effect of narrowing each other in scope. So, at first, climate justice may seem less like a big idea and more like an operational concept or building block for a research project. And yet, one has to wonder about the specifics to which it refers? What empirical phenomena are obviously or necessarily linked to the concept of climate justice—marching in the streets, testifying at the United Nations, or brokering international agreements, or perhaps a multitude of others? Could a reviewer for a funding agency read the term *climate justice* in the title of your proposal and be able to make a good guess about the purpose and scope of your project? No, they really couldn't.

Climate justice, although deriving from a provocative intersection of two big ideas, is still, itself, a pretty big idea and conducting a research project on climate justice, in its entirety, is not feasible. You could, however, conduct a research project on a social setting and set of actions that are being affected by one or more ideas about what climate justice is, and whether and how to achieve it. As you develop a research proposal asking particular questions about the big idea, you will find that you are actually asking questions with and about a suite of more bounded and operational concepts, not about climate justice as such.

A simple ladder diagram, as illustrated and annotated in Box 5.1, can help identify the building blocks of your research questions or project, and how to link

BOX 5.1 A Simple Ladder Diagram

Big Idea	**Climate Justice**	Normative calls for systemic response to address the harm to vulnerable populations caused by climate change
Component Domains	**Climate Politics**	Broad domain of political debate, activist struggle, formal agreements about recognizing and responding to climate change, happening at different political scales
	International Politics	Various mechanisms for international agreements, for example, Paris COP21
	Urban Politics	Political debate, activist struggle, formal agreements about the management of cities; domain of contestation over who gets access to what urban resources
Operational Concept(s)	**Urban Climate Politics**	Bounded domain in which questions are debated and decided about how to respond to climate change at the urban scale, and who will have access to what kinds of resources and protections from the adverse effects of climate change
Social Setting	**Seattle**	Just one of many cities in which political elites and activists are actively working to devise responses to climate change
Social Actions and Artifacts	**Development of Urban Climate Policy**	Processes, artifacts, and outcomes produced in the domain of urban climate politics

Based on Rice (2016)

them to the big idea. The annotations expand on the example of climate justice as the big idea. Remember, there are many entry points into a research project—you may have started by getting excited about a big idea, in which case you can work down the ladder to identify one or more building blocks/operational concepts from which to build your project. Alternatively, you may have questions about an empirical setting you have read about or experienced directly, such that you start filling out the ladder diagram from the bottom up, or from somewhere in the middle. From wherever you start, you can work up and down the ladder multiple times to arrive at a clear sense of your building blocks, and how they relate to big ideas and questions about specific social settings. The example in Box 5.1 draws on Jennifer Rice's (2016) paper on urban climate politics in Seattle, though of course we know that the climate crisis is devastating cities around the world.

Once you have worked back and forth a few times to identify your conceptual building blocks and hone in on your (compelling and answerable) research questions, it will be time to start making sense of your project by drafting a research proposal.

Drafting the Research Proposal

As an artifact, a research proposal is a strongly conventionalized form of writing, consisting of an introduction, a conceptual framework, a background section, research questions, research design and methodology, significance, and bibliography. Each of these sections serves a specific function in relation to the overall arc and logic of the proposal. Doing each section well, meeting its rhetorical demands, calls for you to think carefully through that part of your project.

It is easy to put off the actual writing of the proposal. You may think that you need to read more scholarly literature before you start. It is true that you need competent knowledge of the literature that is shaping your research questions, but it is not true that you need an encyclopedic knowledge of that literature. There is no bright line that marks when you have read enough, so this will be largely up to your own judgment. But resist the urge to read indefinitely before sitting down to start your draft. While squeezing your as yet incomplete idea into the shape of a highly conventionalized form may feel premature and even constraining, the discipline of doing so can reveal both possibilities and limitations to your research project that might otherwise have been overlooked. In writing an effective research proposal, as in any other academic writing, the process of writing *is* the process of thinking (see Chapter 19). The rewards of a careful writing process include refinements to the idea itself, powerfully expressed logical coherency, and a contagious sense of excitement about the conduct of the research itself. In short, the process of writing a research proposal, if done right, will lead to (sometimes surprising) refinements to your original research questions and/or research design. Box 5.2 reviews the rhetorical purpose of each of these sections.

BOX 5.2

A Proposal Is an Argument and a Plan

A proposal is an argument, which is comprised of logically connected parts. A proposal argues that a research project is important, timely, even urgent; that the current knowledge base about a phenomenon of interest is insufficient; and that the gap in existing knowledge must be filled. Further, the proposal argues that the gap should best be filled in a particular way, and that your plan for data collection and analysis will do that work.

The first part of any proposal, the introduction, hooks the reader and piques their interest. It consists of one to three paragraphs that make it clear that your research is important, stating the key issues motivating your research. Generally speaking, the introduction refers both to a set of empirical conditions and to the current state of knowledge about those conditions. In geography, that usually refers to a set of real-world conditions that are curious, paradoxical, timely, urgent, and/or dynamic. The introduction should argue that the current state of knowledge about the these conditions is inadequate and will be addressed in part by the research project at hand.

The introduction to a research proposal should move from a broad issue (e.g., climate change), to a topic (e.g., sea level rise as effect of climate change), to a subject (e.g., the threat that sea level rise in coastal cities poses to real estate), to a question (e.g., how can coastal real estate be protected from sea level rise caused by climate change?) It is important to move steadily and quickly from issue to topic to subject to question, so that by the end of the first paragraph the reader understands what this research is about, what kinds of questions it will address, and why they should care.

The next part, the conceptual framework, is a targeted **literature review** that serves two related purposes. First, it pulls key themes and concepts from the body of literature that has most shaped your research project, in order to locate your project within an intellectual domain or field of inquiry. Second, it sets up research questions by introducing the key building blocks you will use to ask the questions and conduct the research project.

While there is a strong imperative to be innovative in academic research, this imperative is always in tension with a more conservative demand, which is to ground a research project in relation to one or more fields of existing knowledge and scholarship. Broadly, this section of the proposal should demonstrate how your research fits into a given literature or field of

(*continued*)

study, *and* how it offers a new twist. It should offer a reading of the relevant literature that lays the necessary groundwork for the research questions you will pose. What of relevance has and has not been addressed in the literature? Be sure to address how each work you cite influences your own ideas about what is important in this topic area.

More specifically, this section is vital to establishing the foundational concepts that animate your research questions and set the parameters of the research itself. That is, an effective conceptual framework section offers a reading of the literature that identifies and distills for the reader the key building block concepts without which the research at hand could be neither imagined nor executed.

Next, the background (or setting) can be approached almost journalistically. This section briefly highlights those elements of the empirical domain under study that are needed to understand the rationale for or design of the research project. Significantly, it offers an account of the key elements of the story in the language of the conceptual framework established previously in the proposal. Careful choice of such language in this section helps to deepen readers' understanding of both the real-world conditions in question, as well as deepen their understanding of the concepts from which the project is built.

Following on, your research questions should be a crystal-clear statement of the questions you intend to answer with this research. They must grow from and be animated by what you have just laid out for the reader in the foregoing sections of the proposal. These questions must also be answerable. To be answerable, they must be empirically bounded, but not merely empirical—for example, they are set in a place or within a social group but they are not only about the internal dynamics of those settings. Research questions should seem overly narrow, because no matter how narrow they seem, they will become significantly more complicated during the process of qualitative research. Even if you have only one major research question, it will inevitably break down into several smaller questions. These smaller questions point toward your research methodology.

The methodology section explains and describes how you plan to answer these (ever so compelling) research questions. This section should describe and justify both the overall strategy of inquiry, your **epistemological** approach, and the specific research methods you will employ. Identify your methods in detail: what data will you collect, how will you analyze it, what tools (technological or otherwise) will you use during these steps, and what form will your results take? There should be a clear conceptual link between the kinds of research questions you pose, and your strategies for answering them—make sure this link is readily apparent even to a reader or researcher from a completely different field. Justify why these are the best possible methods for answering your particular questions.

The significance section wraps up your proposal by answering the "So what?" question. Here, you should try to state very clearly just what your research will contribute to one or more scholarly literatures and to the larger social world. Your contributions could be conceptual (provide a new way of looking at something), methodological (provide a new way of collecting and/or analyzing particular data), and/or substantial (provide new and valuable information on a topic). Be specific about deliverables: who will get what concrete thing as a result of your research?

Finally, the references section of a research proposal should show that your research is based on the most appropriate literature. Many reviewers will skim the references section first. Often, the references section is included in the overall page count, meaning you need to be quite selective about what to include. Focus on those works that introduce the concepts, tools, and empirical data without which you cannot accomplish your research.

As you work on each of the sections of your proposal, do not feel a need to start at the beginning. Many academic writers find themselves writing the introduction last, once the writing project as a whole has come into clear focus. In this sense, research proposals are not any different from other forms of academic writing. The same moving back and forth and up and down that you did using the ladder exercise to identify your building blocks can be very useful as you write the proposal itself.

As you fine-tune, think seriously about your audience. It is one thing to write a proposal for the narrow circle of subject matter experts comprising your thesis or doctoral committee, but it is another thing to write a proposal for a panel of reviewers for a funding agency, whose research specialization may have nothing to do with your own. It will be important over an academic career (and any career involving writing) to write differently for different audiences. List some of the scholarly attributes of your intended audience. Which of the key terms is most familiar to your intended audience? What level of knowledge can your readers be expected to have in the different literatures from which you are drawing? Share your draft with peers and mentors for feedback and use these reflections to write to your intended audience in ways they find compelling.

While you are imagining the readers of your proposal, you might also consider some attributes of readers beyond their scholarly expertise. For instance, a panel reviewer for a funding agency is likely a mid- to late-career scholar. That matters for how they think about scholarship, especially in periods like the present, when an academic discipline like big-tent geography is particularly dynamic and intellectually diverse. Your readers are not blank slates; they have internalized a set of conventions about how research should and should not be done. They are subject matter experts in something, maybe in your area of expertise, but maybe not. They are also pretty busy with career commitments, and have taken on the added work of reading research proposals just like yours. They are quite likely to be reading your proposal late at night, after other commitments have been met. These are important considerations as you turn to the actual writing of the proposal. Think of that reader as you craft each part of the proposal. What do you need that reader to understand up front? How do you want that reader to react to your words on the page? The answer is that you want that reader to feel a sense of intellectual excitement as they read your proposal, you want them to admire the timeliness of the project, and the tightness of the logic, and you want them to feel pleased to be one of the people giving the project the green light.

Conclusion

Writing an effective research proposal can be among the most stimulating and rewarding projects you undertake. It is indescribably exciting to wield language so as to move material resources, in this case, to move funding and other forms of support toward your research project. Revel in the art and craft of it along the way. Treat the task as a learning process, in which you expect to learn more about the logic and rationale of the research project itself. Look at other examples, and share drafts with readers before submitting it to a funding agency. Ask outright for in depth critique, in order to make it stronger before submitting it for formal review. Allow room for refinements to your research idea in the course of writing the proposal. And remember to savour the privilege of writing a research proposal at all.

Key Terms

audience
conceptual building blocks/ operational concepts
research questions

Review Questions

1. Pick a potential research topic, and fill out the ladder diagram discussed in Box 5.1. Identify the big idea, its component domains, operational concepts, possible social settings, and social actions and artifacts that you think could fit into a research project. Try doing this more than once, and compare and contrast multiple versions of the exercise. The results can help you identify strengths and weaknesses of different approaches.
2. For your hypothetical research project, list the methods and methodological competencies needed to do the research well. Evaluate these against your own knowledge and competency, and make a plan for enhancing any

needed skills before launching the research. One of the best ways to do this is by conducting a pilot project, a mini-version of the research you plan to conduct. You can use lessons learned during the pilot project to refine your approach to actual data collection and data analysis.

3. Consider the data that you hope to collect. How will you collect them, and what form will they take? Will this be a mixed-methods project? Will you need to recruit participants (if so, how and where will you do this)? If it is textual data, whether public records or transcripts of interviews you conducted, what methodological approach will you take for analysis?
4. Much of qualitative research requires ethical review by an institutional board (see Chapter 2). Find the requirements for your institution and create a mock-up of the protocol you would submit for this project.

Review Exercise

Choose a recently published research article from a geography journal and dissect it to identify its components (introduction, literature review/conceptual framework, research questions, data and methodology, findings, and significance). Discuss in a group what the original proposal for the study might have looked like, the author's epistemological framework, alternative ways the work could have been carried out, and different methods that might have contributed to the study. Some articles to consider using for this include Ramírez (2019), Rosenberg (2017), or Spiegel (2020).

Useful Resources

Aoyama, Y., Murphy, J., & Hanson, S. (2010). *Key concepts in economic geography*. Thousand Oaks, CA: SAGE Publications.

Clifford, N., Cope, M., Gillespie, S., & French, S. (Eds) (2016). *Key methods in human geography* (3rd edition). Thousand Oaks, London, New Delhi: SAGE Publications.

DeLyser, D., Herbert, S., Aitkin, S., Crang, M., & McDowell, L. (2010). *The SAGE handbook of qualitative research in human geography*. Thousand Oaks, CA: Sage Publications.

Gallaher, C., Dahlman, C., Gilmartin, M., Mountz, A., & Shirlow, P. (2009). *Key concepts in political geography*. Thousand Oaks, CA: SAGE Publications.

Hay, I. (2012). *Communicating in geography and the environmental sciences* (4th edition). Melbourne: Oxford University Press.

Latham, A., McCormak, D., McNamara, K., & McNeill, D. (2008). *Key concepts in urban geography*. Thousand Oaks, CA: SAGE Publications.

Maxwell, J. (2013). *Qualitative research design: An interactive approach* (3rd edition; Applied Social Research Methods). Thousand Oaks, CA: SAGE Publications.

Rigorous and Trustworthy: Qualitative Research Design

Elaine Stratford and Matt Bradshaw

Chapter Overview

Careful design and rigour are crucial to the dependability of any research. Research that is well conceived results in research that is well executed and in findings that stand up to scrutiny. Thoughtful planning and the use of procedures to ensure that studies are rigorous should therefore be central concerns for qualitative researchers. The questions we ask, the cases and participants we involve in our studies, and the ways in which we ensure the rigour of our work need to be considered; these are hallmarks of any dependable research.

Introduction

In this chapter, we focus on some matters of design and **rigour** that qualitative researchers need to consider throughout the life of a study to ensure that the work satisfies its aims, high standards of scholarly practice, and critical audiences. We outline various principles of qualitative **research design** as well as some specific means by which rigour can be achieved in our work.

The chapter is organized into three main sections. First, we examine what influences us as researchers and consider the effects we have on the conduct of research. This discussion makes a link between the **interpretive communities** in which we work and the sorts of issues that are raised when we begin a study. Second, we elaborate on several steps that map how to select suitable examples and **participants** for study. In qualitative research, important considerations include the people we interview, communities we observe or with whom we participate, and texts we read. Nevertheless, this consideration is secondary to the *appropriateness* of what or whom we involve in our research and subordinate also to *how* we conduct that research. Third, we outline some of the ways in which it is possible to ensure rigour in qualitative research to produce work that is dependable.

Careful research design is an important part of ensuring rigour in qualitative research, because it helps us make and understand complex connections (Roller & Lavrakas, 2015). While scholarship and commentaries about research methods and design often imply that research studies should be conducted in a specific way, no single correct approach to research design can be prescribed (Gould, 1988; Mason, 2004). For certain kinds of work, the order and arrangement of stages may be different, stages can overlap, other stages might well be included, and the combination of qualitative and quantitative techniques in **mixed-method** and **multi-method** research is now commonplace (Anguera et al., 2018). Nevertheless, by the end of the chapter we will have moved through several stages of qualitative research design and summarized this process in three diagrams. We believe that you will find this movement through stages helpful in approaching your own qualitative research work.

Asking Research Questions

Each of us needs to acknowledge that our fellow geographers and other colleagues are *already* involved in our studies (Box 6.1). None of us ever formulates research questions or undertakes research in a vacuum. We are all members of interpretive communities that involve established disciplines with relatively defined and stable areas of interest, theory, and research methods and techniques (Butler, 1997; Fish, 1980). Increasingly, too, we are members of *interdisciplinary* research communities that cross over discipline boundaries in ways that enable collective consideration of **wicked problems** that are difficult to solve because of their complexity. More and more, those communities comprise and seek to engage non-academic others—industry partners, community members, and **more-than-human** agents (Bastian, 2017, Nind, 2017).

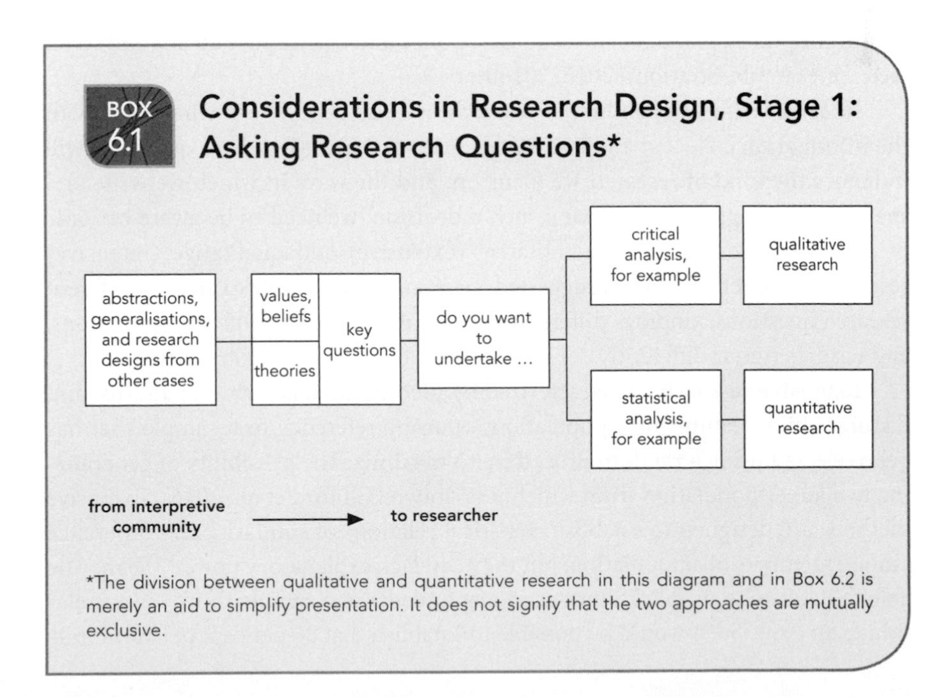

*The division between qualitative and quantitative research in this diagram and in Box 6.2 is merely an aid to simplify presentation. It does not signify that the two approaches are mutually exclusive.

Such collaborative engagements are often deeply rewarding, and that compensates for the challenges they pose, not least among them having to learn each other's distinct vocabularies, methods, values, and conceptual frameworks. These disciplinary and interdisciplinary interpretive communities influence our choices of topic and our approaches to the conduct of our studies; this is because of what Livingstone (2005, p. 395) describes as "the inescapably collective character of interpretation . . . [which, to] the extent that interpretive communities occupy material or metaphorical spaces, they fall within the arc of the cultural geography of reading." To such observations, we would add other geographical actions and analyses. We also fold our own values and beliefs into research, and they can influence both what we study and how we interpret our studies and findings (see Flowerdew & Martin, 2005; Jacobs, 1999; and Chapter 19 for more detail).

From Asking Research Questions to Conducting Research

Research aims affect research design. For example, asking "how many skateboarders frequent a specific public place compared with other types of users?" will involve a research design different from one that aims to answer the question "why do skateboarders use this place, and how do they interact with other types of users?" The first question focuses on quantification and statistical analysis, and the second is more concerned with the qualitative investigation of skateboarders' behaviours and practices, and with questions about how such behaviours and practices inform various socio-spatial relations—with other skaters, peers, members of the public, private security guards, police and municipal officers, and so on (see, for example, Stratford, 2015, 2016b).

In considering the conduct of research, we also need to ask what to do with the information collected and knowledge gained. Answering these questions will influence the kind of research we focus on, and the ways in which we write and share our findings. Before making such a decision, we need to be aware of some of the differences between quantitative (extensive) and qualitative (intensive) research. As Sayer (2010) has suggested, each method helps us to answer different research questions, employs different research methods, has different limitations, and ensures rigour differently.

Extensive research is characterized by identifying regularities, patterns, and distinguishing features of a population, often by reference to a sample that has been selected using a random procedure to maximize the possibility of generalizing to a larger population from which it is drawn (Clifford et al., 2016). Extensive methods are designed to establish statistical relations of similarity and difference among members of a population, but they can lack explanatory power. We may be able to determine that "N" number of respondents in a sample think "P" in relation to an issue. So, it would be possible to establish that 86 per cent of a randomly sampled group of daily users of an inner-city public park are in favour of the infrastructure in that park being redesigned to allow skateboarding to be done safely so that multiple uses of the park could include young people's pursuits.

On such understanding, Wood and Williamson (1996) distributed a standardized questionnaire to a random sample of the users of Franklin Square, an inner-city square in the city of Hobart, Tasmania, in Australia that had been partly claimed through day-to-day use by skateboarders. The data from their study were aggregated, and statements were made about the degree to which these data were likely to reflect the opinions of all the square's users about the presence of skateboarders. This extensive approach produced useful information suggesting the existence of common characteristics and patterns; for instance, skateboarders used certain parts of the square, while other users avoided them.

Yet such findings did not account for the shifting quality of various people's different experiences of Franklin Square and of each other, or the reasons behind their opinions. So how are we to determine *why* respondents held the opinions they did? Such determination is often best advanced by means of **intensive research**, which requires that we ask how processes work or opinions are held, or actions are taken in a specific case (Clifford et al., 2016; Platt, 1988; see also Chapter 9). This kind of open-ended questioning will often illuminate the reasons for something—the *how and why* components. In short, intensive research is a powerful tool when we need to establish what actors do, when we need to understand why they act as they do, and when we need to establish what produces change in actors and the contexts in which they are located. In seeking such understandings, it is crucial to think in nuanced ways about people's identities and ways of being in the world, because we are never *just* an age, a gender, or a member of a specific ethnic, racial, religious, socio-economic, or other group—always, we are complex beings. That complexity needs to be accounted for if rigour is to be maintained, and studies of **intersectionality** are useful in helping us do so (Al-Faham et al., 2019; Hopkins, 2017).

On selecting Franklin Square as a case by which to examine multiple-use conflict in public places, Stratford (1998, 2002) and Stratford and Harwood (2001) used intensive methods such as in-depth interviews and observation to understand various responses to skateboarding in the square and around Tasmania more generally. Indeed, qualitative research methods and intensive modes of investigation characterize most studies of, for example, multiple-use "conflict" in inner cities because researchers and policy makers see pressing need to understand the motivations, values, and positions of all those implicated in the occupation and use of spaces and places. This relationship of research to policy is reflected in, for example, changes in the early 2000s to Australia's traffic rules to legalize skateboarding on certain roads and footpaths (see Australian Transport and Infrastructure Council, 2018; also Stratford, 2016a).

By implication, in various disciplinary and interdisciplinary collaborations wherever such need exists, extensive and intensive methods of work are likely to be highly relevant in varied combinations. Clearly, both approaches have merit and are often used together in mixed-method and multi-method combinations. However, we are pointed in the direction of intensive research if we are mainly interested in working through the elements of structure and process that arise from analyzing responses rather than in generating data that make statistical analysis possible. In opting for a qualitative research design, we are influenced by the theories we are concerned with, by studies undertaken by other researchers in our interpretive communities that we have found interesting, and by the research questions we wish to ask—all of which are interrelated.

Selecting Cases and Participants

Uwe Flick (2018) reminds us that research comes from ideas and experiences, and these are often then transformed into burning questions. We might encounter death and become interested in how people prepare for it or deal with bereavement, he writes, or we might be motivated by societal problems such as homelessness on the basis that it negatively affects people's lifecourse and prospects. We might then work at the meta-level on theoretical discussions about such matters, or we might engage in empirical studies. Either way, our focus on a topic can be described as a focus on a **case**—this differs from a **case study**, explored in more detail in Chapter 7. Sometimes we find a case or example of a larger problem, and sometimes a case or example finds us. In both instances, selection of both case and participants combines purpose and serendipity (Box 6.2).

Selecting Cases

In using the term *cases*, we are referring to examples of general processes or structures that can be theorized. Researchers should be able to ask, "that categorical question of any study: 'What is this case a case of?'" (Flyvbjerg, 1998, p. 8; see also Flyvbjerg, 2006). In the example cited above, Franklin Square was a case or example of multiple-use conflict in a public place, but it embraces a number of general social and spatial processes involving, for instance, the privilege of consumerism, the ways in which citizenship has come to "attach" to acts of consumption, and the relations of government to capital (Stratford, 2015).

On the one hand, we may read about multiple-use conflict in public places in other cities and want to see whether explanations advanced in those studies have merit in—or inform our understanding of—situations with which we are familiar. In such instances, the general or theoretical interest drives the research, and we must narrow the field, selecting examples and participants for research.

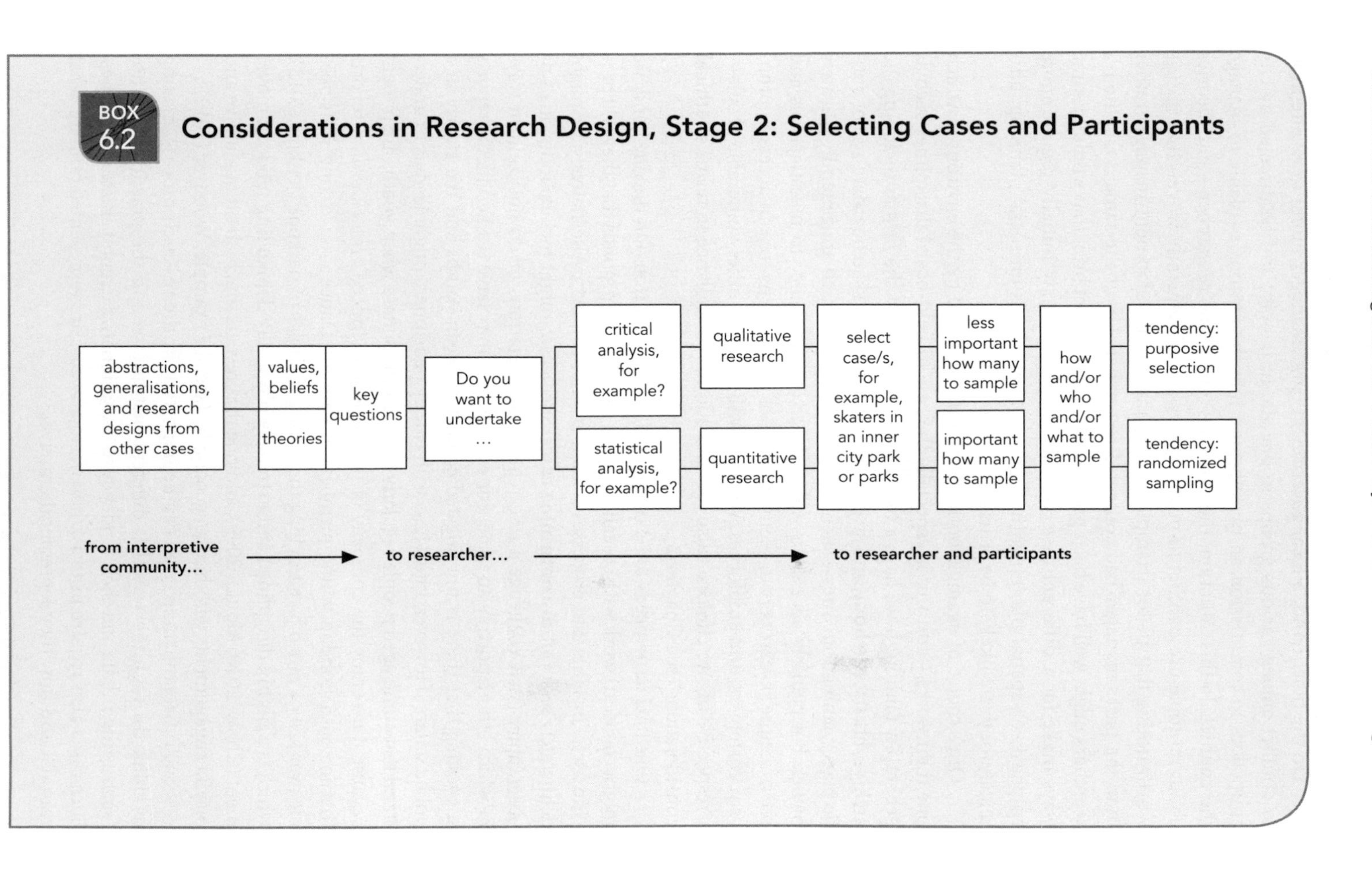
BOX 6.2
Considerations in Research Design, Stage 2: Selecting Cases and Participants
abstractions, generalisations, and research designs from other cases
values, beliefs
theories
key questions
Do you want to undertake ...
critical analysis, for example?
statistical analysis, for example?
qualitative research
quantitative research
select case/s, for example, skaters in an inner city park or parks
less important how many to sample
important how many to sample
how and/or who and/or what to sample
tendency: purposive selection
tendency: randomized sampling
from interpretive community...
to researcher...
to researcher and participants

On the other hand, a local government parks manager might draw attention to conflict among various groups in one site in the city and be directed by city councilors to commission an investigation into the options available to manage the conflict. In this situation, the case has "found" the researcher—and theories about multiple-use conflict in public spaces are subsequently woven into it. It is worth noting that if, for example, the community development manager rather than the parks manager had contacted researchers about the same issue, the researchers might well have been presented with a different brief that would in some ways make for a different case. Broadly, the former may initially require more emphasis on spatial planning; the latter more emphasis on social planning; ultimately, both are likely to be crucial.

One's choice of example both affects and is affected by the environs in which one then works—and this is especially the case in field-based disciplines such as geography (although we have a broad understanding of the *field* as including the archive, library, the body, and so on). In our example of skateboarders' use of public places, ambiguous sites—such as shopping malls, which are generally privately owned but publicly accessible—add a layer of complexity to investigations and often require researchers to secure a range of permissions to gain access to property, customers, and employees. Working with youth or other vulnerable people—irrespective of site choices—also requires additional vigilance in terms of **ethical** considerations (see Chapter 2).

One final issue needs to be considered in case selection. On the one hand, we might choose to work with so-called typical cases on the grounds that they will provide useful insights into processes evident in other contexts. Alternatively, we might deliberately seek out **disconfirming cases**. Such cases might include individuals or observations that challenge a researcher's interpretations or do not confirm ways in which others portray an issue. For example, we may have studied media reports suggesting that there is unmitigated conflict between youths and the elderly in a public square. However, interviews with elderly pensioners might lead us to understand that some aged people regularly frequent the square when youths are present because they enjoy the company of young people. Indeed, *The Guardian* reports that in south London young members of Parkour Dance are working with people over sixty years of age teaching them the principles of parkour (Jenkins, 2013), which is a highly disciplined and extremely physically demanding mode of movement (Chow, 2010; Mould, 2009), and one that reveals much about qualitative and playful engagements with human geography (see, for example, Woodyer (2012)).

Such disconfirming cases can be important in the research process, making us think through the way that different institutions and the practices they use (such as the media and their tendency to sensationalize events) create stereotypes. Such cases also require us to ask how various actors are represented (and for what reasons) and how they represent themselves.

Selecting Participants

Participants are intrinsic to the cases or examples we select, and their identities, values, agency, and practices can be widely divergent. In Franklin Square, the conflict involved the park itself, its heritage values and current uses as well as skateboarders, the elderly, members of the business community, the city council, law enforcement agencies, the international media, health professions, and academics; such an array is not uncommon in other contexts where skateboarding has been or remains controversial (Carr, 2010; Dumas & Laforest, 2009; Kidder, 2012; Stratford, 2014). In some theoretical dispositions, participants can also include non-human elements, or *actants*, the effects of which on the case are profound (Callon, 1986; Murdoch, 2006; Brunner, 2011).

Exploratory and/or background work (for example, reading, observation, viewing documentaries and skaters' YouTube channels, conducting preliminary interviews) will often give researchers the capacity to begin to comprehend the perspectives of participants with whom we think we want to interact. Understanding their perspectives in complex cultural situations usually requires some form of in-depth interviewing (see Chapter 9 for details) or observational method (see Chapter 8 for details) that, though time-consuming, often result in a deep and detailed appreciation of the complicated issues involved (Geertz, 1973; Herod, 1993).

Generally speaking, the more focused our research interest becomes and the more comprehensive our background information and understanding, the more confident we are about who we wish to involve in our research and why. Nevertheless, this confidence needs to be underpinned by a rigorous process of justification. As Mason (2004, p. 129) points out,

> Your answers to questions about which people to sample should therefore be driven by an interpretive logic which questions and evaluates different ways of classifying people in the light of the particular concerns of your study. Underlying all of this must be a concern to identify who it is that has, does or is the experiences, perspectives, behaviours, practices, identities, personalities, and so on, that your research questions will require you to investigate.

In this respect, it is conceivable that conducting in-depth interviews with a small number of knowledgeable **informants** will provide significant insights into a research issue. The choice of technique may be dictated by logistics and financial and human resources, or by the needs of participants, among other factors. That choice has expanded significantly as a result of web-based technologies (see also Chapter 15 on digital geographies). Stratford, one of the authors of this chapter has, for example, undertaken extended and multiple-return interviews using email conversations and Skype as well as telephone and face-to-face modes of engagement, and has supervised several projects using such methods (for example, Fallon, 2006). Participants were able to choose the time at which they responded, and several commented that they valued having that choice. A number also suggested that the email "interview" allowed them to consider their answers, including the use of emoticons, bolded or underlined words, and forms of punctuation for emphasis. This method also provided instant transcription, but the researchers were not privy to important information such as non-verbal cues and intonations that attend "normal" interviews.

Participant Selection

Michael Patton's (2015) work on **purposive sampling** is among the more useful summaries of the topic available to researchers. Patton refers to various forms of purposive sampling, including the following seven commonly employed strategies. *Extreme* or **deviant case sampling** is designed to help researchers learn from highly unusual cases of the issue of interest, such as outstanding successes or notable failures, top-of-the-class students or dropouts, outlier events, or crises. **Typical case sampling** illustrates or highlights what is considered typical, normal, or average. **Maximum variation sampling** documents unique or diverse variations that have emerged through adaptation to different conditions and identifies important common patterns that cut across variations (Williams & Round, 2007). **Snowball** (or **chain**) **sampling** identifies cases of interest reported by people who know other people involved in similar cases (Kirby & Hay, 1997; Stratford, 2008). **Criterion sampling** involves selecting all cases that meet some criterion, such as involvement in natural resource management governance in Australia (Lockwood et al., 2007). **Opportunistic sampling** requires that the researcher be flexible and follow new leads during fieldwork, taking advantage of the unexpected (Clough et al., 2004). **Convenience sampling** involves selecting cases or participants on the basis of access (for example, interviewing passers-by on the street). While this final strategy saves time, money, and effort, it often produces the lowest level of **dependability** and can yield information-poor cases. In practice, much of purposive sampling combines a number of these strategies.

How Many Participants?

In both qualitative and quantitative research, it is usual to study only a subgroup of people or phenomena associated with a case. The size of the group is more relevant in quantitative research because **representative samples** are important. In qualitative research, the "emphasis is usually upon an analysis of meanings in specific contexts" (Robinson, 1998, p. 409) and the sample is not intended to be representative.

The following analogy between a case and an island may help to explain the distinction. Suppose you are looking at a special kind of aerial photograph of an island, so detailed that you can see all its inhabitants.

> Clearly, if the population of the island were ten thousand instead of ten, enumeration would count for a great deal. . . . But this is because of the investigator's limitations: [they] cannot really get to know ten thousand people and the various ways in which each interacts with others. The use of formalist techniques is a second-best approach to this problem because the ideal technique is no longer feasible. Even on this big island, the old technique will count for a great deal, but that is not the main point. The point is that counting and model building and statistical estimation are not the primary methods of scientific research in dealing with human interaction: they are rather crude second-best substitutes for the primary technique, storytelling. (Ward, 1972, p. 185)

Numbers *do* tell us things about the island, and if what interests us happens to be the frequency and geographic distribution of the island's population, then we need no more than the photograph. If we are interested in a particular "story," such as might revolve around an aspect of the cultural geography of the island—for example, multiple-use conflict in public places—then we will need more than the photograph to go on, and might start by thinking about how circumstances on the island compare with others elsewhere (see, for example, Stratford, 2012).

One way to then conduct our specific investigation will be to talk with the island's inhabitants. We could also engage in **participant observation** and consult relevant texts such as submissions to government, letters to the editor of the island's newspaper, or television news stories that might give us an insight into multiple-use conflict in the island's public places. As researchers, however, we are usually resource-limited, both in terms of funding and time, and we must make decisions about what and whom to include and what and whom to exclude from our study. It is clear, however, that we still face the issue of how many people to talk with, how many texts to read, and so forth. While it may seem disconcertingly imprecise, Patton's (2015) brutally simple advice remains accurate: there are few if any rules in qualitative inquiry related to sample size, and it depends on what is needed in the way of knowledge, on the purpose of the research, on its significance and for whom, and on logistics and resources. The richness of information, its **validity** and meaning, is more dependent on the abilities of the researcher than on size of sample. In the final analysis, then, it is you as the researcher who must be able to justify matters of case and participant selection to yourself, your supervisor, your interpretive community, and the readers and users of your work.

Ensuring Rigour

It is no frivolous matter to interpret, share, and represent others' experiences. We need to take seriously "the privilege and responsibility of interpretation" (Stake, 1995, p. 12). This responsibility to informants and colleagues means that it must be possible for our research to be evaluated. It is important that others using our research have good reason to believe that it has been conducted dependably (see, for example, Anfara, Brown, and Mangione, 2002; Barbour, 2001; Filep et al., 2018; Rettke et al., 2018).

Ensuring rigour in qualitative research (Box 6.3) means establishing the **trustworthiness** of our work (Bailey, White, and Pain, 1999a, 1999b; Baxter & Eyles, 1999a, 1999b; Golafshani, 2003; Le Roux, 2017; Morse, 2015). Research can be construed as a kind of **hermeneutic circle**, starting from our interpretive community and involving our research **participant community** and ourselves, before returning to our interpretive community for assessment (Flick, 2018). This circle is a key part of ensuring rigour in qualitative research; our participant and interpretive communities check our work for **credibility** and good practice. In other words, trust in our work is not assumed and must be earned, and there are profound and powerful influences on how we all make meaning, some of which cannot be fully accounted for in the explicit labours of research because they are tacit or unconscious (Peck & Mummery, 2018).

Two steps need to be followed to ensure and defend the rigour of our research for our interpretive communities. First, strategies for ensuring trustworthiness need to be formulated in the early stages of research design and applied at each stage in the research process (see, for example, Forero et al., 2018). These strategies should include *appropriate* **participant checking** procedures in which our work is opened to the scrutiny of interpretive and participant communities (Mason (2004), but see also Bradshaw (2001) on some of the possible perils involved in these procedures). Second, we need to document each stage of our research carefully so that we can report our work to our interpretive community for checking; "we should focus on producing analyses that are as open to scrutiny as possible" (Fielding, 1999, p. 526).

Rigour must be considered from the outset of our research, underpinning the early stages of research design. These procedures were outlined in foundational qualitative research works by Denzin (1978) and Baxter and Eyles (1997) as the four major types of **triangulation**: multiple sources, methods, investigators, and theories (see also Janesick, 2000). For example, as we move through various research stages, we might check (a) our sources against others (research); (b) our process and interpretations with our supervisors and colleagues; (c) our text with our research participant community to enhance the credibility of our research, and (d) our theories with the broader scholarly community. We note that the third check can be problematic if that community has considerable power, such as might be the case with a multinational corporation if its managers refuse permission for us to publish work related to findings derived from the corporation. These forms of checking are related to the practice of **reflexivity**, which is explored in more detail in this volume (Chapters 2, 3, and 4); reflexivity is the critical self-review that ethical researchers must engage in and is an essential part of rigorous research.

As indicated in the research stages in Box 6.3—which often overlap as they become a whole research composition—we also need to document our work fully: how we came to be interested in the research, why we chose to do it, and for what purpose. We may declare our own philosophical, theoretical, and political dispositions, and we will almost certainly review literature dealing with both the general area of our research and the research methods we intend to use. This elaboration

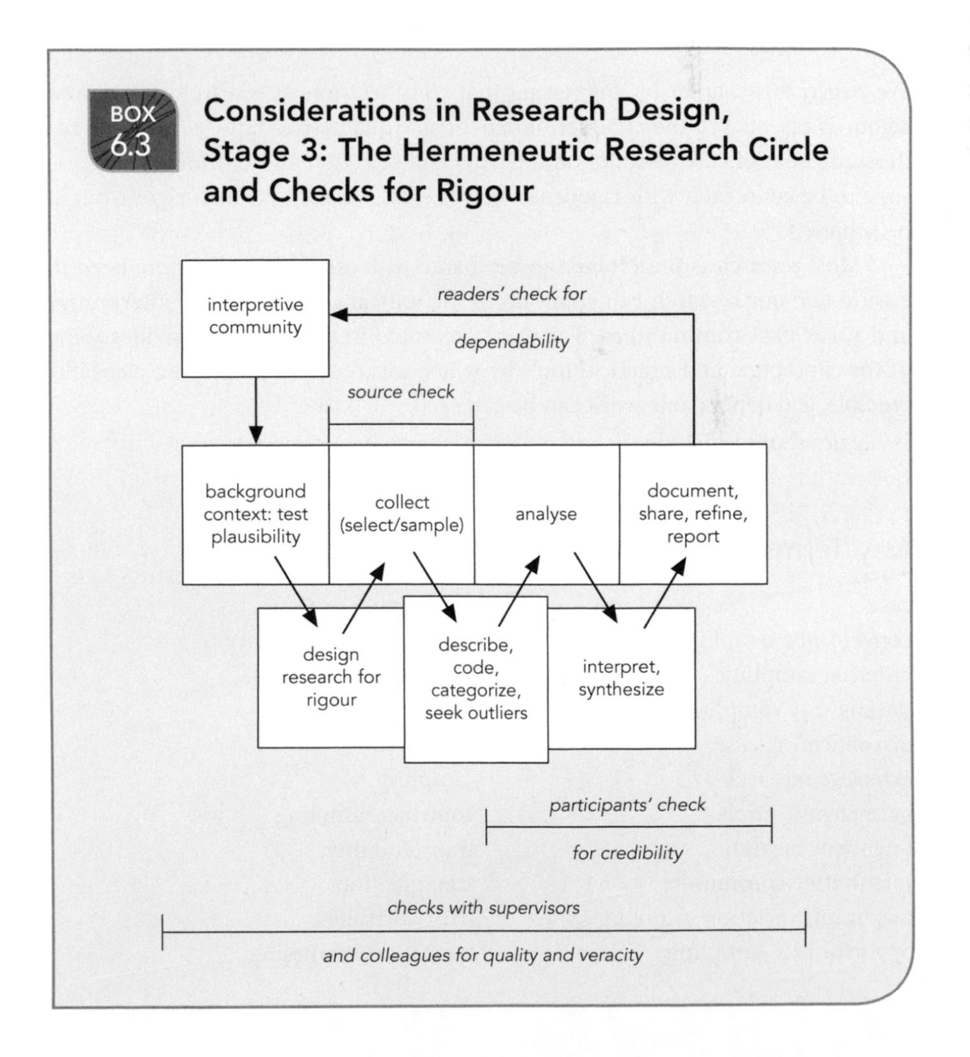

of context permits us to establish the plausibility of our research by demonstrating that we embarked on our work adequately informed by relevant literature and for intellectually and ethically justifiable reasons. We will most likely have checked the plausibility of our research with community partners, supervisors/advisors, and/or colleagues before embarking on detailed research design. At the final stage of reporting research, we should also attempt to acknowledge limits to the **transferability** of our research due to particularities of the research topic, the research methods used, and the researcher. In this way, we confirm that the methods we use and the interpretations we invoke influence our research outcomes. Thus, it is vital that we document all stages of our research process. Such documentation allows members of our interpretive and participant communities to check all of these stages and confirm that our work can be considered dependable.

Conclusion

We began this chapter by suggesting that consideration of research design and rigour is essential to the conduct of dependable qualitative inquiry. We have addressed issues of case selection and participant selection and outlined some reasons to be concerned with rigour, as well as some means by which rigour might be achieved.

Most research is undertaken to be shared with others. We therefore need to ensure that our research can stand up to the critical scrutiny of our interpretive and participant communities. The work presented in this chapter provides some of the conceptual and practical tools by which this outcome of sharing plausible, credible, and dependable work can be achieved.

Key Terms

case
convenience sampling
criterion sampling
deviant case sampling
disconfirming case
extensive research
hermeneutic circle
intensive research
interpretive community
maximum variation sampling
opportunistic sampling
participant
participant community
purposive sampling
research design
rigour
sampling
snowball sampling
transferability
triangulation
trustworthiness
typical case sampling

Review Questions

1. Why is rigour important in qualitative research? Is it an ethical consideration?
2. What is an "interpretive community"?
3. What is meant by the phrase "participant community"?
4. What are some ways we might check our research to establish its dependability to members of our interpretive community?

Review Exercises

1. Imagine that you are working as a geographer/planner in a consulting firm. You have been approached by a municipal government's general manager and asked to investigate ways in which to provide for the civic needs and aspirations of diverse groups and individuals in a central city square. The square has heritage values—including a central fountain dedicated to one of the city founders; one side of it is part of the central bus interchange; it is adjacent to a shopping mall and government and office buildings; and there are large areas of lawn and established trees and benches, as well as intersecting paths throughout the park. Among your first tasks is to determine whether you will use a quantitative, qualitative, or mixed-method approach. Working with your colleagues, draw up a comprehensive table in the form of a SWOT [strengths, weaknesses, opportunities, and threats] analysis to consider the advantages and disadvantages of each approach. Does a clear decision emerge? What do you learn by doing this exercise? How would you then use your research to communicate with policy-makers and demonstrate the work's veracity?
2. Imagine that you and your colleagues in the consulting firm mentioned in Exercise 1 decided that you would adopt a mixed-method approach to your investigation for the municipal government. Before you begin any primary data collection, you have decided to develop a checklist of other types of information you will need in order to ensure rigour. Working together, identify the other forms of information you would seek to draw upon, what methods you would use to collect such data, and justify why you think those forms of information will assist triangulation. How would you explain to the people commissioning your work that the approach is rigorous?

Useful Resources

Bradshaw, M. (2001). Contracts and member checks in qualitative research in human geography: Reason for caution? *Area, 33*(2), 202–11.

Hennick, M., Hutter, I., & Bailey, A. (2010). *Qualitative research methods*. London: SAGE.

Mason, J. (2004). *Qualitative researching* (2nd edition). London: SAGE.

Patton, M.Q. (2015). *Qualitative evaluation and research methods* (4th edition). Beverly Hills, CA: SAGE.

Robson, C. (2011). *Real world research: A resource for users of social research methods in applied settings*. Chichester: Wiley.

Sayer, A. (2010). *Method in social science: A realist approach* (revised 2nd edition). London: Routledge.

PART

Talking, Watching, Text, and Context: The Scope and Practices of Qualitative Research in Geography

Case Studies in Qualitative Research

Jamie Baxter

Chapter Overview

This chapter defines the case study as a broad methodology or approach to research design rather than as a method. Much of the chapter clarifies precisely what a case study is, describes the different types of case studies, and addresses some misplaced depictions (such as N=1) and criticisms (lack of generalizability) of case study research. Although most case study research is cross-sectional, conducted typically on one case at one point in time, this chapter reviews two major variants of multiple case studies that might appeal to geographers in particular: the within-case *temporal* comparison and the *spatial* (place-oriented) cross-case comparison. Much of the how-to of case study methods and fieldwork is covered in companion chapters within this volume. This chapter instead focuses more on broader research design issues specific to case studies.

What Is a Case Study?

Gerring (2004, p. 342) provides a very concise and useful base definition of the **case study** as "an intensive study of a single unit for the purpose of understanding a larger class of (similar) units." However, we must be careful not to conflate sample size with the quality of case study research—a point that is addressed near the end of this chapter. Further, Gerring (2007) later amended his definition to consider the notion of multi-case studies associated with temporal and spatial comparisons. Thus, case study research involves the study of a single instance or small number of instances of a phenomenon in order to explore in-depth nuances of the phenomenon and the contextual influences on and explanations of that phenomenon. Some examples of phenomena researched as case studies might include an event (e.g., a protest rally, a disaster), a process (e.g., immigration, discrimination, risk amplification, deforestation), or a particular place (e.g., a neighbourhood with a high crime rate, a community hosting a hazardous waste facility). Case studies are often used to better understand and sometimes directly resolve concrete problems (e.g., why is uptake of immunization so low in town X?). From an academic point of view, though, case studies are well suited to corroborating existing explanatory concepts (theory), falsifying existing explanatory concepts, or developing new explanatory concepts. Perhaps most important, the case study provides detailed analysis of *why* theoretical concepts or explanations do or do not inhere in the context of the case.

Thus, a case study is perhaps most appropriately categorized as an approach to research design or **methodology** (a theory of what can be researched, how it can be researched, and to what advantage) rather than as a **method** (a mechanism to collect or generate data). It is more of an approach or methodology than a method because there are important philosophical assumptions about the nature of research that support the value of case research. The primary guiding philosophical assumption is that in-depth understanding about one manifestation of a phenomenon (a case) is valuable on its own without specific regard to how the phenomenon is manifest in cases that are not studied. This depth of understanding may concern solving practical/concrete problems associated with the case or broadening academic understanding (theory) about the phenomenon in general, or a case study may do both of these things. Other philosophical assumptions are described below as clarifications particularly concerning N=1 and **transferability/generalizability**.

It is worth examining where case studies fit relative to other approaches to social research. Case studies are often considered equivalent to field research, **participant observation**, **ethnographic research**, or even qualitative research. Although case study research certainly intersects with all of these, it deserves more attention than most social science textbooks provide. That is, case studies are typically mentioned within chapters that mainly concern these other aspects of qualitative research (this volume and chapter notwithstanding). Further, the terms *qualitative* and *case study* are not entirely interchangeable, largely because quantitative researchers also conduct case studies and many use a combination of methods and methodologies within case study research. In fact, much of case study work is indeed qualitative, though some is quantitative or a mixture of the two. Thus, researchers need to specify up front precisely what type of case study is being conducted (quantitative, qualitative, or mixed) so that the quality of the work is fairly judged according to the methodology used (see Chapter 6). However, in keeping with the theme of this text, the focus in this chapter is on the qualitative case study and on qualitative aspects of mixed-method case studies.

The Historical Development of the Case Study

The popularity of case study research has perhaps never been higher. Although the case study has a long history in the social sciences, its popularity faded during the quantitative revolution of the post–World War II era. However, its status has dramatically resurged during the past few decades (Taylor, 2016).

Most writers trace the origin of modern-day case study research to the **Chicago School of Sociology**. The so-called Chicago School emerged in the 1920s and 1930s during a time of rapid industrialization in North America. Industrialization in turn spawned the need to better understand the workings of rapidly growing cities. Major writers from the Chicago School include William Thomas (1863–1947), Robert Park (1864–1944), Ernest Burgess (1886–1966), and Louis Wirth (1897–1952), and their works have been heavily influential on a diverse array of qualitative researchers in areas such as urban and cultural sociology and geography. Much of the work was and continues to be ethnographic, combining quantitative survey work and qualitative participant observation, semi-structured interviews, and unstructured interviews (Platt, 1992). For example, Thomas and Znaniecki (1918) produced a massive four-volume account of Polish immigration to America as a case study of the "immigration experience." Such work and much to follow was heavy on detail and conceptual development that included empathetic accounts of human experience. Thus, the Chicago School approach was appealing to humanistic geographers interested in the manifestation of various phenomena in "places" imbued with contextualized meaning rather than conceptualized simply as "locations" (Tuan, 1977). However, the density and sheer length of the early Chicago School ethnographies made them largely inaccessible except through second-hand interpretation—in geography textbooks, for example. Indeed, the same issue remains a challenge for qualitative case study researchers today who want to provide rich detail but must produce work brief enough to appeal to a wide audience.

During and after the World War II period, quantitative aspects of case study work began to gain more prominence among scholars than its qualitative aspects. Nevertheless, the qualitative/ethnographic methodologies used in case studies continued to be honed throughout this era through such well-known breakthrough case studies as Whyte's (1943) *Street Corner Society* (Boston) and Liebow's (1967) *Tally's Corner* (Washington, DC). Both studies concerned men living in poor neighbourhoods, and the authors produced meticulous accounts of structural social issues about such phenomena as unemployment and gender relations in predominantly Italian and Black working-class neighbourhoods, respectively. The fieldwork was measured in months and years rather than weeks. These works are lauded as much more concise and accessible (though still book-length) accounts than the early Chicago School works while retaining the conceptual/theoretical richness of the Chicago School studies. Further, they helped to develop both participant observation and less structured interview methods more fully.[1] Whyte and Liebow drew out concepts grounded in the case context in concrete terms (e.g., "manliness') rather than academically abstract concepts. Although well-grounded in the case, such concepts did and still do resonate in other contexts that are similar to the ones in the Whyte and

Liebow studies. What is important here is that good case studies are so richly described (theorized) that one generally finds it quite easy to draw parallels with contexts outside the case. Indeed, this detail or richness is one of the best strategies for creating **credible** and **trustworthy** (rigorous) qualitative case study work (see Chapter 6).

It is important to note that one thread of the case study's history is the ongoing use of **multiple methods**, including the combination of qualitative and quantitative methods in the same case studies. Platt (1992) provides a detailed account of the history of the rise, fall, and rebirth of the case study, which is beyond the scope of this chapter. Yet in her account, she carefully shows how the Chicago School studies ushered in a long era during which researchers highlighted and, more detrimentally, exaggerated *differences* between quantitative (**objective**) research and qualitative (**subjective**) research, a tendency that still exists today. In the early years of the Chicago School, however, researchers were more inclined to highlight that quantitative and qualitative approaches were both powerful on their own but also complemented each other (Platt, 1992). That perspective on combining methods has now resurged (Hesse-Biber & Leavy, 2004), including through digital means (Leszczynski, 2018), which means that there is a refreshingly diverse set of research design possibilities from which the case study practitioner can choose. Before we turn to some common ways that geographers design case studies, it is worth clarifying the importance of depth and context as opposed to **sample** size.

N=1 and the Importance of Depth and Context

The term 'N=1' is often used to succinctly describe case study research, but there is far more to case studies than the *number* of units studied. In fact, case study researchers tend not to think in terms of sample size. Thus, it is important to clarify what N=1 means in the context of case study research and to highlight that depth of understanding and contextualized understanding are far more important.

The use of the term *N=1* to describe case studies can be confusing, because it grafts quantitative/statistical terminology onto non-statistical research. In statistics, *N* refers to the **population** (the group about whom conclusions are drawn), while *n* refers to a sample, or subset, of that population. Strictly speaking then, *n=1* is more appropriate statistical terminology—one case from the population of N. Moreover, there is also selection within the case itself that must be accounted for—the sub-units within the case. That is, case study researchers often study large numbers of "things" within their case. Yet qualitative researchers tend to think of their cases as wholes that cannot be fully understood when lumped together with a large number of other cases. That is, the context of the case is important, since it more often than not substantially influences the phenomenon in question (e.g., a change in national employment policy can affect local crime rates). That is why multiple sub-units of things (e.g., people, newspapers, policies) are often studied—to get at these contextual influences. Thus, rather than study a few of each sub-unit across a wide array of communities (e.g., 10 people, one newspaper, and two policies each from 10 or 100 communities), the qualitative case study researcher instead prefers to study one carefully selected community intensively and holistically to understand how the various things studied interact with one another in, for example, one place. There is no statistical notation to adequately account for the importance of context, and any use of N and n does not do justice to the value of case study research.

Further, it is important to underscore that case study research is **intensive** rather than **extensive** research (see Chapter 6). Social scientists use the terms **idiographic** research and **nomothetic** research to describe this difference. Idiographic research is depth-oriented, since it tends to focus on the particular to understand a phenomenon in more detail. Nomothetic is breadth-oriented, since it focuses less on the details and more on investigating a limited number of things across several units (cases) simultaneously. Nomothetic researchers typically use probabilistic sampling to select those units. Although qualitative case study researchers do not typically choose sub-units based on probabilistic sampling, few case study researchers concede that their findings are exclusive *only* to the case. Indeed, qualitative case study researchers working in an idiographic frame expect and look for what might be common *between* cases. A case is viewed as neither entirely unique nor entirely **representative** of a phenomenon (i.e., a "population" of cases, according to statistical terminology). This point is taken up again in the discussion of generalizability and transferability later in the chapter.

To explain the points about sub-unit selection and how to study context more fully, Figure 7.1 offers an example of how a case study can be based on multiples of different sub-units. Suppose a researcher is interested in the phenomenon of community hazard risk perception—specifically, why groups of residents vehemently oppose potentially hazardous or polluting facilities (e.g., industrial, waste, nuclear power) being located near their community.[2] This is a problem of practical importance, since at the very least, failed facility siting efforts cost millions of dollars, not to mention the potential negative impacts of such facilities on local residents and environmental justice concerns. The researcher could look for a community that faces having a facility located in its midst. However, for the purpose of identifying a single valuable case relevant to the problem at hand, the researcher might instead flip the problem on its head and ask the related question "why are some communities *un*concerned and not opposed to these facilities?" Thus, the researcher needs to identify a town where residents seem to be supportive of a local hazardous waste treatment facility—a case study. What the researcher might actually do is use multiple methods whereby they might interview residents in several different types of roles (councillors, facility workers, opposition group members) to understand their risk perceptions and conduct a **discourse analysis** on various types of facility-related media coverage to understand the influences on (a lack of) risk perception. These are some of the many design and selection considerations that are documented elsewhere in this book (e.g., Chapter 6). The intensive, holistic aspect of the study comes in part from trying to understand how the various contexts—which are conceptualized in Figure 7.1 as local, regional, and broad—are involved with facility hazard risk perception. In this example, the analysis of local, regional, and national media coverage helps the researcher to better understand these contextual influences. The analysis could include other influences, such as recent national economic changes and policies at various scales. In the interviews, residents would be asked how they view these external influences, such as the reporting by various types of media. We will revisit this example later in the chapter.

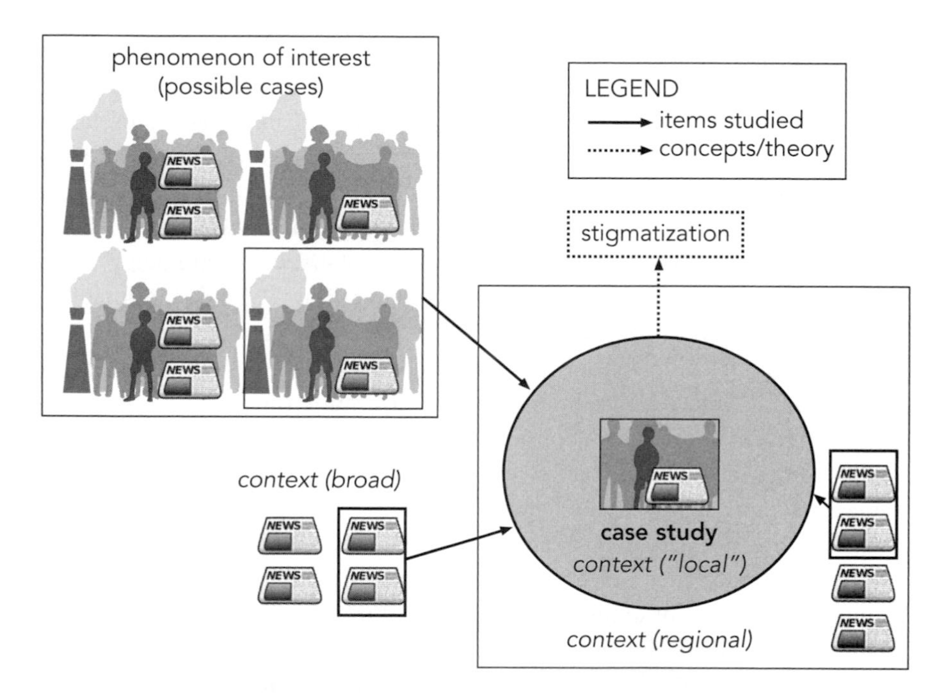

Figure 7.1 Case Study Selection, Sub-units, and Context

A case study is one manifestation of a broader phenomenon. Researchers carefully select the case to understand the practical/concrete aspects of the case itself but also to better understand the broader phenomenon. In this example, one community living with a hazardous facility (n=1) is selected from all possible communities with hazardous facilities (N). Further, multiple people will be studied within the case community, along with multiple newspapers both inside (local) and outside the community (regional and extra-regional). This helps to understand an array of contextual influences on the phenomenon (e.g., low risk perception) within the case to develop contextualized theory (e.g., stigmatization).

BOX 7.1

Studying One Person: A Case Study of "Praxis"

Sometimes cases are not identified as places or institutions with multiple people as sub-units within to be studied. Instead, they may concern multiple manifestations of a phenomenon with one person only. Yet findings from such work can still be transferable beyond the single person. For example, Wakefield (2007) studies the phenomenon **praxis** in critical geography. Praxis concerns the way that researchers may use research itself to make positive change outside academia, commonly using the shorthand "theory + practice." The "case" in this study is Wakefield's own experiences as an activist researcher in the "food movement" operating in such contexts as her own household and Toronto food movement organizations, as well as within broader environmental justice organizations and "globalized, corporate industrial agriculture" (Wakefield, 2007, p. 332). That is, she studies how she uses research in all of these different contexts—how research praxis is manifest in each of these contexts. Wakefield draws on her personal experiences to expand on the underappreciated concept of "activism at home." This concept moves beyond academic teaching and writing, which she argues are the dominant forms of academic praxis. She highlights the value of taking *personal* action in the household and community, partly to set an example for others but also to better understand the challenges of alternative food consumption. Thus, most of her research is **reflexive** in the sense that she systematically studies her own experiences and interactions with others in relation to alternative types and sources of food. She has made a contribution to the general theory of academic praxis in terms of expanding the theoretical concept of activism at home. Further, one can easily imagine that activism at home resonates beyond the case of food movements and Wakefield alone (see Kepkiewicz et al., 2015). That is, it is relatively easy to see how the concept might apply to other cases, such as waste generation or fuel consumption.

Types of Case Studies

Theory Testing and Theory Generating Cases

As outlined above, case studies play two key but not necessarily mutually exclusive roles: to test theory and to generate or expand theory. The former may involve the search for negative or falsifying cases, while the latter concerns cases that are more typical.

For some, what distinguishes case studies from other approaches, such as **grounded theory** and **ethnography**, is that in case studies, theoretical propositions should be stated prior to entering the field (Yin 2003). Yet others tend to view qualitative case studies as primarily **theory generating** endeavours such that ethnography and grounded theory can be easily incorporated within a case study design. In practical terms, both positions are mainly a matter of degree, since most practitioners of grounded theory and ethnography do not commence field research without adequate knowledge of *some* theory. Nevertheless, if one decides to follow Yin's recommendation that formal propositions need to be stated up front, two cautions are necessary.

First, qualitative researchers presume that propositions are contingent or context-dependent such that concepts describing relationships are only "true" *under certain conditions* (Sayer, 2000). That said, concepts are still "true when . . .," and accounting for the context or contingencies within which a truth happens certainly falls within the realm of what most qualitative researchers do in practice. Using our example of the low-risk perceiving, hazardous facility-hosting community depicted in Figure 7.1, the researcher might focus on the concept of "stigma" prior to fieldwork. She might propose that such communities feel that the outside (non-local) media stigmatize the community, especially *when* accidents happen at the facilities and *when* these accidents are widely reported and *when* that reporting does not simultaneously report the social good the community is doing by hosting a facility that solves a widespread social problem (i.e., disposal of hazardous waste). Further, this stigma may minimize or redirect the concern or anger that residents once directed at the facility onto the media instead. The effects of stigma are not always "true," but they seem to be under these specific conditions (and perhaps others).

The second caution is that there is a potential logical flaw in *ever* stating propositions up front at the beginning of a study. That is, formal propositions typically require well-developed theory as their basis. Yet qualitative case studies are often used in exploratory ways to delve into under-explored and thus under-theorized phenomena. Moreover, researchers tend to borrow from related areas of inquiry (see Chapter 5). Often, it is not necessary to re-invent the (conceptual/theoretical) wheel. In the case of our Figure 7.1 example, there might already be a well-developed theory on the negative effects of stigma on hazardous facility-hosting communities, but it might not go as far as suggesting that stigma can redirect or minimize concern about the facility itself. The concept of stigma, which may itself have been developed from case study work, is thus borrowed from the literature and elaborated upon.

This raises the question of how theory is actually *generated* using case studies. This question is covered in greater detail in Chapter 18 on coding and analyzing data, but it is worth pointing out that qualitative research in practice is rarely a purely **deductive** or purely **inductive** endeavour. Rather, it tends to be more cyclical in the sense that theory stated initially either formally as hypotheses or loosely as budding ideas is explored (deductively) by studying the real world of the case and then that information is used to generate new concepts (theory) to explain what is observed (inductively) (Figure 7.2). These refined or new concepts are then further scrutinized by ongoing analysis of the real world of the case. All the while, the good researcher will remain aware of existing academic literature that might contribute to the explanations. Grounded theory and one of its geographical derivatives, **grounded visualization**—described in Chapter 18—are approaches to moving through this deductive/inductive cycle to form in-depth understanding of the entire case.

Case Studies Across Time and Space

Case studies need not be one-off, single studies of one case at one particular point in time. Nevertheless, multiple case studies are generally not approached with the purpose of establishing *statistical* generalizability (see the discussion below). Instead, it is better to view multiple cases in one of two ways. In the first instance, multiple case studies provide a broader basis for exploring theoretical concepts and explanations of phenomena. In the second instance, longer-term study of the case may be useful to **corroborate** and further explore theory as it relates to the case at hand. Both enhance the credibility and trustworthiness of the concepts and explanations by exposing them to different scenarios between cases in the first instance and within the case as it evolves over time in the second. Neither one constitutes **replication** in the statistical or experimental sense of the term, but

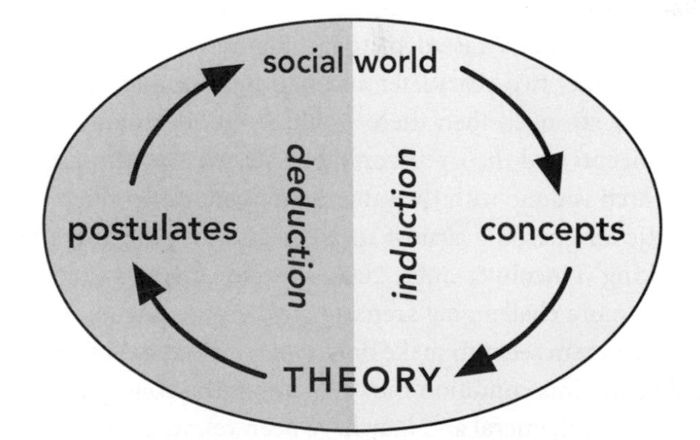

Figure 7.2 Cyclical Modes of Exploration in Case Studies

Often, qualitative case study researchers tend to emphasize the inductive mode of inquiry, moving from empirical observation to concepts/theory. In some cases, the main purpose may instead be to test an existing postulate (hypothesis). However, both inevitably involve multiple loops of reasoning as the researcher tentatively develops concepts (induction) and then compares them to the details of the social world that comprise the case (deduction) they may instead be seen as ways of both deepening and expanding the theoretical concepts (see the examples of stigma and activism at home above). Thus, for multiple cases, little attention needs to be paid to the sequence of the case studies—they can be studied in parallel. However, for **longitudinal cases**, more attention must be paid to the temporal sequencing of the research.

Time: Cross-sectional and Longitudinal Case Studies

The most common form of social research is **cross-sectional**—that is, it is conducted at one point in time (Bryman, 2006). The definition of *one point in time* can be somewhat fuzzy for qualitative researchers since they often spend extended time in the field collecting and analyzing data. Operationally then, a study may be considered cross-sectional if fieldwork is conducted in one block of time regardless of how long it takes. It would only be considered longitudinal if there was a revisit, with the researcher returning to the case after an intervening time period during which no appreciable research was done. Similarly, for studies that do not involve fieldwork or face-to-face interaction (e.g., discourse analysis of media), identifying collections of data at specific time periods and analyzing them for changes from one period to the next would make them longitudinal.

A key advantage of longitudinal research on the *same* case is that it makes it possible for the researcher to address what may be considered the enduring versus the ephemeral by exploring the robustness of the original concepts and explanations (theory). When done as a follow-up study on the same case, longitudinal research amounts to a form of corroboration to determine whether the original explanations have endured over time. Nevertheless, case study researchers should not be so naive as to assume that social settings are invariant over time (see Chapter 1). For example, in our study of risk perception in the hazardous facility-hosting community case depicted in Figure 7.1, if the researcher were to revisit the community two years later and find that stigmatization and its effects were the same or stronger, then there would be good grounds for arguing that the original concepts and theory are credible, are trustworthy, and endure. If the follow-up research is done with the same participants as in the previous study, it serves as a tactic for guarding against such threats to rigour by taking the form of **member checking** (Lincoln & Guba, 2002; see also Chapters 4, 6 and 16).

However, a more challenging scenario is when the phenomena and the concepts that explain them seem to make only *some* sense still, but less so, or perhaps in a changed form. This condition may signal that the concepts or even the phenomenon itself are ephemeral and may have been relevant for only a brief period. For example, residents might talk less about the stigmatizing effect of non-local media during the follow-up. This might threaten the original theory about stigmatization, leaving the researcher to explore the reasons for the apparent change of view. Such an exploration is potentially immensely valuable, because new insights may be gained. Nevertheless, if the changes regarding stigma were due, for example, to the success of the community's legal action against a non-local newspaper, the theory would remain credible. That is, the community might have moved on, knowing that the "stigmatizer" had been appropriately dealt with. If residents were *more* concerned about the facility (e.g., because of a recent accident) and tended to sympathize more with non-local media reporting on the community, it would pose a serious threat to the initial conceptualization. Recall that the original concept of the effect of stigma was that harsh criticism of the facility by outside media stigmatized the community and residents reacted by outwardly expressing little concern about the facility. That is, low concern about facility risks was a reaction to stigma that served to defend the community and enhance community pride. An apparent rise in concern, together with more sympathy toward outside media, go against the original thesis and would require either (1) the reconceptualization of "stigma" or (2) abandoning it as an explanation of low concern altogether.

It is obviously very problematic to assume that social phenomena are static over time. The internal dynamics of a community typically change. For example, new people come into positions of power. Although the research context of the case may change, this does not *necessarily* invalidate (threaten the credibility or trustworthiness) of the original theory in relation to the overall phenomenon. Researchers need to be careful about identifying what aspects of their theory and concepts seem less or more relevant over time and, most important, *why*. Thus, longitudinal case studies are very good for tracking how the phenomenon changes over time in the case in question. Ultimately, this should lead to the development of well-rounded explanations. In this regard, the timing of follow-up research should be theoretically informed. For example, in our community with the hazardous waste facility, if stigmatization of the community by the media is important, then future research might solicit residents' opinions after the publication of key articles and reports that were highly negative—or highly positive—towards the local waste facility and/or the community.

Unlike the *prospective* longitudinal case studies described above, *retrospective* studies go back in time. They constitute the other major class of longitudinal study and may at first appear problematic for the qualitative case study researcher intent on primary data collection. For example, if semi-structured interviews require participants to think back to the way things were and how they felt then, recall bias becomes a serious and unpredictable problem. Nevertheless, in a design that depends on the **triangulation** of methods and data including such techniques as **oral history** (see Chapter 10) and sources such as diaries, letters, policy documents, and meeting minutes, this challenge may be overcome.

Space: Comparative Analysis

Not all comparative analyses are spatial, but most conducted by geographers are spatial in one way or another. For example, there is a long tradition in human geography that emphasizes how phenomena may present very differently from one case to the next because of the place itself—e.g., the meanings derived from the interactions among residents. When conducted at the same point in time, case studies of multiple instances of a phenomenon are commonly known as **comparative analysis, comparative case study**, or **parallel case study**. For example, a comparative analysis might involve a parallel study of three of the communities shown in Figure 7.1. Comparative studies tend to share many of the same advantages as longitudinal case studies in that there are opportunities to generate and modify concepts and theory so that they explain commonalities across cases *despite* being embedded in different contexts. Although research phenomena (e.g., risk perception) need not be place-specific, it is often useful for geographers to focus on place as an initial basis for comparison. It is also important to clarify that we are talking about field comparisons of two or more instances of a phenomenon rather than merely the comparison of a case study with existing, published case studies. Although situating one's research within the literature is essential for good academic work, the focus here is on multiple cases built into a study at the design stage. For example, in a special issue on comparative studies of transnationalism (e.g., remittances of money or visits back to the country of origin), Dunn (2008) suggests that much can be gained by comparing the impact of transnationalism on the same cultural groups in different places or on different cultural groups in the same place, as well as the impact of transnationalism on the places themselves. The emphasis is not on the number of cases *per se* but on understanding how the phenomena are manifest in different contexts.

Castree (2005) underscores the value of comparative analysis by making a call for more explicitly comparative case study work in geography concerning research on "nature's neoliberalisation" (2005, p. 541). He notes that a growing number of such case studies focus too much on the particular and not enough on what is common *across* case studies. He wants to explore some general impacts (manifestations) of neo-liberalization on nature, irrespective of context. This is an ongoing tension in qualitative research generally—balancing the particular with the more abstract when developing explanations. We see this tension in what Castree himself writes in the sense that elsewhere in the same paper, he claims that he is also frustrated by *too much* abstraction. Despite his call for some theory that cuts across case studies, he complains that existing concepts quoted in the literature are "clearly so abstract that [they] fail to tell us how and with what effects otherwise different neoliberalisations work" (2005, p. 543). These over-general concepts include privatization and deregulation. Castree's frustration is directed towards the literature and not towards any particular cross-case comparison. Indeed, a formal cross-case comparison of neo-liberalization's influence on nature in a single study is one solution to Castree's concern, as long as the concepts are sufficiently specific (but of course not *too* specific). Building on Castree's work, similar issues are explored more recently by Jazeel (2016) and González-Hidalgo and Zografos (2019). These ideas are taken up further in the following discussion on generalizability.

Are Case Studies Generalizable?

The short answer is yes. Quite probably, the most common criticism of case study research is its supposed lack of generalizability (Campbell & Stanley, 1966; see also Flyvbjerg, 2006). Yet such concern may be exaggerated. That is, generalizability should not be a problem if case study research is designed appropriately and the analysis is attentive to the tension between concrete and abstract concepts.

Generalizability (or **external validity**) is a term used by quantitative social scientists, but many qualitative researchers prefer the term **transferability** (Lincoln & Guba, 2002). Generalizability or transferability concerns the degree to which findings apply to other cases of the phenomenon in question. It may be interpreted as "the more cases the theory applies to, the better." However, this is only one way to look at generalizability, and qualitative case study researchers tend not to emphasize the more-cases-are-better approach. Instead, qualitative researchers are more concerned that explanations of the phenomenon as manifest in the case are credible. This distinction is partially explained as the difference between statistical generalization and **analytical (theoretical) generalization**. In the former case, generalization is achieved through large probability samples, while in the latter, transferability is accomplished by (1) carefully selecting cases and (2) creating useful theory that is neither too abstract nor too case-specific (Flyvbjerg, 2006; Yin, 2003).

Flyvbjerg (2006) provides some good examples of how generalizable, or transferable, theoretical concepts and explanations can be generated from a single case study. Perhaps the most famous is Galileo's debunking of Aristotle's theory of gravity. Aristotle claimed that heavier objects will accelerate faster than lighter ones. However, with a single experiment (case study) involving only a few different types of balls rolled down inclined planes, Galileo showed that balls of different weights accelerated at the same rate. This case study disproved the theory that weight is a determinant of gravitational acceleration, which paved the way for others like Sir Isaac Newton to suggest that the determinant was something else (the relationship between mass and distance). Thus, the choice of case (in the Galileo example, the choice of a ball and ramp experiment) can be critical. Further, it is important to recognize that falsifying existing theory simultaneously opens up avenues for new theory. Similarly, Popper (1959) argued that it takes only one black swan to falsify the theory that all swans are white. Keep in mind that although these examples serve to show both the value of single cases and the strategy of **falsification**, social scientists rarely expect any concept or theory to apply in all cases—that is, to constitute a theoretical law. What is important is to describe *why* a theory does or does not apply in a particular case.

There are numerous examples of case studies in geography that falsify, particularly in areas like political economy and political ecology, which use case studies to show how social and environmental problems are connected to flaws in the global socio-economic system. For example, Weis (2000, p. 300) provides an explanation of unsustainable deforestation in the blue mountains of Jamaica that shifts blame for the problem away from the "poor stewardship" of the peasants towards the "grossly inequitable land regime" within a globalized food system. As with much of the move toward **decolonizing research**, Weis's empathetic account of these Jamaican farmers is a form of falsification as it attempts to supplant a long-standing and traditional explanation of deforestation with a more compelling alternative one.

Finding cases that falsify existing theory is but one of many examples of the role case studies can play in terms of addressing generalizability or transferability. For most qualitative researchers, the development of a coherent theory itself is of primary concern, not necessarily whether the findings challenge hegemonic wisdom or whether findings adhere in *all* or even *most* cases at the time they are studied (George & Bennett, 2005). In practical terms, it is too great a burden for one study not only to understand the context, contingencies, and details of the case but at the same time to know all contingencies and contexts for all other cases of the phenomenon in question—regardless of whether the phenomenon is risk perception, academic praxis, transnationalism, neo-liberalization, or deforestation.

Strictly speaking, the only way to truly know whether a theory applies in other cases is to study those other cases—but perhaps not in as much detail as the cases that were used to develop the theory in the first place. The true value of a qualitative case study, then, may not be known until several years after the research or policy community has had a chance to digest the concepts and theory. The theory may only *eventually* prove useful for explaining similar phenomena in different cases. Indeed, this is one of the main reasons for the enduring fame of Whyte's (1943) and Liebow's (1967) studies. Although they focused on Italian and Black men in Boston and Washington, respectively, the studies' insights apply in numerous cities among a wide array of ethnic and minority groups. Their concepts have resonated in several contexts over space and time. They seemed to strike the right balance between describing concepts that explain the concrete details of the case and creating concepts that are sufficiently abstract to apply to similar phenomena in different places and times (i.e., living in a poor ethnic neighbourhood in a large urban centre). The take-home message is to generate theoretical concepts and explanations that potentially resonate in other (as yet unstudied) contexts.

Conclusion

Case studies have a long and rich history in the social sciences and geography. Case study methodology is a powerful means by which to both (1) understand the concrete and practical aspects of a phenomenon or place and (2) develop theory. That is, case studies may be used to understand and solve practical problems relating to the case alone, *and* they may be used to test, falsify, expand, or generate explanatory theoretical concepts. Case studies are valuable because, when done well, they produce deep, concrete explanations of social phenomena that are attentive to a variety of contextual influences at various scales. In fact, most of what qualitative researchers do involves a case study methodology. This research often uses various combinations of qualitative (and quantitative) *methods* to support data collection (e.g., interviews, focus groups, participant observation) as well as a variety of analytical *strategies* (e.g., grounded theory, discourse analysis). These methods and analytical strategies may be used in longitudinal analyses of the same case, cross-case comparisons of different cases, or one-time analysis of a single case. It is important to recognize that although a case study may only involve a sample of one, a carefully chosen and well-studied case can be used to produce very robust, credible, and trustworthy theoretical explanations. These explanations are generalizable, or transferable, in the analytical sense rather than in the statistical sense. Good theoretical explanations are those that are well rooted in the concrete aspects of the case yet sufficiently abstract that others in similar situations can see how they might apply to their own context.

Key Terms

analytical generalization
case study
Chicago School of Sociology
comparative analysis
comparative case study
corroboration
cross-case comparison
cross-sectional case study
ethnographic research
external validity
falsification
generalizability
idiographic
longitudinal case study
member checking
method
methodology
nomothetic
parallel case study
participant checking
praxis
theoretical generalization
theory generation
theory testing
transferability

Review Questions

1. How can a qualitative study based on N=1 produce useful research results? What role does generalizability (transferability) play in evaluating the quality of such results?
2. How have multiple methods been usefully employed in case study research?

3. Outline some of the key advantages and challenges of temporal or spatial comparative case studies.
4. What are the roles of induction and deduction in case study research? How do they relate to nomothetic and idiographic research?

Review Exercise

Break into groups of three to four and design a qualitative case study of a phenomenon important to students at your university. Discuss what you will measure and consider how you will select units to study. Try designing your project with a mix of qualitative methods—what does each one bring to the research? Share your design with the entire class.

Using all of the designs from the small groups (above), collectively identify some possibly theoretical insights that might be abstracted from the case studies.

Useful Resources

Colorado State University. (2020). Writing guides—Case studies. https://writing.colostate.edu/guides/guide.cfm?guideid=60 A helpful guide to the history and conduct of case studies as well as to the presentation of their results.

Flyvbjerg, B. (2006). Five misunderstandings about case-study research. *Qualitative Inquiry, 12*(2), 219–45.

George, A., & Bennett, A. (2005). *Case studies and theory development in the social sciences*. Cambridge, MA: MIT Press.

Gerring, J. (2007). *Case study research: Principles and practices*. Cambridge: Cambridge University Press.

Hamel, J., Dufour, S., & Fortin, D. (1993). *Case study methods*. Newbury Park: SAGE.

Hesse-Biber, S., & Leavy, P. (2004). *Approaches to qualitative research: A reader on theory and practice*. New York: Oxford University Press.

Stake, R. (2006). *Multiple case study analysis*. New York: Guilford Press.

Taylor, L. (2016). Case study methodology. In N. Clifford, M. Cope, S. Gillespie, & S. French (Eds), *Key methods in human geography* (pp. 581–95). Thousand Oaks, London, New Delhi: SAGE Publications.

Yin, R. (2003). *Case study research: Design and methods*. Los Angeles: SAGE.

Notes

1. At the time, structured surveys were by far the most popular means of collecting information in social research.
2. The example is modelled on my own research largely to avoid any misinterpretations about research intentions and interpretations. See Baxter and Lee (2004).

"Placing" Participant Observation

Annette Watson

Chapter Overview

Participant observation is a method to turn witnessed actions and experiences into qualitative data: researchers watch social and environmental phenomena, take notes on these phenomena, and record their own experiences, and, to varying degrees, participate in the minutiae of daily life. This chapter provides an overview of the history of the evolution of this method, how to conduct participant observation, and how to use the data in geography. The chapter provides examples of the "field" and "home" spaces in which geographers have used the method, ranging across urban, rural, and global landscapes. The chapter also reviews how to engage participant observation to incorporate techniques of knowledge co-production in research practices, addressing the critical turn in geography.

Introduction: The Most Social of Methods

Participant observation is perhaps the most social of methods. Certainly, my greatest joy in doing participant observation is that I get to hang out with people, and learn from them. **Participant observation** is a method whereby the researchers themselves are the instrument processing and recording observations directly as data, with the researcher(s) participating in daily life while observing (and writing down) those practices. Participant observation can be used as a stand-alone method, but it is also employed to begin or as a part of any multi-method project, including those with or without a field component.

I want to begin this chapter stating that it is important not to conflate participant observation with ethnography, even though ethnographers often rely on participant observation. **Ethnography**, aiming to qualitatively describe the interior lifeworlds of cultures and situations, is a kind of academic writing. Yet in this era of "post-normal" and interdisciplinary sciences that engage with multiple communities, many kinds of academics (qualitative and quantitative, new students and experienced professors) can generate and employ the data from participant observation or find use in its practice (e.g., Pierce and Lawhon, 2015). In this chapter I focus on practices and data because participant observation is a skill that teaches you how to observe and document the minutiae. Whether your study is of a singular event with a short time span, or encompasses a larger spatio-temporal frame, anyone can draw on participant-observation data for their academic work as geographers.

Geographers have used participant observation to understand many different places and spaces: urban, rural, or suburban, and even virtual spaces. And while the discipline of anthropology initially forged the method of participant observation (Malinowski, 1922), it is through the discipline of geography that one can ably theorize the "place" of participant observation amongst many possible research methods.

The Spectrum of Places, Practices, and Uses

As Robin Kearns points out, "participant observation for a geographer involves strategically placing oneself in situations in which *systematic understandings of place* are most likely to arise." (emphasis added, Kearns, 2016). Participant observation can be employed as a method to understand the places of a researcher's lived world ("home"), as well as places unfamiliar to the researcher, or accessed through longer distance travel ("the field") (England, 1994; Kearns, 2016; Mukherjee, 2017). It can be a method that is utilized just one time for a project, or as part of a long-term research strategy.

Doing participant observation includes a wide spectrum of practices that can emphasize either the *participation in* or the *observation* of a phenomena, if not an equal mix. For example, a researcher who wants to understand the geography of protest might choose to *observe* the crowd without being the one chanting and holding signs. Public places are all open to more observation-heavy approaches that do not require much interaction between the researcher and the people/places being observed. To observe a protest—or religious service, or a conference, or a restaurant/bar, or a street intersection, or a classroom—you describe in **field notes** what you heard, what you saw, the interactions between the different social groups that meet in that observed space, and the details of the sites of interaction (the shape of the room, the architecture, the sounds, smells, etc.). Many geographers, for instance, are interested in the social rules of place that shape human actions in different material and symbolic settings; people behave differently in a café, a health centre's waiting room, and a children's playground. The rules of place reflect cultural priorities, social differences (gender, age, race, etc.), and the interaction of people with their natural and built environments. Your recorded observations in any of these contexts will include what you sense externally as well as the feelings/emotions that come with your witnessing.

For those who focus on the other end of the spectrum, emphasizing greater *participation* in their participant observations, it is a method of seeing and feeling, but also a method of doing. This involves trust-building and often living with other people—socializing, friend-making—as much as it does the component of writing down and analyzing what you observed. **Trust**, in the context of participant observation, refers to a belief in the honesty of a research relationship, and mutual respect between parties of that relationship. Trust is the currency through which a researcher obtains information from others—if you want to put research in terms of an exchange. Without trust, it is difficult to engage participants in any study, and without trust, it is difficult to continue working in an area with a particular group of people. Without trust no one will talk to you.

Regardless of whether a researcher is more participant-heavy or observant-heavy in their approach—or, as is quite common, occupies a spot somewhere in the middle – participant observation is a valuable way to **triangulate** or cross-check analyses with other methods, such as those I use: **interviewing**, **focus groups**, workshops, and **participatory mapping**. I would argue that most geographic field methods, whether qualitative or quantitative, include participant observation, even when the researchers are not explicit in their use of this method. This chapter explains how to do it in a conscientious and **rigorous** way, in a variety of situations. I also highlight what participant observation techniques bring to understanding different geographies.

Positionality—The Place of the Researcher and the Practice of Reflexivity

Because the data from participant observation is derived from a researcher's observations, a rigorous practice of participant observation requires reflection on one's **positionality**. Positionality refers to a researcher's social, locational, and ideological placement relative to the research project or to other participants in it. Positionality may be influenced by biographical characteristics, such as class, race, and gender, as well as various formative experiences, and these identities persist within a power-laden context that must be unpacked by the participant observer (see Chapters 2, 3, and 4).

The need to discuss one's positionality as part of a rigorous practice of participant observation emerged as a result of a critique called the **crisis of representation**, an academic debate that peaked in the 1980s and 1990s, which questioned who had a right to represent a particular culture (see for example, England, 1994). This was especially relevant to the writing of ethnographies because, historically, the method of participant observation and ethnographic writing was forged in the context of colonial expansion and nascent Western scientific disciplines, when the written observations of "far away" places by Euro-American military officers and natural historians gave way to a professionalized corps of anthropologists and geographers who described and mapped the non-Western world for Western audiences. Extensive **post-colonial** and feminist literature stemming from the critical turn in geography has recounted the numerous ways that these depictions of **Other** places and societies reflected racism and embedded justifications for territorial and resource exploitation of places and people by the West (see Chapter 4). And more broadly, many marginalized groups within Western spaces, such as Indigenous peoples, LGBTQ+, and people of colour, have questioned the knowledges being produced by **outsiders** regarding their own communities (IPSG, 2010; Kobayashi, 1994; 2014; Sultana, 2007). For example, a key directive to non-Indigenous researchers in an ethics statement authored by the Indigenous Peoples Specialty Group (IPSG) of the American Association of Geographers asks:

> How could you explore options for research that do not include "studying" Indigenous peoples? If non-Indigenous communities or institutions are a primary obstacle or barrier to Indigenous self-determination, would it be more helpful to the Indigenous community for you to study non-Indigenous policies or attitudes? If the Indigenous community does not want you to conduct research within their community, how could you flip or shift your research to focus more on your own community, or on broader social relations, policies and institutions that affect Indigenous peoples? (IPSG, 2010)

The problem, as post-colonial and feminist researchers confront it, is that when a participant observer writes up their results, the image of the culture or event is unavoidably distorted because of the outsider perspective of the researcher—and yet the method of participant observation seemed to give them licence to represent that society as if they were neutral, with "expert" knowledges always superseding the **insider** knowledges of peoples in those communities.

In terms of understanding the research process, this critical perspective suggests that a researcher's identity *matters* in the collection of data as well as the interpretation of that data, and thus assumes that all knowledge is produced with a perspective, referred to as **situated knowledges** (Haraway, 1988). Such arguments characterize the emergence of the research worldview of **social construction**, which assumes human identities and cultures are produced through socially agreed-upon meanings and are highly changeable, rather than fixed and **objective** (Katz, 2009). All people have life experiences that can be connected to their gender, class, age group, culture/subculture, sexuality, education, whether they are a parent, whether they grew up in a rural or urban place, and so on—and people occupying these identities experience and exercise **power** differently. Because of the **intersectional** (Crenshaw, 1991) or compounded nature of our identities, people experience a range of social advantages and disadvantages, and understanding these power dynamics is not so clear cut (Whitson, 2017). Therefore the positionality of the researcher can both provide insight into, and affect, the kinds of data being collected; it can also affect the interpretive framework of the experience. This goes for any kind of research, whether qualitative or quantitative.

This is a critique that Sundberg (2003) launched at Latin American geographers who she claimed practised a masculinist **epistemology**, or *way of knowing*, which assumes that fieldwork is free of power relations and knowledge is produced objectively. Proceeding instead from a feminist epistemology, which assumes that categories such as gender always influence the production of knowledge, Sundberg deployed a survey to Latin American researchers and found that her "respondents highlighted the gendered nature of fieldwork. Negotiating gender relations also means negotiating unequal relations along other axes of power, including class position, race, and geographic location," and yet she found that few publications about Latin America were **transparent** about these politics (Sundberg, 2003, p. 186).

As a result of these issues, participant observers engage in a critical **praxis** of **reflexivity**, which refers to the act of thinking critically through one's relationships and affects within the world being studied. The participant observer ought to practise reflexivity at multiple points during a research project, and deeply reflect on their positionality. It is essential that researchers examine their identity vis-à-vis the research participants, since the researchers themselves constitute the instrument in the method of participant observation. Reflexivity is a means to achieve greater transparency, being clear about one's motives and influences on the research endeavour and the kinds of analyses and conclusions to be made; reflexivity is also a means to query the overall ethics of a research project (see Chapter 2).

In participant observation methodology the terms **emic/etic** emerged to talk about how the perspective of an insider (emic) differs from that of an outsider (etic) (England, 1994). In a rigorous practice of reflexivity, you first reflect on your insider or outsider status relative to the research context, the contours of your emic or etic perspectives, and otherwise place your proximity to the social interaction being studied. As an exercise to unpack your positionalities you might sketch out your own key characteristics, identify places and social settings where you fit in (and those where you do not), then progress to deeper reflections on the power dynamics inherent in your daily life and how your positionalities situate you in those. Such reflections allow you to query the ethics of the research agenda and establish a strong **subjectivity**.

Of course, absolute boundaries of emic/etic are not clear cut. Importantly, Mukherjee (2017) writes about how her work in urban Bangalore technology spaces underscored how it was problematic to understand positionality as a static characteristic of insider and outsider:

> As a female student of Indian origin affiliated with a U.S. university, I enjoyed many privileges as a research scholar, although my position vis-à-vis the professional elites in the software industry was rather tenuous. Throughout my research, I confronted how my multiple subject positions (including my feminist politics) emerged at particular moments

> along with and in response to those of my interviewees, as we were all implicated in the wider political economic transformations in urban India. (Mukherjee, 2017)

As Mukherjee indicates, neither the insider nor outsider position exists in a stark binary because of the intersectional nature of our identities (see also, Pacheco-Vega & Parizeau, 2018). Other scholars indicate that because of their place-based research, participant observation methods can cultivate intimate networks where such boundaries become less clear over time, sometimes due to increasing ethical responsibilities that come with the research network (e.g., Cuomo & Massaro, 2013; Smith, 2015; Watson, 2012).

The key to a rigorous practice of reflexivity is to repeat the exercise of unpacking your positionality throughout a single project, and throughout the lifespan of a research relationship, which for some might be just as long as researcher's career. Box 8.1 highlights some of my writing on my positionality for a recent participatory mapping project that drew on my experiences doing participant observation. As my own example shows, the act of reflexivity is not to be understood as absolving or erasing these power-laden relationships (Rose, 1997), but rather to acknowledge and work with transparency across these positionalities of researcher and researched.

BOX 8.1

Writing About Positionality

The following are extracts from my field note memos for a project I led in partnership with eight Alaska Native tribes—a cultural resource study that reviewed 10,000 years of land use for an area now called Gates of the Arctic National Park and Preserve, and which combined the ethnography with a participatory GIS (geographic information system) database of contemporary subsistence regimes (Watson, 2018). It is an example of a long-term research relationship that I have maintained in a region where I did not grow up: I am not an Indigenous person, which is the largest marker of my difference between myself and my collaborators on the project. Throughout my years of collaborating on this and other projects in the region, my approach has been more participant-heavy in its use of participant observation because of my positionality. The United States National Park Service (NPS) required this particular project to be completed through extensive consultation with the tribes, and it included some areas where I did not have prior field experience. Thinking about the IPSG directive, I of course continually asked myself: why should I "study" these Alaska Natives at all? I wrote about my positionality in relation to this particular project at the beginning of this study period, as well as toward the end:

> [June 2016] The purpose of this project is to produce "baseline data" over a large spatial and temporal scale . . . over an area at the boundary between two different cultural groups—Inupiat and Athabascan—and eight sovereign tribes. Why am I, as a non-Native, the one privileged with this task? . . . [I]f appropriate I might formally share authorship with a researcher who is Indigenous. . . . I don't want to do this contract if it means that I'm taking an opportunity away from someone who grew up in these communities.
>
> . . . Interestingly though, my positionality cannot be fixed across the study boundaries. I'm non-Native, but practically married into one of the Athabascan families; thus for this project, some of the people I am to consult for the study consider me part of the family, even while I am a "stranger" to some who live in the other communities. But just because I now have familial relations to people in the region, doesn't give me any "right" to accept the contract from the NPS to complete the study, nor does that relationship position me to represent that culture as an insider. What I need to be vigilant about is to ensure that no one feels pressure to participate in the study because of the relationship I have with the Olin family. And I need to ensure that things shared with me through that familial network, especially by my partner's father, are not shared in this ethnography without his and the tribe's explicit permission.
>
> But . . . it is hoped that this data will be used in an upcoming environmental impact statement for a proposed road, the Ambler Road. This Ambler Road might travel hundreds of miles of heretofore roadless territory, and across these communities' traditional hunting grounds to a mining area. The people in the region are aware of this potential future, and in every meeting I've attended . . . the people asked that subsistence land use be better understood by the agencies. Residents want this study to happen, to represent their subsistence in a way that could be used to drive land use policy. And I want to facilitate the narrative(s) they wish the State to know.
>
> [July 2018] . . . I found that the political context was an overwhelming motivation for tribes to each consent to participate in

(*continued*)

the study, as well as an overwhelming motivation for individuals choosing to share with me their lifetime subsistence use areas: where they hunted, fished, and gathered. When in other eras the trust had been broken between Indigenous communities and "outsider" researchers, in this case, given prior history of development in the State, tribal leadership did not question the logic of mapping these spaces, though they had particular ways they wanted their subsistence use mapped, and skepticism on whether such maps could affect the road development. I think they trusted the State less than they trusted me with their data, and I think I was able to include a few more people in the database because some trusted me already, knowing me for years . . .

Because I wasn't an "insider" to most of the villages consulted in the study, my ostensibly "neutral" and "expert" positionality could also be an advantage when it comes to ensuring that this information will be used for the Ambler Road impact studies. The expertise of someone from "outside" the region might be . . . more trusted by other non-Natives as an "objective" account of how subsistence functions. After all, it seemed obvious to me in the public meetings regarding the Ambler Road planning, that elders and community leaders' statements were being categorized as part of a general "public opinion" rather than an expert opinion. My positionality might be the best advantage the tribes have in getting their arguments about subsistence considered by the State; some community leaders told me as much.

Participant Observation in Practice—Expanded Positionalities and Knowledge Co-production

The crisis of representation contributed to fundamental changes in **knowledge production** but not just through new methodological practices of reflexivity. The crisis also diversified the positionalities, questions, subjects, and practitioners of the method of participant observation; it changed processes of consent, and opened spaces for collaborative models of research that we see in greater frequency today.

For example, England (1994) suggested the positionality of the "supplicant" could productively shift power dynamics in the research context. I have long noticed that a "student" positionality is ideal, as far as learning an honest perspective of individuals in a community, and I have often taken a role of *apprentice* in many fieldwork settings: students are often conceived as the "empty vessel" that strangers desire to talk to, and broad cultural assumptions about the ignorance of youth (and the wisdom of age) can actually work in their favour. Further, students are thought to be "learners" rather than "knowers," more open to the ideas of community members (England, 1994). I now tell my students that this could be the easiest time they will have in their careers to forge community networks in places where they are otherwise conceived as an "outsider." Trust is less a barrier in the sharing of information when you are there, in earnest, to learn.

Participant observers have also greatly diversified the places in which to conduct their research, consciously exploring positionalities within spaces where they are insiders to the culture. For example, participant observers can adopt the stance of **studying up**, akin to the supplicant positionality but employing a language that underscores the power-laden research relationship. The term *study up* (which is not without controversy) refers to the act of a researcher trying to understand a cultural group that exhibits *more* power than the identity from which the researcher comes—the supposed opposite of the researcher studying marginalized communities. Studying up allows the participant observer to elide the crisis of representation, with some examples of research questions turning toward urban governance or the institution as the site of research (e.g., Billo & Mountz, 2016; Henderson, 2016).

Many participant observers today site their research within "home" spaces, asking research questions that contribute to political change within their own polities. Indeed, Nancy Hiemstra argues that the metaphor of the periscope describes how the researcher's gaze "can facilitate the study of state institutions, policies, capitalist corporate endeavors, and power-laden social relations broadly speaking," such as in her work understanding US international migration and detention centers (Hiemstra, 2016). Additional examples include covert studies of the beef industry or in spaces where heteronormativity forces researchers in the closet; such participant observers aim to expose either illicit or researcher-defined immoral behaviour (Jansson, 2010; Maguire et al., 2019; Thiem & Robertson, 2010). Other participant observers have developed study designs that are multi-sited, such as to articulate the phenomenon of globalization through observing the global flows and nodes in a commodity chain (e.g., Cook, 2004).

In many of these examples, authors state that they are not describing "in toto" a culture or event, but instead are writing in ways to highlight the partial, situated perspective of the participant observer. Some choose a team-based or collaborative model of fieldwork in recognition of the partiality of that knowledge, with attention to how the intersectional identities of participant observers might engage the research context differently in collaborative ethnography (Brosius & Campbell 2010; Mountz et al., 2003).

Geographers have also diversified their use of participant observation to query the contours of **embodied knowledge**, conduct digital geographical ethnography in new frontiers of the virtual world (see Chapter 15), and contribute to the emergent fields of non-representational geography and the **more-than-human** (Basnet, Johnston, and Longhurst, 2018; de Jong, 2015; Duggan, 2017; Pitt, 2015; Vannini, 2015). Many of these projects demonstrate ethical explorations of communities by insiders; for example, de Jong (2015) wrote about her positionality as follows:

> I'm a 26-year-old woman of middle class and European background. I use Facebook most days in varying ways, both for work and leisure. Being a fulltime PhD student with limited alternative time constraints, I not only have access but also the power of flexibility to use Facebook, blurring the boundaries between leisure and work time. Becoming "friends" with participants on Facebook and sharing festival stories was not a premeditated decision; it unfolded as a result of my own everyday use of this site, alongside the converging ways many participants also used this space. "Knowing" Facebook, I was able to use this space in culturally appropriate ways. (de Jong, 2015, p. 213)

Importantly, the crisis of representation encouraged scholars using participant observation to value the insider position in the production of knowledge; as an insider you have the opportunity to access many insights from your own culture, and the ability to better understand the insights that people within your culture present to you. Being perceived as an insider allows you to begin a research project with some level of currency of trust from your research collaborators (Pacheco-Vega & Parizeau, 2018), and thus has the potential to produce work for greater mutual benefit with more efficiency in terms of time. Of course, special challenges also arise in insider research, such as difficulty noticing deeply embedded cultural practices and worldviews with the same acuity that an outsider might have.

The crisis of representation also opened spaces in the academy for people from non-Western or marginalized cultures to be valued as experts, with and without academic training. Today many critical scholars argue that, covert studies notwithstanding, academic questions need to be better aligned with the goals of the community engaged for research. This task is easier for a researcher who is already an insider to the community, but accessible to outsiders via **participatory research** practices (see Chapter 16): methods to share the power and responsibility of research design with non-academic community partners and individuals, increasingly known as the **co-constitution of knowledge** (also called knowledge co-production). One way this can be achieved by participant observers and qualitative researchers more generally is co-authorship with those more often construed as research **subjects** (e.g., Benson & Nagar, 2006; Watson & Huntington, 2008).

Doing Participant Observation—Learning the Rules of the Game and Recording These as Data

Doing participant observation anywhere along the continuum from participant to observer can be described in four stages: (1) preparation; (2) start/arrival; (3) observing phenomena; (4) recording data.

1. Preparation

A researcher prepares for doing participant observation by obtaining the equipment to record data, and obtaining any necessary permissions to conduct the research. Different research contexts will present a variety of unique logistical challenges for every researcher to consider—you need to carefully consider the logistics of "being there" even if you are an insider doing local work.

The actual tools I gather in preparation to record data for participant observation include my laptop, a small notepad/pen, and digital audio/video/still camera equipment—it is important to ensure redundancy in my data collection toolkit. Make sure to practise using all devices if you are unfamiliar with them, prior to your period of participant observation.

Consider whether having such tools present during your observation will alter the kinds of observations you might make. When will you be writing down the notes you take (on scene, or after an observed event)? Are you able and allowed to take photographs or video? Usually, a participant observer wants to be able to record phenomena that might happen to people in places whether or not the researcher is there to document; especially in situations where the researcher is doing more observing than participating, it is probably best to remain unnoticed while you observe behaviour in public places, if not completely covert. Sometimes the choice is purely practical. I've observed public meetings in which it was not out of place for me to be typing incessantly on my laptop, but in other cases, such as a bird surveying group whose work I observed, it was more practical for me to keep a small notepad where I only had the time to jot down phrases or words for later write-up. Likewise, in places where others are found often using cameras, such as in tourist or event spaces, I can record the scene visually without much concern about whether the people around me react to being recorded.

The other main preparation for participant observation, whether or not one is an insider, is to obtain permission to conduct that research, also known as **consent**. Covered more extensively in Chapters 2 and 4 of this volume, consent is a process of transparently informing potential research participants of the risks and benefits associated with the research, and may also include obtaining administrative permissions and negotiating expectations for knowledge co-production, if warranted. These permissions could come from a variety of sources, institutional and community-level, formal and informal, and the researcher needs to ascertain what kinds of permissions would be required or beneficial for that particular research project. Sometimes the exact procedures will vary due to the context of the researchers themselves, and, because universities have differing expectations regarding informed consent, conversations with the institution's ethics board are often the first step. Rules can also vary between using participant observation as a class assignment or for a larger project for publication; academic institutions may or may not require going through formal review by their review boards to ensure the ethics of the participant observation.

In some locales, if you are aiming to observe activity in a public location—such as pedestrian and vehicle traffic patterns at a particular urban crossroad or social interactions at a farmer's market in a public park—there may be no prior permissions needed other than those of the ethics review committee for that research. However, you may need to consider how you will respond if any of the people whose activity you are observing asks you to cease. You are encouraged to check your own region's rules on public observations. Observations in primary and secondary education classrooms will require the permission of, for example, the relevant municipal education department, school administrative agents, principal, and teacher, the students' parents, and, usually, a protocol for students themselves to opt in. Similarly, observations on private property typically require the permission of the landowner and any residents. Explaining the purpose of your observations, including what will be done with the data, is essential to communicate when asking permission, and the ways to open communication for permissions can vary. Some constituencies are comfortable with social media or email, while others might prefer formal letters or phone calls. In most cases I get these permissions in writing, specifying a particular range of time in which my observations will be taking place, summarizing and repeating the purpose of the research, use of data, and my contact information, similar to any individualized consent form.

Specific to the Indigenous context in the US (and potentially relevant in other places), the IPSG declaration instructs a diplomatic stance; recognizing the sovereignty of Indigenous peoples, they request that researchers formally gain permission from a community's leadership to conduct the work, including a negotiation of the purposes of that research (see Chapter 4). Thus, to ethically employ participant observation in Indigenous communities today, as well as among other marginalized groups, regardless of whether the researcher is an insider or not, you must leave room for a participatory praxis that begins with the process of a community level of consent, even before individual protocols for consent are broached.

Once I ensure I have obtained the appropriate institutional and community permissions, then I prepare for potential interactions with individuals at the event I'm observing. Even the researcher who chooses to observe more than participate might have to answer questions like, "what are you doing here?". In the case of such encounters, I bring with me a stack of business cards that I'm prepared to share with people who wish to talk to me further, or as a way of showing my institutional affiliation. For my university, I'm not required to have signed consent forms for participant observations when I'm talking one-on-one or with a small group, but have an approved script that I've memorized and verbally communicate about the project, which obtains consent to participate. You have to be prepared to engage with people, even if you are aiming to do more observing than participating. What are you going to say when someone chooses to talk to you? A concise, truthful script briefly describing your position and your project will always be useful.

2. Start/Arrival

Data collection for participant observation always has a starting point, and there are some things to consider when you arrive at your research site and/or start recording observations. For the covert participant observer, or those who wish to observe more than participate, the largest concern would be to find ways to hide the fact that you are observing, including practicing memorization and creating a system to write field notes in private. For researchers with no need to be covert, you must be prepared to explain your work if asked, and clearly communicate your role as a researcher to the individuals you are interacting with when you are thinking of the exchange as data. The script you made in preparation will come in handy to describe your project, and as de Jong (2015) notes for her project as an insider in virtual space, a re-introduction of your project is often necessary to obtain individual consent, even when you are considered a member of that community.

For those who are outsiders to a given research context, and especially for those doing longer distance fieldwork, once you have arrived at your field location you need to be equally prepared to discuss the kinds of phenomena you want to observe. The additional challenge in this context is the need to perceive the informal leadership and interpersonal politics of the research context, if you are not already familiar with that community. For example, the first person who decides to talk to you in a place that is new for you may be a community leader, but also could be an outsider to that community, and befriending them might affect how others see you, even in a negative way. Be perceptive through observing that person with others, as a way to start to understand the local power dynamics of an unfamiliar place. You don't want to befriend the "wrong" crowd right off the bat—don't be seen talking to a lot to police at first, if the aim of your project is to be inclusive of groups that feel persecuted by police. An insider has much less need to spend time on mapping the landscape of community dynamics and build trust, but for an outsider this period of arrival can be key to the success of a project, and thus might take much more time.

3. Observing Phenomena—The Emotional Labour of Participant Observation

How do you observe *and* participate? When I explain what I "do" when I collect data through participant observation I use the analogy of learning or documenting how to play a sport. Football, baseball, cricket, or surfing: I'm aiming to articulate the rules of the game. You can read a book on the rules, you can spectate, listen to the commentators, interview people involved in the sport, but if you want to understand the world of the players and the embodied experience of the game, you ought to play. What body language do you read in your opponent in the heat of the game? What did the crowd sound like when the outcome was assured? What does the callus on your foot feel like at the end of the game, or the end of the season? Through participant observation you become aware of what you experience with all five senses, plus what you feel—one's emotional experiences of a situation are sometimes incredibly telling. You watch, you partake, you converse, you think and reflect on those experiences.

A geographer has many research tools to choose from, but few of these can access the human subconscious. Interviews and **surveys** are tools that ask participants to tell you what they think; their responses then become data. But sometimes, people don't know why they do what they do. The social construction of meanings are not codified in text or law, but rather, are *performed*; participant observation can be a tool to turn these performances into data.

Like all methods, the skill of observing is built with practice, but a skilled participant observer is particularly sensitive to social cues. Across innumerable cultures and places, people are often non-confrontational, but being reflexive allows you to be more sensitive to non-verbal communication. Reading body language and social customs might seem easier if you are already an insider to the broader cultural context, but even insiders encounter subcultures of which they might not be a part (Mukherjee, 2017) or they fail to recognize something important because it is perceived as normal (and is therefore unseen). Those moments when your behaviour is being watched or informally policed can offer important observations, too, during a research project; through these moments you become consciously aware of how to be a member of that group in that setting.

There is a lot of emotional labour (that is, the work required to control how one is feeling in order to complete the research) that goes into doing participant observation so maintaining a support network of friends, advisors, and others is an important (often unspoken) facet of doing research. One's position in the research relationship is not just given but actively cultivated throughout the process of participant observation, and the boundaries you set and maintain come with emotional costs. Policing your own boundaries and continually cultivating self-awareness of your positionality and effect on others can be an exhausting, but necessary, component of participant observation. Further, managing your role in the research context so as not to alienate or unduly influence anyone may require avoiding conflict over things for which you have deep emotional resonances. This may mean taking a stance of non-judgment as to what you are witnessing, whether it is a fight between people, illicit activity, or what you might consider animal cruelty. Maintaining a non-judgmental stance also takes an emotional toll when the setting challenges one's identity and experiences, such as the covert research performed by researchers identifying as LGBTQ+ or as an ally within heteronormative spaces (Maguire et al., 2019).

Yet any good participant observer develops and expresses empathy, or the ability to understand another's perspective. Having empathy does not mean the researcher adopts an emic perspective, but rather respect for one's differences as much as one's similarities. This, too, takes emotional labour, because a participant observer needs to evaluate or read people to gauge whether they are dissembling, masking the truth, or just plain lying—nothing can be accepted at face value if you are doing research. And any elation or disappointment a researcher can feel after a participant observation event can be just as taxing as physical exertion. And then you still have to write about it.

4. Recording Data

There are many ways to record what is learned through participant observation. Some use video, or digitally record soundscapes, while others record their experiences in text, or a combination of approaches. This chapter focuses on textual recording of data. There are two kinds of textual record: observational field notes, and **memos**, though practitioners sometimes just refer broadly to their field notes without separating out these types of writing and data recording. While observational field notes aim to be the textual record of an event observed with the five senses and emotional experience, usually articulated in a chronological fashion (that is, what happened), memos record the reflections that the researcher confronts in the process of doing participant observation, both emotional and intellectual, and are also a place to write out preliminary analyses (see also Chapter 18). Sometimes I create stand-alone memos, such as a memo on one facet of my positionality, while other times I embed my memos in the observational field note either as a separate paragraph(s) or even as bracketed statements. It depends on what plans you have for analyzing the notes.

In general, for each observational field note you want to set the scene, keeping in mind that the setting is not merely about location, but indicative of power relationships. What do you see, hear, smell? What are people saying? What are you doing throughout the event you write about? Are you interacting with others, or taking a more observational stance? Are there ways that your presence and positionality are affecting these interactions and information being communicated? What are you learning from the experience? The appendix at the end of this chapter highlights excerpts of some observational field notes I took as part of a project on subsistence land use in Alaska, including my own insights on how I am writing, which I annotated for the purpose of this chapter.

Usually I am unable to take notes when I am conducting field observations, unless it is at an event where others have out their papers and pens. But if you are able to, write down particular quotes that you hear people say, because the verbatim quote, even a partial sentence, is an essential piece of data for the participant observer. Write your notes as soon as you can; but memory is a practice honed through participant observation. When I completed my first 15-minute observation, I thought it would take me the whole day to write up what I saw and experienced in that quarter of an hour. But the more I wrote field notes, the more I found that I could go longer periods of time before writing notes with just as much detail, and it became much faster to write down all the necessary details. Do not despair the first time you attempt to record field notes as data: as with observation and participation, writing field notes is a practised skill.

For some team-based participant observation projects, there will be some standardizing of the observations on which to take notes, but for many people, the kinds of notes that emerge are simply chronological. But creating the observational field note is not a once-and-done sort of activity, even if I am able to write notes in the moment. Field note writing is iterative rather than linear. Even for a participant observation that took place for 15 minutes or just an hour, I might begin by writing a few quotes I remembered that I didn't want to forget, and then I'll sketch out more description. As my memory unfolds, iteratively, I in turn move around in the document as I remember new details. I eventually ensure that I fill in the details chronologically; I find that it is less important to write linearly, even though the observational note will have a linear chronological progression.

Sometimes it makes sense to create one day as one entry, while other times, a short five-minute interaction can stand on its own, as a lesson that I learned. Or I might write other preliminary analyses that consider the extent of fieldwork to date—maybe for just a few sentences. My most important recommendation is this: always take observational notes in a way that you can read them weeks later and be right back in that scene—even if you never aim to share these notes with another person. Your future self does not want to find a jumble of half-sentences and bulleted lists, but rather, a fully explicated, full-sentence description of the observed event, the setting, and who was involved. Your field notes are your data: treat them with respect and safeguard them. Similarly, field notes are sensitive documents that often include private information; they should be backed up digitally when possible and carefully secured with password protections, hardcopies should be stored in locked facilities, and you should consider developing pseudonyms for respondents that mask individuals' identities.

My final piece of advice to researchers starting to write notes is to try to have a focused purpose for what you observe, tied to your research question, rather than taking notes on *everything* you witness. In turn, the more specific your research questions, the less you will have to write in terms of observational field notes. While you should not narrow the scope too much in case something becomes relevant at a later date, if you are studying children's food practices, a detailed accounting of adults' music choices is probably extraneous. The farther along you are in a project, the fewer notes you will probably take, because the experiences will be less new for you and you will develop short-hand tricks to record familiar happenings.

Analyzing Participant Observation Data and Writing Up the Results

As with the practices of participant observation, there is a variety of ways to use and analyze fieldnote data. Some scholars write traditional descriptive ethnographies, but I have argued in this chapter that any qualitative or quantitative researcher can (and maybe should) employ the techniques of participant observation for all their work. The key for each researcher is to anticipate the use of the field notes and have a plan for their analysis.

Field notes are often analyzed via coding strategies to further assess the patterns in the recorded data. **Coding** is a system to identify themes and patterns in your observational notes and memo writing, just as you might in reviewing participant diaries or interviews. Sometimes coding will inspire the creation of additional analytical memos, and sometimes the patterns can surprise you (see Chapter 18 for detailed guidelines on coding and analysis). Many times participant observational field notes provide a rich basis for creating vivid depictions of a **case study** site or event, or an illustrated vignette that demonstrates an element that emerged through other research methods in the project.

In other examples, participant observation can be important for critical work that tries to see lesser-known perspectives, such as non-representational ethnographies that attempt "to rupture rather than merely account, to evoke rather than just report, and to reverberate instead of more modestly resonating," thus taking a more creative and less report-like approach to writing up participant observations (Vannini, 2015, p. 318).

But for me, it is important to note that there have been many instances in which I chose not to write "results" at all. This has to do with negotiating my positionality, and being vigilant about the politics of representation. For example, when geographers were first writing about theories of the more-than-human, I asked one of my collaborators from a Gullah Geechee community (descendants of enslaved Africans on the South Atlantic coast of the US) to co-write an ethnography engaging this theoretical framework; my colleague quickly pointed out to me that Humanism was a problematic theory for her, because African peoples had often been left out of definitions of the human. In another instance, I had asked one of my Alaska Native collaborators to co-author a paper about a spiritual figure in

Key Terms

coding
consent
co-constitution of knowledge
crisis of representation
emotional labour
empathy
ethnography
emic/etic
feminist epistemologies
field notes
Indigenous methodologies
insider/outsider
intersectional
knowledge co-production
memo
masculinist epistemology
participant observation
participatory research
positionality
reflexivity
situated knowledges
social construction
"study up"
transparency
trust

Review Questions

1. Explain the concept of the crisis of representation, and how it affected the practices of participant observation.
2. Explain the role of trust in participant observation.
3. Explain how to record the data from participant observation.
4. What are some ethical considerations that arise from doing participant observation? You may wish to consider issues that arise from various researcher positionalities (e.g., as an outsider; as someone joined to the study community by marriage).
5. What are some ways that our presence as participant observers might influence interactions in research settings?
6. In what ways can participant observation complement other research methods?

Review Exercises

1. Identify a theme you are interested in (e.g., gender relations, consumer purchasing activities, people's screen use) and spend 15 minutes observing in a public place such as a café or university library, taking notes as possible. Share your field notes in pairs and refine them to be descriptive and as complete as possible.
2. Identify a hypothetical research project or adopt one from a recent article in a geography journal and consider your positionality relative to the chosen research.
3. Write a letter of introduction about yourself and your research topic to a community leader.

Useful Resources

The following are helpful introductions to participant observation in research.

Cook, Ian. (2005.) Participant observation. In R. Flowerdew & D. Martin (Eds). *Methods in human geography*. London: Taylor & Francis.

Laurier, Eric. (2016). Participant and non-Participant observation. In N. Clifford, M. Cope, S. Gillespie, & S. French (Eds), *Key methods in human geography* (pp. 169–81). Thousand Oaks, London, New Delhi: SAGE Publications.

Watson, Annette, & Till, Karen. (2010.) Ethnography and participant observation. In D. De Lyser, S. Herbert, S. Aitken, M. Crang & L. McDowell (Eds), *The SAGE handbook of qualitative geography* (pp. 121–37). London: SAGE.

Also see:

Hiemstra, Nancy. (2016). Periscoping as a feminist methodological approach for researching the seemingly hidden. *The Professional Geographer, 69*(2), 329–36.

Appendix 8.1

Field notes Kitlaska January 18 2017

Kitlaska is the name of the City; Kitta is the name of the Tribe

About zero degrees Fahrenheit in Kitlaska for the daytime; all day cloudless, calm and crisp. Snow is piled high in the village, and packed down on the roadways.

On the plane ride from Fairbanks were two female aides that worked for the local nonprofit, going to another village on our route today. Also a passenger who also got off at Kitlaska; pilot knew him so he invited him to sit up front for the best view. I knew this to be a route taken by Athabascans from the villages. We left Fairbanks at 8am, and even with arriving in Kitlaska at 9am, it was still dark outside. Sun will not rise until after 10 . . .

I will only be highlighting sections of this much longer set of notes that summarize my first day visiting this community.

This place name is a pseudonym. This is a village near the Arctic Circle, where I traveled to conduct participatory mapping with multiple tribes.

The place and tribal name are pseudonyms, but demonstrate one way I use field notes is to ensure I get the "story" and use of terms and names properly. In many places in Alaska, the place names are not the same names as the tribal entities.

You always set the stage in taking field notes.

(*continued*)

No one picked me up at the air strip even though I had arranged to meet a tribal official [typical, ha ha], but I hitched a ride on the van driven by the new owner of the Lodge. . . . He learned my purpose of being there, and said that he was "for" the road, because "this place is dying." He and his wife just relocated to Kitlaska to operate the lodge, from "down south," he said, and he said he was far happier in Kitlaska, even though he wasn't earning very much. They were in town since 2014.

I often bracket some phrasing like this, which gives more information—that this is a pattern, but also indicates my attitude about the pattern. Everything in field notes is observations, but I bracket observations I make about my own thoughts as a mnemonic device.

Whenever I am sure, I use exact quotes. Even if it is only for a word or a phrase, if you can recall that exact use of words, your notes need to show they are quotes or partial quotes.

. . . I got dropped off at the tribal office . . . and spoke with Chief Mark Case . . . He might have been my age; and I knew a little about his Athabascan family history. He spoke to me about the road; he was insistent that the Ambler road was not going to ever happen, and that the money being spent in planning is really for nothing, and that all these waves of researchers (the folks "from Anchorage" had already been to Kitlaska to do their interviews, and now me) were studying things really "for nothing." I tried to say that I hoped we could create some maps that would be useful to his community, like looking at climate change and how that might impact future subsistence . . . but I do not know if I changed his mind about the utility of our project together. Or about me, as he initially posed me as another arm of "the government." Mark certainly didn't approach this conversation as if it were "our" project.

All people's names are pseudonyms.

Even though I didn't learn that bit of information right at that moment, in the notes it was appropriate to put that information there.

You will find that when you conduct research with communities not your own, there will be variable responses to you. Even though this community's Tribal Council approved of the research, this Chief remained skeptical, and was sure to express that to me. Writing about these interactions keeps me honest and humble.

Mark said their ice road—having problems at the moment—could be open from December until April. As long as December 1st to mid-April. Reliably about three months. So it was really, he said, all their community needed in terms of being connected to a road system; they didn't need a road year-round.

This was the same conversation, but I didn't spend the time in field notes to create a transition between topics. Writing a transition is not important in field notes. But knowing about the Chief's assessment of their ice road, in a place without any paved roads in and out of town, is an important thing to note about a community's land use. Use notes to remember details about statistics or timelines people tell you.

Sometimes the sentence structure of field notes is not so pretty or easy to read. Because you try to write the field notes fast, as you remember. And THEN you may edit them later on, if you want. For me, as long as I understand the information contained in the notes, I do not edit for further clarity. If I end up using the information in work that I publish, then I just write the info how I want to write it.

. . . I spoke quite a bit with Justin, my research assistant, once we were introduced in the tribal office. Justin described how he had fairly recently (last 5 years? Need to check) moved back to town permanently—he was born here, but lived in Fairbanks for a long while, with shorter (often summer) visits to Kitlaska. . . . The tribal administrator Marcy had given me a community list, "with more information than you need," and Justin and I went through it, talking about who is an elder and who practices subsistence. Then after a round of phone calls to schedule interviews, Justin and I went for a walk around town. Things I learned from both venues of our chat: most people in town do not hunt moose, but get that meat from the sport hunters that come through in fall time. Justin said just a few people go out hunting, though many go out for berries. On our walk around town, he described how there was an increasing amount of erosion in recent years, cutting drastically toward town—he showed me a cabin that got moved three years ago, and pointed toward another that also was recently relocated (I have pics on iphone of that cabin, and the space it once occupied). . . .

Some other important things I learned today:

the Old Kitlaska is what keeps showing up downriver on our maps in GIS, but not the same place as I am walking around today.

It's important to make notes to yourself within your field notes, because if you go to quote this later, you know this is not a for sure number, phrased like this.

For this project, each research assistant was an essential source of data, and provided interpretation for my participant observations while I was in town, taking me for guided walks and sometimes longer overnight trips to understand land use.

I will often note exactly what I have pictures of, so I can coordinate my textual and visual data.

While this set of field notes for this particular day often reads like a novel, ultimately you don't care about taking care of the narrative in those ways. This ends abruptly. Many of my field notes do. You have to get some essential information down. By the time I wrote these last few lines, I was already madly writing down my observations from the second day in this community. . . .

Engaging Interviews

Kevin Dunn

Chapter Overview

This chapter provides advice on engaging interview design, practice, transcription, data analysis, and presentation. I describe the characteristics of each of the three major forms of interviewing and critically assess what I see as the relative strengths and weaknesses of each. I outline applications of interviewing by referring to examples from economic, social, and environmental geography. Finally, some of the unique issues regarding digital, web-based interviews are reviewed.

Interviewing in Geography

Interviewing in geography is so much more than "having a chat," though good social skills and the right atmosphere are certainly helpful in conducting interviews. Indeed, interviewing is one of the most common qualitative research methods in the discipline, perhaps because it is both familiar to people and it generates rich data.

Successful interviewing does require careful planning and detailed preparation, however. An audio-recorded hour-long interview can involve considerable preliminary background work and question formulation. Interviewing also requires diplomacy in contacting **informants** and negotiating various institutional hurdles for permissions, access, and human subject protections, such as privacy. A 60-minute interview will typically require approximately four hours of transcription if you are a fast typist, and verification of the record of interview could stretch over a week or more. After all that, you have still to analyze the interview material. These are time-consuming activities. Is it all worth it? Yes! In this chapter, I outline some of the benefits of interviewing and provide a range of tips for good interviewing practice.

An interview was once defined as "a face-to-face verbal interchange in which one person, the interviewer, attempts to elicit information or expressions of opinion or belief from another person or persons" (Maccoby & Maccoby, 1954, p. 499), that is, a data-gathering method in which there is a spoken exchange of information. While this exchange has traditionally been face-to-face, researchers have also used telephone interviews (Groves, 1990). However, the current era has seen the emergence of **digital interviewing**, a mode in which there is no direct access to the informant, which includes techniques such as email exchanges and **video call** interviews. Video call interviews (e.g., Skype, Zoom, FaceTime) have become increasingly popular in the past decade, and they do resemble face-to-face interviews in many ways.

Types of Interviewing

There are three major forms of interviewing: structured, unstructured, and semi-structured. These three forms can be placed along a continuum, with the **structured interview** at one end and the **unstructured interview** at the other. Structured interviews follow a predetermined and standardized list of questions. The questions are asked in almost the same way and in the same order in each interview, and are much more like questionnaires (see Chapter 13). At the other end of the interviewing continuum are unstructured forms of interviewing such as **oral histories** (discussed fully in Chapter 10), in which the conversation is actually directed by the informant rather than by set questions. In the middle of this continuum are **semi-structured interviews**. This form of interviewing has some degree of predetermined order and topical prompts but maintains flexibility in the way issues are addressed by the informant. Different forms of interview have varying strengths and weaknesses that we review here.

Strengths of Interviewing

Interviews are an excellent method of gaining access to information about places, events, opinions, and experiences, which vary enormously among people of different class, ethnicity, age, gender, sexuality, and disability. Because of the variety of perspectives uncovered by interviews, researchers should be careful to resist claims that they have discovered *the* truth about a series of events or that they have distilled *the* public opinion (Goss & Leinbach, 1996, p. 116; Kong, 1998, p. 80). Interviews can also be used to counter the claims of those who presume to have discovered the public opinion. This can be done by seeking out the opinions of different groups, often marginalized or **subaltern** groups, whose opinions are insufficiently heard.

In general, research interviews are used for four main purposes (see also Krueger & Casey, 2014; Minichiello, Aroni, & Hays, 1995, 2008; Valentine, 1997):

1. to investigate complex behaviours and motivations;
2. to collect a diversity of meanings, opinion, and experiences. Interviews provide insights into the differing opinions or debates within a group, but they can also reveal consensus on some issue;

3. to show respect for and empower the people who provide the data. In an interview, the informant's view of the world should be valued and treated with respect. The interview may also give informants cause to reflect on their experiences and the opportunity to find out more about the research project than if they were simply being observed or if they were completing a questionnaire; and
4. to fill a gap in knowledge that other methods, such as **participant observation** or the use of census data, are unable to bridge efficaciously.

Most of the questions posed in an interview allow for an **open** response as opposed to a **closed**, yes/no set of response options (see also Chapter 13 on questionnaires). In this way, each informant can describe events or offer opinions in her or his own words. One of the major strengths of interviewing is that it allows you to discover what is relevant to the *informant* and respond accordingly.

Because the face-to-face verbal interchange is used in most interviewing, the informant can also tell you if a question is misplaced or confusing (Box 9.1), though ideally a **pilot study** of the questions would reveal weaknesses prior to the launch of the full project. Furthermore, your own opinions and tentative conclusions can be checked, verified, and scrutinized. This may disclose significant misunderstandings on your part as the researcher, or reveal issues that you had not previously identified (Longhurst, 2016).

BOX 9.1 Asking the Wrong Question: A Tale from Cabramatta

On 23 June 1990, I began my first formal research interview. The informant was a senior office-holder from one of the Indo-Chinese cultural associations in New South Wales, Australia. My research interest was in the social origins of the residential concentration of Indo-Chinese Australians around the outer Sydney suburb Cabramatta and the experiences of these immigrants. The political context of the time was still heavy with the racialized and anti-Asian overtones of the 1988 "immigration debate" in which mainstream politicians and academics such as John Howard (later Australian prime minister from 1996 to 2008) and Geoffrey Blainey (professor of history) had expressed concern about "Asian" immigration and settlement patterns in Australian cities. Specifically, Vietnamese immigrants were accused of congregating in places like Cabramatta (Sydney) and Richmond and Springvale (both in Melbourne) and of purposefully doing so in order to avoid participating with the rest of Australia. I had hypothesized that Vietnamese Australians did not congregate voluntarily but that they were forcibly segregated by the economic and social constraints of discrimination in housing and labour markets. Indeed, the geographic literature supported my assertion at the time. I was a somewhat naive and **colonialist** investigator who saw his role as "valiant champion" of an ethnic minority.

But back to my first interview. One of my first questions to these Indo-Chinese-Australian leaders was: "Please explain the ways in which discrimination has forced you, and members of the community you represent, to reside in this area?" Their answer: "I wouldn't live anywhere else." This informant, and most subsequent informants, described the great benefits and pleasures of living in Cabramatta. They also explained how residing in Cabramatta had eased their expanding participation in Australian life (Dunn, 1993). I had asked the wrong questions and had been told so by my informants. I decided to focus the project on the advantages and pleasures that residence in "Cab" brought to Indo-Chinese Australians. The face-to-face nature of the exchange, and the informed subject, make interviews a remarkable method. The participants can tell the researcher, "You're on the wrong track!"

Interview Design

It is not possible to formulate a strict guide to good practice for every interview context. Every interview and every research issue demands its own preparation and practice. However, researchers should heed certain procedures. Much of the rest of this chapter focuses on strategies for enhancing the **credibility** of data collected using **rigorous** interview practice. In the next section, we look at the organization of interview schedules and the formulation of questions.

The Interview Schedule or Guide

Even the most competent researcher needs to be reminded during the interview of the issues or events they had intended to discuss. You cannot be expected to recall all of the specific questions or issues you wish to address, and you will benefit from some written reminder of the intended scope of the interview. These reminders can take the form of **interview guides** (less structured) or **interview schedules** (more structured).

An interview guide is a list of *general* issues you want to cover in an interview and are usually associated with semi-structured forms of interviewing. The guide may be a simple list of key words or concepts intended to remind you of discussion topics, often drawn from existing literature on an issue. The identification of key concepts and themes is a preliminary part of any research project (see Babbie, 1992: and Chapter 6).

One of the advantages of the interview guide is its flexibility. As the interviewer, you may allow the conversation to follow as "natural" a direction as possible, but you will have to redirect the discussion to cover issues that may still be outstanding. Questions can be crafted *in situ*, drawing on themes already broached and from the tone of the discussion. The major disadvantage of using an interview guide is that you must formulate coherent question wordings on the spot. This requires good communication skills and a great deal of confidence, so a guide is inadvisable for first-time interviewers. Guides are more appropriate for skilled interviewers and for particular forms of interview, such as **oral history** (see Chapter 10).

By contrast, interview schedules are used in structured and sometimes semi-structured forms of interviewing. An interview schedule is a list of carefully worded questions (see Box 9.2) and proceeds with less fluidity. The benefits of using interview schedules mirror the disadvantages of interview guides. They provide greater confidence to researchers in the enunciation of their questions and allow better comparisons between informant answers. However, questions that are prepared before the interview and then read out formally may sound insincere, stilted, and out of place.

I have found that a half-hour interview will usually cover between six and eight **primary questions**. Under each of these central questions I nest at least two detailed questions or **prompts**. In some research, it may be necessary to ask each question in the same way and in the same order to each informant. In others, you might ask questions at whatever stage of the interview seems appropriate.

BOX 9.2

Formulating Good Interview Questions

- Use easily understood language that is appropriate to your informant.
- Use non-offensive language.
- Use words with commonly and uniformly accepted meanings.
- Avoid ambiguity.
- Phrase each question carefully.
- Avoid leading questions as much as possible (i.e., questions that encourage a particular response).

A mix of carefully worded questions and topic areas capitalizes on the strengths of both guides and schedules. Indeed, a fully worded question can be placed in a guide and yet be used as a topic area. The predetermined wording can be kept as a fall-back in case you find yourself unable to articulate a question on the spot. I find it useful to begin an interview with a prepared question. It can be damaging to your confidence if an informant asks, "What do you mean by that?" or "I don't know what you mean" in response to your first question.

Unless there is a compelling reason for uniformity in how questions are asked, interview design should be dynamic throughout the research (Tremblay, 1982). As a research project progresses, you can make changes to the order and wording of questions or topics as new information and experiences are fed back into the research design. Some issues may be revealed as unimportant, offensive, or silly after the initial interview and can be dropped from subsequent interviews. The interview schedule or guide should also seek information in a way that is appropriate and tailored to each informant, such as when interviewing people across a wide age span; record such factors in a **research diary** to help you recall them later.

While the primary purpose of interview schedules and guides is to jog your memory and to ensure that all issues are covered as appropriately as possible, they can also be used to provide informants with a copy of the questions or issues before the interview to prompt thought on the matters to be discussed.

Types of Questions

Interviews utilize primary (or original) questions and **secondary** or **follow-up questions**. Primary questions are opening questions used to initiate discussion on a new theme or topic. Secondary questions are prompts that encourage the informant to follow up or expand on an issue already discussed. An interview schedule, and even an interview guide, can have a mix of types of original questions, including descriptive questions, storytelling prompts, structural questions, contrast and opinion questions, and devil's advocate propositions (see Box 9.3). Since different types of primary questions produce very different sorts of responses, a good interview schedule will generally comprise a mix of question types.

On an interview guide or schedule, you might have a list of secondary questions or prompts (Box 9.5). There are a number of different types of prompts, ranging from formal secondary questions to nudging-type comments that encourage the informant to continue speaking (Longhurst, 2016). Sometimes prompts are listed in the interview guide or schedule, but often they are deployed, when appropriate, without prior planning.

Primary Question Types

Type of Question	Example	Type of Data and Benefits
Descriptive (knowledge)	What is the full name of your organization? What is your role within the organization? How many brothers or sisters do you have?	Details on events, places, people, and experiences; easy to answer opening questions
Storytelling	Can you tell me about the formation and history of this organization and your involvement in it?	Identifies a series of players, an ordering of events, or causative links; encourages sustained input from the informant
Opinion	Is Canadian society sexist? What do you consider to be the appropriate size for a functional family?	Impressions, feelings, assertions, and guesses
Structural	How do you think you came to hold that opinion? What do you think the average family size is for people like yourself?	Taps into people's ideologies and assumptions; encourages reflection on how experiences and social expectations may have influenced opinions and perspectives
Contrast (hypothetical)	Would your career opportunities have been different if you were a man? Or if you grew up in a poorer suburb?	Comparison of experience by place, time, gender, and so forth; encourages reflection on advantage and privilege
Devil's advocate	Many practising town planners are voicing concern about the lack of transparency in the new development assessment process.	Controversial/sensitive issues broached without associating the researcher with that opinion

See also Box 9.4.

Ordering Questions and Topics

It is important to consider carefully the order of questions or topics in an interview guide or schedule. Minichiello, Aroni, and Hays (1995) advise that the most important consideration in the ordering of questions is preserving **rapport** between you and your informant. An interview might begin with a discussion of the general problem of homophobia or racism: how widespread it is, how it varies

BOX 9.4 Asking the Tough Questions without Sounding Tough

My work with Indo-Chinese communities in Cabramatta occurred within a political context in which Vietnamese Australians were being publicly harangued by academics and politicians (Dunn, 1993). I felt it was important for informants to respond to their critics. My interview schedule had the following two devil's advocate propositions. I also used a preamble to dissociate myself from the statements.

> In my research so far, I have come across two general explanations for Vietnamese residential concentration. I would like you to comment on two separate statements that to me represent these two explanations:
>
> First: that Vietnamese people have concentrated here because they don't want to participate in the wider society.
>
> Second: that the Vietnamese are segregated into particular residential areas through social, economic, and political forces imposed upon them by the wider society.

The aim was to gather people's responses to both statements. In most cases, informants were critical of both views. Some informants took my question as a request for them to select the explanation that they thought was the most appropriate. Those people selected the second statement. Others were in no doubt that I disagreed with both views. Either way, devil's advocate propositions are often leading. My political views were noticeable in the question preamble and wording as well as in the preliminary discussions held to arrange the interviews. It is fairer that the researcher's motives and political orientation are **transparent** to the informant rather than hidden until after the research is published (see Chapters 2, 4, 6, and 19 of this volume).

from place to place, and how legal and institutional responses to those forms of oppression have emerged. Following a **funnel structure**, after broaching these general or macro-level aspects of oppression, the interview might then turn to the particular experiences of the informant.

In an interview with a **pyramid structure**, the more abstract and general questions are asked at the end. The interview starts with easy-to-answer questions about

Types of Prompt

Prompt Type	Example	Type of Data and Benefits
Formal secondary question	Primary Q: What social benefits do you derive from residing in an area of ethnic concentration?	Extends the scope or depth of treatment on an issue
	Secondary Q: What about informal child care?	Can also help to explain or rephrase a misunderstood primary question
Clarification	What do you mean by that?	Used when an answer is vague or incomplete
Nudging	And how did that make you feel? (Repeat an informant's last statement.)	Used to continue a line of conversation
Summary (categorizing)	So let me get this straight: your view, as just outlined to me, is that people should not watch shows like The Bachelor?	Outlines in-progress findings for verification; elicits succinct statements (for example, "quotable quotes")
Receptive cues	Audible: Yes, I see. Uh-huh. Non-audible: nodding and smiling	Provides receptive cues, encourages an informant to continue speaking

an informant's duties or responsibilities or their involvement in an issue. This allows the informant to become accustomed to the interview, interviewer, and topics before they are asked questions that require deeper reflection. For example, to gather views on changes to urban governance, you might find it necessary to first ask an informant from an urban planning agency to outline their roles and duties. Following that, you might ask your informant to outline the actions of the informant's own agency and how those actions may have changed in recent times. Once the "doings and goings-on" have been outlined, it may then make sense to ask the informant why agency actions and roles have changed, whether that change has been resisted, and how the informant views the transformation of urban governance.

A final question-ordering option is to use a hybrid of the funnel and pyramid structures. The interview might start with simple-to-answer, non-threatening questions, then move to more abstract and reflective aspects, before gradually progressing towards sensitive issues. This sort of structure may offer the benefits of both funnel and pyramid ordering.

When thinking about question and topic ordering, it can be helpful to have key informants comment on the interview guide or schedule (Kearns, 1991). Key informants are often initial or primary contacts in a project. They are usually the first informants, and they often possess the expertise to liaise between the researcher and the communities being researched. Key informant review can be a useful litmus test of interview design, since these representatives are "culturally qualified." They have empathy with the study population and can be comprehensively briefed on the goals and background of the research (Tremblay, 1982).

Structured Interviewing

A structured interview uses an interview schedule that typically comprises a list of carefully worded and ordered questions (see Boxes 9.3 and 9.5 and the earlier discussion on ordering questions and topics). Each respondent or informant is asked exactly the same questions in exactly the same order. The interview process is question-focused.

It is a good idea to **pre-test** a structured interview schedule on a subset (say two to four) of the group of people you plan to interview for your study to ensure that your questions are not ambiguous, offensive, or difficult to understand. Though helpful, pre-testing is of less importance in semi-structured and unstructured interviews in which ambiguities (but not offensive questions!) can be clarified by the interviewer.

Structured interviews have been used with great effect throughout the sub-disciplines of geography, including economic geography (Box 9.6).

BOX 9.6 Interviewing—An Economic Geography Application

In 1991, Erica Schoenberger argued that most industrial geography research had been on the outside looking in, deducing strategic behaviour from its locational outcomes rather than investigating it directly (1991, p. 182). One of the assumptions challenged by the use of structured interviews was that the location of firms was strongly associated with proximity to industry-specific inputs and markets. Using structured interviews with managers, Schoenberger was able to show that the location of foreign chemical firms in North America was as much, if not more, related to historical and strategic contingencies than to contemporary location preferences.

(continued)

For example, one of Schoenberger's case studies was a German-owned chemical firm. Her interviews revealed that the firm's board of directors had decided on a major expansion in the US market. But the board had been split between establishing a "greenfields" site, which would be purpose-built to company needs, and acquiring an already established chemical plant, which would hasten their expanded presence in the market. Plans to establish a greenfields facility were foiled by organized community opposition. The directors who argued for an acquisition then gained the upper hand, and at about the same time a US chemical firm came up for sale. For many decades, German chemical firms had agreed among themselves to specialize in certain parts of the chemical sector. These agreements were about to end, and Schoenberger's case-study firm was keen to expand horizontally into another speciality. The US firm that came up for sale happened to specialize in that area. A host of historical and strategic events had combined to produce a particular locational result.

The historical and strategic contingencies that accounted for the location of the German chemical company were revealed through structured interviews. Their location was in fact quite at odds with the apparent preferences of the firm and revealed nothing about the firm's location preferences (Schoenberger, 1991, p. 185). The US chemical sector has a high level of foreign ownership, and most of it was established through acquisition. Replicated interviews with managers and directors across the chemical sector revealed the prevalence of location choices being determined by historical and strategic contingency. Interviewing was therefore an essential method for unravelling the location determinants of chemical plants in the North America.

Semi-structured Interviewing

Semi-structured interviews typically employ an interview guide. The questions asked in the interview focus on content and deal with the issues or areas judged by the researcher to be relevant to the research question. Alternatively, an interview schedule might be prepared with fully worded questions for a semi-structured interview, but the interviewer would not be restricted to deploying those questions. The semi-structured interview is organized around ordered but flexible questioning. In semi-structured forms of interview, the role of the researcher (interviewer or facilitator) is recognized as being more interventionist than in unstructured interviews. This requires that the researcher redirect the conversation if it has moved too far from the research topics.

Unstructured Interviewing

Various forms of unstructured interviewing exist. They include oral history, **life history**, and some types of group interviewing and in-depth interviewing. Unstructured interviewing focuses on personal perceptions and personal histories. Rather than being question-focused like a structured interview or content-focused as in a semi-structured format, the unstructured interview is informant-focused. Life history and oral history interviews seek personal accounts of significant events and perceptions, as determined by the informants and in their own words (Chapter 10 and also McKay, 2002). Each unstructured interview is unique. The questions you ask are almost entirely determined by the informant's responses. These interviews approximate normal conversational interaction and give the informant some scope to direct the interview. Nonetheless, an unstructured interview requires as much, if not more, preparation than its structured counterpart to gain a solid understanding of past events, people, and places related to the interview through **archival research** and other historical background reading. But through these interviews we can find out about events and places that had been kept out of the news or that had been deemed of no consequence to the rich and powerful (Box 9.7).

Oral Environmental Histories

Oral history interviews can collect data about environmental history. This type of interviewing helps produce a broader picture of the cause and process of environmental change than is available solely through physical methods of inquiry. Data collected might include people's memories of changes in local land use, biodiversity, hydrology, and climate.

Lane (1997) used oral history interviews to reveal changes in watercourses, weeds, and climate in the Tumut Region high country of the Australian Alps. Interviews were conducted with five main informants, first in their homes and then while driving and walking through the countryside where they had resided. The informants told of the waterholes and deep parts of creeks where they would fish and swim and where they and their children had learned to swim. One informant commented that one of the creeks used to be almost a river—and now you could step over it (Lane, 1997, p. 197). The same informant noted the change in colour and quality of the water. Lane's informants described how the water level and

(continued)

quality had steadily degraded since pine plantations had been planted in the 1960s. This description was consistent with "scientific" understandings of the impact of pine plantations in which there is an ever-decreasing level of run-off as the pines grow.

Such specific observations from local residents may often be the only detailed evidence on environmental change that is available. Oral history can fill gaps in the "scientific record," or it can be used to complement data gathered using physical or quantitative methods. More important, with the use of oral history, environmental change can be set in a human context and related to the history of people who live in the region (Lane, 1997, p. 204).

Interviewing Practice

Rapport with another person is basically a matter of understanding their model of the world and communicating your understanding symmetrically (Minichiello, Aroni, & Hays, 1995).

Achieving and maintaining rapport, or a productive interpersonal climate, can be critical to the success of an interview. Rapport is particularly important if you need to have repeat sessions with an informant. Even the first steps of arranging an interview are significant, including the initial contact and other preliminaries that might occur before the first interview. Interviews in which both the interviewer and informant feel at ease usually generate more insightful and more valid data than might otherwise be the case. In the following paragraphs, I outline a set of tips that can help you to enhance rapport before, during, and while closing an interview.

Contact

Informants are usually chosen purposefully on the basis of the issues and themes that have emerged from a review of previous literature or from other background work. This involves choosing people who can communicate aspects of their experiences and ideas relevant to the phenomena under investigation (Minichiello, Aroni, & Hays, 1995). Researchers should, of course, seek a diversity of respondents and views, as suitable within the frame of the topic. Your recruitment materials and protocols, how you will include diverse respondents, and a rough list of questions will need to be part of your ethics review and approval by the board at your institution; addressing such issues in the research design phase is therefore imperative (see Chapter 2 on ethics and Chapter 6 on design). Once you have identified a potential informant, you must then negotiate permission for the interview. This means getting the consent of the informants themselves, and in some circumstances, it will also involve gaining the sanction of gatekeepers like employers, parents, or teachers. This might occur, for example, if you wanted to interview schoolchildren, prisoners, or employees in some workplaces.

Your first contact with an informant will often be by telephone or email. In this preliminary phase, you should do at least four things (Robertson 1994, 9):

1. Introduce yourself and establish your bona fides. For example: "My name is Juan Folger, and I am an honours student from Java State University."
2. Make it clear how you obtained the informant's name and telephone number or address. If you do not explain this, people may be suspicious and are likely to ask how you got their name or email address. If you are asked this question, rapport between you and your informant has already been compromised.
3. Outline why you would like to conduct the interview with this particular informant. Indicate the significance of the research, and explain why the informant's views and experiences are valued. For instance, you may believe that they have important things to say, that they have been key players in an issue, or that they have experienced something specific that others have not. On the whole, I have found that most people are flattered to be asked for an interview, although they are often nervous or hesitant about the procedure itself.
4. Indicate how long the interview and any follow-up is likely to take; review ethics board conditions regarding consent, privacy of data, and other related issues.

All of these matters can be outlined in a letter of introduction, which may be sent to an informant once they have agreed to an interview or while agreement is still being negotiated. This formal communication should be under the letterhead of your organization (for example, your university) and should spell out your qualifications, the topic of the research, the manner in which the interview will be conducted, and any rules or boundaries regarding confidentiality. Students should seek permission from their supervisors to use the letterhead of an organization such as a university, although ethics procedures within the institution (see Chapter 2) are likely to have made this mandatory. In the absence of a letter of introduction, informants should be made aware of their rights during the interview, an arrangement that can be set out in written form. (See Box 9.8 for some of the rights of informants that can be established. See also Hay (2016) and Israel and Hay (2006).) These preliminary discussions are important to the success of an interview. Indeed, they set the tone of the relationship between interviewer and informant.

The Interview Relationship

The relationship established between interviewer and informant is often critical to the collection of opinions and insights. If you and your informant are at ease with each other, then the informant is likely to be communicative. As previous chapters have outlined, relations of power, knowledge, and identity all saturate qualitative research and that is particularly the case in such a personal method as interviewing. Dowling, Lloyd, and Suchet-Pearson (2016) provide an excellent round-up of recent developments in interviewing practices among geographers.

Codifying the Rights of Informants

In their research on the Carrington community in Newcastle, New South Wales, Winchester, Dunn, and McGuirk (1997) decided to codify informants' rights in the oral histories and semi-structured interviews that were to be conducted. They included the following list of informants' rights on university letterhead, and they gave a copy to each of the informants:

- Permission to record the interview must be given in advance.
- All transcribed material will be anonymous.
- Files and transcripts will be made available to informants who request them.
- Informants have the right to change an answer.
- Informants can contact us at any time in the future to alter or delete any statements made.
- Informants can discontinue the interview at any stage.
- Informants can request that the audio recorder be paused at any stage during the interview.

To this list one might add further statements (for example, that informants can expect information about the ways in which their contributions to the research may be used). A codification of rights was deemed necessary for two reasons. First, it was done to empower the informants and assure them that they could for example pause or terminate the interview process whenever they deemed it necessary to do so. Second, the researchers had employed a local resident to conduct the interviews, and so it was important that the interviewer was also constantly reminded of the informants' rights.

Decisions about the interview relationship will vary according to the characteristics of both the informant and the interviewer. The cultural nuances of a study group will at times necessitate variations in the intended interaction. However, it is wise to remember that despite any empathy or relationships that are established, the interview is still a formal process of data gathering for research. Furthermore, there is usually a complex and uneven power relationship involved: information, and the power to deploy that information, flows mostly from the informant to the interviewer, but the power of the interviewer as an "expert" and interlocutor makes for complicated relations (see Chapters 2 and 4). Feminist geographers have been particularly engaged with critically examining further power relations in interviews, noting the **intersectional** influences of gender, race, sexuality, age, nationality, and other **axes of difference** (see Dowling, Lloyd, and Suchet-Pearson (2016) for a review).

Establishing good rapport may increase the level of understanding you have about the informant and what they are saying. There are a number of strategies for enhancing rapport. The first is through the use of respectful preliminary work. The second involves the use of a **warm-up** period just before an interview commences, which could be a chat about the weather, matters of shared personal interest, or "catching-up" talk. In their surveys and interviews of Vietnamese Australians in Melbourne, Gardner, Neville, and Snell (1983, p. 131) found that "[t]he success of an interview (when measured by the degree of relaxation of all those present and the ease of conversation) generally depended on the amount of 'warm-up' (chit-chat, introductions, etc.)."

My own warm-up techniques for face-to-face interviews have included giving the informant an overview of the questions I plan to ask and presenting relevant diagrams or maps, as well as discussing historical documents. Maps, diagrams, tables of statistics, photos, and other documents can also be used as references or stimuli throughout an interview. If an informant offers you food or drink before an interview, it would be courteous to accept. As we shall see in more detail later in this chapter, many of these forms of warm-up are not available in digital interviews but the principle goal of making everyone comfortable remains.

You should also have acquainted yourself with the cultural context of the informants before the interview. As Robin Kearns pointed out, "If we are to engage someone in conversation and sustain the interaction, we need to use the right words. Without the right words our speech is empty. Language matters" (Kearns, 1991, p. 2). For instance, you must be able to recognize the jargon or slang and frequently used acronyms of institutions or corporations as well as the language of particular professions or cultural groups.

Listening strategies can improve rapport and the productivity of the interview. Your role as interviewer is not passive but requires constant focus on the information being divulged by informants and the use of cues and responses to encourage them. Your role as an active participant in the interview extends well beyond simply asking predetermined questions or topics. You should avoid "mental wandering"—otherwise, you may miss unexpected leads. Moreover, it is irritating to the informant, and a threat to rapport, if you ask a question they have already answered.

Adelman (1981) advises researchers to maintain a **critical inner dialogue** during an interview. This requires that you constantly analyze what is being said and simultaneously formulate the next question or prompt. You should be asking yourself whether you understand what the informant is saying. Do not let something pass by that you do not understand with the expectation that you will be able to make sense of it afterwards. Minichiello, Aroni, and Hays (1995, p. 103) provide a demonstration of how critical inner dialogue might occur: "What is the informant saying that I can use? Have I fully understood what this person is saying? Maybe, maybe not. I had better use a probe. Oh yes, I did understand. Now I can go on with a follow-up question."

Strategies to enhance rapport continue throughout the interview, through verbal and non-verbal techniques that indicate that the responses are valued. When undertaking repeat interviews it is essential that the researcher does not appear to have forgotten the informant's story, as told so far. In some professional fields, such as police, forensic, or cognitive interviewing, it is recommended not to re-interview because of the enhanced opportunity for "interviewer introduced" perspectives and suggestibility (Burgwyn-Bailes et al., 2001). However, repeat interviews can be used to successfully build rapport and enhance the communicativeness of the informant, expanding the breadth and depth of the data collected (Hershkowitz & Terner, 2007). Vincent (2013) found from interviews with pregnant schoolgirls and schoolgirls' mothers in Britain that these advantages were particularly salient with vulnerable groups and when dealing with sensitive topics. Researchers should review the information provided in the previous interview so that they can be ready to probe for additional information and note any changes in respondents' perspectives.

Informants may sometimes recount experiences that upset them or stir other emotions. When an informant is becoming distressed, try pausing the interview or changing the topic and possibly returning to the sensitive issue at a later point. If the informant is clearly becoming very distressed, you should probably terminate the interview. As mentioned in Chapter 2, most research ethics boards will require you to arrange for counseling information that can be passed along to distressed participants.

There may be a stage in an interview when your informant does not answer a question. If there is a silence or if they shake their head, the informant may be indicating that they have not understood your question or simply do not know the answer. They might be confused as to the format of the answer expected: is it a "yes–no" or something else (Minichiello, Aroni, & Hays, 1995, p. 93)? In these cases, try restating the question, perhaps using alternative wording or providing an example. You should always be prepared to elaborate on a question. It is important to remember, however, that choosing not to respond is the informant's right and you should not usually press them. They may have chosen not to answer because the question was asked clumsily or insensitively, or for some other reason—if the question dealt with sensitive matters, for instance. This issue also requires cultural sensitivity and familiarity on your part: recall from Chapters 4 and 8 that some Indigenous cultures avoid saying "no" and merely move on. If you prepare your questions carefully, you should avoid this sort of problem and the consequent loss of empathy and data.

As an interviewer, you should also learn to distinguish between reflective silence and non-answering. Robertson cautions, "Do not be afraid of silences. Interviewers who consciously delay interrupting a pause often find that a few seconds of reflection leads interviewees to provide the most rewarding parts of an interview. . . . There is no surer way of inhibiting interviewees than to interrupt, talk too much, argue, or show off your knowledge" (Robertson, 1994, p. 44). It is important to allow time for informants to think, meditate, and reflect before they answer a question (see Box 9.9). It is also important to be patient with slow speakers or

BOX 9.9

Finishing Sentences, Interrupting, and "Rushing-on"

During February 2001, Minelle Mahtani (then of the University of British Columbia) joined me in Sydney to undertake joint, and comparative, research on the media representations of ethnic minorities in Australia and Canada. This involved interviews with managers and employees within newsrooms. They were powerful and confident informants. Dr Mahtani had a wealth of expertise in such environments, having been a producer with the flagship Canadian Broadcasting Corporation's *The National*, a news television program. Our first field interview was with a network news editor for one of the commercial networks in Australia. Our questions included themes such as media representations of ethnic minorities, attempts by the organization to improve the portrayal of ethnic minorities, the presence of minority journalists, and circumstances in which they or their staff had challenged stereotypical storylines. The questions had been developed and agreed to in advance, but what very different styles we had. The informants would sometimes provide very short and dismissive responses to some questions. When it was clear that they had answered, I would probe or move to another question. Dr Mahtani would wait, however. The silence would hang heavy over the interview. I felt uncomfortable, but these powerful informants got the idea that we wanted a fuller response. They would then justify the view they had briefly given, or they would admit that there were alternative viewpoints to the one they had expressed. My rushing-on was a strategy vastly inferior to the "sounds of silence" for uncovering richer insight into ethnic minority representations and the dynamics of the newsroom (Dunn & Mahtani, 2001).

people who are not entirely fluent in the host language. Resist the temptation to finish people's sentences for them. Supplying the word that an informant is struggling to find may seem helpful at the time, but it interrupts them and inserts a term they might not have ordinarily used.

Closing the Interview

Do not allow rapport to dissipate at the close of an interview. Because an interview establishes a relationship within which certain expectations are created, it is better to indicate a sense of continuation and of feedback and clarification than to end the interview with an air of finality. Indeed, near the end of an interview, or after the recording device is switched off, an informant may continue to divulge very interesting information. The ethical and political protocols of the interview should be maintained in these closing moments.

Try not to rush the end of an interview. At the same time, do not let an interview drag on. There is an array of verbal and non-verbal techniques for closing interviews (Box 9.10). Of course, non-verbal versions should be accompanied

Techniques for Closing Interviews

Four types of verbal cue:

- direct announcement: "Well, I have no more questions just now."
- clearinghouse questions: "Is there anything else you would like to add?"
- summarizing the interview: "So, would you agree that the main issues according to you are . . . ?"
- making personal inquiries and comments: "How are the kids?" or "If you want any advice on how to oppose . . . just ring me."

Six types of non-verbal cue:

- looking at your watch
- putting the cap on your pen
- stopping or unplugging the audio recorder
- straightening your chair
- closing your notebook
- standing up and offering to shake hands

Source: Adapted from Minichiello, Aroni, and Hays, 1995, p. 94–8.

by appropriate verbal cues; otherwise, you could appear quite rude. The most critical issue in closing an interview is to express not only thanks but also satisfaction with the material that was collected. For example, you might say, "Thanks for your time. I've got some really useful, insightful information from this interview." Not only is gratitude expressed this way, but informants are made aware that the process has been useful and that their opinions and experiences have been valued.

Recording and Transcribing Interviews

Interview recording, transcription, and field note assembly are referred to as the mechanical phases of the interview method. These are the steps through which the data are collected, transformed, and organized for the final stages of analysis.

Recording

Audio recording and note-taking are the two main techniques for recording face-to-face and telephone interviews. Audio files are captured using audio recorders and smart phones, and increasingly they are also captured using video calls via Skype and Zoom. Both audio recording and note-taking have associated advantages and disadvantages, as will become clear in the discussion to follow. Therefore, a useful strategy of record-keeping is to combine note-taking and audio recording.

The records of an interview should be as close to complete as possible. Audio or video recording can allow for a natural conversational interview style because the interviewer is not preoccupied with taking notes and can be a more attentive and critical listener. Audio recording is also preferable to note-taking alone because it allows you more time to organize the next prompt or question and to maintain the conversational nature of the interview. An audio recording will help to compile the fullest recording, while the primary aim in note-taking is to capture the gist of what was said, as well as non-audible cues such as gestures and facial expressions.

On the other hand, some respondents will prefer not to be recorded for various reasons. The presence of the mobile telephone or other recording device may serve as a reminder of the formal situation of the interview, or informants may feel vulnerable if they think that someone might recognize their voice if the recording were aired publicly. Opinions given by the informant on the spur of the moment become fixed indelibly on the hard drive (or memory stick) and can become a permanent public record of the informant's views. This may make the informant less forthcoming than they would have been if only note-taking had been used. Some informants become comfortable with an audio recorder as the interview progresses, but others do not. If you find the latter to be the case, consider stopping the recorder and reverting to note-taking. The downside to solely note-taking is that researchers can be so engrossed in taking notes that they may find themselves unprepared to ask the next question, or might miss important movements, expressions, and gestures of the informant while they are scribbling at a furious pace. This may undermine rapport and detracts from attentive listening (though sometimes the pause while notes are written can be a natural moment of reflection for the respondent).

If you use an audio recorder, place it somewhere that is not too obvious without compromising the recording quality. Modern digital audio recorders and phone apps have long recording capacities and can be turned on and then left alone. But take care when using an audio recorder not to be lulled into a loss of concentration by the feeling that everything is being recorded safely. Regardless of which technology is used, every experienced interviewer has a story of a failed recording! To maintain concentration and avoid the problem associated with recorder failure take some written notes. I find this particularly important when I am conducting the second or third interview in a long day's fieldwork.

Because an audio recorder does not keep a record of non-verbal data, non-audible occurrences such as gestures and body language will be lost unless you are also using a video call or taking notes. If an informant points to a wall map and says, "I used to live there," or if they say, "The river was the colour of that cushion," the audio recording will be largely meaningless without some written record. These written notes can be woven into the verbal record during the transcription phase (described below). Overall, a strategic combination of audio recording and note-taking can provide the most complete record of an interview with the least threat to the interview relationship.

Transcribing the Data

The record of an interview is usually written up to facilitate analysis. Interviews produce vast data sets that are next to impossible to analyze if they have not been converted to text. **Transcripts** are "digital text documents that can originate from . . . recorded verbal accounts, expressions, exchanges, and interactions that occur in research activities such as focus groups, interviews, and oral histories." (Cope, 2020, p. 357). The transcript should be the best possible record of the interview, including descriptions of gestures and tone as well as the words spoken (although see Box 9.13). The name or initials of each speaker should precede all text in order to identify the interviewer(s) and informant(s). Line numbers at the top and bottom of each page of the transcript enable quick cross-referencing between the transcript and digital recordings of the interview. Converting interviews to text is done either through a reconstruction from handwritten notes, a transcription of an audio or video recording, or editing digital interview correspondence.

Interview notes should be converted into a typed format preferably on the same day as the interview. If there were two or more interviewers, it is a good idea to compile a combined reconstruction of what was said using each researcher's notebook. This will improve the breadth and depth of coverage. The final typed record will normally comprise some material recalled verbatim as well as summaries or approximations of what was said.

Recorded interviews should also be transcribed as soon as possible after the interview. Transcription is a time-consuming and therefore resource-intensive task (Cope, 2020). On average, most interviews take four hours of typing per hour of interview. Transcription rates vary according to a host of variables, such as typist skill, the type of interview, the informant, and the subject matter. Transcription apps are freely available to help manage digital files and allow you to pause, review, and even slow down the audio while you type (a favorite as of 2020 is *Descript* www.descript.com). The recent expansion of video communications due to the COVID-19 pandemic has rapidly improved the captioning functions of Zoom and similar tools through automatic talk-to-text technology; however, the resulting transcripts still need significant editing to correct mistakes, particularly for respondents with varied accents.

While there are paid transcription services, many researchers prefer to transcribe their own interviews for two main reasons. First, since they were present at the interview, they are best placed to reconstruct the interchange, including non-audible occurrences and where such events should be inserted into the speech record. Interviewers are also better able to understand the meaning of what was said and less likely to misinterpret the spoken words. Second, transcription, although time-consuming, does enable the researcher to engage with the data again, providing a preliminary form of analysis.

There is no accepted standard for symbols used in transcripts, but some of the symbols commonly used are set out in Box 9.11. Emailed interviews do not require transcription, since each informant's answers are already in a text format. The text may well include some emoticons that are popular in digital communications. Many of the texting symbols below are now represented by emoji and emoticons. As of March 2019, the world had amassed 3,019 emoji in the Emoji Unicode Standard, whereas in 1999 the were only 176 such pictograms (Buchholz, 2019).

Once completed, the transcript should be given a title page stating the informant's name (or a code if there are confidentiality concerns), the number of the

BOX 9.11

Symbols Commonly Used in Interview Transcripts

Symbol	Meaning
//	Speaker interrupted by another speaker or event: // phone rings//
:	Also used to indicate an interruption
KMD	The initials of the speaker, usually in CAPS and bold
—	When used at the left margin, refers to an unidentified speaker
Ss	Several informants who said the same thing
E	All informants made the same comment simultaneously
. . .	A self-initiated pause by a speaker
. . . . or	Longer self-initiated pauses by a speaker
-	Speech that ended abruptly but without interruption
( )	Sections of speech, or a word, that cannot be deciphered
(jaunty)	A best guess at what was said
(jaunty/journey)	Two alternative best guesses at what was said
*	Precedes a reconstruction of speech that was not recorded
(. . .)	Material that has been edited out
<u>But I didn't want to</u>	Underlined text indicates stressed discourse
I got nothing	Italicized text indicates louder discourse
[sustained laughter]	Non-verbal actions, gestures, facial expressions
[hesitantly]	Background information on the intonation of discourse
Emoticon	(see http://en.wikipedia.org/wiki/List_of_emoticons)
:) or ☺ or :-) or :-D	Smiling, joke marker, happy, laughing hard
:(or :-(	Frowning, sad
;) or *)	Wink
:'(or :: or :,(	Tears, shedding a tear, crying
lol	Laughing out loud

*The complete list of computer recognized emoji is maintained by the Unicode Standard, see: https://unicode.org/emoji/charts/full-emoji-list.html

interview (for example, first or third session), the name(s) of the researcher(s) (i.e., who carried out the interview), the date of the session, the location, duration of the interview, and any important background information on the informant or special circumstances of the interview. Quotations that demonstrate a particular point and that could be presented as evidence in a final report on the research might be circled or underlined.

The transcript can be given to the informant for vetting or authorizing. This will normally improve the quality of your record (see Box 9.12). This process of **participant checking** continues the involvement of the informants in the research process and provides them with their own record of the interview.

BOX 9.12

Debates about Changing the Words: Vetting and Correcting

In general, it has been thought that a transcription should be a verbatim record of the interview. This would include poor grammar, false starts, "ers," and "umms." There are a number of good reasons advanced for this position. A verbatim record will include the nuances of accent and **vernacular** speech, it will maintain any sense of hesitancy, and it could demonstrate an embarrassment that was present. For example, Sarah Nelson (2003, p. 16) reflected on how the "humming and hawing" of some Ulster politicians when asked about sectarian killings was reflective of their hesitancy and hypocritical stances on sectarianism in Northern Ireland. Transcripts that are not exact textual replications of an interview will lose the ethnographic moment of the interview itself. However, it may be difficult to search for key terms if they are misspelled in a transcript (that is, misspelled as a means of indicating accent or mispronunciation), so proceed with caution.

A range of researchers working in different disciplines and countries have expressed some concern about the political effects of exact transcription. Many have reflected on the embarrassment that many informants articulate when they receive the transcript of their interview. They express anxiety about the grammar, the false starts to their sentences, repetition, and the "ers" and "umms" and "you knows." This anxiety is even more strongly felt by informants who live in societies where the dominant language is not their first language, or where language use indicates class status or other dimensions of marginalization. Informants might be so concerned as to withdraw their interview and avoid any future ones

(continued)

(McCoyd & Kerson, 2006). Moreover, research reports on the less powerful in society (people living in poverty, the elderly, young people, etc.) that use the real language of informants, even those that are largely sympathetic to those people, can often portray them in a way that reproduces negative images and stereotypes. Nelson (2003) reflected on the way such quoted material reconstructs images of illiteracy, powerlessness, and inferiority. As bell hooks (1990, p. 152) famously stated, "I want to know your story. And then I will tell it back to you in a new way . . . . Rewriting you, I write myself anew. I am still author, authority." Ultimately, informants are usually more interested in the impact of their words than in the nuances of expression, so some researchers send informants summaries or interpretations of the interview rather than transcripts. It is certainly a good idea to send informants the eventual publications and reports.

Assembling Field Note Files

Assembling interview records marks the beginning of the analysis proper. It begins with a critical assessment of the interview content and practice and is followed by formal preparation of interview logs. By creating wide margins in the transcript file, you can make written annotations. Comments that relate to the practice of the interview, such as the wording of questions and missed opportunities to prompt, should be placed in the left margin. These annotations and other issues concerned with contact, access, ethics, and overall method should be elaborated upon in a **personal log** (Box 9.13). The right margin of the transcript file can be used for

Comparing Transcript Files, Personal Logs, and Analytical Logs

Transcript File	Personal Log	Analytical Log
Includes the record of speech and the interviewer's observations of non-audible data and intonation. Also includes written annotations in the margins on the practice and content of the interview.	Reflection on the practice of the interview. Includes comments on the questions asked and their wording, the appropriateness of the informant, recruitment and access, ethical concerns, and the method generally.	Exploration of the content of the interview. A critical outline of the substantive matters that have arisen. Identification of themes. Reference to the literature and theory. In-progress commentary on the research aims and findings.

Source: Adapted from Minichiello, Aroni, and Hays, 1995, p. 214–46.

annotations on the substantive issues of the research project. These comments, which generally use the language and jargon of social science, are then elaborated upon in the **analytical log**. The analytical log is an exploration and speculation about what the interview has found in relation to the research question, similar to **memos** discussed in Chapter 18. It should refer to links between the data gathered in each interview and the established literature or theory.

Analyzing Interview Data

Researchers analyze interview data to seek meaning from the data. We construct themes, relations between variables, and patterns in the data through various kinds of analysis (see Chapter 18 for additional discussion). A simple, but essentially *quantitative*, technique to start with is content analysis, which can be based on a search of either manifest or latent content (Babbie, 1992). **Manifest content analysis** assesses the visible, surface content of documents such as interview transcripts. An example would be a tally of the number of times the words *cute* and *cuddly* are used to describe koalas in interviews with members of the public. This might be important to understanding public opinion and the politics of culling in areas of koala overpopulation (for example, Muller, 1999). Searching interview data for manifest content often involves tallying the appearance of a word or phrase, which can easily be done with the find function in word processing or spreadsheet programs. More complex **computer-aided qualitative data analysis software** (CAQDAS) programs such as NVivo or Atlas.ti are particularly effective at undertaking these sorts of manifest searches, as well as much more sophisticated analysis operations (see Chapter 18), but they require software licences and training to use.

Latent content analysis involves searching the document for **themes**. For example, you might keep a tally of each instance in which a female has been portrayed in a passive or active role. Latent content analysis of interview texts requires a determination of the underlying meanings of what was said. This determination of meanings within the text is a form of **coding**.

A coding system is used to sort and then retrieve data, which help build the analytical processes of identifying themes and explanations. For example, the text in transcripts of interviews with urban development authorities could be coded based on the following categories: structures of governance (for example, legislation, party political shifts), cultures of governance (with sub-codes like "managerialist perspective" and "entrepreneurial perspective"), coalitions and networks (of various types and agendas), the mechanisms through which coalitions operate, and the various scales at which power and influence emanate and are deployed (see McGuirk, 2002). Once the sections of all the interviews have been coded, it is then possible to retrieve all similarly coded sections. These sections of text can be amalgamated and reread as a single file (Box 9.14). This might allow a researcher

Coding Interview Data: Five Suggested Steps

Coding Step	Specific Operations: Computing/Manual Versions
Develop preliminary coding system	Prepare a list of emergent themes in the research. Draw on the literature, your past findings, as well as your memos and log comments. Amend throughout.
Prepare the transcript for analysis	Meet the formatting requirements for the computing package being used. / Print out a fresh copy of the transcript for manual coding.
Ascribe codes to text	Allocate coding annotations using the "code text" function of computing packages. / Place handwritten annotations on transcript.
Retrieve similarly coded text	Use the retrieve text function of computing packages to produce reports on themes. / Extract and amalgamate sections of text that are similarly coded.
Review the data by themes	Assess the diversity of opinion under each theme. Cross-referencing themes allows you to review instances where two themes are discussed together. Begin to speculate on relations between themes.

to grasp the varying opinions on a certain issue, unravel the general feeling about an issue, or arrive at an explanation that was previously hidden.

Not every section of text needs to be coded. An interview will include material that is not relevant to the research question, particularly warm-up and closing sections and other speech focused on improving rapport rather than on gathering data. By contrast, some sections of text will be coded multiple times. For example, in one sentence an informant may list a number of causes of fish kills, including open-cut mine run-off, super phosphates, acid sulphate soils, and town sewage. This may require that the sentence be given four different coding values. Coding is discussed more fully in Chapter 18.

Presenting Interview Data

Material collected from interviews is rarely presented in its entirety. The emerging exceptions include the increasing trend for transcripts to be available for review in digital repositories. Some journals and some international learned societies and unions are mandating that data files (e.g., interview recordings or transcripts) be "published" as an appendix to published papers or in online repositories managed by the discipline groups. Such open access files should be accessible and searchable, using meaningful file names. These affirm the importance of the records being good transformations of the interview, and the importance of those vignettes on ordinary life being respectful, though they also raise new concerns for privacy and ethics procedures. An excellent review of such issues is provided by Elman and Kapiszewski (2017).

Most interview data must be edited and (re)presented selectively in research publications. While it is difficult to locate a genuinely *representative* statement (Minichiello, Aroni, & Hays, 1995), it is usually possible to indicate the general sense and range of opinion and experience expressed in interviews. One way to indicate this is to present summary statistics of what was said. CAQDAS packages such as NVivo can help you to calculate the frequency with which a particular term or phrase appeared in a document or section of text (see Chapter 18). However, the more common method is through a literal description of the themes that emerged in the interviews (see, for example, Boxes 9.6 and 18.1 and the discussion in Chapters 19 and 20 on presenting results).

When describing interview data, you must cite transcript files appropriately. For example, in her interview-based honours research on the changing identity of the Australian industrial city of Wollongong, Pearson (1996, p. 62) noted that "Several respondents asserted that elements excluded by the new identity were of little significance to the overall vernacular identity of Wollongong (Int. #1, Int. #6, and Int. #7)." The transcript citations provided here indicate which of the informants expressed a particular type of opinion. In research publications, the transcript citations can indicate the informant's name, number, code, or recorder count. Whenever a direct quotation from an informant is presented, then a transcript page reference or recorder count can be provided.

Transcript material should be treated as data. A quotation, for example, ought to be treated in much the same way as a table of statistics by being introduced and then interpreted by the author. The introduction to a quotation should offer some idea of where an informant is coming from; information about their role, occupation, or status is important in this regard. Also important, as Baxter and Eyles (1997, p. 508) point out, is "some discussion of why particular voices are heard and others are silenced through the selection of quotes," which researchers can use to reflect critically on their own practices. Quotations should be discussed in relation to, and contrasted with, the experiences or opinions of other informants. Statements of opinion by an informant should also be assessed for internal contradiction. Finally, a quotation cannot replace a researcher's own words and interpretation: as the author, you must do the interpretive work of explaining clearly what theme or issue a quotation demonstrates.

In terms of data presentation, it will sometimes be important that an informant's identity be concealed. Pseudonyms or interviewee numbers have been used by geographers to disguise the identity of their informants when it has been thought that **disclosure** could be harmful. Informants can be given the opportunity to select their own pseudonym. For example, Robina Mohammad (1999) used this technique with some success in her interviews with young Pakistani Muslim women in England. The interviews included discussion of patriarchal authority, "English cultures," and the cultures and dynamics of the Pakistani Muslim community in Britain. Some informants selected Pakistani Muslim pseudonyms for themselves, others chose very English names. These selections were themselves very interesting and provided further insight into the cultural perspectives and resistances of these women. Other researchers allocate pseudonyms that reflect the ethnicity of the informants (McCoyd & Kerson, 2006). Gill Valentine (1993) felt it necessary to disguise the name of the town in which her interviews with British lesbians had taken place. Similarly, Mariastella Pulvirenti (1997) disguised the street names that were mentioned by female Italian Australians when discussing their housing and settlement experiences in Melbourne.

Naming an informant (or locating them in any detailed way) and directly associating them to a quotation could be personally, professionally, or politically harmful. Researchers must be very careful when they deploy data they have collected. Interviewers are privileged with insights into people's lives. Some researchers recommend instituting an alias or pseudonym for informants very early in the mechanical phase so that no digital records will bear the informant's real identity. However, it can prove difficult to remember who the real people behind the aliases are, and some researchers only impose the pseudonym in the presentation phase of the research.

The presentation of interview-based research must contain an accessible and transparent account of how the data were collected and analyzed (Baxter and Eyles, 1997). This account should outline the **subjectivity** of the researcher, including their **positionality** (see Chapters 2, 3, 4, and 20). As we have already seen from the discussion in Chapters 4 and 6, it is only through **transparent** accounts of how interview-based research is undertaken that the **trustworthiness** and wider applicability of the findings can be assessed by other researchers.

Interviews Using Digital Communication

Digital interviewing can include interviews with individuals or groups and can be either **asynchronous** or **synchronous** (see Mann & Stewart, 2002; Meho, 2006). Asynchronous digital interviewing has most often taken the form of email exchanges. Indeed, most early uses of electronic communications for social research involved attaching questionnaires (and sometimes interview schedules) to emails.

And, of course, video-conferencing software now provides a noteworthy medium for the conduct of synchronous interviews with the person next door, in the next city, in the next state, or in another country. While such software has been available for some time (e.g., Skype since 2003, Apple FaceTime since 2010, Microsoft Teams since 2017) the COVID-19 pandemic made Zoom, in particular, central to many people's lives. In the next section, I review advantages and disadvantages of digital interviewing and then provide a series of tips for better practices specific to this mode.

Advantages of Digital Interviewing

The term digital interviewing here encompasses any online means of conducting an exchange with a participant (e.g., sending questions via email) and the rapidly emerging realm of real-time video interviews. Digital interviews offer five general sets of advantages: (1) an expanded sample; (2) reduced **interviewer effects**; (3) enhanced convenience; (4) more reflective informant responses (for email exchanges); and (5) cost savings (see Chapter 13 for additional discussion of some of these issues).

First, online delivery of questions allows a researcher to overcome spatial, temporal, and social barriers that would restrict access to informants for face-to-face interviews. Interviews can be more easily facilitated with people living overseas or in inaccessible locations (remote places or war zones) or who have mobility limitations, and the asynchronous quality of emailed or posted exchanges can make them much more convenient for people who are shift workers or those who are based at home with small children (Bampton & Cowton, 2002; Mann & Stewart, 2002; Meho, 2006). While video interviews offer these benefits as well, language and other cultural barriers may still be present. Email interviews can also allow researchers to surmount social hurdles to gain access to informants: they will suit people who are introverted, who are cautious about their identity being revealed, or whose cultural context is disparate from that of the researcher; they are also appropriate when the topic is very sensitive. Good examples in the literature include interviews with political and religious dissidents, people who are incarcerated, oppressed minorities, and people in subcultures viewed as radical or deviant. McCoyd and Kerson (2006) found from their research with women in North America who had had medical pregnancy terminations (after learning of a fetal anomaly) that email interviews were very effective for discussions on a topic that generated strong emotions among the informants and broached issues that are not normally discussed in public. Meho's (2006) review of the uses of email interviewing indicates that response rates among these "difficult to reach" samples were on average about 40 per cent, which is impressively high.

One advantage of email interviews is a reduction in interviewer effects as a result of visual anonymity. Many of the cues we use to make judgments about people are based on visual appearance (e.g., dress, body shape, skin colour, jewellery, hair styles), and these cues are much more limited in email-based interviews in which informants (and researchers) can share only what they want about their embodied selves. This anonymity comes at the cost of the non-audio data that are usually gathered in a face-to-face interview and some contextual effects may dissipate. At the instrumental level, this means that the interview is not interrupted by telephone calls, colleagues, or family members. A more complicated matter is the absence of the researcher from the informant's own ethnographic setting and the loss of those observations. However, email interviewing offers great convenience to informants, allowing them to choose the time of their responses, to consider their answers at their leisure, and to do so in the comfort of their own home (Bampton & Cowton, 2002; Mann & Stewart, 2002; Meho, 2006). An informant in McCoyd and Kerson's (2006, p. 397) research with women who had medical terminations reflected that "I'm looking forward to doing the interview . . . it is a much more relaxed and productive way to do it [through email]. This way, I can do it when things are quiet and I'm in the right frame of mind." The informant has much more control over the pace and flow of the interview, more so than if it were a telephone or face-to-face interview during which they might feel rushed to offer an answer.

Another advantage of email interviewing is that that informants' answers can be more detailed, reflective, and well-considered than those in other formats. James and Busher's (2006) email interviews with people in British university settings (and beyond) found that there was a "richness of reflection among the participants." Informants who took their time in responding to a question "tended to generate more thoughtful answers" (2006, p. 414). James and Busher gave the example of an informant who began one of their answers with "I didn't email you straight back, because I was thinking about my answer. So my responses were more carefully thought through and probably longer than if I'd tackled the whole thing in a face-to-face interview" (university-based informant, quoted in James and Busher [2006], p. 415). Informants can rethink, proofread, and re-craft (and even spell-check) their responses so that they most accurately represent their views and experiences (Bampton & Cowton, 2002).

Another advantage of online interviews (though not video call) is in the mechanical phase, because the answers are already in text form the issues of transcription error and interpretation are removed (Chen & Hinton, 1999; Mann & Stewart, 2002). Informants have "cleaned" their own responses to a level they are satisfied with before they post them (McCoyd & Kerson, 2006). Of course, researchers do have to edit out the unnecessary email symbols, signature sections, and line returns. A final set of advantages of all forms of digital interviewing is the reduced cost relative to face-to-face interviewing. The obvious savings are associated with travel costs, time, and carbon footprints.

Challenges of Digital Interviewing

The weaknesses or limitations of digital interviews, relative to face-to-face interviewing, stem mostly from the spatial and temporal displacements between the informants and the researcher. These issues include concerns about the authenticity of the informant, the loss of visual cues that assist rapport-building, and the "clunkiness" of the interview relationship. There are also issues of uneven internet access and comfort with the medium, as well as ethical issues associated with privacy and anonymity.

In email interviews, the advantages that stem from the visual anonymity and the use of pseudonyms have the negative effect of reducing our ability to know who we are interviewing (though feminists and others have wondered how participants are ever "known"—see McDowell, 2005), which means that people can make misleading claims about who they are (James & Busher, 2006). However, Mann and Stewart (2002) noted that maintaining a false identity in a substantive correspondence is not so easy to do, though researchers ought to be wary of inconsistencies and contradictions that reveal such false personae. More subtly, informants are more able to embellish or be bombastic in emailed answers, because the researcher does not have the visual cues to help detect and address such tendencies (Meho, 2006). Similarly, prompts are delayed until the next email interchange, at which point they may be considered less important than new primary questions, crowding out space for such clarifications (Chen & Hinton, 1999; Meho, 2006). Researchers using email interviews must instead rely heavily on reading between the lines of answers or participant checking for clarity.

The absence of **paralinguistic clues** in email interviews raises broader issues. It is much more difficult for researchers to tell whether an informant is becoming distressed or uncomfortable when reflecting on or writing an answer to a question (Bampton & Cowton, 2002). Their answers are in a "narrower bandwidth," and this has dramatic consequences on the potential for rapport. For example, it is difficult to communicate empathy and sympathy (e.g., in regard to grief) to informants via email without it being perceived as banal or insincere (Mann & Stewart, 2002). The generally truncated process of email interviewing means that interviewing a single informant can extend over weeks and months. There will be substantial gaps at times between responses, linked to the personal availability of the informants and to the ability of the researcher to properly assimilate previous answers and to consider necessary probes and relevant primary questions. These gaps could be interpreted, incorrectly, as disinterest on the part of the informant or as a sign that the researcher is underwhelmed or disgusted by previous responses. It is also difficult for researchers to know why an informant might be delaying a response (Bampton & Cowton, 2002).

Digital interviewing of all types is really only appropriate for study groups with widespread internet access and digital literacy, which may limit recruitment.

Participating in an interview through email, blogs, or online in a chat format may be demanding, requiring a lot of time on a phone or computer, which may be more onerous for the informant than answering a question orally over the telephone or face-to-face. One early observation was that digital interviews may be most appropriate for technologically savvy study groups (Mann & Stewart, 2002), though as a greater portion of the general population engages in digital cultures, that is less and less relevant (see Chapter 15).

Decisions about good ethical practices surrounding digital methods continue to take shape (Ash, Kitchin, & Leszczynski, 2019). One concern surrounds the privacy of informant comments, especially as anonymity has become nearly impossible to preserve. A bigger issue concerns the usual promises that researchers offer to informants about the confidentiality of their comments; institutional server back-ups, hacking, and other concerns suggest using a secure file exchange service rather than email would be a wise choice for documents. Although many universities currently require informants to sign consent forms before they participate in interviews, the digital environment has different protocols for signatures and approvals, and some researchers consider that an email from the informant detailing their informed consent or an e-signature on a secure document should be sufficient. Checking on such questions with the information technology and research ethics experts at your institution well before you begin is highly suggested.

Because of the limitations and challenges of digital interviewing, some have concluded that it is best seen as a complementary mode and that in-person interviewing remains the gold standard method (see Meho, 2006). According to this view, digital interviewing is appropriately used for informants who are difficult to physically reach or for those who already have a presence within some form of online community or network. However, the mode has special advantages of its own, such as the opportunity for more reflective responses, and it represents an appropriate method for the digital age—indeed, Chapter 15 examines emerging methodologies such as app-based inquiries that are rapidly upstaging email as a method.

The Rise of Video Call Interviewing

Video calls facilitate the reception and transmission of audio-video signals by users at different locations in real-time. Video call interviews have the benefits of email interviews, but they also have the benefit of being a synchronous, visual medium. Common video call apps include Skype, Zoom, and FaceTime, all of which have free, account-based interfaces.

There is only a relatively small set of literature on the benefits and challenges of video call interviewing, including some notable contributions from geographers (Adams-Hutcheson & Longhurst, 2017; Deakin & Wakefield, 2014). The benefits surround access, rapport, and ethics. As with email interviews, video extends the reach of a sample to informants globally, to those who are less mobile, more remote, socially isolated or living in dangerous places, and who might otherwise be excluded from a sample (Deakin & Wakefield, 2014). Reaching such people is also less expensive in terms of time and resources (both financial and environmental), generating advantages for students and time-constrained academics. It is also much less taxing on informants to prepare for an interview using video than it is if they were to be interviewed in person in their office or at home. Weller's (2017) use of Skype and FaceTime interviewing found that 83 per cent of the youth said it was as good as the face-to-face interviews they had had previously, and many were more comfortable with the video call. This was part of a longitudinal study, following the lives of 50 young people from across Britain:

> there's less of a pressure of presence if you like . . . nothing against you or anything (laughs). . . . when you're sitting in a room with someone opposite you, you feel a lot more under pressure than when it's over the computer, so I guess it does give you the freedom to sit back and actually think so in that way I think it was quite nice actually as pressure does get to me a little. (Carl, Britain, quoted in Weller, 2017, p. 618)

It has also been noted that video call interviews provide the informant with a vista into the researchers' domain, whether that is an office, meeting room, or home study, thereby humanizing the relationship (Weller, 2017). Callegaro (2011) also noted that video facilitates the exchange of files, such as photographs and screen shares, and these can be useful mechanisms to build and enhance rapport. And, unlike email or telephone interviews, video calls allow the researcher to better monitor paralinguistic cues. Finally, the ethical benefits of video call interviews include the limited privacy invasion (from not being in peoples' physical domains) as well the relative ease with which informants can withdraw consent and end an interview if they so wish (Deakin & Wakefield, 2014).

The challenges of video call interviews mostly surround the risks of technological failures, uneven access, and the lack of ethnographic context relative to face-to-face interviews. Technical failures can disrupt the start of the interview or seriously interrupt it through screen freezing and audio drop-outs. Varied expertise with new technology and uneven access to broadband streaming are challenges. These will have a more dramatic effect on rural residents, lower income communities, and people in precarious social circumstances, who often have poorer access to fast internet. Further, Weller (2017, p. 619) reflected on the ethnographic context that she no longer had access to through video call interviews compared to her face-to-face interviews: "[I] have gained much from wandering around the areas in which participants live, absorbing myself in the scenery of their homes, experiencing customs and hospitality, and meeting family and friends. The valuable contextual material that enriched my understanding of participants' lives [was missing]." Video call interviews offer only glimpses into the ethnographic context of the informant's environments. A final ethical concern with video call interviews is that—like audio-recorded interviews—a very powerful record is generated, and the actual recoding equipment is not overtly visible to the informant; researchers must inform participants that the call will be recorded, provide details on how the digital file will be kept secure, and offer to share the transcription. Caution is particularly recommended with free versions of apps as both the privacy parameters and researchers' access to files may not be as secure as required.

Most reviewers of video call interviews have concluded that they are generally as effective and beneficial as face-to-face interviews, with some claiming it as "favoured choice." The benefits of video call interviews can be enhanced through proper netiquette, focus, and preparedness (see Box 9.15). In general, video call interviews ought not be used at all for informants who are uncomfortable with these technologies, or who do not have reliable streaming and hardware.

Tips for Effective Video Call Interviews

Category of action	Tips for effective video call interviews
Netiquette and ethics	Restate request for consent, and remind informant that the interview is being recorded. Make visible your user identity. Use the informant's own identity in the meeting topic box or subject, not a generic meeting ID. Allow host and participant to turn their webcams on and off at any time. Mute when you are not talking.
Focus and rapport	Have warm up tactics that leverage the capabilities of video call. Avoid multitasking. Disable notifications on your phone and computer to minimize distractions. Watch carefully for paralinguistic cues. Allow "audio only" if requested by informant. Ensure eye-to-eye is facilitated, and a "head and shoulders" set placement, especially of the researcher. Include physical gestures at closing and exit, such as a wave.
Preparedness	Allow the informant access before the meeting start time (click "enable join before host," etc.), in case they want to check out the interview space. Approach the technical check-in moments ("can you hear me okay?") as ice breaker opportunities. Be prepared for technical hiccups, using the chat functions or other media to maintain contact if necessary.

Sources: Deakin & Wakefield, 2017; Seitz, 2016; Weller, 2017.

Conclusion

The rigour of interview-based research is enhanced through adequate preparation, diverse input, and verification of interpretation. Being well-informed and prepared will give you a deeper understanding of the "culture" and discourse of the group(s) you study. You can then formulate good questions and enhance levels of rapport between you and your informants. You should also purposely seek out diversity of backgrounds and opinions. By interviewing more than one informant from each study group, you can begin to draw out and invite controversy or tensions, or, alternatively, generate multiple explanations for a phenomenon. An opinion from one informant should never be accepted as demonstrative of a broader group opinion unless it is shown to be so from interviewing multiple informants. Finally, some means of verifying your interpretations of interview data are necessary (for example, participant checking, peer checking, and **triangulation** with documentary material).

Interviews bring people "into" the research process. They provide data on people's behaviour and experiences. They capture informants' views of life. Informants use their own words or vernacular to describe their own experiences and perceptions. Kearns (1991, p. 2) made the point that "there is no better introduction to a population than the people themselves." This is what I find the most refreshing aspect of interview material. Transcribed interviews are wholly unlike other forms of data. The informant's vivd text reminds both researcher and reader of the lived experience that has been divulged. It reminds geographers that there are real people behind these data.

Key Terms

analytical log
asynchronous interviewing
critical inner dialogue
digital interviewing
disclosure
emoticon
funnel structure
informant
interview guide
interview schedule
interviewer effects
latent content analysis
life history
manifest content analysis
oral history
paralinguistic clues
participant checking
personal log
pre-testing
primary question
prompt
pyramid structure
rapport
secondary question
semi-structured interview
structured interview

synchronous interviewing
transcript
unstructured interview
vernacular
video call
warm-up

Review Questions

1. Match the three categories of interviewing (unstructured interviewing, structured interviewing, semi-structured interviewing) with the following three descriptions (question focused, informant focused, content focused). Explain what is meant by each of these three descriptions.
2. List some rapport strategies you could use if you were to conduct a face-to-face interview with an older relative not well known to you. Provide a rapport strategy for the preliminary, contact, warm-up, and closing phases of such an interview.
3. Identify five benefits and five disadvantages or challenges of digital interviewing.

Review Exercises

1. Select one of the three statements below, and spend about 10 minutes constructing an interview schedule for a hypothetical five-minute interview with one of your peers. Use a mix of primary question types and prompts. Think about the overall structure of your schedule, and provide a sense of order in the way the issues are covered. Try to imagine how you will cope if the interviewee is aggressive, very talkative, or non-communicative. Will your schedule still work?
   a. The climate emergency we are in necessitates a rapid transition away from personal cars toward public transit.
   b. Beach activity is decidedly spatial. Performances are expressive, and behaviour is at times territorial.
   c. The local environment plan (LEP) of every city/town should allocate a specific area for sex industry uses.
2. Construct two interview guides on the research topic of your choice, of which one uses the funnel method of framing questions and the other uses a pyramid model. Share your plans with a peer group and decide which of the two (or perhaps a hybrid model) would work best.
3. The absence of direct contact in digital interviewing has implications for rapport. Map out a list of strategies specifically tailored to maintain rapport in an interview comprised of email exchanges then compare it to ways to maintain rapport in face-to-face and video call interviews.

Useful Resources

Callegaro, M. (2011). Videophone interviewing. In P.J. Lavrakas (Ed.) *Encyclopedia of survey research methods*. Thousand Oaks: SAGE Publications.

Deakin, H., & Wakefield, K. (2014). Skype interviewing: reflections of two PhD researchers. *Qualitative Research, 14*(5), 603–16.

Dowling, R., Lloyd, K., & Suchet-Pearson, S. (2016). Qualitative methods 1: Enriching the interview. *Progress in Human Geography, 40*(5), 679–86.

Longhurst, R. (2016). Semi-structured interviews and focus groups. In N. Clifford, M. Cope, S. Gillespie, & S. French (Eds), *Key methods in human geography* (pp. 143–56). Thousand Oaks, London, New Delhi: SAGE Publications.

Weller, S. (2017). Using internet video calls in qualitative (longitudinal) interviews: Some implications for rapport. *International Journal of Social Research Methodology, 20*(6), 613–25.

Listening Sensitively: Oral Histories

Darius Scott

Chapter Overview

This chapter describes how oral history recordings elicit emotional and informative accounts from research participants. After defining the method and considering its place in geography, the "Doing Oral History" section details preparatory aspects such as finding participants and creating a guide, as well as strategies for collecting, analyzing, and preserving oral histories. In the final section, the chapter breaks down excerpts from actual oral histories to show how they are useful for geographic research. This chapter is meant to inspire a firsthand exploration of this method.

Understanding Oral History

What is Oral History?

Oral histories are recorded dialogues between a researcher and one or more participants. These co-productions aim to chronicle some series of participants' past experiences. In a single recording, a person may recount their first day working on the family farm as child in addition to the calm they felt after selling that farm decades later. For human geographers, oral histories foreground emplaced stories of experience, emotion, defeat, triumph, and mundanity. The method can also support collaborative and politically-engaged research.

Oral histories are distinct from in-depth **interviews** (see Chapter 9) by being conversational and focused on the life histories of individuals. Participants are invited to recollect and express themselves with minimal prompting. The method typically elicits narratives, or stories set in participants' past, which may be interspersed with recounts of day-to-day activities. Geographer Kyle Evered, for instance, conducted oral histories with retired opium poppy farmers in Turkey who described both the intimate and laborious routine of milking poppy plants during harvest time as well as the moments their farms closed following international drug policy (Evered, 2011). The farmers' oral histories provided localized, personal perspectives of a large-scale process.

Oral histories are often poignant. Participants may recount difficult experiences of loss, transition, and finality. For instance, the *Remembering Black Main Streets* oral history project focused on thoroughfares in the US South that were loci of Black cultural emplacement. These important geographies were razed by inequitable twentieth-century urban redevelopment. The oral histories in the collection capture some sense of what was lost to mutually constituted processes of racial injustice and capital wealth accumulation. The excerpt below illustrates a participant recounting his view of West Broad Street in the city of Savannah, Georgia, which he recalls beginning to undergo a fateful transformation in the 1940s:

> Everything was nice back then, now that's when West Broad was West Broad. We don't have any West Broad now. We don't have anything now. Not a thing. The Blacks don't have anything. When I came here, West Broad Street was known from . . . Broughton Street all the way to Henry was Black business, Black undertakers, Black taxi service, Black theatres, Black hotels, Black restaurants, Black bars. . . . Right down on down West Broad. We had it. Urban renewal came through here. You know what was going to happen. . . . That urban renewal came through here and cleaned us out. (Interview R-0171, University of North Carolina at Chapel Hill's Southern Historical Collection)

Beyond providing an informative, on-the-ground perspective of the street's demise, the participant mourns it for the Black population of Savannah in saying "[w]e don't have anything now. . . . The Blacks don't have anything." In this way, his recorded recollection is an *emotional* one that offers a grim survey of Black geographic assets in Savannah.

Oral history provides insight into several geographic themes including contested places, place attachment, historical inequities, production of space, emotional perspectives of place, local experiences of large-scale processes, and experiences of injustice. Through these themes, oral history research may be able to privilege the firsthand narratives of participants in the analysis of both material and **affective** geographic issues.

Oral History as a Research Method

Oral history research aims to centre the **situated knowledge** of participants alongside those of the researcher. Oral history participants are invited to openly share what is important to them, and recordings often veer away from research topics to accommodate the narratives. The method, then, can constitute a "coproduced archival practice" between participants and researchers (Maharawal & McElroy, 2018, p. 5). For example, the community-based Anti-Eviction Mapping Project, discussed in Maharawal and McElroy (2018), used oral history to map and historicize the "social worlds" of people resisting the recent eviction crisis in the Bay Area of California. The oral history recordings centre participants' narrated *place attachment* or longstanding residential relationships to the city, which are ignored in socially decimating processes of urban development (see also www.antievictionmap.com/). Similarly, Rogaly's *Stories from a Migrant City* (2020) foregrounds the cosmopolitan working class narratives of Peterborough in England, which are marginalized by debates concerning Brexit.

Critics of oral history contend that participants' recollections may be influenced by nostalgia and false memories, which, they argue, makes the data less **reliable** than that of, say, archived documents and records. However, psychological research has found that people tend to remember what is important to them, which may coincide with topics of oral history recordings (for more on this debate, see Raleigh Yow, 2014, pp. 41–76). Further, oral histories can relay perceptions of places over time where memory is less important than the *emotional* tenors the method may highlight. In Aotearoa New Zealand, Egoz (2000) recorded the frustrated narratives of organic farmers whose practices countered the longstanding maintenance of manicured "clean and green" agricultural landscapes. The farmers shared evocative recollections of disputes with neighbours and reflections on their earnest devotion to the land.

Emotion-focused oral history research can be politically pressing. Mei-Po Kwan (2008), for instance, led an oral history project considering how Muslim women's day-to-day movements in Columbus, Ohio, were restricted by fear of anti-Muslim hatred following the terrorist attacks of 9/11 in the US in 2001. The research considers how actual and anticipated violence imbued the city with a pervasive air of threat for the women. Kwan's work also highlights oral history's purchase in exploring *local experiences of large-scale processes*. The women negotiated global currents of anti-Muslim sentiment within the space of Columbus. However troubled by memory issues, oral history offers emplaced and potentially sensitive perspectives of geographic matters.

Recent pushes for community-engaged work bolsters the acceptance of oral history as a research method. Initiatives like those of the aforementioned Anti-Eviction Mapping Project and the community-development focused Oral History Centre (www.oralhistorycentre.ca) at the University of Winnipeg illustrate how the oral history collection process can accommodate politically-engaged collaborations between the researcher and participants. One way this engagement manifests is through **participatory research** (see Chapter 16). Interested participants may be prompted to collect oral histories from one another in which case they may steer what questions are asked. Oral history also stands to support community-engaged work in related disciplines such as environmental history (see Williams & Riley, 2020). For instance, Lane (1997) consults the oral recollections of New South Wales, Australia, residents to deepen scientific understandings of environmental change.

Most social scientists collect a number of oral histories from multiple participants for a single research project. Evans recorded the histories of several locals in Blaxhall, Suffolk, UK, to write his influential work, *Ask the Fellows Who Cut the Hay* (1965). However, other researchers rely on just one interlocutor. Historian Theodore Rosengarten's *All God's Dangers* (1974), for instance, was created with several recordings of just one person—Ned Cobb (aka Nate Shaw), a farmer and storyteller from Alabama. Indeed, no set number of participants or recordings are requisite, and the numbers range: a researcher may write a journal article based on just two richly textured oral history recordings. When beginning with oral history as a student or newcomer, it is appropriate to focus on first recording one session with an enthusiastic narrator.

Doing Oral History

The relaxed dialogue of oral history recordings belies the planning and preparation the method calls for. This section draws from my own personal experiences including work on the Southern Oral History Program's *Back Ways* project, which used oral history to explore the social experience of rural segregation in the US. Long before any recording, the *Back Ways* team determined our aims, conducted background research, and, most importantly, identified potential participants who could offer well-rounded accounts. Beginner oral historians can engage with local community history organizations to identify potential participants. Forthcoming potential participants may also be found through academic departments. Indeed, scholars have explored the disciplinary formation of geography using the oral histories of academic geographers (e.g., Matless, Oldfield, & Swain, 2007).

Students and newcomers are advised to explore options for preserving recordings and **transcripts** before embarking on oral history projects.

Preserving Oral Histories

Preservation is a major topic among oral historians. This is the process by which the recordings and accompanying materials are saved for future review. Some researchers, such as Sommer and Quinlan (2018), contend that preservation distinguishes the method: "For oral historians, preservation of intact, complete interview recordings sets oral histories apart from other types of interviews" (p. 3). Free websites, such as Soundcloud, can be used to preserve digitally recorded oral histories and have them be publicly accessible (when that is appropriate). However, libraries and similar institutions are usually preferable for preservation as they have measures in place to preserve materials in perpetuity, with approved levels of public access. By contrast, websites sometimes close with little notice. Administrators at local libraries and preservation organizations should be contacted directly to start the process before oral histories are recorded.

Each organization may have requirements regarding recording formats (e.g., WAV or MP3) along with particular consent or permissions forms that must be used.

Preservation often means making the oral histories publicly accessible online and/or via physical access. For instance, when the University of North Carolina at Chapel Hill's Southern Historical Collection preserves oral histories, the audio recordings and transcripts are accessible to the general public via its website. Such public-facing preservation could lead to more attention being paid to the oral history work and any pertinent issues the communities they come from may be facing. However, it also necessitates having each participant consent to having their histories preserved in such a fashion.

Finding Participants

Connecting with oral history participants often requires **rapport** and vouching. Rapport means forming a dialogue with participants prior to recording. In the prior edition of this volume, George and Stratford (2016) soundly suggest using preliminary meetings for this purpose. Vouching, or identity confirmation, may require the help of community **gatekeepers** or individuals such as local church and non-profit leaders who have experience connecting communities to outsiders. Such people are ideal first contacts and may be willing to introduce researchers to potential participants. They may also be willing to sit for recordings themselves. At the onset of the *Back Ways* project, a local public historian sat for an oral history recording and notified potential participants about the project before they were contacted.

Connections to oral history participants may also be established via the **snowball** method. That is a process in which one participant connects a researcher with other potential participants. The way this works in practice is that after a recording, a participant is asked if they know other people who might be good potential candidates. The participant may then facilitate introductions or provide names and contact information. However they are come by, personal references are excellent means of connecting with oral history participants.

There are many more means of recruiting participants beyond community gatekeepers and the snowball method. Researchers can also email listservs or place flyers in public locations. See Chapter 6 of this volume for more recruitment advice.

Preparing for a Recording

Preparing for an oral history recording typically begins with background research, or the review of pertinent records from the internet, **archives**, and libraries. When I worked on the *Back Ways* project, the process involved reviewing archived road surveys, county planning documents, and secondary literature on North Carolina segregation. These documents inspired specific questions about how planning and development shaped experiences of rural segregation that could be answered in my own oral history recordings.

After conducting some measure of background research, a **guide** is created. The guide is a document that lists questions or **prompts** to be asked during the sessions. To be sure, the document is meant to *guide* rather than dictate what is ultimately covered in a given oral history. It is important that researchers remain very flexible to accommodate the conversational tone of the research method.

Guides should be arranged as chronologically as possible. That means starting with topics where anticipated answers reference the earlier parts of participants' lives and concluding with topics where answers might reference the later part. It may be helpful to start with a general question about childhood experiences (e.g. "how do you remember childhood?") before getting into specific questions (e.g. "what sorts of games would you play?"). Arranging topics chronologically is not a perfect process as there is generally some overlap. However, doing so as much as possible yields some structure and a narrative arc for oral histories.

A working guide with **common questions** should be the starting point, but then researchers can personalize the guide for each participant. Personalizing the guide stimulates rich narrative accounts from each participant and typically means conducting additional background research tailored to the life history of an individual participant to create **specific questions** for them. Factors that influence the extent to which a guide may be personalized include the participant's notability and how much information is already known about them. For instance, one *Back Ways* participant was a well-known community organizer who had also run for county sheriff. For this participant, there were news articles to review that were not available for other participants who had lived more private lives. If some aspects of a participant's life history are known, such as their long-term place of employment, it is also possible to conduct background research about those related specifics. Asking participants to bring pertinent photographs or mementos can also help prompt personal, specific narratives. Such relics may be tied to particular stories.

Finally, before the recording, oral historians secure the proper equipment and ensure they have a place to record. It is ideal to use a standalone handheld audio recorder device such as a Xoom H4N to record the oral history. While a standard smartphone may be adequate for recording a first oral history, good quality recording equipment is often available for loan at university libraries and media centres. Further, it is best to record in a quiet place where you will not be interrupted.

Acquiring Consent

Acquiring clear consent is a critical aspect of oral history research. As the subject matter is typically personal, the participants must understand what will happen with their accounts, especially regarding public access. Professional oral history projects typically have a standardized consent form that each participant signs. The form outlines the purposes of the recording as well as information about how it will be used and preserved.

University ethics review requirements for oral histories vary from country to country. Data protection legislation in the UK requires that researchers provide explicit documentation to participants that discloses how their information will be used. Further, UK university research ethics committees typically require that all human subjects research undergoes review. In the US, oral histories will typically be exempted by a university's institutional review board (IRB). However, each oral history project must be exempted by the IRB on a case-by-case basis. Readers should check with their ethics review boards and university policies and national laws and err on the side of institutional review and transparency. The collection of any personal material to be used for research and publication (including theses and dissertations) should invoke high standards of protection of people's privacy. Signed consent or agreement forms are promoted by leading oral history organizations including Oral History Australia, Oral History Society (UK), and the National Oral History Association of New Zealand.

It is ultimately the responsibility of the researcher to ensure that the research process is verbally explained to each participant before any recording. Also, care should be taken to be sure that participants do not feel coerced and remain willing to share their accounts throughout the actual recording process. While difficult recollections are bound to occur in oral history recording, this does not permit coercion or insensitivity. After explaining the consent process to the participant and briefly going over the purpose of the recording, it is ethically compulsory to let the participant know they may stop the recording at any time.

Conducting the Oral History

Once the recorder is turned on, a researcher might state the date, their own name, the name of the participant, and the location of the session before starting at the top of the guide. Interruptions are minimal to maintain a story-telling tone. If the answer to a question is not understood, the oral historian can say so, politely. In an oral history, curiosity steers the conduct of the recording within the confines of an established research area. Care must be taken to allow for unexpected directions and stories, even if they seem tangential in the moment.

Sometimes participants cannot (or prefer not to) recount an experience, in which case the researcher might propose a short break, redirect the topic, or, in some cases, sit with silence while the participant mulls over what to say. Relatedly, there may be moments when participants cry or get upset while recounting difficult experiences. If this happens, be open to pausing the recorder and checking in with the participant. They may wish to conclude the recording or continue at a later time. Be patient and respectful of participants' boundaries.

There are two steps to take just after a recording concludes: (1) ensure the participant understands the research process; explain any information about preservation and where the recording may be publicly accessed and (2) record field notes describing any unusual circumstances as well as where the oral history was recorded (e.g., a living room) and what the environment was like (e.g., neat and cozy). Such notes help with future research writing. To that end, field notes can also reflect immediate thoughts given whatever research questions are being approached.

Analyzing Oral Histories

The process of **transcription** is critical to using oral histories for text-based research products such as essays. Some oral historians transcribe recorded pauses, repetitions, and shorthand colloquialisms to retain as much of the participants' voice as possible. For detailed recommendations on transcribing, see Chapter 9 of this volume.

Before **coding** (see Chapter 18), time should be taken to sit with each oral history transcript and answer the big question: "What is the story the participant shared?" This is the part where the transcript is simply read through, the big events are pieced together, and the defining aspects of the overall account are identified. It is important to remember that oral history recordings are not just arranged clips of data. The transcript may not be perfectly linear—starting from childhood or some other point and moving straight ahead to end with the present day or some other designated time. Instead, it might be imagined as a thread within part of the tapestry that is a person's life story. It is important to honour that. Further, oral history researchers can compare how experiences across the lives of participants compare to one another by creating coding categories such as "childhood experiences" and "early work experiences" even if research questions are not particularly focused on childhood or work. Coding is not the only analytical technique suitable for oral histories—**narrative analysis** (Cope & Kurtz, 2016), **narrative mapping** (see Chapter 14), and other approaches can also yield valuable insights in rigorous ways.

Oral histories are useful for several creative and intellectual projects including public history exhibits as well as audio and video documentaries. For examples see the Columbia Oral History Master of Arts website (http://oralhistory.columbia.edu/), which features a number of public-facing thesis projects. Such public productions draw more attention to research topics. Also, they are useful ways to share the results of oral history work with the participants themselves. Share any public projects with the local press—they may be enthusiastic about covering multimedia work and publicizing research. Further, public-facing projects and related press coverage are highly valued by academic institutions that are charged with public scholarship and strengthening community–university relations.

Oral History in Practice

Concerning Spatial and Historical Inequities

> *Researcher*: And how did y'all get around as children?
> *Participant*: Well, we walked. And if it was somewhere like we were going to church, we had to go on the wagon. That's the only transportation we had. My daddy couldn't afford—we couldn't afford a car. I'm sure he was making mortgage payments, I'm sure. We had to go on the wagon. And everybody—that's the way all Black people traveled then. Nobody had a car. It was a horse and wagon, or mule and wagon, whichever one you had, for church or going into town to get your groceries. Wherever you had to go, that was your transportation. (Interview X-0028, University of North Carolina at Chapel Hill's Southern Historical Collection)

Injustice affects people intimately. The co-constitutive power of American racism and capitalism produces spatial injustices across time and down to the smallest of scales. This is acutely evident in the American South where Black communities settle amidst former sites of forced agricultural labour. Localized sites of tenant farming, forced labour, and institutionalized segregation populate the region's history of racialized economic discrimination. Oral history recordings elicit firsthand recollections of such emplaced *historical and spatial inequities* (Woods, 2002).

The participant above is a Black eighty-one-year-old retired seamstress who spent her life living on family farmland in rural North Carolina. She shared her oral history with me in 2014 when cars had come to dominate, but she recalls a time when travel was slower for her community—done by foot or wagon. Economic disenfranchisement prevented most rural Black Americans at the time, the 1930s and '40s, from owning a car. She recalls the transportation norm as being racially distinct—"that's the way all Black people traveled then." In the rest of the oral history, the participant deftly points out, on maps and in photos, the paths and wagon roads people in her community would use.

The excerpt comes from the *Back Ways* project, a collection of oral histories focused on segregation in rural North Carolina. Black participants were able to recount inequitable economic opportunities and uneven infrastructural development in their communities. A number of the *Back Ways* participants recounted the pains of still relying on unpaved roads due to discriminatory local, state, and federal governments. Participants also recounted how twentieth-century institutionalized segregation affected their individual lives. The same woman quoted above continued:

> *Participant:* As you got older, you began to resent some things.
> *Researcher*: Was that because you started to see how it was for other people?

Participant: Uh-huh, and see how it was for the white people. Like, we couldn't go to the University when we graduated. We had to go to Durham, to North Carolina College, at that time. But a lot of us couldn't go, because we had to catch the bus, and the bus station was on the other side of town, and you had to walk from the bus station to the college, which was about four miles. So, it was so many things that you saw that you realized was a block for you. And that's the way it was. Some of the kids, when we graduated, had relatives up north, and they were able to go up there and go to school. But here, see, it was nothing nearby. I don't reckon I should say it, because you're there, but I'm not in love with [University of] Carolina now because of that. [Laughs] After all these years, I'm not. Because I would have loved to have gone to college, but it wasn't there. (Interview X-0028, University of North Carolina at Chapel Hill's Southern Historical Collection)

The forces of institutional segregation, economic disenfranchisement, and transportation inequity coalesced to prevent this participant from pursuing higher education. She had already mentioned that walking and wagon travel were the norm for Black people in her community, and in this excerpt she describes the infeasibility of commuting to the closest segregated institution, North Carolina College. In the early 1950s, the nearby university forbade admittance of both women and African Americans, which rendered college inaccessible for the participant. She mentions not being "in love with Carolina" in the present day, which highlights the personal importance of the issue and her regard of a local geographic landmark. The participant's narrative locates instances of racial inequity and reveals their lingering impacts.

Recent spatial scholarship tackles how Black place-making in the American South manifests against the weight of plantation afterlives (Purifoy, 2014; Roane, 2018). These projects figure localized Black agency into critiques of relentless racial capitalism. For example, Williams, 2018 has used oral history to assess agrarian racism in the Mississippi Delta of the US. Narratives from pilots and activists reveal the recklessness of aerially applied herbicides at commercial cotton farms. The crop dusters knowingly exposed nearby Black communities to the toxic substance without fear of legal retribution. Oral history supports assessing such intimate encounters with historical racial inequities, and in this example could serve as both a spark for further investigation and a way to confirm what had been suspected using other research methods (e.g., archival records on herbicide use).

Concerning Contested Places

Oral history is a particularly valuable complement to rural geography research (Woods, 2010). Away from urban centres, people in rural regions and localities confront biased, contentious changes to their "home" space including deforestation, resource extraction, waste site planning, and more. Previously, geography researchers recorded oral histories with farmers to explore their perspectives of changing rural landscapes of the UK, US, and Aotearoa New Zealand (Egoz, 2000; Evered, 2011; Riley & Harvey, 2007). Indeed, farmers may offer compelling oral history recordings predicated on their economic and cultural interest in rural landscapes. In the excerpt below, an American dairy farmer vividly recalls a meeting between his agricultural community and the local government regarding the imminent construction of an ill-conceived water reservoir. Such contested places are often tied to memories of significant emotional experiences.

> Well there was a meeting, it was in 1976—it was during when the [baseball] World Series was going on—on Wednesday night, we had a meeting at the community building over there—I remember Cincinnati was playing somebody. And anyway, they started the meeting out, "Well, we sure are glad everybody is here tonight and glad you're here coming out, supporting us in the building of the lake." And, he kept saying how happy he was that everybody was there, you know, in favor of the lake and everything. Finally a guy just got up and said, "Now wait a minute. I think y'all are misunderstanding, mistaking; there is not anybody here in favor of the lake." And then it broke loose you know. They, the head had just introduced a surveyor, and one guy said, "Yeah, we want to get a real good look at him so we'll know who to shoot." I mean they were serious you know. I'll be honest with you, if this had happened back when my father was a kid, and my grandfather and several of the ones coming up, that lake would never have been built. (Interview K-0013, University of North Carolina at Chapel Hill's Southern Historical Collection)

This oral history comes from a series with farmers, politicians, and scientists about a particular place—the Cane Creek Reservoir in rural North Carolina. Even this excerpt alone highlights a number of geographic themes including a *local experience of a large-scale process*—the county government's top-down placement was opposed by the participants' community. Reflecting on the sway of his grandfather, the participant reveals a *place attachment* to the area near the reservoir. The *emotional* weight of the issue is reflected in his recounting of a meeting attendee vowing violence. The importance of the memory is further revealed by the participant recalling the day of the meeting—occurring the same day as a specific game of the 1976 baseball World Series. The meeting and subsequent reservoir's construction had an impact on the community the farmer's family had resided in for generations with longstanding residents being bought out to make space for the construction. Oral history is useful for eliciting accounts of particularly contentious sites such as the Cane Creek Reservoir.

There are myriad ways oral histories may be useful for geography research in addition to offering intimate accounts of contested places and historical inequities. The two participants above both felt personally connected to the topics being discussed, which brings out more detailed and deeply textured stories. The recordings evoked ardent memories and perspectives. At their best, oral histories are experience-based, emotional recordings that centre the concerns of participants.

Conclusion

Oral history recordings mobilize intimate accounts of individuals' lives and experiences. They accommodate not just the memories themselves but also the emotional registers that accompany them. Indeed, oral histories elicit descriptions of what people remember of their experiences as well as how such recollections are tied to fear, joy, and anxiety. Often, these emotions are matters of multigenerational triumph and defeat alike.

In addition to emotional ties, oral history narratives elicit multi-scalar perspectives. In a single recording, a participant may offer personal reflections on large-scale processes such as international policy change as easily as they may recount a family dispute. Narrated recollections may include firsthand perspectives of space production and obliteration as with the *Remembering Black Main Streets* collection. Indeed, the temporal and personal nature of oral history means the method can aid in assessing lived impacts of "bygone" inequities *and* ongoing racial capitalism.

While a process of intense evocation, the work of conducting geographic oral history research is accessible. It can be as simple as turning on a recorder and asking a few questions about someone's life experiences. However, an oral history is only an opportunity—a vehicle for rich, intimate stories. With this in mind, the value of the oral history depends on preparation and rapport. Further, it depends on the site of the recording and several other factors that are beyond the researcher's control such as the participant's mood. Maintaining some patience is as critical as following the stories. Follow-up responsibilities, including preservation and some form of analysis, help share the insights broadly and contribute to public scholarship.

Key Terms

background research
common questions
gatekeepers
guide
narratives
preservation
snowball method

Review Questions

1. How are oral histories useful for geographers despite concerns with the reliability of memory?
2. What aspects of any individual's personal life history are interesting for geographical research?
3. How can participants consent to their oral histories being used as part of a research project?

Review Exercises

1. Listen to an oral history or, if this is not available, read through the full transcript of an oral history and try to determine how the oral historian arranged the guide used to prompt conversation.
   a. How structured does the oral history seem? In what ways do you suspect the oral historian allowed the recording to move beyond the predetermined questions?
   b. How did the oral historian negotiate questions that yielded short responses?
2. Imagine you have been asked to create an oral history project about the local fire department/service.
   a. Make a list of potential gatekeepers you should contact to get started.
   b. Brainstorm some ways the project might be interesting to the gatekeepers and ways you could share the project's results with the public.
   c. Draft an email that you could send to the gatekeepers introducing yourself and the potential project. This should be based on any university template you might be required to complete to satisfy institutional ethics requirements.

Oral Histories Referenced (in order of appearance)

Interview R-0171 with Leroy Beavers Sr by Kieran Taylor, 7 August 2002, in the Southern Oral History Program Collection (#4007), Southern Historical Collection, Wilson Library, University of North Carolina at Chapel Hill.

Interview X-0028 with Hattie McCauley by Darius Scott, 31 October 2014, in the Southern Oral History Program Collection (#4007), Southern Historical Collection, Wilson Library, University of North Carolina at Chapel Hill.

Interview K-0013 with Bobby Kirk by Karl E. Campbell, 28 October 1985, in the Southern Oral History Program Collection (#4007), Southern Historical Collection, Wilson Library, University of North Carolina at Chapel Hill.

Useful Resources

Columbia University Oral History Master of Arts Program. http://oralhistory.columbia.edu/

Lorimer, H. (2003). Telling small stories: Spaces of knowledge and the practice of geography. *Transactions of the Institute of British Geographers, 28*(2), 197–217.

National Oral History Association of New Zealand (NOHANZ). www.oralhistory.org.nz/

Oral History Australia. www.oralhistoryaustralia.org.au/

Oral History Society. www.ohs.org.uk/

Sommer, B.W., & Quinlan, M.K. (2018). *The oral history manual* (3rd edition). Rowan & Littlefield.

Williams, B., & Riley, M. (2020). The challenge of oral history to environmental history. *Environment and History*, 26(2), 207–231, 26. https://doi.org/10.3197/096734018X15254461646503.

Woods, C. (2002). Life after death. *The Professional Geographer, 54*(1), 62–66.

Focusing on the Focus Group

Jenny Cameron

Chapter Overview

An investigation of options for the post-mining uses of industrial peatlands in Ireland (Collier & Scott, 2010), an exploration of emotions experienced by volunteers participating in conservation programs in South Africa (Cousins, Evans, & Sadler, 2009), and a study of the barriers to Latinos using parks in Los Angeles (Byrne, 2012)—all are examples of research projects that employ focus groups to disentangle the complex web of relations and processes, meaning and representation, that comprise the social world. With the shift to more nuanced explorations of people–place relationships in geography, the focus group method has been increasingly recognized as a valuable research tool.

Focus groups can be exhilarating and exciting, with people responding to the ideas and viewpoints expressed by others and introducing you, the researcher, and other group members to new ways of thinking about an issue or topic. This chapter discusses the diverse research potential of focus groups in geography, outlines key issues to consider when planning and conducting successful focus groups, and offers strategies for analyzing and presenting results.

What Are Focus Groups?

The focus group method involves a small group of people discussing a topic or issue defined by a researcher. Briefly, a group of between six and ten people sit facing each other around a table; the researcher introduces the topic for discussion and then invites and moderates discussion from group members. To get conversation started, participants might go around the circle, sharing an experience or viewpoint relevant to the discussion topic; interaction often proceeds naturally from that point, with occasional prompts or questions from the researcher/facilitator. A session usually lasts between one and two hours (you might see parallels here with university tutorial group meetings or seminars).

Figure 11.1 Relationship between Focus Groups, Group Interviews, and In-depth Groups

The focus group is one of the group techniques used in research. As shown in Figure 11.1, these techniques range from group interviews in which each participant is asked the same question in turn and there is little or no interaction between participants (Barbour, 2007) to in-depth groups in which the emphasis is on the interaction between participants, with participants sometimes even deciding on discussion topics (Kneale, 2001). In-depth groups also meet regularly for extended periods of time (sometimes months).

As with in-depth groups, interaction between participants is a key characteristic of focus groups. The group setting is generally characterized by dynamism and energy as people respond to the contributions of others (see Box 11.1). One comment can trigger a chain of responses. This type of interaction has been described as the **synergistic effect** of focus groups, and some propose that it results in far more information being generated than in other research methods (Berg, 1989; Stewart, Shamdasani, & Rook, 2007). In the focus group excerpt in Box 11.1, for example, three social enterprise practitioners discuss their motivations. It starts with Practitioner A discussing how they are motivated by the people that their social enterprise works with. Practitioner B then discusses the challenge of balancing the social goals with the business operations. Practitioner C brings the discussion back to the personal motivations and makes the point that they are "in it for me." There's then discussion about what this means with Practitioners A and C refining what being "in it for me" means in the context of a social enterprise.

The interactive aspect of focus groups also provides an opportunity for people to explore different points of view and to formulate and reconsider their own ideas and understandings. Kitzinger (1994, p. 113) describes this form of interaction in the following terms: "[p]articipants do not just agree with each other. They also misunderstand one another, question one another, try to persuade each other of the justice of their own point of view and sometimes they vehemently disagree." For researchers who are interested in the socially constructed nature of knowledge, this aspect of focus groups makes them an ideal research method; the multiple meanings that people attribute to places, relationships, processes, and events are expressed and negotiated, thereby providing important insights into the practice of knowledge production.

As in group interviews, the researcher plays a pivotal role. They promote group interaction and focus the discussion on the topic or issue. The researcher draws out the range of views and understandings held within the group and manages—sometimes even encourages—disagreement among participants (Myers, 1998).

BOX 11.1

The Synergistic Effect of Focus Groups

Practitioner A: When I first started [the social enterprise] we didn't have anyone who could look after the business side and because the other guy was even less good at it than me, I ended up falling into that role. I learnt. To me it's not about money it's about "Why are you doing it?". I'm not doing it for me. I'm doing it for them. To me that's the fundamental distinction between what I do and what a business man does. The business man is doing it ultimately for themselves. I'm doing it for the people that I'm working with. That's my motivation.

Facilitator: What you're saying is that you're clear about what your motivation is. You are adapting yourself to the context in order to make sure you can act on that motivation.

Practitioner A: The money doesn't scare me.

Practitioner B: Social enterprise should have a social mission as its ultimate purpose for existing. Sometimes the only way you can make that mission happen is by having the business feed money into it. But the more money you make the less social mission. I can remember someone saying we could make more money but we would probably have to replace some of the students with more full-time chefs. We could make more, but then we can put less students through. Then our social mission drops and our profits go up.

Practitioner A: Why are you doing it? Why are you doing it?

Practitioner B: So it's that very fine balance.

Practitioner C: Yeah, but just going back to what you said. I'm actually in it for me too. I have no bones about that.

Facilitator: Because of the challenge?

Practitioner C: Absolutely. I don't want to sound like I think I'm good. You've really have got to have your heart in it and you really have to love it. You have to be really selfish like that. I'm in it for me.

Practitioner A: Yeah, yeah, in it for you—in terms of your own motivations. You're not in it for you to see how many bucks you can make for yourself so you can retire at whatever age or buy the big house or whatever.

Practitioner C: That's right.

Source: Audio excerpt from focus group conducted by Jenny Cameron and Sherelle Hendriks, Newcastle, Australia, 23 July 2012 (see Cameron & Hendriks, 2014).

Initially, focus groups can be extremely challenging for researchers who are new to the process. They are, however, well worth it. In focus groups, the diversity of processes and practices that make up the social world and the richness of the relationships between people and places can be addressed and explored explicitly. Furthermore, group members almost invariably enjoy interacting with each other, offering their points of view, and learning from each other. Researchers also find the process refreshing (for example, see the discussion by two sceptical anthropologists in Agar and MacDonald [1995]).

Using Focus Groups in Geography

Focus groups were originally used by sociologists in the US during World War II to examine the impact of wartime propaganda and the effectiveness of military training materials (Merton, 1987; Morgan, 1997). Although this work resulted in several sociological publications on the technique, focus groups were neglected by social scientists in the post-World War II period in favour of one-to-one **interviews** and **participant observation** (Johnson, 1996). It was in the field of market research that the focus group method found a home. Since the 1980s, there has been renewed interest in the technique among social scientists, and this has led to considerable diversity in focus group research, including in geographic research.

One way that focus groups are used in geography is in an *exploratory* mode. This usually involves gathering information on a topic or collecting insights into an issue. The discussion in Box 11.1, above, is from exploratory research investigating the motivations of social enterprise practitioners. Another example is research by Zeigler, Brunn, and Johnson (1996) which used focus groups to explore people's responses to a major hurricane during the pre-impact and post-impact phases. The researchers claim that they garnered insights that would not have been revealed through methods such as **questionnaires** or individual interviews. One finding was that after the hurricane people were keen to begin recovery efforts almost immediately, and that rather than having emergency workers lead the recovery efforts they wanted assistance from emergency workers. This resulted in a series of recommendations for initiatives that could help to support people's strongly-expressed desire for self-reliance in the aftermath.

In a different setting, Cousins, Evans, and Sadler (2009) explored the emotional responses of conservation volunteers from the UK as they participated in wildlife projects in South Africa (such as lion and rhino monitoring projects). They found that that the volunteers journeyed through emotional highs and lows, from awe and exhilaration to anguish and disappointment, as they were confronted with the practicalities of working with wildlife. As a result of the research, the authors recommended changes in the preparation of volunteers so that their expectations were better matched to the realities they would face.

A second way that focus groups are used in geography is in an *explanatory* mode. This involves researchers trying to understand why people think and act in certain ways. Barbour (2007, p. 24) notes that focus groups are particularly well-suited to studies that seek explanations. In focus groups, "participants are given scope to justify and expand on their views in a non-judgemental environment," giving researchers a chance to learn how people rationalize what might seem illogical to researchers. For example, Barr et al. (2010) used focus groups to explain why environmentally-aware behaviour at home may not be replicated while "away." They explored the beliefs of people who were environmentally-conscious at home and were aware of the contribution of air travel to greenhouse gas emissions but were not prepared to reduce holiday air travel (though some would consider paying for carbon offsets). Participants explained and justified their use of air travel for holidays for a range of reasons that included "trading-off" sustainable practices when at home for less sustainable practices when on holidays, claiming that people living in other parts of the world were more of a problem than they were, and even expressing skepticism about the contribution of air travel to climate change. The research highlighted just how difficult it is to change air travel behaviour, especially as low-cost airfares have become so readily available.

A third way that focus groups are used in geography is in a *generative* mode. Here there is an emphasis on how people's views or insights may not pre-exist the focus group. Instead of accessing what is already known to participants, the focus group becomes a place for producing new knowledge. For example, Gibson, Cameron, and Veno (1999) were concerned not just with reproducing knowledge of the problems and difficulties confronting rural and non-metropolitan communities in Australia but with reshaping understandings so that new responses might be engendered. The seemingly isolated instances of innovation that several focus group members could readily recall provoked other participants to think of additional examples and come up with new ideas. The beginnings of a body of knowledge on regional initiative began to emerge through these discussions.

In their research on individual and household strategies for the allocation of land, labour, and capital in Indonesia, Goss and Leinbach (1996) also highlight how focus groups can contribute to new **knowledge production**. By interacting with other focus group participants, Javanese villagers developed new understandings of their social conditions. Indeed, Goss and Leinbach argue that "the main advantage of focus group discussions is that both the researcher and the research subjects may simultaneously obtain insights and understanding of particular social situations *during* the process of research" (Goss & Leinbach, 1996, pp. 116–17, emphasis in original). Similarly, Carolan (2017) used focus groups to explore strategies for shifting consumers' feelings not just about what they eat but the system of production that provides the food that seems to just appear in the supermarket. Carolan's study was conducted in a context in which claims were being put forward about immigrant farm labourers supposedly stealing jobs from citizens and residents. Through their involvement in the study, participants were in their words "moved" (p. 68) and they developed new insights into how food was produced (for more on this study see the discussion below). For geographers who are committed to the idea that research can be used to effect social change the potential for focus groups to create and transform knowledges and understandings of researchers and participants is compelling (see also Johnson, 1996).

Another dimension to how focus groups are used in geographic research is whether they are used as a stand-alone method or in combination with other methods. This dimension applies to focus groups across the exploratory, explanatory, and generative modes. For example, focus groups were the only method used by Zeigler, Brunn, and Johnson (1996) in their exploratory research on disaster responses; whereas the focus groups exploring the emotional responses of conservation volunteers by Cousins, Evans, and Sadler (2009) formed part of a larger study that also used participant observation. In this study, the researchers started with participant observation and waited until the third week to hold the focus groups to give the participants time to be exposed to the volunteering experience and to feel comfortable with each other so they would share their views. In this example, focus groups were used with another qualitative method, but focus groups can be used with quantitative methods. For example, focus groups have been used to generate questions to be tested in surveys (Goss & Leinbach, 1996), to refine the design of survey questionnaires (Jackson & Holbrook, 1995), and to follow-up the interpretation of survey findings (Goss & Leinbach, 1996), particularly when there are contradictory or unexpected results (Morgan, 1996).

The focus group method has made important contributions to geographic research. It is a highly effective method that can be used alone or in combination with other methods to investigate the nuances and complexities associated with people–place relationships. Focus groups can access what people already know to provide insights that describe, document, and explain the social world. Focus groups also provide opportunities for researchers and participants to jointly develop new knowledges and understandings (see also Chapters 4 and 16).

Planning and Conducting Focus Groups

Given that the focus group method can be used for a range of research purposes in geography, there will be some variation in how groups are organized and conducted. There are, however, basic principles and methodological issues that need to be considered. To be sure, the success of a focus group depends largely on the care taken in the initial planning stage.

Selecting Participants

Selecting participants is critically important. Generally, participants are chosen on the basis of their experience related to the research topic. Burgess's (1996) study is a good example of this **purposive sampling** technique (see Chapter 6 for a discussion of participant selection). In work intended to ascertain the perceptions of crime and risk in woodlands among different social and cultural groups, she selected women and men of varying age, stage in the lifecycle, and ethnicity to participate in focus groups. In their study of home-based and travel-based environmental behaviour, Barr et al. (2010) recruited participants who displayed what could be called strong to medium to weak environmental behaviours at home in order to investigate each group's attitude to air travel.

Composition of Focus Groups

Should people with similar characteristics participate in the same group, or should groups comprise members with different characteristics? This decision will be largely determined by the purpose of your research project.

Holbrook and Jackson (1996), for example, sought to address issues of identity, community, and locality by grouping together people with characteristics like age and ethnicity in common. In their research on environmental responsibility, Bedford and Burgess (2002) had people with similar experiences in each focus group but a range of different focus groups—suppliers, retailers, regulators, consumers, and advocates. They describe this as "ensur[ing] homogeneity within the group and heterogeneity between them" (2002, p. 124). Other researchers have noted that discussion of sensitive or controversial topics can be enhanced when groups comprise participants who share key characteristics (Hoppe et al., 1995; O'Brien, 1993). In some projects, it may be more appropriate to have groups made up of different types of people. Goss and Leinbach (1996) were interested in the social relations involved in family decision-making and deliberately chose to conduct mixed-gender groups. The different knowledges, experiences, and perspectives expressed by women and men became an important point of discussion. By contrast, in the context of forest fringe communities in Ghana, Teye (2012) conducted separate focus groups with women and men as the patriarchal nature of these traditional communities meant that women would be unlikely to discuss and debate issues with men.

Another consideration is whether people already known to each other should participate in the same group. Generally, it is best not to have people who are acquainted in the same group, but in some research, particularly place-based research, it may be unavoidable. Researchers need to be aware of the limitations that this can produce. One is peer pressure, with participants not wanting to appear out of step with their acquaintances. Similarly, some participants may **under-disclose** or selectively disclose details of their lives, as Pratt (2002) found in her research.

A different problem arises when participants **over-disclose** information about themselves. One strategy for dealing with this is to outline fictional examples and ask group members to speculate on them. In groups they ran in Indonesia, Goss and Leinbach (1996) provided details of three fictional families and asked group members to discuss which of the families would be most likely to accumulate capital. Participants did not have to disclose information about their own situations but could still discuss family strategies. Participants are often asked to treat discussions as **confidential**. Since this confidentiality cannot be guaranteed, it is appropriate to remind people to disclose only the things they would feel comfortable about being repeated outside the group.

Of course, you should always weigh whether a topic is too controversial or sensitive for discussion in a focus group and would be better handled through another technique such as individual in-depth interviews. (Universities have ethics committees, institutional review boards, or research ethics boards to ensure that researchers carefully manage material from focus groups and other qualitative research methods. For more on this, see Chapter 2.)

Size and Number of Groups

The size of each group and the number of groups are other factors to be considered. Too few participants per group—fewer than four—limits the discussion, while too many—more than 10—restricts the time available for individual participants to contribute.

In terms of the number of groups, one rule of thumb is to hold three to five groups, but this guideline will be mediated by factors such as the purpose and scale of the research and the heterogeneity of the participants. A diverse range of participants is likely to require a larger number of groups. For instance, Burgess (1996) was interested in perceptions of crime and risk in woodlands among different social and cultural groups so conducted 13 focus groups with people of varying age, stage in the lifecycle, and ethnicity, whereas in Byrne's (2012) study of why Latinos do not use parks in Los Angeles only two focus groups were conducted with Latinos who were non-park-users.

The structure of the focus group is also a factor to consider. When fewer **standardized questions** are used and when there is a relatively low level of researcher intervention and moderation, more groups are needed, since both these factors tend to produce greater variability among groups (Morgan, 1997). Time, cost, and availability of participants may also limit the number of groups that can be held. The overall research plan—especially whether focus groups are the sole research tool or one of a number of tools—will also affect decisions about the number of groups convened. Finally, another guide to the number of focus groups is to use the concept of **saturation** (Krueger, 1998, p. 72). This means that you continue to conduct focus groups until you can gather no new information or insights.

Recruiting Participants

The strategy used to recruit participants will depend on the type of participants you require for your study. Gibson, Cameron, and Veno (1999) recruited business and community leaders in two regions by initially contacting local people who featured in local newspapers and targeting managers of key government and non-government agencies. These initial contacts were asked to suggest other people who would make interesting contributions to the study, including those who would have a different point of view from them (this **snowball** recruitment technique is also discussed in Chapter 6). A preliminary phone conversation quickly established whether nominees were interested and able to attend. This was followed by a letter with more information about the project. A few days before the focus groups were held, participants were telephoned again to reconfirm their participation. Twelve people were invited to attend each group to allow for cancellations due to illness, last-minute change of plans, and so on (several people from each group did drop out).

After an unsuccessful attempt to recruit participants through strategies such as web-based postings, Byrne (2012) worked directly with two community-based organizations that recruited participants on his behalf. When Eden, Bear, and Walker (2008) wanted a more diverse range of food consumers, they made contact with a whole-food retailer and a vegan group, among others. Like these researchers it is important to think strategically about how best to locate potential participants.

Questions and Topics

Before conducting focus groups, give thought to the questions or topics for discussion. This involves not only the general content of questions or selection of issues for discussion but also the wording of questions and issues, identification of key phrases that might be useful, the sequencing and grouping of questions (see Chapter 13 for additional material on question order), strategies for introducing issues, and the links that it might be important to make between different questions or issues. Also give thought to using other sorts of stimuli to generate discussion. Barbour (2007, pp. 84–8), for example, discusses the use of cartoons, snippets from television shows, photographs, newspaper clippings, advertising materials, and vignettes. However, it is important to select such material carefully and, where possible, to check that it will not offend participants. Other strategies that combine focus groups with other stimuli include, first, **participatory mapping**, in which participants annotate a large paper map and use it to generate stories and various perspectives; and, second, **arts-based approaches** in which materials are provided (paper, markers, glue, photos, etc.) to engage participants in tangible activities, often simultaneously with dialogue. While some arts-based methods have been more typical in work with children and youth, they are easily adapted to adult participants (Carter & Ford, 2013).

One way to proceed is to devise a list of questions. Hares, Dickinson, and Wilkes (2010) had a list of six open-ended questions that acted as **probes** and that moved from the general to the specific. Holbrook and Jackson (1996) identified six themes related to the experience of shopping and then used these themes to develop questions that were raised spontaneously and that fitted with the flow of the discussion. Burgess (1996) first took the participants in each focus group on a walk through a woodland and then introduced for discussion five primary themes related to elements of the walk. As part of their recruitment process, Cameron and Hendriks (2014) asked each participant to prepare a brief two to three-minute statement about their background and their current role working in a social enterprise. The themes that emerged from these statements provided the initial basis for discussion.

Take care when letting people know in advance what the questions or topics will be. If attendance or discussion is likely to be enhanced by providing this information, then it may be appropriate. Sometimes, however, it might be necessary for you to paint a very broad picture. For example, even though Hares, Dickenson, and Wilkes's (2010) research was on people's awareness of the impacts of tourism on climate change, they told potential participants that the research was about climate change and everyday lives. They deliberately avoided mentioning tourism in the recruitment process as they did not want to create a connection in participants' minds between tourism and climate change. (See Chapter 2 for a consideration of the ethical dimensions of this sort of approach.) When recruiting and explaining the research it is also important to use language that participants will understand.

Generally, questions or topics should be suited for a discussion of between one and two hours. With very talkative groups, it may be necessary to intervene and move the discussion on to new topics. Alternatively, if you have planned a hierarchy of questions or themes, then it may be appropriate to allow the group to focus on the more important areas of discussion. With less talkative groups, you may need to introduce additional or rephrased questions and **prompts** to help draw information out and open up the discussion. You should think about such questions and prompts during the preparation stage.

Another issue to consider is whether questions and topics should be standardized across all focus groups involved in your study or whether new insights from one group should be introduced into the discussions of the next. In many qualitative research situations, it may be appropriate to incorporate material from earlier groups, but this issue should be determined with reference to the project aim. Information that might identify people who attended earlier groups should not be revealed to subsequent groups.

As well as conducting meetings with several groups, you may find it useful and appropriate to have each group meet more than once. Multiple focus groups may be a particularly useful strategy when participants are being asked to explore new and unfamiliar topics or to think about an apparently familiar topic in a new way. Multiple groups may also be appropriate as a way of developing trust between the researcher and research participants. For instance, when researching the experiences of single mothers, I met several times with one group of teenagers who were very wary of talking with people associated with educational, medical, and media institutions (Cameron, 1992).

When focus groups are used in combination with other methods it is important to pinpoint the purpose of each method and the relationship between the various methods, and to ensure this is reflected in the questions and topics. In his multi-method "strawberry study," Carolan (2017, pp. 66–9) sought to test different learning approaches that might shift consumers' feeling about the food they eat and the system of food production. He started with individual interviews to find out what consumers knew about how strawberries were produced. This was followed by a focus group during which participants ate their way through several kilograms of strawberries while watching a documentary about the industrial production of strawberries in California and discussing what they had seen. A week later, the same participants spent seven hours picking strawberries and photographing (with their mobile phones) anything that seemed important to capture. This was followed several days later with individual debriefings during which each participant discussed with the researcher the photographs they had taken, their experiences picking strawberries, and how the project had shaped their feelings about strawberry and food production more generally. Because of his carefully designed study, Carolan was able to conclude that *experiential* approaches (especially the experience of a seven-hour day on hands and knees picking strawberries), rather than the *representational* mode of listening to and talking about a topic, are crucial to shifting attitudes and feelings and thereby producing a different type of knowledge about, in this case, how the food system operates.

Conducting Focus Groups

Generally, focus groups are best held in an informal setting that is easily accessible to all participants. The rooms of local community centres, libraries, churches, schools, and so on are usually ideal. The setting should also be relatively neutral: for example, it would not be advisable to convene a focus group about the quality of service provided by an agency in that agency's offices. Food and drink can be offered to participants when they arrive to help them relax, but alcohol should never be provided. It is also helpful to give out nametags as participants arrive.

Much has been written about the ideal focus group **facilitator** or **moderator** (for example, Morgan, 1997; Stewart, Shamdasani, & Rook, 2007). In academic research, the researcher, who is familiar with the research, is often best positioned to fill this role. To gain some confidence and familiarity with the process, a less experienced researcher might initially take the role of note-taker while a more experienced researcher facilitates the first groups. Focus groups can also be run with more than one facilitator, and a less experienced researcher might invite a more experienced researcher to take the lead. But beginning researchers should not be afraid to try facilitation; as Bedford and Burgess (2002, p. 129) note, "the desired qualities are those possessed by the average undergraduate—the ability to listen, the ability to think on your feet, and a knowledge of and interest in the subject the group is discussing."

When a note-taker is present, that person should sit discreetly to one side. The notes, particularly a list of who speaks in what order and a brief description of what each talks about, as well as non-verbal cues such as smiles, nods, and head shakes, can be helpful when transcribing audio recordings of the discussion. A seating plan is also essential. Because the facilitator has to attend to what participants are saying and monitor the mood of the group, they should not take extensive notes, although the facilitator may want to jot down a point or two to return to in discussion.

It is highly advisable to audio-record focus groups. The group will usually cover so much material that it will be impossible to recall everything that was discussed. In addition, because presentation of focus group results generally includes direct quotes to illustrate key points, a transcribable audio recording can be very helpful. The quality of the recorder and microphone is crucial. (See Chapters 9 and 10 for a fuller discussion.) Most recorders come with a built-in microphone, but several flat desk microphones placed around the table will ensure much better sound quality and that quieter voices are recorded. Several high-quality mobile phones can also capture the audio. Ensure that you test the equipment long before the focus group as well as later in the room before group members arrive. Make sure that any devices you need are fully charged (or that you have spare batteries). Take care that the setting for the group meeting is quiet enough for discussion to be recorded clearly.

The facilitator usually initiates discussion by giving an overview of the research and the role of the focus group in the project. The themes or questions for discussion can then be introduced. Since group members may be unfamiliar with the focus group technique, a brief summary of how focus groups operate should also be given. Box 11.2 provides an example of a focus group introduction.

The facilitator moderates discussion by encouraging exploration of a topic, introducing new topics, keeping discussion on track, encouraging agreement and disagreement, diplomatically curbing talkative group members, and encouraging quiet participants. Examples of the sorts of phrases used by facilitators are outlined in Box 11.3.

Some aspects of facilitation require special comment. Expressing and exploring different points of view is important in focus groups, yet research shows that groups have a preference for agreement (Myers, 1998). The facilitator plays a central role in creating the context for disagreement. This can be done by stating in the introduction that there is no correct answer and that disagreement is normal and expected, by asking directly for different points of view, and by making explicit implied disagreement and introducing it as a topic for discussion (Myers, 1998, p. 97). Watch for non-verbal signs of disagreement such as folded arms,

- Keeping on track:
  "There was an important point made over here a moment ago, can we just come back to that?"
- Inviting agreement:
  "Has anyone else had a similar experience?"
  "Does anyone else share that view?"
- Inviting disagreement:
  "Does anyone have a different reaction?"
  "We've been hearing about one point of view, but I think there might be other ways of looking at this. Would anyone like to comment on other sorts of views that they think other people might have?"
  "There seem to be some differences in what's been said, and I think it's really important to get a sense of why we have such different views."
- Clarifying:
  "Can you give me an example of what you mean?"
  "Can you say this again, but use different words?"
  "Earlier you said that you thought . . . now you're saying . . . can you tell us more about what you think/feel about this topic/issue?"
- Curbing a talkative person:
  "There's a few people who've got something to add at this point. We'll just move onto them."
  "We need to move onto the next topic. We'll come back to that idea if we have time."
- Encouraging a very quiet person:
  "Do you have anything you'd like to add at this point?"

Source: Drawn from discussions in Carey (1994), Krueger (1998), and Myers (1998) and from personal experience.

movement away from the table, and a shaking or downcast head. You might ask the whole group or target the disagreeing member to give a different point of view. Of course, as facilitator, never state that someone is wrong or display a preference for one position. In the unlikely event that the discussion becomes heated, then intervene immediately, suggest that there is no right answer, and move the group on to the next question.

Very talkative or very quiet participants can be a challenge. Talkative people need to be gently curbed, while quiet ones need to be encouraged to participate. Along with the sorts of phrases listed in Box 11.3, your non-verbal signals can be useful. Pointing to someone who is waiting to speak indicates to the talkative person that there are others who need to have a turn. Making frequent eye contact with the quieter person and offering encouraging signs, like nodding and smiling when they do speak, is important. Remember, though, that silence gives people time to reflect and gather their thoughts. Do not feel that you have to fill silences; give people time to respond.

At the conclusion of a focus group, you might review key points of the discussion, providing a sense of completion and allowing participants to clarify and correct your summary. Group members should always be thanked for taking the time to attend and for their contributions. You can do this again with a personal letter to each participant.

Online Focus Groups

With developments in digital technology, some researchers find **online focus groups** useful, particularly for bringing together people who are extremely busy or geographically dispersed (even in different parts of the world). Moore, McKee, and McLoughlin (2015) provide an excellent discussion of using online focus groups to research young people's housing opportunities in the UK.

Analyzing and Presenting Results

Krueger (1998, p. 46) importantly reminds us that "analysis begins during the first focus group." Listen carefully to responses, and clarify any unclear or contradictory contributions, since this information may be critical later when you present the results. For example, if young people say that they would watch television news and current affairs if the coverage were more relevant to them, it is probably important to get them to explain or give examples of how news items could be made more relevant (see also Box 11.4).

Since there is always a richness of material, analyzing focus group discussions can be as time-consuming as it is interesting. The first step involves transcribing the audio recording. A complete **transcript** of the entire discussion takes a great deal of time, since one hour of recording usually requires more than four hours to transcribe. When a detailed comparison of groups is to be undertaken, full transcripts may be necessary. Generally, a partial or abridged transcription (which involves transcribing only key sections of the discussion) will suffice. This is best done as soon as possible after the focus group, with the facilitator(s) and note-taker working in collaboration to decide which sections should be transcribed. A record of the seating plan and running order of speakers and a brief description of what was said are extremely helpful at this point. If you as the researcher have the time, it is also advantageous for you to transcribe the audio recording, because it will give you an opportunity to become more fully immersed in the content (and if you are new to focus group research, you will be able to reflect on your facilitation style and identify your strengths and weaknesses). (For a full discussion on transcribing interviews, as well as a notation chart, see Chapter 9.)

It is advisable to transcribe and undertake a preliminary analysis of the first focus group before conducting any others. This is a way of checking that your questions are understood by participants and are eliciting the type of information you need for your research. It is also a way of checking that you understand and can interpret the responses of participants. For example, in an initial focus group, you might not think to ask young people to clarify what they mean by "relevant news coverage," but by carefully reading the transcript, you are likely to pick up this omission.

Once you have the complete set of focus group transcripts available, read the material over several times to help make yourself very familiar with the discussion. One relatively straightforward strategy for proceeding draws from the questions or **themes** on which the discussion focused. Write each question or theme on the top of a separate sheet of paper (or electronic document), and then list on each sheet the relevant points made (see Chapter 18 on coding and analyzing strategies). Finally, make a note of key quotes that might be used in written material (Bertrand, Brown, & Ward, 1992). This is an approach that works well when the discussion did not deviate widely from the questions or themes set by the researcher or when comparisons are to be made between focus groups (Bertrand, Brown, & Ward, 1992). For example, in a research project comparing the land management strategies for dealing with salinity preferred by farmers, policy-makers, and researchers, the sheets with the responses of the different groups to each question or theme can be easily compared.

When the purpose of the research project is to identify key themes or processes associated with a particular issue or topic, it may be more appropriate to use margin **coding** (Bertrand, Brown, & Ward, 1992). To do this, read through the transcripts, identify key themes or categories, and devise a simple colour-, number-, letter-, or symbol-based coding system to represent the themes or categories. You should then reread the transcripts and highlight the words, sentences, and paragraphs related to each category or theme by writing the appropriate code in the margin. Once transcripts have been coded, a cut-and-paste technique—completed either on a computer or manually—can be used to group the discussion related to each theme or process (see Chapters 9 and 18 for more information on this). Always keep an original of the transcripts for future reference. A variation of this thematic analysis is to develop a list of key words and, in a word-processing package, type two or more key words beside each comment. Using the search function, you can then locate related points of discussion. Computer programs specifically designed for qualitative analysis, like **NVivo**, can also be used and are particularly helpful when you have a large amount of transcribed material to analyze (see Chapter 18 for a discussion of this).

The ability to find material quickly is an important consideration, since analysis and writing rarely proceed in a linear fashion. During the writing process, new insights unfold (see Chapter 19), and frequently you may find it necessary to return to the original transcripts to refine and reformulate ideas. Sometimes it will be necessary to listen to and make additional transcriptions of sections of the recordings.

When reporting on focus group research, present your results only in terms of the discussion within the groups. Focus groups do not produce findings that are **generalizable** to a wider population, but are important for revealing explanatory, experiential, and process-based insights that can be tied to specific contexts. Focus group results are also expressed in impressionistic rather than numerical terms. Instead of precise numbers or percentages, the general trends or strength of feeling about an issue are typically given. As Ward, Bertrand, and Brown (1991, p. 271) have noted, focus group reports are "replete with statements such as 'many participants mentioned . . .', 'two distinct positions were observed among the participants', and 'almost no one had ever . . .'." Reporting on their study into people's responses to emergency procedures, Zeigler, Brunn, and Johnson (1996), for example, noted that the people in their focus groups generally responded with either compliant behaviour or under-reaction. Zeigler, Brunn, and Johnson then used direct quotes to illustrate the different ways that the responses were expressed (see Box 11.4 for an example of a focus group analysis, and see also the ways that Byrne [2012], Barr et al. [2010], and Cousins, Evans, and Sadler [2009] discuss focus group findings).

BOX 11.4 Example of How Focus Group Results Can Be Written Up

The following is an extract from a journal article reporting on findings from Australian focus group research on the impact of television news and current affairs on young people's political participation and active citizenship. The analysis of the focus group discussion starts by highlighting the way that young people find the reporting of news and current affairs too complex, particularly because of the sophisticated language and the absence of background information. The analysis continues:

> Political current affairs television was viewed by respondents as too complex to incorporate into their everyday viewing habits, but young people also feel it is not worth investing time in television current affairs because any political information received from the programs is usually trivialized and played for entertainment value. For example,

(continued)

> *A Current Affair* [a news magazine show] was described by Debra, a nineteen-year-old university student, as "Hey Hey It's Ray" after the celebrity of its host Ray Martin, and was seen by her focus group as a form of populist emotional exploitation. As the following responses suggest, there was a strong feeling amongst the groups that television current affairs portrays politicians as being "full of it."
>
> Bianka: The whole politics thing. They're all liars; they're all full of it.
>
> Craig: All the media carry on with is stuff like when they asked Hewson [a former politician] if [the] GST [Goods and Services Tax] would be applied to a birthday cake and they just blew that up. Who gave a shit?

The respondents felt news and current affairs did not help them develop a political identity. They also expressed distrust in politicians who attempt to "persuade" them to choose a lesser evil. As Bianka points out: "They all change their minds when they get what they want. I mean what's the point?" What eventuates is a distrust of not only politicians but also the media that is supposed to decipher the positive and negative elements of each candidate's actions.

Note how the findings are reported only in terms of the focus groups and in tentative terms with the use of phrases such as "responses suggest," "respondents felt," and "[w]hat eventuates." Main themes that emerged from the focus groups are summarized by the authors, and quotes from participants are used to illustrate and elaborate these themes.

Source: Evans and Sternberg, 1999, p. 105.

In some projects, it might not be the general trends but the ambiguous or contradictory remarks that the researcher particularly wants to explore. The development and presentation of an argument may refer not only to what was talked about but the way it was talked about in the group setting. This became a significant aspect of the focus group research conducted by Gibson, Cameron, and Veno (1999, p. 29):

> The stories of success and hope that emerged when the discussion was shifted onto the terrain of community strengths and innovations were numerous. They came stumbling out in a disorganised manner suggesting that these stories were not readily nor often told. In the face of dominant narratives of economic change perhaps such stories are positioned

> as less important or effective. It is clear that there is a lack of a language to talk about this understanding of community capacity; yet, as we will argue, this understanding has the potential to contribute to the ability of a region to deal effectively and innovatively with the consequences of social and economic change.

In research on peat mining in Ireland, Collier and Scott (2010) also comment on the significance of shifts in tenor of the discussion. They found that when presented with photo-montage images of mined and restored landscapes "there was a palpable change in participant attitude" and that the focus groups became so animated the researchers had to ask participants to slow down so that individual comments could be heard (p. 310). Collier and Scott note that at this point the focus groups became a setting for social learning with participants being much more willing to put forward and discuss different ideas, and to ask the researchers for more information that might help them explore options (p. 311).

One important element in the process of writing up (or "writing-in," as Mansvelt and Berg call it in Chapter 19) is to find a balance between direct quotes and your summary and interpretation of the discussion. When too many quotes are included, the material can seem repetitive or chaotic. Too few quotes, on the other hand, can mean that the vitality of the interaction between participants is lost to the reader. Morgan (1997, p. 64) recommends that the researcher should aim to connect the reader and the original participants through "well-chosen" quotations (see Chapter 20).

Conclusion

Focus groups demand careful preparation on the part of the researcher. The selection and recruitment of participants, the composition, size, and number of groups, and the questions and topics to be explored are all key points to consider during the planning stage. Even the apparently mundane details of appropriateness of venue, provision of refreshments, and quality of audio equipment are critical to the success of focus groups. A well-prepared researcher also gives thought beforehand to the process of facilitation, including the points to cover in the introduction, the wording of key questions, topics, and phrases, the probes and prompts that might be useful in further exploring a theme or topic, and strategies for drawing out different points of view, keeping the discussion on track, and dealing with both the more talkative and quieter members of the group. As soon as possible after the focus group, start the process of analysis, beginning with transcribing the audio recordings, followed by reading and rereading the transcripts, summarizing main points, and identifying central themes.

Although they require careful planning beforehand and a great deal of reflection afterwards, focus groups are an exciting and invaluable research tool for geographers to use. Participants almost invariably enjoy interacting with each other, and the discussion can generate insights and understandings that are new to both participants and researchers. The interactive element makes focus groups ideally suited to exploring the nuances and complexities of people–place relationships, whether the research has a primarily data-gathering function or is more concerned with the collective practice of knowledge production.

Key Terms

coding
facilitator
moderator
online focus groups
over-disclosure
probe
purposive sampling
recruitment
saturation
standardized questions
synergistic effect
transcript
under-disclosure

Review Questions

1. What is meant by the "synergistic effect" of focus groups and why might this effect be important for geographers interested in qualitative research?
2. Focus groups are useful for both collecting data and generating knowledge. Brainstorm some examples of focus group research in which collecting data might be the main aim and focus group research in which generating knowledge might be the main aim.
3. Why and how might focus groups be used in combination with quantitative research methods?
4. What are some key issues that researchers need to consider when planning focus group research, particularly in terms of the composition of focus groups; the size and number of groups; and the organization of questions and topics?
5. What are some strategies that researchers can use to ensure that focus groups run as smoothly as possible?
6. What are some characteristics of how focus group findings are reported?

Review Exercises

1. Find a research project from a recent issue of a geographical journal that you think could have been conducted using focus groups. Why do you think focus groups would be appropriate? Discuss the participants you

would select, the composition of the focus groups, the size and number of groups you would use, the questions you would ask or themes you would use, and strategies for recruiting participants.

2. A local university is redesigning the "For Current Students" section of its webpage. Your research company has been commissioned by the university to conduct a focus group study on what students think about the current design and what they think is important in the redesign. Devise a series of questions that you would use to canvass students' views. Select two students to co-facilitate the focus group and select seven to eight students to participate in the focus group. Other students observe, noting how the content of the discussion flows, and the verbal and non-verbal interactions between participants and the facilitators. You could do this exercise around other topics. For example, building on Evans and Sternberg's (1999) research, discussed above, into the impact of television news and current affairs on young people's political participation and active citizenship you could explore the media young people use to find out about news and current affairs and how this impacts their view of politics.

Useful Resources

Barbour, R. (2007). *Doing focus groups*. Los Angeles: SAGE.

Bryman, A. (2012). *Social research methods* (4th edition). Oxford: Oxford University Press.

Cry, J. (2019). *Focus groups for the social science researcher*. Cambridge: Cambridge University Press.

Krueger, R.A., & Casey, M.A. (2014). *Focus groups: A practical guide for applied research* (5th Edition). Los Angeles: SAGE.

From Dusty to Digital: Archival Research

Michael Roche

Chapter Overview

Earlier iterations of this chapter began with a muted plea for human geographers to realize the value of having a historical strand to their research and to recognize that the examination of archival sources fell within the realm of qualitative methods. Historical geographers for whom archival research is virtually inescapable had tended to treat the task as an in-house craft, although more recent widespread use of digital devices for copying in the archive and digitization projects providing online access serendipitously has brought archival research closer to mainstream human geography. In a significant reworking of the field geographers have provided archival research as craft or technique with a much richer critical theoretical foundation as well as, under the heading of "animating the archive" (Dwyer & Davies, 2010), have offered new possibilities for the outputs of archival inquiry over and above the generation of text. To some extent the surer theoretical footing for archival research in the literature has tended to crowd out "how to do it" advice. Consequently, the central core of this chapter still contains advice for novices about conducting archival research.

Introduction

After nearly 40 years of working as a historical geographer, I still relish the opportunity to undertake archival research (and I am not alone; see Lovell, 2001). There is a continuing sense of delving into the unknown, of engaging in academic detective work trying to understand inevitably fragmentary and partial surviving **records**, of striving to make sense of the sometimes highly partisan evidence you are scrutinizing. My own introduction to **archival research** was orchestrated only at the graduate level. To a large degree, I was able to learn by trial and error, following the tendency of historical geographers at the time to regard archival research skills as something to be acquired on the job rather than by reading about how to do it. Good archival scholarship was to be inferred from reading its products in the form of journal articles or books by leading historical geographers and from discussions with supervisors. In many ways, this was a laudable model, one that allowed me to develop my skills and understanding at my own pace, but geography students of today wishing to use archival sources can benefit from a more overt discussion about the nature of archives in a theoretical sense as well as the fundamentals of archival research. This is further the case because archival researchers still need to engage with wider disciplinary theory and the research ethics that are also a part of historical inquiry. Even so, like other methods, archival skills can to some extent only be learned by doing archival research. Expertise improves with experience, and this is not easily reduced to a checklist of best practice.

What Is Archival Research?

> Archival scholarship at its best, it seems to me, is an ongoing, evolving interaction between the scholar and the voices of the past embedded in the documents. (Harris, 2001, p. 332)

Archival sources are a subset of what historical geographers and historians refer to as **primary sources**. They include non-current records of government departments held in public **archives** but can be extended to include company records and private papers. As well as **documents**, handwritten and typed, these sources can embrace personal letters, diaries, logbooks, and minutes of meetings, as well as reports, plans, maps, and photographs. More recently, they have included records created in electronic format, which brings new challenges (Tibbo, 2012). This chapter concentrates on official papers, including manuscript and typescript files although most of the following comments are also applicable to company archives and private papers.[1] With the target readership of this book in mind, the chapter concentrates largely on government archives, on the late nineteenth and twentieth centuries and is written from an Aotearoa New Zealand perspective, that hopefully is transferable at least to other "settler colonial" contexts. A late nineteenth- to twentieth-century focus means that most of the records are typewritten and comparatively easy to read, and the prose of the recent past remains relatively easy to comprehend. Ease of reading and limited amounts of handwritten documentation should not, however, obscure the fact that official records, in terms of what was originally created and what has been retained in archives, do not offer impartial or full insights into matters of class, gender, and race. As a collection of unique, single documents, created contemporaneously with the events they discuss, the materials lodged in archival repositories provide a particular window into the geography of earlier times. Historical approaches applied to archival sources will not allow all of the research questions of human geography to be addressed; however, they do provide a means of answering questions about the recent as well as the more distant past that are not recoverable by the other techniques or from other sources available to human geographers and they can provide a way of helping to "verify" and contextualize other sources of information such as **oral testimony** and official statistics.

Researchers will find archives in a variety of settings: public archives are typically housed in a government agency charged with the preservation of non-current records; small regional or specialized collections are often located in museums, historical societies, or universities' special collections departments; while the records of private organizations may be accessed through the organizations themselves. Regardless of their physical location, increasingly for the researcher, the internet provides the initial contact point with archival collections, whether through a portal such as Hathi Trust (www.hathitrust.org) or through repositories' own sites. Box 12.1 lists some major repositories and their web addresses.

The digital era has enabled various specialized archives to make the most of digital technologies, by bringing together material on frequently overlooked groups, topics, or sources (Mason & Zanish-Belcher, 2007). For example, the Aluka Project focuses on liberation struggles in southern Africa and broadens the notion of "archive" as conceived in this chapter to contain published and oral material (see www.aluka.org) (Isaacman, Lalu, & Nygren, 2005). Other examples are provided by sites that contain original and interpretive material relating to the Tantramar marshlands of New Brunswick in Canada (www.mta.ca/marshland) and the "Mapping our Anzacs" site relating to Australian soldiers who served in World War I hosted by National Archives of Australia (https://discoveringanzacs.naa.gov.au). The capabilities of digital media have been taken up in countless online preservation projects, digital humanities sites, and research portals. Some of these are associated with museums or other externally funded research projects, such as *Digital Harlem* (www.digitalharlem.org), or are projects of universities' digital scholarship centers, such as Mapping Inequality (https://dsl.richmond.edu/panorama/redlining), while others have been set up beyond an institutional framework, for example, the Palmyra Atoll Digital Archive (www.palmyraarchive.org) (Johnston, 2018). These sites are examples of what has been termed "animating the archive," which is returned to later in the chapter. A challenge for such endeavours is assuring that they stay active and do not fall into abeyance. More broadly, regarding digital projects, Hodder (2017) also points to questions around unreadable formats, information being easily altered, duplicated, and lacking marks of origin, and little clarity over who owns it (and has responsibility for preservation). For curators of such non-institutional digital platforms there are issues of maintenance of the digital technology, training, use, and related costs as well as potentially others relating to the ownership of materials (Beel et al., 2015) that will need to be frequently revisited.

Identifying the Archives

Archives New Zealand: www.archives.govt.nz/index.html
Library and Archives Canada: www.collectionscanada.ca/index-e.html
National Archives of Australia: www.naa.gov.au
National Archives of Ireland: www.nationalarchives.ie
National Archives of South Africa: www.national.archives.gov.za
National Records of Scotland: https://discovery.nationalarchives.gov.uk/details/a/A13531995
Public Record Office (England and Wales): www.nationalarchives.gov.uk
US National Archives and Records Administration: www.archives.gov

Advice on Conducting Archival Research

> Archives are constructed, shaped, produced and manipulated by those who choose to create them. (Beel et al., 2012, pp. 204–5)

So where to begin? Good archival research practice is difficult to reduce to a list of bullet points. However, it is useful to explore sequentially the sorts of things you might need to do and the obstacles that you might encounter in undertaking archival research. At the start, like any other research project, your work ought to be informed by in-depth reading on current scholarship around your topic but accompanied by an openness about the ultimate direction of the research. This point is well made by the eminent Canadian historical geographer Cole Harris (Box 12.2).

Archival research begins before you arrive at the archive. You ought to be familiar with the existing **secondary literature** on your research topic before beginning any search for archival material. With larger official archives this may involve registering as a user and then searching for and ordering files through an online system similar to but not the same as a library catalogue. This may even be possible at distance and in advance of arriving at the archives. Do check the registration requirements on the institution's website before arriving and ensure that you have the appropriate sorts of identification. For smaller regional archives it may be advisable to email or write ahead of time outlining your research topic. Even experienced archival researchers may wish to seek some further advice about their research project from the **archivists** on duty.

Archivists can provide helpful, expert advice about record sets that you may not have considered useful. Typically, as a new user you will be given the

BOX 12.2

Approaching Archival Research

The first point to make about archival research is that it cannot be contained within a single methodology. Any sizable archive holds a vast array of material, and even if one's research questions are fairly specific, the chances are good that there will be far more potentially relevant documents than there will be time to examine them . . .

. . . [A]rchival research tends to gravitate towards one of two polar reactions—neither, I think particularly helpful. It is easy enough to be taken over by the archives, to attempt to read and record all their relevant information. In this way months and perhaps years go by, and eventually the investigator has a vast store of notes and usually, rather weak ideas about what to do with them. A fraction of the archives have been transferred from one location to another, while the challenges of interpretation have been postponed . . .

In effect the archives have swallowed the researcher. At the other pole are those who come to the archives with the confidence that they know precisely what they want. They have conceptualized their research thoroughly in advance. They pretty much know how they will argue their case and what their theoretical position is. But they do need a few more data, which is why they return to the archives. As long as they cleave to their initial position, either they will find that data they need and leave fairly quickly or they will not find them and also leave. Fair enough for certain purposes. But they are imposing their preconceptions on the archives. They have solved the problem of archival research by, in effect, denying the complexity of the archives and the myriad voices from the past contained in their amorphous record (Harris, 2001, pp. 330–1).

opportunity to explain what you are researching and why. The archivist will tell you how the **finding aids** work and can offer suggestions about where to start looking. While their experience and expertise can often prove invaluable, it is important to remember that they may have limited time available to assist individual researchers. Sometimes ingenuity is called for in searching and you may only be able to address research questions obliquely.

A crucial difference between a library and an archive lies in the way that each organizes and stores material. Libraries catalogue books and journals typically by either the Dewey Decimal or the Library of Congress classification systems. These group together all books on similar subjects. In contrast, public archivists seek to maintain the integrity of the record sets they receive from government departments, organizations, or individuals in terms of preserving the place of specific files in the broader record set, maintaining the original ordering of documents in the file, providing storage conditions that will ensure the long-term survival of the records, and making them available to the public. There is no universal agreement over the terminology archivists use for describing their holdings but in some countries **fonds** is used as an overarching term to mean all the material created by a single administrative entity. Maintenance of the relationship between materials as originally and organically created within a fonds is an essential task for archivists. Archivists place great emphasis on the **provenance** of the **files**; the actual order of the material within the files in itself tells the researcher something about the situation that prevailed when the record was being created.[2] Thus, whereas in a library you can refer to a catalogue to find a book on a particular subject on an open shelf, in an archive basic finding aids take the form of sequential **series lists** of all the files held by particular agencies. These lists itemize all the files created by an organization using the original description system (usually numerical but sometimes alpha-numerical).

Fortunately, in most archives electronic searching of the collections is now possible. This means that you can look for specific items in the same way, superficially at least, that you would use a library catalogue. Some have more advanced search systems that also allow you to look for specific files by number or for selected file series. This can be useful if you need to look systematically through a batch of material where the individualized files names do not provide clues about their contents.

Several additional challenges await in archival discovery. What survives in the file is likely to be only a fragment, and it may be quite partial in terms of providing any insights about the past. With small archives, you may have only the series list of files to guide you. Inevitably, you will find that some of the originating records staff have been more thorough and less idiosyncratic than others in managing their filing systems. The name of a file may not always be a clear guide to its contents, material may have been misfiled, and some files may have been lost or destroyed. For instance, only after requesting found files marked "railways accommodation" did I realize that they had nothing to do with railway department housing, the topic I was working on, but actually in a technical sense referred to the number of passenger carriages and freight wagons that could be "accommodated" at particular railway station yards.

In some national collections, precious and fragile originals that have previously been available only on microfilm are now increasingly scanned and accessible on-line (e.g., Hackel & Reid, 2007). As Summerby-Murray (2011, p. 117) has noted "an unexpected side effect of digitalization" has been "a dramatic improvement of legibility" as well as making items more widely and conveniently available to researchers. Notably, optical character recognition (OCR) and better machine-learning have also allowed extensive transcription of large volumes of archives, such as the manuscript US Census, which is now available both as images of the enumerator sheets and in transcribed form at the bottom of each file (www.archives.gov/research/genealogy/census/online-resources).

One of the most fundamental technical difficulties relates to the ability to actually *read* the documents retrieved in the archive. During the first half of the nineteenth century, many official documents were handwritten in **copperplate script**. This script looks elegant, but it can take a little time to learn to read it proficiently, a situation that may be exacerbated when officials wrote both across and along a page in order to save paper. Perseverance will pay off. Archives from the later part of the nineteenth century are commonly written in a **modern hand**. They are generally readable with a bit of effort.[3] In any case, all kinds of handwritten documents made in the past tend to be difficult to decipher, especially when the investigator is trying to read a faint letterbook copy of the original. Not only do spellings change as you move back in time, but so does the very construction of the English language.[4] This makes it more of a challenge to understand the world view of these earlier times. A fundamental trial can be posed by difficult-to-read handwriting; where the document was particularly important, I have on occasions been reduced to identifying how each letter of the alphabet was being written and then attempting to transcribe the document letter by letter. For those who essentially learned to write on keyboards this may be an extra but hardly insurmountable challenge. From around the 1880s, typewritten material becomes more common in government files, but important annotations in the margins will be handwritten and often cryptic in meaning. These annotated comments are particularly important for the insight they can give into discussion within an organization about the issue to which the larger document relates.

Anyone contemplating the use of archive material as a source for qualitative research in human geography must be prepared to be patient and resourceful; using documentary evidence is rarely easy and may require a considerable amount of time. The units of land area and currency may also be different from those in use today (for example, acres rather than hectares). This raises the issue of whether to convert every measurement to the current system or to give a general conversion factor and use the units of the period (generally, I prefer the latter). Some facility with the original units is useful. Appreciating that there are 640 acres to a square mile makes it possible to recognize, for example, that the apparently precise data on the forest areas in Otago Province in Aotearoa New Zealand in 1867 are in reality estimates to the nearest quarter square mile, or 160 acres. You may also need to understand more specialized measures, depending on your field of research. For example, throughout much of the British Empire in the nineteenth century, quantities of sawn timber are often given in superficial feet (colloquially referred to as *superfoot*)—that is, 12 inches by 12 inches by 12 inches (30 cm by 30 cm by 30 cm), but in North America the equivalent term *board foot* was used.

You may also find that there is restricted access to some files. Personnel files from corporations, as well as records from institutions such as orphanages, prisons, and insane asylums, often fall within this category. The period during which restricted access applies varies from one country to another, but 30 years after the closing of the file is typical. In some circumstances a lesser degree of restriction applies, and permission to look at files may be granted by a senior archivist, government official, or someone associated with the organization that created them. A formal written request outlining your research project may result in the granting of access; however, some conditions may be attached—for example, you may be permitted to read only a specific portion of the material while the remainder of the file remains physically sealed. In other cases, the researcher has no choice but to wait patiently until the time embargo ends and material is released.

Photocopying material is usually possible, but it can be comparatively costly, and you may have to pay in advance. Some material may be deemed too fragile or too difficult to photocopy. Many archives now offer digital scanning using machines that are designed to protect delicate bindings, and often will deliver images right to your email free of charge. It is therefore advisable for you to find out what the policy is beforehand. Plans, maps, and larger charts can be scanned or copied by other means, but this is sometimes quite expensive or requires a special request to the staff. Most archives now permit researchers to make their own digital copies of documents. Various conditions apply, sometimes including registering your camera or phone, completing associated documentation, and agreeing not to use flashes or tripods or to fold documents. The use of digital images does raise issues regarding labelling and storage if you are to make effective use of such materials (Box 12.3).

When the archivist gives you the file to work on, you will find in most cases that new items are on top of the older material, particularly if the material is secured by paperclip or in a folder. You may want to work from back to front. Will the material answer any of your research questions? It may be immediately obvious that the material is relevant to your inquiry, or it may appear only tangentially relevant or even irrelevant. Sometimes it is difficult to make a judgment at first glance, and you may have to recall material you have examined previously but whose significance you did not appreciate at the time. Alan Baker, a British historical geographer, has offered some guiding thoughts on evaluating primary sources, including archival materials (Box 12.4).

Digital Images

In creating and storing digital archival files, be mindful of Harris's (2001) comment about merely transferring the archive from one location to another. Ease of copying can create other difficulties if reference details are not kept meticulously. As a set of suggestions, resting on painful personal experience, I would suggest the following points:

1. If using a camera or phone check that you have adequate storage space (or upload directly to a cloud-based, secure server) and that the battery is charged (take spare batteries or charger).
2. When copying lengthy documents, be wary of making blurred images and of missing pages.
3. Recognize that you may require some maps and images to be reproduced with greater clarity than you can obtain with your hand-held camera or phone and be prepared to pay for high-quality scanned images.
4. Ensure that you have a reliable system for linking the digital image to the source file (my low-tech approach to this has been to include a slip of paper with the file details on it alongside the photographed page so that I have a visual reference on each digital image).
5. Name and store the digital images so that they can be located and retrieved easily.
6. Post-archive editing of the digital images may be necessary to improve readability but in so doing take care not to crop off any key details or the file reference information.

Baker's words seem to me to be crucial for those using archives as qualitative sources in human geography. It is essential to understand as fully as possible the original purpose of the document, who created it, what position they held, and how and when it was made. Some generic questions to pose when assessing documentary sources are laid out in Box 12.5.

The questions raised in Box 12.5 provide a useful start, although I would make three qualifying points. First, it is possible to extend "document" to include maps and plans (see Harley, 1992). Second, this approach tends to privilege the ideas behind actions. Underpinning at least some of Black's questions is an acceptance of the importance of ideas as the key to understanding actions. This means causation is found in the thoughts of individual actors at the time and has been termed an

Assessing Evidence in Historical Geography

No source should be taken at face value: all sources must be evaluated critically and contextually. The history and geography of a source needs to be established before it can legitimately be utilized and incorporated into a study of historical geography. The historical sources we use were not compiled and constructed for our explicitly geographical purposes; they were more likely to have been prepared, for example, for the purposes of taxation and valuation, administration and control. We also have to understand not only the superficial characteristics of a specific source but also its underlying motivation, background, and ideology of the person(s) who constructed it. In order to make the most effective and convincing use of a source we must be aware of its original purpose and context and thus its limitations and potential for our own project.

Source: Baker, 1997, p. 235.

idealist approach (Guelke, 1982). This is not the only point of view from which to understand historical geographies of the past (Baker, 1997). Third, the documents themselves cannot be read in isolation but must be understood in their wider context. Even in the most voluminous archives the files were created for specific purposes and do not necessarily capture a wider worldview. Some topics and subjects

Questions to Ask of Documentary Sources

1. Can you establish the authenticity of the source—is it genuine? Are you looking at the original?
2. Can you establish the accuracy of the document—how close is it to the source of events or phenomena? How accurately was the information recorded? (Cross-check it with other sources, if possible.)
3. What was the original purpose for collecting the information? How might it have influenced what information was collected?
4. How has the process of archiving the information imposed a classification and order upon historical events?

Source: After Black, 2006.

of importance to historical geographers of today may be absent and others silent in the official record. Any conclusions will be provisional and revised and adjusted in the light of new evidence and changing interpretations.

To some extent, all archival researchers develop individualized approaches to note-taking from archival materials. There are two basic and contrasting strategies. The first involves collecting material by *topic*, noting specific details and suitably referenced quotations. New topics can be added as more files are read and new research questions formulated. The alternative approach is to record *chronologically* any pertinent information from each file and then subsequently identify **themes** that emerge from across the files. Both strategies have advantages and disadvantages. The former depends on identifying key topics at the beginning of the project within which to collect information. Such an approach still allows you to add new topics or identify dead ends and see how themes merge or diverge. My personal view is that while this approach means that many diverse sources are brought together, it can blur a researcher's capacity to make good inductive judgments when in the archive. The second method is more sensitive to the provenance of files and can give a clearer sense of the role of specific individuals, officials, or departments. It does, however, involve a degree of double-handling in that evidence that has already been collected by the researcher needs to be reorganized after each visit to the archives and perhaps annotated further. It is important to follow up other research questions that may emerge from this re-sorting process. The latter approach is one that I have used over many years. It suits me, and as a full-time academic, I can incorporate it into my way of working. But I would acknowledge that it probably works best when a researcher has been working in a familiar area over a longer period of time, even if the specific contents of the files are unknown. Students with limited time for archival research may prefer to adopt the first strategy and develop digital file systems that support storage and analysis by topic.

After you have located and extracted archival evidence, it must be adequately cited in the written products of the research. The first step is to carefully record the specific document description and file reference. For example, the personnel file of L.M. Ellis, a Canadian who was first Director of Forests in New Zealand (1920 to 1928), is located amongst the Forest Service files at Archives New Zealand in Wellington. The specific item reference is R21098142, the agency code is ADS Q and Series 185/38, the accession is W607 and the item is 4/4i. Accuracy is crucial especially if you have to submit a written request for the item. Fortunately, the Ellis file can be identified and ordered online more or less at the click of a button, though locating it illustrates some of the other challenges of archival work. Ellis' given first name was Leon but he never used it and searching for him under this heading reveals two unrelated files. To further complicate matters, Ellis preferred the spelling MacIntosh, but on his official file it was recorded as McIntosh—thus MacIntosh Ellis only generates his original job application and a draft report, but not the personnel file. Be prepared to think around a problem rather than immediately accepting that the material you are looking for is not held. Record the details carefully: it enables you as a researcher to keep track of where you found specific information and it enables a subsequent researcher to relocate the material. The idea is simple enough, but given the nature of archival material, it is somewhat more exacting than, for example, the standard bibliographic requirements of author, date, title, and publisher/place for a book in the reference list of a thesis. Citing archival materials correctly can also pose problems in that human geography has tended to adopt versions of in-text citation systems, such as Harvard and Chicago styles. Most archival sources sit uncomfortably within this framework and are generally better referenced in footnotes or endnotes, typically used by historians. Students undertaking archival research may need to negotiate a variation from social science–oriented referencing formats favoured by most human geographers.

In 2006, Ogborn (2006, p. 111) wrote with qualified confidence that "You may have to go to the data rather than having them come to you." While his essential point remains, digital scanning of selected files breaks down a traditional assumption of archival research, of the researcher travelling to the archive (approximating "the field" in other branches of human and physical geography). Not only can the file material be viewed at a distance via your computer, it can be recopied digitally as the basis of future writing and visual projects. As Beel et al. (2015, p. 202) pithily note, "digitisation changes the relationship between people, place and archive." This can mean a dramatically altered time budget for a research project. Travel can be offset against more time searching files but Harris' warning about being "swallowed by the archive" still applies. Also, keep in mind that other pertinent materials may not have been digitized and, in any case, collecting alone is insufficient—time must be set aside to analyze the materials.

Born Digital Records in the Archive

In contrast to paper records that have been scanned as part of a digitization project, **born digital records** have no original paper-based (or equivalent) form. This may include but is not limited to emails, text-based documents (e.g., Word documents), PowerPoint presentations, videos, spreadsheets, PDFs, social media, and geotagged data (see Poorthuis et al., 2016; and Chapter 15). Increasingly archives and research libraries are **accessioning** such materials. They pose some new challenges for researchers and archivists. A 2017 survey of Aotearoa New Zealand institutions identified a lack of staff expertise, funding, technical infrastructure, and time for planning as the major impediments to collecting born digital content. Immediate issues include the development of catalogues or databases of digital material and transferring of the content to safer storage medium (Moran, 2017). For the moment, searching born digital holdings is somewhat uneven depending on the capacity and priorities of the archives in question.

Born Digital and the Digital Novice

To date my own experience with born digital records in the archive has been slight. What I did notice was that a particular set of documents I looked at were preserved as a PDF that I was able to download via my computer at work, thus saving a day's trip, a real time bonus. On the other hand, the document in question was the final version of a report completely shorn of any of the surrounding memoranda, file notes, and indeed draft versions with marginal annotations that you would typically find in a paper copy file. In that sense the born digital material is dislocated from its context—while that can still be reconstructed to some extent, in this case it will require some careful newspaper and periodical searching plus some interviews with key informants to do so. Born digital material may soon represent a formidable part of archival collections. My immediate concern is less about technical questions of readability and permanence of the electronic records than about searchability of holdings and with what may amount to be a reduced capacity to make sense of the context of much of this material.

The archive—paper-based or digital—does not constitute the only source for historical research. For instance, newspapers, private papers, and unpublished memoirs may provide valuable material for cross-referencing with the archival record. Once archival work is completed, the researcher may need to follow up on unfamiliar key actors by checking old editions of *Who's Who* or newspaper obituaries, as well as finding out about unfamiliar organizations or period issues; here newspaper accounts can be invaluable (and some are now available electronically by searching "historical newspapers" and the region you are interested in). This "post-archive" work can of course help to shape and inform the purpose of subsequent trips to the archive.

Moreover, files are not the end-point of research, as American cultural geographer Carl Sauer reminded historical geographers 80 years ago:

> Let no one consider that the historical geographer can be content with what is found in archive and library. It calls, in addition, for exacting fieldwork. One of the first steps is the ability to read the documents in the field for instance of an account of an area written long ago and compare the places and their activities with the present, seeing where

> the habitations were and the lines of communication ran, where the forests and the field stood, gradually getting a picture of the former cultural landscape behind the present one. (Sauer, 1941, p. 13)

Sauer's words may indicate nostalgia for a rural past while your focus could just as easily be urban and social and much closer to the present day, but his challenge remains pertinent. For example, Keighren's (2012) case study of archives-based inquiry into "moral politics" in early twentieth-century New York undertaken as part of an undergraduate field excursion to the city shows how the experience enabled the students to better contextualize the city's current social geography.

Changing Archival Practice: A Cautionary Tale

The first decade or so of my archival work was characterized by assiduous note taking from files, occasionally even including making simple sketches of key portions of maps and plans. On a few occasions I ordered photocopies of some important documents and maps. At the time this was both comparatively expensive and some weeks would elapse before I received the copies. This detailed transcription of the primary material and incidental notes about interpretation and other lines of inquiry certainly helped imprint the contents on my memory.

The second decade revealed some changes as supported by research funding but, facing greater time pressures and restricted time for out-of-town archival work, I began to order photocopies of larger amounts of material, although still maintaining a fairly substantial set of written notes and comments on the original material. This behaviour was rationalized on the grounds of photocopying being cheaper than a return airline fare. It was sometimes advantageous when portions of documents not originally thought to be significant proved so and were available in my office.

In recent years, particularly once digital devices were permitted in the archive, my practice has further changed, and not necessarily for the better. I find that I am now tending to digitally copy much of the contents of files that look as if they may be of use. This has some advantages; many original plans and maps are coloured and this can now be captured on the camera and help with the interpretation, although as Rekrut (2011, pp. 152–4) notes, digital copying can also result in the loss of some information. By way of compensation I have tended to read the file more fully before copying anything in it. I have also kept a log of items digitally copied and write a commentary on the nature of the material in the file, what is of interest and why (especially if this is not necessarily reflected in its title). Because a large part of my research collection is paper based, I find am caught with a hybrid system that risks assuming the worst features of both. This is not something the digital natives of today will have to face.

That said it is not all bad news; various **digitization** projects are making primary source material available online so that researchers are freed from being physically in the archive. While this is disadvantageous in terms of the loss of mental space for focused inquiry without interruption it is advantageous in that it helps relocate historico-geographical research to a position alongside and much closer to a range of other qualitative approaches used in human geography.

Challenges of Archival Research

> Archives always have a *creator* and these makers of memory need to be considered carefully and archives can contain hidden voices of both the powerful and powerless—but often one or the other is absent. (Hodder, 2017, pp. 703 and 707)

When dealing with the files contained in an official government archive, it is important to bear in mind the sorts of **power** relations inherent in the surviving materials. This is rather more than just acknowledging that the archived files are fragmented and partial. Records were often created by elites and officials and, as such, reflect the outlooks and understandings of the dominant groups in the context at the time they were created. A particular challenge lies in disentangling under-recorded lives of women, minority groups, children, and Indigenous peoples, who may be hidden within or omitted entirely from colonial archives (see Chapter 4). For example, Francesca Moore (2010, p. 263), in her work on illegal abortions in late nineteenth- and early twentieth-century Lancashire, UK, suggests "drawing force from absence and incompleteness." Similarly, Sarah Mills (2013) confronts the "fragments" and "ghosts" of the archives when constructing a historical geography of childhood. A number of authors have explored various archives related to the British Empire in a theoretically informed and empirically rich fashion, such as Ruth Craggs's (2008) work on the Royal Empire Society library. Duncan (1999) writes of these concerns in terms of complicity stemming from use of the "colonial archive." In settler societies, there has been "an under theorization of indigenous experience in contemporary archives studies" (Adams-Campbell, Glassburn and Rivard, 2015, p. 109; see also Chapters 2 and 4). To recover something of the historical geography of marginalized people the archives may need to be read against the grain, recognizing the internal organization and structure of the collections and "looking for traces of the least powerful within the papers of the wealthy and the state" (Falzetti, 2015, p. 137). For much of the nineteenth century and well into the twentieth century, these records were created largely by men in the upper echelons of society, and in states such as Australia, Canada, and Aotearoa New Zealand, they are predominantly the records of colonizing British settlers. Summarizing the contents of files from the archives as the "truth" merely reproduces these uneven power relations rather than interrogating them.

The records of non-governmental, community, or even sporting groups, as well as personal records such as letters and **diaries**, may provide a way into understanding the concerns and aspirations of those who had subaltern positions in the public political sphere. In the same way, **oral histories** from the recent past may provide insights into lives otherwise ignored by "official" records (see Chapter 10). Furthermore, an awareness of the power relations *within* the archival material may allow the researcher to reinterpret surviving materials. For instance, what I once mapped as examples of illegal felling of forest in Aotearoa New Zealand in the 1870s I would now be inclined to understand as resistance on the part of Māori forest owners to the imposition of authority by the Crown and as the flouting of government regulations by European timber-cutters who had limited alternative means of supporting themselves.

Ethics and Archives

It is all too easy for archival researchers to dismiss **ethical** issues as something relevant only to geographers working on present-day topics using other qualitative or quantitative methodologies. Yet archival researchers also have ethical obligations, accentuated by the fact that the individuals who created—or are the subjects of—the records in question are, in all likelihood, now deceased and unable to represent themselves. Other ethical dimensions of archival research have less to do with safety, harm, and risk to the researcher and more to do with retaining the integrity of the material contained in the files and the preservation of the archives themselves. After all, those using archives should regard access to the material as a privilege. Historical records are precious and often irreplaceable. All researchers are under an obligation to look after archival material and to ensure that it is preserved in good order for any subsequent scholars.

The US National Council of Public History identifies three guidelines for using archive materials.[5] We can reasonably substitute *geographer* for *historian* in each of these guidelines.

1. Historians work for the preservation, care and accessibility of the historical record. The unity and integrity of historical record collections are the basis of interpreting the past.
2. Historians owe to their sources accurate reportage of all information relevant to the subject at hand.
3. Historians favour free and open access to all archival collections. (National Council on Public History 2003)

Another situation in which ethical issues may arise is when files that contain classified or otherwise restricted material are issued to you by mistake. While it may be tempting to capitalize on an archivist's error in issuing a file before a time embargo or other restriction has elapsed, in the longer term this is counter-productive; it is equivalent to an unsanctioned questionnaire or an interview in which the participant does not know the true purpose of the research. Such behaviour can result in tighter lending conditions being imposed on all subsequent archival users.

Public archives typically specify conditions to which users must adhere when they sign in or request a reader's card. Not all primary documents are in public archives, however, and having access to such documents can present practical and ethical issues. Finer (2000) recounts an episode in which, after she initially received unlimited access to the records of a prominent Italian social reformer, she learned partway through the project that new conditions governing access, the scope of the research, and its objectives were being imposed (Box 12.7).

This is not the full extent of ethical issues and the archive. Moore (2010, p. 264) has further considered these as they relate to "research on sensitive topics."

Researching in a Private Archive

For researchers using public archives, the protocols are fairly well-established and reinforced in documentation that is part of the user registration process. On occasions, researchers will have access to private papers or records of small organizations. On the basis of a particular research project, Finer (2000) puts forward four "negotiations" that ought to be undertaken on those occasions to ensure the smooth running of the project. They are:

1. To insist on and ideally participate in the drafting of a detailed written agreement regarding precisely what is to be attempted in the research and to what end.
2. To draft a timetable agreed to in advance by staff members who are in a position to affect access to records or to other facilities such as photocopying.
3. To reach agreement in advance on the handling of sensitive material and the extent to which and on what terms it is to be cited.
4. To reach agreement on matters of faith/ideology—that is, the extent to which it is or is not considered necessary for the researcher to be of the same persuasion as the person(s) being researched.

While Finer's project had a biographical orientation, her points have a more general utility for geographical research. Her point about ideology and faith is further explored by Bailey et al. (2009).

These embrace the realms of sexual, illegal, and political behaviour, as well as experiences of violence. As she notes **informed consent** in historical research is "at the worst meaningless" (Moore, 2010, p. 267). Archival research may unearth "revelatory materials;" in her research on abortion at a time when it was illegal in the UK, Moore took the position, influenced by emerging practice about social science on living human subjects, that, "Arguably, the impracticality of consent in historical work means researchers have greater responsibilities to research subjects in relation to their privacy." Thus, following practices common in ethnographic and anthropological inquiry, Moore used pseudonyms to protect the identity of individuals even though they had been deceased for two generations in an effort to protect, "the dignity, rights and welfare of research participants" (Moore, 2010, p. 267). This suggests a larger ethical issue for historical geographers, which lies in negotiating the terrain between accurate reportage from archival sources and social science driven concerns for the privacy of individuals.

Presenting the Results of Archival Research

There is no single correct way of presenting archival research. The theoretical foundation of the research project, the sorts of empirical information retrieved, and the writing style of the researcher all shape how the research project or thesis is expressed.[6] Typically, however, archival researchers will make use of direct quotations from key documents to demonstrate their case. They will also be mindful of the actions of key actors within organizations (and sometimes the importance of the role of obscure players as well) in shaping decisions and policy that may have had far reaching geographical significance. They also make use of **case study** material to illustrate points (see Chapter 7). Identifying key themes inductively from the archival records can be a valuable way to organize the central argument of your paper—issues of territorial control and governmentality, migration, border conflicts, social/spatial inequality, symbolic place meanings, landscape change, and uneven development are useful geographic themes that are present in many historical projects and can help spark new insights (see also Chapters 10 and 18). Frequently, there is integrated use of cartographic and pictorial material, which not only serve as evidence, but also greatly enrich the visual component of a finished product.

Adept researchers are often able to move easily from specific points of detail to a much larger picture and can connect to what is known about related topics. I would recommend critical perusal of recent issues of the *Journal of Historical Geography* and *Historical Geography* and significant books by recognized figures in the field. However, it is not just a matter of identifying key quotations but rather of building an argument (see Chapter 5). This obliges you to select ideas in a logical way from the pre-existing literature and then to use them to provide an informed discussion based on what you have found in the archives. The desirable end point, however, is to be in command of the source material. Rather than merely reproducing a chronicle of part of what is contained in the archive, strive to make your writing a synthesis of specific detail and informed interpretation.

While the fragmentary nature of archives and power relations involved in their creation and maintenance continues as a point of discussion, in the archive the researcher, paradoxically, can feel rather overwhelmed by the abundance of material. At a time when human actors attract more attention in human geography than structural arguments "biography as method" offers "another way of managing the tasks of shifting, sorting and scoping that have become increasingly central to historical research in geography" (Hodder, 2017, p. 455). By "biography as *method*" Hodder means studying "the external factors that have shaped lives" compared to biography as *subject*; the study of "internal motivations" (Hodder, 2017, p. 453).

The centrality of text in the form of an essay or research report has shaped the way in which evidence and insights from the archive are drawn on. More recent writing in human geography has pointed to new possibilities encapsulated in phrases such as "animating archives" (Dwyer & Davies, 2010). Mills (2013, p. 702) explains that she uses the "term 'animating' to cover a range of creatively engaged and enlivened archival practices that I believe has defined this recent research—its approach to reading the archive, the materials used to 'bring the past to life,' and the relationships and responsibilities of 'playing' with the archival record." In practice this may mean use of "digital images and the opportunity to zoom into fragments, turn pages on a tablet and piece together screen shots" (Mills, 2013, p. 710). It can also mean creative use of some of the ephemeral memorabilia, especially "tactile stuff" (e.g., tickets, programmes, badges), preserved in files, which may provide an illuminating way of linking events, places, and people together.

Conclusion

Although it has much to offer human geography in general, archival research has tended to be neglected by other than historical geographers; this may be shifting as digital projects and GeoHumanities expand. As a research method, historical research using archival sources:

- calls for creative thinking in identifying source materials relevant to your research problem;
- needs patience, precision, and critical reflection in collecting and evaluating material;
- requires a sense of historical-geographical imagination in interpreting source material whereby theorization does not outstrip the evidence;
- is partial and requires that you relate archival material to other contemporary sources of a textual and pictorial sort that may be held in other collections;

- increasingly asks researchers to access digital rather than original items; and
- asks researchers to continually negotiate between the theoretical and the empirical.

Archival work can be time-consuming and, superficially at least, frustrating in that the information retrieved may offer only partial answers, particularly when you find yourself under time pressure to complete a research project. Archival work done properly requires patience. Rarely will the surviving archival material provide full answers to the questions you pose. In the case of public archives, the surviving material typically says more about politics, economics, the concerns of elites, and men than it does about social and private spaces, women, and minority groups. It is, however, still possible to use these records to recreate something about the lives of ordinary people. But surrendering to the temptation to merely summarize the content of files, a trap into which inexperienced archival researchers can fall, is another way of being—as Harris terms it "swallowed by the archive."

It is all too easy in discussing archival research to create the impression that there is no room for novices when in fact more human geographers need to be encouraged to incorporate archival work into their research programs. I would simply describe archival research as somewhat akin to confidently accepting the challenge of working on a jigsaw puzzle even though you can be reasonably certain that pieces are missing and that the box cover with the picture of the completed puzzle will never be found. Good archival research can be extremely satisfying, both in learning the skills to conduct it and in the presentation of results.

Key Terms

archival research
archives
archivist
copperplate script
digitization
document
files
finding aids
fonds
modern hand
primary sources
provenance
records
secondary literature
secretary hand
series lists

Review Questions

1. What sorts of research questions can be addressed using archival sources?
2. What are the problems of a researcher being, as Harris terms it, "swallowed by the archive"?

3. What is meant by *provenance*, and why it is important to archival researchers?
4. What are the potential benefits and challenges of digital archives?
5. What are the four steps that Black (2006) recommends for assessing historical evidence?
6. What approach would you adopt to organize and store digital images from archival collections?

Review Exercises

1. Make an appointment for a tour of a local historical archive—this could be a governmental facility (e.g., a state or provincial archive), a historical society, or your university's special collections department. Identify their specializations, learn about how finding aids work, and ask (ahead of time) to see a sample box of records to familiarize yourself with both the intellectual and embodied experience of being in an archive.
2. On your own or with a partner, visit your country's national archives website. Most will have rotating special digital exhibits—choose one of these and dive in, spending about 30 minutes perusing the collection. Then do a little "forensic" work on the exhibit by identifying how the archivists put it together. Answer the following questions: what are the themes and topics addressed? What primary sources do they draw from? What is the provenance of the records that are activated in the exhibit? How do they engage the public (photos, maps, interactive elements, storytelling, other means)?

Useful Resources

Adamson, G.C.D. (2014). Institutional and community adaption from the archives: A study of drought in Western India, 1790–1860. *Geoforum*, 55, 110–19.

Keighren, I.M. (2012). Fieldwork in the archive. In R. Phillips & J. Johns (Eds), *Fieldwork for human geography* (pp. 138–40). London: SAGE.

Lorimer, H. (2010). Caught in the nick of time: Archives and fieldwork. In D. DeLyser, S. Herbert, S. Aitken, M. Crang, & L. McDowell (Eds), *Handbook of qualitative geography* (pp. 248–73). Los Angeles: SAGE.

McLennan, S., & Prinsen, G. (2014). Something old, something new: Research using archives, texts, and virtual data. In R. Scheyvens (Ed.), *Development fieldwork: A practical guide* (pp. 81–100). London: SAGE.

Moore, F.P.L. (2010). Tales from the archive: Methodological and ethical issues in historical geography research. *Area, 42*(3), 262–70.

Smiley, S. (2006). Field notes: Historical geography research at the Tanzania National Archives. *African Geographical Review, 35*(1), 107–16.

Notes

1. Much of this chapter also relates to photographs, but this sub-field has a literature of its own—for example, Schwartz and Ryan (2003) and Quanchi (2006).
2. A discussion of the behind-the-scenes work of the archivist in appraising, arranging, and describing records is provided by Harvey (2006).
3. Original manuscripts concerned, for example, with the early European settlement of North America before 1700 may be written in **secretary hand**. This was the script of professional scribes of the time, and it is difficult to read without specialized instruction and its translation requires additional palaeographical skills, discussion of which is beyond the scope of this chapter.
4. For instance, in the mid-nineteenth century, the long *s* written much like an *f* was frequently used in official correspondence. Words with a double *s*—*lesson*, for instance—are rendered as what looks to us like *lesfon*. Indeed, you may be dealing with records in another language. For instance, there is good deal of correspondence in Te Reo Māori in Archives New Zealand, much but not all of which is accompanied by an English version prepared by official translators.
5. These guidelines are expanded on in the American Historical Association's Statement of Standard of Professional Conduct, which is available online at www.historians.org/jobs-and-professional-development/statements-standards-and-guidelines-of-the-discipline/statement-on-standards-of-professional-conduct.
6. Given that this volume focuses on qualitative methods, I have omitted discussion of how a researcher might extract and present in tabulated form quantitative information derived from archival sources. For examples of this genre see Holland and Olson (2016, 2017) and Holland, Olson, and Garden (2018).

Using Questionnaires in Qualitative Human Geography

Pauline M. McGuirk and Phillip O'Neill

Chapter Overview

This chapter deals with questionnaires, an information-gathering technique used frequently in mixed-method research that draws on quantitative and qualitative data sources and analysis. We begin with a discussion of key issues in the design and conduct of questionnaires. We then explore the strengths and weaknesses for qualitative research of various question formats and questionnaire distribution and collection techniques, including online techniques. Finally, we consider some of the challenges of analyzing qualitative responses in questionnaires, and we close with a discussion of the limitations of using questionnaires in qualitative research.

Introduction

Qualitative research seeks to understand the ways people experience events, places, and processes differently as part of a fluid reality, a reality constructed through multiple interpretations and filtered through multiple frames of reference and systems of meaning-making. Rather than trying to measure and quantify aspects of a singular social reality, qualitative research draws on methods aimed at recognizing "the complexity of everyday life, the nuances of meaning-making in an ever-changing world and the multitude of influences that shape human lived experiences" (DeLyser et al., 2010, p. 6). Within this epistemological framework, how can questionnaires contribute to the methodological repertoire of qualitative human geography? This chapter explores the possibilities.

Commonly in human geography, questionnaires pose standardized, formally structured questions to a group of individuals, often presumed to be a **sample** of a broader **population** (see Chapter 6). Questionnaires are useful for gathering original data about people, their behaviour, experiences and social interactions, attitudes and opinions, and awareness of events (McLafferty, 2016; Parfitt, 2013). They usually involve the collection of quantitative and qualitative data. Since such **mixed-method** questionnaires first appeared with the rise of behavioural geography in the 1970s (Gold, 1980), they have been used increasingly to gather data in relation to complex matters like the environment, social identity, transport and travel, quality of life and community, work, and social networks.

While there are limitations to the qualitative data that questionnaires are capable of gathering, they have numerous strengths. First, they can provide insights into social trends, processes, values, attitudes, and interpretations. Second, they are one of the more practical research tools in that they can be cost-effective, enabling **extensive research** over a large or geographically dispersed population. This is particularly the case for questionnaire surveys conducted online where printing and distribution costs can be minimized (Sue & Ritter, 2012). Third, they are extremely flexible. They can be combined effectively with complementary, more **intensive** forms of qualitative research, such as **interviews** and **focus groups**, to provide more in-depth perspectives on social process and context. For instance, Waitt and Knoble's (2018) investigation of inner-city dwellers' experiences and use of urban parks combined key informant interviews with city planners, questionnaires with local residents, and follow-up in-depth interviews with volunteers who had participated in the questionnaire. Data from the questionnaire provided a framework for the in-depth interviews, allowing key **themes**, concepts, and meanings to be teased out and developed (see Gibson et al. [2013] and McGuirk and Dowling [2011] for similar examples). In this mixed-method format particularly, questionnaires can be both a powerful and a practical research method. Showing these advantages, Beckett and Clegg (2007) report on the success of qualitative research into women's experiences of lesbian identity using only postal questionnaires to gather rich accounts from respondents. This process allowed respondents the privacy and time to consider and develop their responses to sensitive questions. The questionnaire as a research instrument, then, seems to have nurtured rather than constrained the data collection exercise.

Questionnaire Design and Format

While each questionnaire is unique, there are common principles of good design and implementation. Producing a well-designed questionnaire for qualitative research involves a great deal of thought and preparation, effective organizational strategies, and critical review and reflection, as an array of literature suggests (for example, de Vaus, 2014; Dillman, 2007; Fowler, 2002; Gillham, 2000; Lumsden, 2005; see also the relevant chapters in Babbie, 2016; Bryman, 2016; Clifford et al., 2016; Flowerdew & Martin, 2013; Hoggart, Lees, & Davies, 2002; and Sarankatos, 2012). The design stage is where a great deal of researcher skill is vested, and it is a critical stage in ensuring the worth of the data collected.

Notwithstanding the quality of the questionnaire devised, we are beholden as researchers to ensure that we have sufficient reason to call on the time and energy of the research participants. The desire to generate our "own" data on our research topic is insufficient justification (Hoggart, Lees, & Davies, 2002). As with any study, the decision to go ahead with a questionnaire needs to be based on careful reflection on detailed research objectives, consideration of existing and alternative information sources, and appropriate **ethical** contemplation that is attuned to the particular cultural context of the research (see Chapters 2, 3, 4, and 6).

The content of a questionnaire must relate to the broader research question as well as to your critical examination and understanding of relevant processes, concepts, and relationships. As a researcher, you need to familiarize yourself with relevant local and international work on your research topic. This ensures clarity of research objectives and will help you to identify an appropriate participant group and relevant key questions. You need to be clear on the intended purpose of each question, who will answer it, and how you intend to analyze responses. You also need to be mindful of the limits to what people are willing to disclose, being aware that these limits will vary across different social and cultural groups in different contexts. Public housing tenants, for instance, might be wary about offering candid opinions about their housing authority for fear of retribution. Respondents might be cautious about what they are willing to disclose in questionnaires administered via email because of the loss of anonymity that occurs when email addresses can be matched with responses (Van Selm & Jankowski, 2006). Every question, then, needs to be carefully considered with regard to context and have a clear role and purpose appropriate to the social and cultural norms and expectations of the participant group (Madge, 2007).

Begin by drawing up a list of topics that you seek to investigate. Sarantakos (2012) describes the process of developing questions for a questionnaire as a process of translating these research topics into variables, variables into indicators, and indicators into questions. Identify the key concepts being investigated and work out the various dimensions of these concepts that should be addressed (see also Chapter 5). Then identify indicators of the dimensions and use them to help you formulate specific questions. Doing this will ensure that each question relates to one or more aspects of the research and that every question has a purpose. De Vaus (2014) suggests that it is helpful to think about four distinct types of question content:

1. *attributes:* Attribute questions aim at establishing respondents' characteristics (for example, age or income bracket, dwelling occupancy status, citizenship status); some researchers call these *demographic* questions.
2. *behaviour:* Behaviour questions aim at discovering what people do (for example, recreation habits, extent of public transport use, food consumption habits).
3. *attitudes:* Questions about attitudes seek to discover what people think is desirable or undesirable (for example, judgment on integrating social

housing with owner-occupied housing, willingness to pay higher taxes to fund enhanced social welfare services).
4. *beliefs:* Questions about beliefs aim at establishing what people believe to be true or false or preferred (for example, beliefs on the importance of environmental protection, beliefs on the desirability of social equity).

A guiding principle for question types, however, is to ensure that your target participant group will understand the questions and has the knowledge to answer them (Babbie, 2016). As is the case in writing for the popular media, it is recommended that unless you are targeting a specialized and homogenized group, you phrase questions to accommodate a reading age of approximately 11 years (Lumsden, 2005). Rather than dumbing down your questionnaire, this tactic helps with clarity and direction. It also encourages respondents to answer the questions: for instance, a complexly worded question asking whether government planning policies contribute to local coastal degradation may lead them to abandon the questionnaire, but asking about changes local residents have noticed in erosion or flooding is much more answerable.

Apart from the typology of question content, there is a range of question formats from which to draw. We commonly make a distinction between **closed** and **open questions**, each of which offers strengths and weaknesses and poses different challenges depending on the mode through which the questionnaire is being administered (e.g., mail, face-to-face, email). Closed questions may seek quantitative information about respondent attributes (for example, level of educational attainment) or behaviour (for example, how often and where respondents buy groceries). You should provide simple instructions on how to answer closed questions (e.g., how many responses the respondent can tick). Some examples are set out in Box 13.1. Closed questions can ask respondents to select categories, rank items as an indicative measure of attitudes or opinions, or select a point on a scale as indicative of the intensity with which an attitude or opinion is held (see Sarankatos, 2012, ch. 11). A major benefit of closed questions is that the responses are easily coded and analyzed, a bonus when interpreting a large number of questionnaires. Indeed, for web-based questionnaires, a data file can be assembled automatically as respondents type in their answers. Closed questions are demanding to design, however, because they require researchers to have a clear understanding of what the range of answers to a question might be. Respondents' answers are constrained to the range of categories designed by the researcher, and this can be a limitation. It has also been found that when respondents are asked to "tick all appropriate categories" on a list (see the category list question in Box 13.1), they can turn to **satisficing behaviour**; that is, they keep reading (and ticking) until they feel they have provided a satisfactory answer and then stop. Relatedly, a significant limitation of closed questions is that they rest on the assumption that words, categories, and concepts carry the same *meaning* for all respondents, which is not always the case. For example, how a respondent answers the question "How often have you been a victim of crime in the past two years?" will depend on what the respondent sees as a crime (de Vaus, 2014). It is worthwhile to be aware, too, that the ways particular questions are posed or how they relate to preceding questions can influence respondents' answers. Babbie (2016) shows that greater support in questionnaire surveys is indicated habitually for the phrase "assistance to the poor" rather than for "welfare" and for "halting rising crime rate" rather than "law enforcement." A further critique of closed questions is the lack of spontaneity in respondent's answers and the removal of the possibility of "interesting replies that are not covered by the fixed answers" (Bryman, 2016, p. 249). This limitation might be overcome by offering an answer option such as "other (please specify)" or by using **combination questions** that request some comment on the option chosen in a closed question (see Box 13.1).

In general, open questions have greater potential to yield in-depth responses that match the aspiration of qualitative research: to understand how meaning is attached to place, process, and practice. Open questions offer less-structured response options than closed questions, inviting respondents to recount understandings, experiences, and opinions in their own style. Rather than offering alternative answers, which restrict responses, open questions provide space (and time) for free-form responses. Open questions also give voice to respondents and allow them to question the terms and structure of the questionnaire itself, demonstrate an alternative interpretation, and add qualifications and justifications. This capacity acknowledges the **co-constitution of knowledge** by researcher and research participant (Beckett & Clegg, 2007). For instance, Mee (2007) used open questions in her questionnaire-based research exploring public housing tenants' experiences of "home" in medium-density unit dwellings in Newcastle, Australia. Despite normative cultural assertions that link ideas of home to home ownership and detached (single-family) housing, respondents used the open-questions to describe their rented apartment homes as "heaven," "a blessing," and as "wonderful" and "beautiful." Open questions, then, are capable of yielding valuable insights, many of them unanticipated, and they can open intriguing lines of intensive inquiry in scenarios where extensive research is the main focus or where a more intensive person-to-person approach is not possible (Cloke et al., 2004b). Such scope, however, means that open questions can be effort-intensive for respondents to answer and time-consuming to **code** (Bryman, 2016). An open format can also generate responses that lack consistency and comparability. Certainly, respondents answer them in terms that match their interpretations. So open questions and the responses they yield are certainly more challenging to analyze than are their more easily coded closed counterparts (see Chapter 18 for details on coding).

BOX 13.1

Types of Questionnaire Questions

Closed questions

Attribute information

How often do you have a conversation with a neighbour? (please tick the appropriate box)

- Less than once a week ❑
- Once a week ❑
- Twice a week ❑
- More than twice a week ❑

Category list

What was the *main* reason you chose to live in this neighbourhood? (please tick the appropriate box)

- Close to work ❑
- Close to family and friends ❑
- Close to schools or educational facilities ❑
- Close to shopping centre ❑
- Close to recreational opportunities ❑
- Environment ❑
- Housing costs ❑
- Good place to raise children ❑
- Other (please specify) ❑

Rating

Please rank the reasons for buying your current house (please rank all relevant categories from 1 [most important] to 6 [least important]).

- Price ❑
- Location ❑
- Size ❑
- Close to job/family ❑
- Investment ❑
- Children's education ❑

Scaling

Please indicate how strongly you agree/disagree with the following statement (please tick the appropriate box):

(continued)

Having a mix of social groups in a neighbourhood is a positive feature.

Strongly disagree ❑
Disagree ❑
Neutral ❑
Agree ❑
Strongly agree ❑

Grid/matrix question

Think back to when you first got involved in environmental activism. What initially inspired you to get involved? (please tick the appropriate box for *each* reason)

	Very influential	**Fairly influential**	**Not very influential**	**Not influential**
Spirituality/religious beliefs				
Fear/anxiety about ecological crisis				
Nature/ecology experiences and care for the environment				
Political analysis				
Felt like you could make a difference				
Influential person, book or film (please specify)				
Key event (please specify)				
Contact with an organization, campaign, or issue (please specify)				
Sense of personal responsibility				
Other (please specify)				

Combination question

Have changes in the neighbourhood made this a better or worse place for you to live? (please tick the appropriate box)

Changes have made the neighbourhood better ❑
Changes have not made the neighbourhood better or worse ❑
Changes have made the neighbourhood worse ❑

Please explain your answer below.

Open questions

What have been the biggest changes to the neighbourhood since you moved in?

What, if any, are the advantages for civic action groups of using the internet, email, and cell phones?

Please describe any problem(s) you encounter using public transport.

In summary, using open questions makes it possible to pose complex questions that can reveal people's experiences, understandings, and interpretations of social processes and circumstances. as well as their reactions to them. Closed questions are not capable of such in-depth explorations. Answers to open questions can also tell us a good deal about how wider processes operate in particular settings. Thus, they enable research that addresses the two fundamental questions that Sayer (2010) poses for qualitative research: what are individuals' particular experiences of places and events? And how are social structures constructed, maintained, or resisted? (see Chapter 1). We would add to this the need to dissolve the idea of general processes into the messy contexts of everyday life.

Beyond choice of question content and type, getting the wording, sequence, and format of a questionnaire right is fundamental to its success. Guidance on these is given in Box 13.2 with discussion revolving around clarity, simplicity, and logic. In question wording, you need to be sure that questions are sufficiently precise and unambiguous to ensure that the intent of your question is clear and well communicated. It is advisable to be familiar with the **vernacular** of the participant group. In online contexts, this may include becoming familiar with the jargon, abbreviations, and grammatical rules commonly used within the online community being approached, for instance, the language styles of specific blogger or social media groups (Madge, 2007; and see Chapter 15). Remember that the language of a questionnaire is not just textual. Graphical and numerical modes might also be present. These modes work together to affect respondents' perception of the survey and are perceived in ways that are influenced by cultural context (Lumsden, 2005). The web's capacity for global reach also means that online questionnaires may target international participants, not all of whom communicate expertly in English. There are software programs that allow the researcher to convert a questionnaire written in English into other languages (see www.objectplanet.com/opinio/howto/translation.html) as well as commercial providers (e.g., QuestionPro; www.questionpro.com/features/multi-lingual.html), though proofreading by a native speaker is always recommended. Beyond issues of logic, clarity, and comprehension, questions should avoid threats or challenges to respondents' cultural, ethnic, or religious beliefs, which may arise from a researcher's insensitivity, ignorance, or lack of preparation, even in the absence of overt prejudice. The need for concern about respondents' "cultural safety" (Matthews, Limb, and Taylor, 1998, p. 316) is part of the researcher's broader ethical obligations. See the section below for suggestions on how to pre-test your survey for logic, inadvertent offenses, and language comprehensibility.

The flow and sequence of the questionnaire are fundamental to respondents' understanding of the purpose of the research and to sustaining their willingness to offer careful responses and, indeed, to completing the questionnaire in full. Grouping related questions and connecting them with introductory statements will help here. In general, open-ended questions are

Guidelines for Designing Questionnaires

- Ensure questions are relevant, querying the issues, practices, and understandings you are investigating; that is, make sure the questions you ask will actually help you answer your project's research questions.
- Keep the wording concise (about 20 words maximum), simple, and appropriate to the targeted group's vernacular.
- Ensure that questions and instructions are easily distinguishable in format and font.
- Avoid double-barreled questions (for example, "Do you agree that the Department of Housing should cease building public housing estates and pursue a social mix policy?").
- Avoid confusing wording (for example, "Why would you rather not use public transport?") and be alert to alternative uses of words (for example, for some people "dinner" implies an evening meal while for others it implies a cooked meal, even if eaten at midday).
- Avoid leading questions (for example, "Why do you think recycling is crucial to the health of future generations?"), and avoid loaded words (for example, *democratic, free, natural, modern*).
- Avoid questions that are likely to raise as many questions as they answer (for example, "Are you in favour of regional sustainability?" raises questions of what sustainability means, how a region is defined, and how different dimensions of sustainability might be prioritized).
- Order questions in a coherent and logical sequence.
- Ensure the questionnaire takes no more time to complete than participants are willing to spend. This will depend on the questionnaire *context* (for example, whether it is conducted by telephone, face-to-face, or online) and its *format* (mostly closed questions or many open questions). Generally, 20 to 30 minutes will be the maximum, although longer times (45 minutes) can be sustained if the combination of context and research topic is appropriate.
- Ensure an uncluttered layout with plentiful space for written/typed responses to open questions.
- Use continuity statements to link questionnaire sections (for example, "The next section deals with community members' responses to perceived threats to their neighbourhood.").
- Begin with simple questions, and place complex, reflexive questions or those dealing with personal information or sensitive or threatening topics later in the questionnaire.

better placed towards the end of a questionnaire, by which time respondents are aware of the questionnaire's thrust and may be more inclined to offer fluid and considered responses. In terms of layout, aim for an uncluttered design that is easy and clear to follow. Where you use closed questions, aligning or justifying the space in which the answer should be provided will contribute to clarity and simplify coding. With open-ended questions, particularly in hard copy, you need to be conscious of the need to leave enough space for respondents to answer without leaving so much as to discourage them from offering a response altogether.

All of these questionnaire design principles need to be observed regardless of how the questionnaire is being distributed: whether by mail, face-to-face, by telephone, by email, or online. However, there are additional design factors that are important to consider when using an online environment (Dillman, 2007). Web-based questionnaire delivery makes it possible to incorporate advanced features such as split screens, drop-down boxes, multimedia elements, and branching functions that present different questions based on answers to earlier ones. Some of these features and functionalities do require higher-level software (e.g., a paid licence rather than the free version). As more people use phones as their primary device, you may also want to check that the survey software is responsive, that is, legible and scaled to interactivity on various devices.

You need to consider whether the participant group has the ability and the capacity to receive and respond to the questionnaire and its mode of delivery. Web surveys with advanced multimedia features, for example, have high bandwidth requirements, though this is of diminishing concern (Vehovar & Manfreda, 2017). You also need to remember that online questionnaires require respondents to think about how to respond to the questionnaire while simultaneously thinking about technical options, a matter that is particularly important if your target participant group is less computer-literate. Keeping things simple and limiting the number of actions a respondent has to undertake is sensible. Finally, you need to take account of whether you will administer your questionnaire solely online or through other modes as well, in which case you need to be mindful of how questions will be posed in those other modes. Box 13.3 outlines additional key principles for the design of online questionnaires (adapted from Dillman & Bowker, 2001).

Finally, whether developing a conventional or online questionnaire, you should include a cover or introductory letter or email. Box 13.4 offers examples. The letter or email needs to provide general information about the purpose of the study and the questionnaire as well as information about **confidentiality** and ethical review of the study, how the respondent has been selected, how long the questionnaire will take to complete, and when relevant, instructions on how and when to return the questionnaire.

Guidelines for Designing Online Questionnaires

- Introduce web based questionnaires with a welcome page providing basic instructions and information and encouraging completion.
- Ensure the first question is interesting to respondents, easily answered, and fully visible.
- Use conventional formats for questions, similar to those normally used on self-administered paper questionnaires.
- Provide clear instructions including technical advice on how to respond to each question, and position them at the point where they are needed.
- Limit the length of the questionnaire. Remember that a typical print page can take up several screen pages and thus may seem excessive.
- Keep the design simple to aid navigational flow and readability and ensure the format is maintained across different browsers and screen set-ups.
- Allow respondents to move to the next question without having to answer a prior question.
- Allow respondents to scroll from question to question without having to change screen pages.
- Wherever possible allow all answer choices for a given question to be viewed on one screen.
- Include advice that indicates how much of the questionnaire the respondent has completed.
- Close with a thank-you page with your email address for any questions.

Sampling

Before administering a questionnaire, you will need to make a decision about the target audience, or sample. In quantitative research, questionnaires are used commonly to generate claims about the characteristics, behaviour, or opinions of a group of people (the population) based on data collected from a **representative sample** of that population. The population might be, for example, tenants in public housing, the residents of a given local government area, or people living with diabetes. The sample—a subset of the population—is selected to be representative of the population such that the mathematical probability that the characteristics of the sample are reproduced in the broader population can be calculated

(May, 2011). In such cases, a list of the relevant characteristics of the population, the sampling frame, is required so that a sample can be constructed. A **sampling frame** might be, for example, the tenant list of a given public housing authority, a local electoral register, or a health register of all people in a given geographical area receiving treatment for diabetes). The rules surrounding sampling are drawn from the central limit theorem used to sustain statistical claims to representativeness, **generalizability**, and **replicability** (see McLafferty, 2016; Parfitt 2013).

Examples of Invitations to Participate in Questionnaire Studies

Sample cover letter

School of Geography
Geography Building
East Valley University
East Valley, NB E1E 1E1
Telephone: (506) 898-9778
Facsimile: (506) 898-9779
Email: E.saunders@evu.edu.ca

High-density residential living in Port Andrew, East Valley

I am Edith Saunders, a research student with the School of Geography at East Valley University. As part of my research on high-density residential environments in East Valley, I am investigating how people understand and create feelings of home in high-density neighbourhoods. The research is being conducted in collaboration with East Valley Council and is aimed at informing its policy and planning decision-making. The work is focused on the Port Andrew area, and you have been selected to receive this questionnaire as a local resident.

The questionnaire asks about the ways you understand and use your home and the ways you interact with your local neighbourhood spaces and services. The questionnaire will take approximately 30 minutes to complete, and participation is voluntary. The questions ask primarily about your experiences and opinions. There are no right or wrong answers. All answers will be treated confidentially and anonymously, and individuals will not be identifiable in the reporting of the research.

It would be appreciated if you could complete the questionnaire at your earliest convenience and no later than July 30. Please return the completed questionnaire in the reply-paid envelope provided. Return of the questionnaire will be considered as your consent to participate in the survey.

Your participation is greatly appreciated. Your opinions are important in helping to build understanding of high-density residential living and how it can be supported through local government planning and provision of neighbourhood spaces and services.

Questions about this research can be directed to me at the address provided.

Thank you in advance for your participation.

Yours faithfully,
Edith Saunders

The university requires that all participants be informed that if they have any complaints concerning the manner in which a research project is conducted, they may be directed to the researcher or, if an independent person is preferred, to the university's Human Research Ethics Officer, Research Unit, East Valley University, E1E 1E1, telephone (506) 898-1234.

Sample email invitation to participate in an online questionnaire

From: kanchana.phonsavat@EVU.edu.ca
To: [email address]

Subject: Survey on high-density residential college living

Dear Student,

I am a research student with the School of Geography at East Valley University (EVU). As part of my research, I am investigating how students understand and create feelings of home in high-density residential college environments. The research is being conducted in collaboration with EVU and East Valley Council. You have been selected to receive this invitation to participate as a student resident of one of EVU's residential colleges.

We are interested in the ways you understand and use your college accommodation and the ways you interact with your local neighbourhood spaces and services. The questionnaire will take approximately 30 minutes to complete and is completely voluntary and confidential. The data will be used to evaluate university and council policies and their support of high-density residential environments.

(*continued*)

To complete the questionnaire, please click on the following link:

www.newurbanliving.evu.org.ca/surveys.html

It would be great if you could complete the questionnaire in the next two weeks. If you have any questions or need help, please email me at kanchana.phonsavat@EVU.edu.ca.

Thank you in advance for your participation.

Kanchana Phonsavat

The university requires that all participants be informed that if they have any complaints concerning the manner in which a research project is conducted, they may be directed to the researcher or, if an independent person is preferred, to the university's Human Research Ethics Officer, Research Unit, East Valley University, E1E 1E1, telephone (506) 898-1234.

On the other hand, questionnaires in qualitative research are usually a part of mixed-method research aimed at establishing trends, patterns, or themes in experiences, behaviours, and understandings. Important to the analysis, then, is uncovering the influence of a *specific context*, rather than making generalizable claims about whole populations (Herbert, 2012). A more appropriate sampling technique for qualitative research is non-**probability sampling** where generalization about a broader population is neither possible nor desirable. Sampling frames may not, in any case, be available. Some web surveys, for instance, involve self-selection by respondents where anyone who agrees to complete the questionnaire can be included in the sample. For example, Drozdzewski and Klocker's (2019) research into parents' experiences of balancing their childcare responsibilities and careers, in Australian workplaces, used an online questionnaire. Respondents were invited to complete the questionnaire through university-based online list servers, a nation-wide survey distribution company, and the researchers' industry contacts. A total of 1,047 people from across the country participated in the questionnaire.

Specifically, **purposive sampling** (see Chapter 6) is commonly used where invitation to participate is made according to some common characteristic, be it a social category (for example, male single parents), a behaviour (for example, women who use public transport), or an experience (for example, victims of crime). There are no specific rules for this type of sampling. Rather, the determinants of the appropriate sample and sample size are related to the scope, nature, and intent of the research and to the expectations of your research communities.

As in all research, these considerations are overlain by resource constraints (time and money). Nonetheless, a lack of hard-and-fast rules and a need for pragmatism do not imply the absence of a systematic approach—quite the opposite.

Complex and **reflexive** decisions need to be made about how to approach sampling. For instance, in research on what motivates Australian "sea-changers," like many other retirees around the world, to abandon city life and relocate to regional, coastal areas, researchers would need to take into account whether they should seek respondents in all age groups, all household types, and all income categories. Research on people living with HIV/AIDS would need to take into account whether the researchers should target, say, early-stage individuals only, a range of sexual identities, people of any sexual orientation, only individuals infected from a particular source, and so on. Each decision will have ramifications for how sample recruitment proceeds and what mode of questionnaire distribution is suitable. Questionnaires administered online, for example, may be well suited to research on factors shaping environmental advocacy where the target respondents are likely to have digital skills and access to computers as part of their work. By comparison, this mode of distribution may be poorly suited to research on perceptions of displacement among low-income elderly or immigrant populations in gentrifying areas. These cases illustrate the importance of research scope, purpose, and intent in shaping the sampling approach and in determining appropriate sample size. Bryman (2016) provides details of various types of purposive sampling, along with a discussion of sample size, and Chapters 6 and 7 in this book provide an extended treatment of further questions regarding selecting cases and participants. In the end, decisions about samples are shaped by compromises between cost, need for targeting, the nature of the research, and the limits of possibility.

Pre-testing

We highly recommend trying out a questionnaire before it is distributed. **Pre-testing** is when a questionnaire is piloted or road-tested with a sub-sample of your target population to assess the merits of its design, its appropriateness to the audience, and whether it does in fact achieve your aims. For web-based questionnaires, rigorous testing of the questionnaire on a range of platforms and browsers should be undertaken to identify and weed out potential technical problems. In online contexts, technical bugs are very likely to result in respondents abandoning the questionnaire. Getting feedback from those with extensive questionnaire-design experience and from those who might use the data generated (for instance, in the example in Box 13.4, a local authority and a university) will allow possible problems to be identified and improvements made. Scheduling a pre-testing stage provides the opportunity for post-test revisions that might dramatically increase the questionnaire's effectiveness.

Both individual items and the overall performance of the questionnaire need attention at this stage. Are instructions and questions easily understood? Would any of them benefit from the addition of written prompts? Do respondents interpret questions as intended? Do any questions seem to make respondents uncomfortable? Discomfort and sensitivity (perhaps the question is considered too intrusive) might be indicated by respondents skipping or refusing to answer a question or section. Alternatively, such outcomes could mean that respondents do not understand the question or do not have the knowledge or experience to answer it. Consider too how respondents react to the order of the questions. Does it seem to them that the questions flow logically and intuitively? Are there parts where the questionnaire seems to drag or become repetitive? Technical aspects can also be tested: is there enough space for respondents to answer open questions? How long will the questionnaire take to complete? Do the data being generated present particular problems for analysis? If you plan to conduct the questionnaire face-to-face with respondents, the pre-test stage can also be a useful exercise in training and confidence-building for the researcher.

Modes of Questionnaire Distribution

Consideration of the *mode* of questionnaire distribution should be one of the earliest stages of your questionnaire design. This has implications for design, layout, question type, and sample selection. The main distribution modes are mail, face-to-face, telephone, and online survey platforms. Each mode has distinctive strengths and weaknesses, and our choice depends on the research topic, type of questions, and resource constraints. The best choice is the one most appropriate to the research context and target participant group, while the success of any particular mode is dependent on a design appropriate to context and participant group. So the question is: what should researchers interested in qualitative research be aware of to guide them in the choice of mode?

Mailed questionnaires have clear advantages of targeted coverage. They can be distributed to large samples over large areas (for example, an entire country or province). The anonymity they provide may be a significant advantage when sensitive topics are being researched—for example, those dealing with socially disapproved attitudes or behaviours, such as racism or transgressive sexual behaviour, or topics involving personal harm, such as experience of unemployment or crime. Respondents may also feel more able to take time to consider their responses than they might if an interviewer were present (see below). Clearly, too, the absence of an interviewer means responses cannot be shaped by how an interviewer poses a question, interacts with the respondent, or interprets cues in the conversation in culturally specific ways.

Nonetheless, mailed questionnaires are generally the most limited of the four modes in terms of questionnaire length and complexity. The scope for complex open questions is particularly limited by the need for questions to be self-explanatory and brief, and this may be a significant consideration for qualitatively oriented research. Once the questionnaire is sent out, there is little control over who completes it or, indeed, over how it is completed; respondents may choose to restrict themselves to brief, unreflective, or patterned responses. A response to the question "what do you value about living in this community?" might yield a response of several paragraphs from one respondent and the comment "friends and neighbours" from another. There is no opportunity to clarify questions or probe answers. Nor is there control over the pattern and rate of response. Some parts of the target participant group may respond at a higher rate than others. It is common, for instance, for mailed questionnaires to achieve significantly higher response rates in wealthy neighbourhoods than in less socially advantaged neighbourhoods. Finally, mailed questionnaires can be subject to low response rates unless respondents are highly motivated to participate. While follow-up steps can increase a response rate somewhat (May, 2011), non-response rates on mail questionnaires have been reported at 80 per cent (Patten, 2015), raising questions about their utility.

Distributing questionnaires electronically adds potential for innovation and experimentation (Babbie, 2016). There are three main means of electronic distribution: (1) sending the entire questionnaire to respondents as an email attachment, (2) posting or emailing respondents an introductory letter with a link to a web-based questionnaire, and (3) distributing a general request for respondents (for example, via social media) to complete a web-based questionnaire. You might also use a mix of these distribution strategies (Bryman, 2016, p. 232). A major benefit of electronic distribution is that it "compresses" physical distance and expands enormously the reach of the questionnaire, including enhancing access to difficult to reach populations (McInroy, 2016; Sikkens et al., 2017). Furthermore, people practising covert or illegal behaviours—for example, graffitists or drug users—may be more easily recruited through the internet. The internet is also a powerful way of gaining access to self-organized groups—those with common interests, lifestyles, or experiences organized into chat-rooms, social media groups, and online forums. For example, Banaji and Buckingham's (2010) study on internet activism and young people sought out specific activist websites and conducted a questionnaire with 3,000 users. Mailing lists, online forums, and social media such as Facebook pages can be used for circulating the questionnaire or inviting participants to complete an online questionnaire. However, some online groups are sensitive to the intrusion of researchers (Chen, Hall, & Johns, 2004). Many discussion groups state their privacy policy when you join, so researchers should check the welcome message of public discussion lists for guidelines before using them to recruit potential participants (Madge, 2007; see also Chapter 15).

Regardless of the specific means of electronic distribution used, the recruitment of participants will be affected by the age, class, and gender biases that shape computer, email, and social media use (see Gibson, 2003). For instance, online delivery of a questionnaire investigating the leisure habits of elderly people is likely to confront participation problems, given that elderly people may be less likely to complete online surveys. Low-income groups may similarly have restricted access (Babbie, 2016).

Other benefits of electronic distribution include cost-savings and efficiencies. Digital dissemination enables the use of attractive formats and colour images without associated printing costs. Digital technologies also open up opportunities for flexibility in question design, for more complex questions, for incorporating adaptive questions with encoded skip patterns (thus removing the need for complex instructions and filter questions), for **geotagging** data using mobile devices' GPS, and for increasing the potential to generate rich and accurate qualitative data with fewer unanswered questions (Bryman, 2016). Researchers who have deployed digital questionnaires report lower response rates than conventionally distributed questionnaires; although rates can be comparable when pre-notification and follow-up emails are used (Fan & Yan, 2010). Online respondents characteristically submit lengthy commentaries on open questions (Van Selm & Jankowski, 2006), a plus for qualitative research. In addition to saving on printing and postage costs, the electronic collection of data offers the major advantage over paper questionnaires of eliminating the need for a separate labour-intensive phase for data entry and coding of closed questions (Van Selm & Jankowski, 2006).

Mailed and online questionnaires do, however, present a particular set of challenges surrounding hidden costs, ethical issues, and technical capacities and failures (see also the discussion in Chapter 9 regarding online interviews). The cost and labour savings of avoiding coding and data entry through electronic data capture can be offset by the costs of design and programming (Hewson et al., 2003). Online web survey hosts (e.g., Survey Monkey, Lime Survey) are quite easy to learn and use, and they usually offer free or entry-level versions for a small monthly fee. Universities increasingly hold subscriptions for these services available for use by faculty/staff and students. When it comes to ethical issues, the challenge of obtaining **informed consent** digitally has been largely addressed by universities allowing for digital versions of signed consent forms to be provided to participants. In terms of privacy, the identity of web-based questionnaire respondents can be protected if they withhold their names, although technically adept researchers could collect data about web-based participants using, for example, user log files or Java Applets (Bryman, 2016; Lumsden, 2005). Anonymity cannot be provided to email questionnaire respondents when the returned questionnaire attaches an email address, which bears consideration. Responses stored on computer files, and online, could be accessible to hackers, and this may be a particularly important concern if the study being conducted involves sensitive and personal data. However, many universities now provide secure cloud-based storage to faculty/staff and students for the secure storage of such data, where security standards meet the requirements of university ethics committees.

Qualitative research is often very effective if questionnaires are administered face-to-face, in what becomes close to an interview setting. This can be a costly option due to the human labour required, but also has some important benefits. The major benefits of this mode flow from the fact that an interviewer's presence allows complex questions to be asked (see also Chapter 9). As well, an interviewer can take note of the context of the interview and of respondents' non-verbal gestures, all of which add depth to the data collected (May, 2011). As an interviewer, you can motivate respondents to participate and to provide considered, informative responses. Moreover, people are generally more likely to offer long responses orally than in writing. However, as Beckett and Clegg's (2007) work on lesbian identity suggests, this outcome is context-dependent. Perhaps more crucially, face-to-face questionnaires give an interviewer the opportunity to clarify questions and probe vague responses (see Chapters 9 and 10 for related discussions). For example, adding probes like "why is that exactly?", "in what ways?", or "anything else?" can elicit reflection on an opinion or attitude. Long questionnaires can also be completed because direct contact with an interviewer can enhance engagement. The ability to pose complex questions and elicit more in-depth and engaged responses is a major benefit for qualitative research. Moreover, this high level of engagement can also secure high response rates with a minimal number of blank responses and "don't know" answers (Babbie, 2016). However, the level of interviewer skill and reflexivity required to secure optimal outcomes should not be underestimated.

As Kevin Dunn discusses more fully in Chapter 9, the presence of an interviewer can be a powerful means of collecting high-quality data, but it introduces limitations as well. Interviewer/respondent interaction can produce **interviewer effects** that shape the responses offered. People filter their answers through a sense of social expectation, especially when interviewed face-to-face (Lee, 2000). They may censor or tailor their answers according to perceived social desirability. That is, they may avoid revealing socially disapproved behaviours or beliefs (such as racism or climate change skepticism) or revealing negative experiences (for example, unemployment). Beckett and Clegg (2007) chose postal questionnaires specifically to ensure the *absence* of an interviewer. Their argument was that participants should be allowed to recount their stories on their own terms, without any identification with the researchers' associations with particular geographical spaces or social and cultural attributes and without fear of judgment by the researcher. When interviewers are used, one means of dealing with respondents' self-censoring is to incorporate a self-administered section in the questionnaire or to reassure respondents through guarantees of anonymity.

Moreover, the interviewer's presence (as an embodied subject with class, gender, and ethnic characteristics) can also affect the nature of responses given. For instance, Bryman (2016) suggests that the gender, ethnicity, and social background of the interviewer can introduce significant variations; indeed feminist geographers have been leaders in raising the issues of **positionality** and **subjectivity** in research relations (see Whitson [2017] and the review of recent interview-related geography publications by Dowling et al. [2016]). So, while distinct benefits arise from using face-to-face distribution, there are drawbacks. Perhaps the most limiting is the practical consideration of cost. Interviewer-administered questionnaires are expensive and time-consuming and tend to be restrictive both spatially and with respect to population coverage. However, as we suggested before, this factor may not be a significant drawback if a particular, localized participant group is targeted.

While the opportunities for personal interchange are more restricted in telephone than in face-to-face questionnaires, the telephone mode still offers the possibility of dialogue between researcher and respondent and can provide some of the benefits of an actual face-to-face interview but with a level of anonymity that may limit problematic interviewer effects. Conducting questionnaires over the phone may encourage respondent participation because it may be seen as less threatening than being stopped on the street or opening the door to a stranger wanting to administer a questionnaire. However, telephone delivery constrains the scope for lengthy questionnaires, with about 30 minutes being the maximum time respondents are willing to participate (de Vaus, 2014). Furthermore, because the mode relies on a respondent's memory, the question format must be kept simple and the number of response categories in closed questions needs to be limited. However, the advent of **computer-assisted telephone interviewing** (CATI) and **voice capture** technology is significantly enhancing telephone questionnaires (see Babbie, 2016) and extending their potential. Moreover, they can be administered with great convenience and at relatively low cost.

Several drawbacks for the telephone mode are worth considering, however. First, telephone questionnaires often rely on a telephone directory as a sampling frame, and this can introduce specific class and gender biases among respondents as well as ruling out people whose numbers are not listed. Moreover, as cell phones become ubiquitous and more people abandon having landlines altogether, landline directories have largely lost their usefulness as a sampling frame. Historically, telephone surveys have had good response rates. However, growing public annoyance with unsolicited marketing calls means approaches by telephone face automated blocking of unidentified numbers, screening via voicemail, or call rejection (Dillman et al., 2009; Guthrie, 2010).

Maximizing Questionnaire Response Rates

Questionnaire response rates are shaped by the research topic, the nature of the sample, and the quality and appropriateness of questionnaire design as much as by the mode of distribution. In any case, questionnaire response rates tend to be higher when using a purposive sample—as is common in qualitative research—wherein interest in the research topic may be strong. There is good evidence that response rates for online questionnaires are stronger if the questionnaire is relatively brief, taking no longer than 20 minutes to complete, is not complex to complete, is simple in design, and does not require participants to identify themselves (Lumsden, 2005). Regardless of the mode of distribution, response rates can be improved by undertaking a series of strategies before questionnaire distribution and as follow-up (Bryman, 2016; Dillman, 2007). Box 13.5 summarizes the strategies that enhance questionnaire response rates according to the different modes of distribution.

BOX 13.5

Strategies for Maximizing Response Rates

Strategy	Face-to-Face	Telephone	Mail	Online
Ensure mode of distribution is appropriate to the targeted population and research topic.	√	√	√	√
Send pre-notification letter/email introducing the research and alerting to the questionnaire.		√		√
Place advertisement in local community newspaper/magazines or online forums/social media sites introducing the research and alerting to the questionnaire.	√	√	√	√
Ensure questionnaire is concise.	√	√	√	√
Ensure appropriate location of approach.	√			
Ensure appropriate time of approach.	√	√		
Vary time if no contact is made initially or pre-arrange time/location, if appropriate.	√	√		
Ensure pre-paid envelope is included in mail-out.			√	
Send follow-up postcard/email thanking early respondents and reminding others (one week after initial receipt).			√	√
Send follow-up contact and additional copy of questionnaire (two to three weeks after initial receipt).			√	√
Avoid abrasive manner.	√	√		
Dress appropriately to the target population.	√			

Analyzing Questionnaire Data

Analyzing questionnaires used in mixed-method research that blends qualitative and quantitative data requires an approach that distinguishes between closed questions in which responses are provided in an easily quantified format and open questions that seek qualitative responses. Quantitative data arises primarily from closed questions that provide counts of categorical data (for example, age and income bands, frequency of behaviour) or measures of attitudinal or opinion data (see Box 13.1 for examples). Questions such as these are relatively easy to code numerically and analyze for patterns of response and relationships between the variables that the questions have interrogated (May, 2011). Indeed, as noted above, response categories can be pre-coded on the questionnaire, simplifying matters even further (see de Vaus [2014] for more detail), while digital survey tools generate data spreadsheets automatically and typically enable quite substantial statistical analysis very easily. The analysis of qualitative responses is more complex. The power of qualitative data lies in its uncovering of a respondent's understandings and interpretations of the social world, and these data, in turn, are interpreted by the researcher to reveal the understandings of structures and processes that shape respondents' thought and action (for elaboration, see Crang [2005b]). Chapters 9 and 18 discuss the techniques and challenges of coding and analyzing qualitative data in detail. Nonetheless, it is worth raising some important points specific to analyzing qualitative data arising from questionnaires.

While quantitative results may help establish some broad *patterns*, in qualitative responses, the important data often lie in the *processes* revealed in detailed explanations and precise wording of respondents' answers. For qualitative research, then, it is best to go beyond classifying qualitative responses into simple **descriptive** categories so as to confine reporting to quantitative dimensions, stating, for example, that "49 per cent of respondents had positive opinions about their neighbourhood." There are two problems here. First, such reporting may well be statistically misleading given the data might have been derived from a relatively small purposive sample and could be used incorrectly to frame generalizations. Second, this approach involves "closing" open questions so that the richness of how respondents constructed, in this example, their positive understandings and experiences of their locality, is lost. Certainly, classifying qualitative responses into descriptive categories allows us to simplify, summarize, compare, and aggregate data. Yet, in so doing, we should be careful not to forfeit the nuance and complexity of the original text, which was collected as a qualitative exercise to further our understanding of the meanings and operations of social structures and processes and people's interpretations and behaviour in relation to them. Analysis that is more attuned to the strengths of qualitative research will approach questionnaire data with techniques such as "sifting and sorting" to identify key themes and broader concepts that connect them (see the discussion of analytical coding in Chapter 18). Reporting the findings in terms that recognize and elevate people's experiences, social processes, and human–environment interactions is much more meaningful than falling back on awkward attempts at quantification.

Further, in analyzing qualitative responses, we need to make no assumption that respondents share a common definition of the phenomenon under investigation (be that quality of neighbourhood, experience of crime, understanding of health and illness, and so on). Rather, we should assume that variable and multiple understandings coexist in a given social context. We need to incorporate this awareness into how we make sense of respondents' answers. Indeed, one of the strengths of using questionnaires in qualitative research is their ability to identify variability in understanding and interpretation across a selected participant group, providing the groundwork for further investigation through additional complementary methods such as in-depth interviews.

Finally, keep in mind that qualitative data analysis is sometimes referred to as more of an art than a science (Babbie, 2016) in that it is not reducible to a set of neat techniques. Although useful procedures can be followed (see Chapter 18), they may need to be customized to the distinctive concerns and structure of each questionnaire and the particular balance of quantitative and qualitative data it gathers. For this reason, and others, at all stages of the process of analysis we need to be mindful of engaging in **critical reflexivity**, especially when considering how our own frames of reference and personal positions shape the ways in which we proceed with analysis (see Chapters 2 and 19).

Conclusion

In seeking qualitative data, questionnaires aim not just to reveal attitudes and opinions but to identify and classify the logic of different sets of responses, seek patterns or commonality or divergence in responses, and explore how they relate to concepts, structures, and processes that shape social life. This is no easy undertaking, and researchers using questionnaires struggle with the tensions of seeking explanation while being generally limited as a result of the instrument's form and format to obtaining concise accounts.

Hoggart, Lees, and Davies (2002) argue that the necessarily limited complexity and length of questionnaires prevent them from being used to explain action (since this requires us to understand people's intentions), the significance of action, and the connections between acts. Compared with the depth of information developed through more intensive research methods such as in-depth interviews, focus groups, or participant observation, questionnaires may provide only superficial coverage. Nonetheless, they go some way in offering explanations in that they are useful for identifying regularities and differences and highlighting incidents and trends (see de Vaus [2014] for an extended critique). Indeed, as Beckett and Clegg's (2007) work shows, in some contexts they can enable the collection of full and frank, thoughtful and detailed accounts in ways that more intensive methods involving interviews and interviewers' presence may inhibit.

There are ways of constructing and delivering effective questionnaires that are largely qualitative in their aspirations, being mindful of the possibility of acquiring deep analytical understandings of social behaviours through careful collection of textual materials. Certainly, the interview, through its record of close dialogue between researcher and respondent, provides a particularly powerful way of uncovering narratives that reveal the motivations and meanings surrounding human interactions, and questionnaires can only ever move incompletely in this direction. However, by not requiring close and prolonged engagement with the research subject, the questionnaire offers opportunities to reach a wider range and greater number of respondents, in particular through online applications, and to collect data on people's lived experiences. This extensiveness and diversity makes questionnaires an important, contemporary research tool.

Key Terms

closed questions
co-constitution of knowledge
combination questions
computer-assisted telephone interviewing (CATI)
cultural safety
mixed-method research
open questions
population
pre-testing
probability sampling
purposive sampling
sample
sampling frame
satisficing behaviour
voice capture

Review Questions

1. Why are open questions more suited to qualitative research than closed questions?
2. Why is the choice of the mode of questionnaire distribution specific to the nature of the sample and the scope of the research topic?
3. Why should we avoid merely "closing" open question responses for the purpose of reporting findings?
4. What are the limitations of the use of questionnaires for qualitative research?
5. What are the particular benefits of administering questionnaires online?

Review Exercise

Sydney, Australia, continues to experience population growth. As a counter to its sprawling suburbs and congestion, the city is proposing to redevelop public housing in Waterloo, an area very close to the Central Business District, as a high-density urban community. By 2040 it is expected to include nearly 7,000 new dwellings in 17 new residential blocks, including 35 per cent of dwellings in the affordable housing category. The Social Housing Minister says the completed masterplan "will make Waterloo one of the most attractive suburbs in inner Sydney for people to live, work, and play."

Imagine it is now 2040 and the residents are in place. Your task is to design a six-question questionnaire to explore the ways social relations have evolved in Waterloo and how these are expressed. Do this in three steps:

i. Make a list of the types of social interactions that might be evident at the scale of the neighbourhood (e.g., friendship networks, acts of care, attitudes to strangers, use of public and shared spaces, security of private property). Then, think of the varied ways these relations or interactions might be expressed. For example, indicators of friendship networks might be counted using simple categorizations such as location, age, and length of residence; gauging residents' attitudes to strangers might use Likert scales to assess positive and negative emotions; identifying acts of care may involve eliciting evidence based on direct experience and observations; and assessing the security of property might use secondary crime data as prompts for statements requiring agreement or disagreement.
ii. Design the questionnaire using a variety of types of questions (e.g., open questions, closed questions, scaled questions).
iii. Outline the methods you would use to analyze data collected from the questions you have devised.

Useful Resources

Babbie, E. (2016). *The practice of social research* (14th edition). Belmont, CA: Wadsworth.

Bryman, A. (2016). *Social research methods* (5th edition). Oxford: Oxford University Press.

Cloke, P., Cook, I., Crang, P., Goodwin, M., Painter, J., & Philo, C. (2004). *Practising human geography*. London: SAGE. See Chapter 5.

de Vaus, D.A. (2014). *Surveys in social research* (6th edition). Sydney: Allen and Unwin. See Chapters 7 and 8.

Duke University's Initiative on Survey Methodology. An interdisciplinary initiative on survey methodology containing extensive tips and resources on survey research methods. https://dism.ssri.duke.edu/survey-help/tipsheets

Fielding, N.G., Lee, R.M., & Blank, G. (2017). *The SAGE handbook of online research methods*. London: SAGE. See Chapters 10–13.

Hoggart, K., Lees, L., & Davies, A. (2002). *Researching human geography*. London: Arnold. See Chapter 5.

Parfitt, J. (2013). Questionnaire design and sampling. In R. Flowerdew & D. Martin (Eds), *Methods in human geography: A guide for students doing a research project* (2nd edition) (pp. 78–109). Harlow: Pearson/Prentice Hall.

Sarankatos, S. (2012). *Social research* (4th edition). New York: Palgrave Macmillan. See Chapter 11.

———. 2012. *Social research* (4th edition). Palgrave MacMillan. www.macmillanihe.com//companion/Sarantakos-Social-Research/workbook/. This is a companion website for Sarantakos's book *Social Research*. It offers a digital workbook on questionnaire surveys.

Sue, V., & Ritter, L. (2012). *Conducting online surveys* (2nd edition). London: SAGE.

SurveyMonkey.com. SurveyMonkey.com—create surveys. www.surveymonkey.com. This is a commercially available web-based interface for creating and publishing custom web surveys and then viewing the results graphically in real time.

SurveyMonkey—The Monkey Team. n.d. "Smart survey design." s3.amazonaws.com/SurveyMonkeyFiles/SmartSurvey.pdf. SurveyMonkey's guide to effective design and question-writing for online questionnaires.

"Where I Went Today . . .": Solicited Journals and Narrative Mapping

Sarah Turner

Chapter Overview

Solicited journaling and narrative mapping offer excellent opportunities for the richness, plurality, and fluidity of everyday experiences to be explored and represented in human geography. This chapter introduces a range of ways that solicited journaling has been used by geographers, before focusing on factors to consider when undertaking this method, and outlining possible analysis and writing-up strategies. I also introduce different journaling techniques, before raising a range of potentially sensitive and ethical concerns for consideration. In the final section of the chapter I briefly focus on narrative mapping as one approach to present data gained from solicited journaling or from other qualitative methods, and some tactics for creating such maps.

Introduction

Women's experiences of violence in South Africa, street vendor livelihoods in Vietnam, and Antarctic researchers' emotions during fieldwork—these are all topics that have been studied by human geographers utilizing **solicited journals**. Such journals can help to reveal a participant's experiences, emotions, and daily personal geographies, supporting nuanced understandings of the relationships between people and the environment over time. Participants write their solicited journal entries—on a theme guided by the researcher—in the context of their own space and time, and decide what they want to include and omit. This ability of participants to reflect upon and contemplate their entries allows them to control the narrative; to decide upon the meaning and weight that they attach to different events. Due to these benefits, solicited journaling has been increasingly acknowledged as an important method for understanding the thoughts and experiences of disempowered and marginalized communities.

As one specific way to represent the findings from solicited journaling research, as well as from other projects that produce narratives, **narrative mapping** is gaining popularity amongst human geographers. Narrative maps allow the provided rich insights into participants' embodied and emotional spaces, and in particular into the ebbs and flows of their daily lives.

Other examples of solicited journaling being drawn upon by geographers include McGregor (2005) who asked male fishers in camps around Lake Kariba, Zimbabwe, to keep journals regarding their daily experiences, and Thomas (2007) who collected solicited diaries from individuals affected by HIV/AIDS as well as their main caregivers, to better understand how HIV and AIDS shape rural livelihoods, vulnerability, and support networks in Namibia. In addition, Spowart and Nairn (2014) used a diary-interview method to explore the emotions and **subjectivities** of snowboarding mothers in Aotearoa New Zealand, while Filep et al. (2015) asked researchers in Antarctica to journal their feelings regarding their physical and social environments while undertaking fieldwork.

Planning and Directing Solicited Journal Activities

Selecting Participants and Recruitment

During the initial planning stages, it is important to think about how you will find willing participants, and what sampling approach makes the most sense for your chosen topic (see also Chapters 6 and 13). Given that solicited journaling essentially removes the researcher from a central role in data collection, rapport and trust between the researcher and participants are vital so that the journals contain relevant information (and are completed). Your personal contacts can be an effective channel by which to gain participants, although other approaches such as sending advertisements through social media, email lists, radio broadcasts, or flyers in local shops, community organizations, and meeting places are all possible. Journaling usually involves a fairly limited number of participants, therefore **purposeful sampling** is common with potential participants selected deliberately due to the fact that they might be willing participants and rich sources of data in relation to the chosen topic. Often, recruited participants will drop out of the process for a variety of reasons, especially if the journals are to be kept over a long time period, hence over-recruitment can be helpful.

It is also important to consider, given the context in which you are working, what the literacy levels of your potential participants are likely to be. For example, Eidse and Turner (2014) found that some of the street vendors whom they had asked to keep journals in Hanoi, Vietnam, felt uncomfortable trying to write down their thoughts and asked someone else to do the writing for them as they dictated, or they passed the journal onto another household member entirely. Similarly, Meth (2003, 2004) found that some of the women with whom she worked in South Africa passed their journals on to their children to write up, as they dictated. Clearly this approach can raise **ethical** and **confidentiality** problems. Asking participants to audio-record their journals is a possible option in such cases, although this comes with its own set of concerns, as noted below.

Setting Up the Journals

The degree to which you provide participants with specific journaling guidelines can vary, ranging from requesting very specific limited responses to free-form entries. However, it is important that you carefully think through what you wish to achieve with the journaling exercise in advance, including how participants' responses will help answer your research questions (see Chapter 5), and provide clear guidelines as to your expectations, so as not to waste participants' time and energy (see Box 14.1. for a possible example, and also Meth [2003] for her written diary instructions). Generally, provide instructions that give participants a sense of direction and an understanding of what your expectations are, while at the same time allowing them to express their creativity and what is important or valuable to them (Hayman, Wilkes, and Jackson, 2012). Solicited journals can also incorporate approaches other than writing, such as asking participants to add photographs or sketches to the journals.

Example of Journaling Instructions for Participants—Motorbike Taxi Drivers in Bangkok, Thailand

Thank you for agreeing to keep a journal! I am specifically interested in your daily experiences as a motorbike taxi driver in Bangkok. For the next two weeks, please try to fill in your diary each evening, reflecting on your day's journeys and other related activities (e.g., meal breaks, motorbike repairs, etc.) including times, places, and any events that impacted your ability to carry out your day's work (e.g., police check points, traffic jams). It would be interesting to also learn about any emotions and any concerns you have about your job. You may write as little or as much as you like.

For each entry please add:

- The date and time of day when you're writing your journal
- Name relevant locations, such as the places where you drove that day, or places you waited for customers.

If there are any experiences that you would like to write about, but that did not take place during this two-week period, please feel free to include them as well, and when they took place.

Please do not worry at all about grammar, spelling, or handwriting!

(continued)

After 4–5 days I will contact you again to make sure everything is going ok. You can also call me anytime if you have questions or concerns. At the end of the two weeks I will collect your journal. At that time I would also appreciate if you would participate in an interview, which will last no more than 40 minutes, at a time of your convenience.

If you know of anyone else who may be interested in taking part in this project, please let me know.

Thank you for your participation!

Note: A separate consent form also covered what the drivers' journal data would be used for, their right to stop journaling at any time and withdraw from the project, and the confidentiality of their writing.

The length of time that you give participants to complete their journals can also vary greatly from a few days to a few months, although providing participants with a clear timeline and checking in with them periodically helps to maintain participation rates. Likewise, the frequency with which you wish participants to journal can range from a daily request, to writing when a specific event occurs.

It is also important to think through the actual structure of the journal—if participants are presented with a notebook with far more pages than they are likely to be able to fill in their allocated time, this might create stress and anxiety. Alternatively, a notebook with too few pages might curtail writing and creativity. You also need to consider practical elements such as whether to have lines on the pages or leave them blank—lines help with written forms of expression but might hinder participants who also want to draw; one solution is for each page to have half blank space and half lines. Consider if you want participants to use a pen that you provide, so their text or drawings will be dark enough to be reproduced later in a possible representation of your data (if you gain their permission to do so, of course). While the traditional journal is a physical booklet in which participants record their entries with pen or pencil, it is also possible that some may wish to write up their journals on a computer or phone if they have access to one. Sometimes participants can find this more private and reassuring, since a file can be easier to password-protect and encrypt than hiding a physical journal.

Pre-journal Activities, Checking In, and Debriefing

Journals are often integrated with other qualitative methods as part of the overall research process. An interview or focus group just before the journaling process can provide a good opportunity for you to explain the exercise and expectations, while allowing participants to ask any clarification questions. This provides participants with the opportunity to decide whether to proceed or not after they are fully informed of the process. It is ethically imperative to let potential participants know how their data might be used at this stage. This is also when **informed consent** should be gained, if you have not already done so (see Chapter 2 on research protections). A pre-journal activity can also be used to collect background and demographic information about the journaling participants, such as age, gender, socio-economic background, education or work experience, or a range of other factors, as relevant to your project. A check-in meeting or interview half-way through the journaling exercise can be helpful, if time and resources permit, for you to confirm that the journaling is proceeding as hoped, and to answer any further questions. This can also help maintain participants' levels of enthusiasm and commitment.

An exit interview, focus group, or **survey** can allow you to start to reflect on the main themes emerging from the journals. Because the journal entries are usually written in isolation, this provides participants with a fair degree of freedom to decide what to include in their entries. While this is usually seen as one of the strengths of this method, it also means that information self-selection occurs, without you necessarily knowing why (for example, why a certain event is focused upon more than another). An exit interview therefore provides a great opportunity for you to ask questions regarding this selection process and to put participants' writing in a larger context. Such interviews also allow you to clarify the meanings of certain segments of what participants have written, or to confirm that your initial interpretations of participants' writing are correct. These exit activities can also provide a chance to ask participants broader questions with regards to journaling itself and their experiences with the method and should be reflected upon carefully.

Analyzing and Presenting Results

The analysis of data collected from solicited journals often follows thematic coding procedures, which can help you to draw out the main themes emerging in the data through an iterative process (see Chapter 18 for more on coding). For example, Singer et al. (2000), who were analyzing neighbourhood drug use and HIV risk, coded journal material that they received by locations and drug-use behaviours of interest. Alternatively, Morrison (2012, p. 71) used an "iterative rather than linear process" allowing her to be flexible and change her interpretations as she dug into the journals, so as to "describe, classify, and connect data". It is also possible to draw upon **narrative analysis** approaches that highlight the narrative structures (such as a story-line) within journal entries and that help maintain the participant's voices throughout the analysis process (see Wiles, Rosenberg, and Kearns [2005] and Cope and Kurtz [2016] for more on this approach).

It is important in your write-up to be **transparent** about how the data were gathered and analyzed. You should consider including information outlining your **sampling** procedures and decision-making, as well as the steps that the journaling process took. Also consider adding details regarding the selection criteria used to decide which data were included in your analysis, and how the data were transformed into an interpretative discussion. You need to be able to show readers "whose meanings are represented and why" (Baxter & Eyles, 1997, p. 509). How you approach these different stages of data collection, analysis, and interpretation rest in part on your **positionality** as a researcher, so reflecting carefully on your own positionality and **subjectivity** is important throughout this process (see Chapter 2).

As part of this reflexivity, throughout the analysis and writing-up processes it is important to consider the **power** relations inherent in this method and how best to enable the voices and opinions of participants to be heard. As for many other qualitative methods, while the journal writer has power over what they include in their contributions, the researcher usually retains the power to decide what information is revealed to a broader audience through analysis and writing (see Chapter 19 on writing qualitative geographies and Chapter 20 on communicating with wider audiences). To help in this regard, as well as paraphrasing participant responses where appropriate, it can be useful for you to include direct quotations from journals in your final write-up. This helps readers to see how the evidence and data you collected substantiate your interpretations, and reassures readers that your conclusions are **credible** (Filep et al., 2018).

The written and visual presentation of your findings can be supplemented by scanned images of participants' journal entries to highlight individual voices and narratives while providing visual depth and context to your interpretations. However, it is important that you gain the permission of the journaling participant first, as they might prefer to remain more **anonymous**. A distinctly geographic route to presenting journal results (as well as other qualitative results) can also be narrative mapping, introduced below.

Other Forms of Solicited Journals

Letter Correspondence as Journaling

A slightly different form of solicited journaling from what I have outlined above is letter correspondence. For example, Jennifer Harris (2002) completed a research project with women who self-harm by drawing on letters as a form of journaling. She notes that face-to-face interviews would have been very difficult to arrange given the degree to which self-harm is stigmatized, as well as raising ethical concerns regarding the possible impact on participants. Via a national organization for women who self-harm, Harris was able to contact possible participants and invite them to be part of her study. The women were told that they could write as much or as little as they wanted in their letters, while Harris (2002) reflected that the letters turned out to be extremely detailed and included intensely personal accounts. Harris responded to her participants' letters, attempting to show empathy but avoiding providing any advice; a strategy that she noted was frustrating at times but important with regards to ethical considerations. It is important to note that many Institutional Review Boards/Research Ethics Boards would also require you to provide participants with the contact information of a local support organization or professional, if you decide to undertake research on such sensitive topics.

Email Journals

With increasing access to technology in the Global North, and more and more in the Global South, it is not surprising that email journals are now being considered a useful method to generate research data. The benefits of this approach include the ease by which researchers can check in with participants to make sure the journaling is proceeding smoothly, and the lack of need to **transcribe** data. Nonetheless, depending on your research topic and location, you need to consider "digital divides," since different groups within societies might lack the skills, experience, and/or access to complete such journals with ease and comfort. As with other online approaches discussed in this volume (see Chapters 9, 13, and 15), concerns over confidentiality, privacy, and security are also important, with possible legal ramifications if you are working in a country where emails can be collected and used in court by authorities, or if a participant responds from a work email address. You also need to make sure that participants are who they say they are.

Jones and Woolley (2015) completed an email-diary research project regarding the impacts of the London 2012 Olympic games on commuters in the city. The authors sent participants daily emails with a maximum of four questions each time, asking them to "produce a log of their journey, detail anything they thought was different or significant, and to reflect on issues and news stories provided by the researchers" (Jones & Woolley, 2015, p. 712). Over time, these responses created a journal-like set of data for each participant. They also asked participants to complete a pre- and post-journal **questionnaire** on the research themes and method. Interestingly, Jones and Woolley (2015) sent follow-up emails to only some of their participants to gauge the differences this made to enthusiasm and completions. In the post-journal questionnaires, participants noted that they were encouraged by receiving such feedback.

Sound Journals

Sound journals or diaries can be drawn upon to help understand the nuances of individual engagement with sound, music, and place. Duffy and Waitt (2011) employed this method at the Four Winds music festival in New South

Wales, Australia. They noted "how individuals each respond to particular sounds is informed by life histories and what they value" (Duffy & Waitt, 2011, p. 121–2). In this particular study, participants were given mobile recorders and asked to record sounds they deemed meaningful while at the festival. The participants then took part in a conversation with one of the researchers during which they listened to the recordings together and the researcher asked questions regarding what was recorded and why. The authors found the use of sound journals particularly useful to capture the emotions participants experienced momentarily. Or as they put it, the journals were "successful in facilitating access to some aspects of these more difficult-to-articulate, in-the-moment feelings and affects of participants as they recorded, listened attentively and let the festival unfold around them" (Duffy & Waitt, 2011, p. 128). Such sound journals can be used to record people's relationships to a number of specific places, social settings, and events.

Other Kinds of Journals

There is a range of possible journal approaches used in geography that are more functionalist than those discussed above, such as travel journals in which transport geographers ask individuals to track their movements across a city. These journals, often more quantitative in nature, can help geographers to find patterns with regards to pollution exposure, public transit use, travel times, and specific traffic environments, but may also explore participants' qualitative experiences of transportation, sometimes in combination with other methods (e.g., Hansson and Roulston, 2017).

Economic journals, or household income and expenditure journals, can also provide quantitative data with regards to household income and spending patterns. These journals have also been used very effectively by human geographers to gain qualitative, in-depth information. For example, Magalie Quintal-Marineau (2016) asked Inuit women in Kangiqtugaapik, Nunavut, in northern Canada to track and detail their economic contributions to their own households as well as to other households in and outside their community. Participants completed two journals, each a week long. One of these was a week when government transfers were received, the other when no specific income was expected. The participants were asked to record all their income and expenditures, as well as more general information about recurring payments and government funds. Quintal-Marineau visited the women nearly every day to encourage them to complete their entries, sometimes filling in the journal with the participants. This allowed her to gain further information on themes that were of interest to the women at that time. She also completed an exit-interview in which she asked about specific interesting or unusual entries in the journals to gain further clarifications. These discussions, combined with the journals, provided rich insights into women's economic contributions at various scales and the sociocultural meanings of their contributions (see also Wiseman et al. [2005]).

Potentially Sensitive and Ethical Concerns

> Sometimes I was too tired to write. I didn't have a lot of time. Usually I get back to my place at 8 pm, then many women take turns to use the only bathroom. The landlord doesn't like us keeping the light on too long in the evening [because of electricity costs]. So by 10 pm, I need to turn off the light. I often used the time when I couldn't sell to write a bit here and there, because I can't really write in the evening. But I get anxious if I spend too much time in one place—I feel like I am missing potential sales in other places I haven't covered, so I take very little time to rest and write during the day. (Street vendor journaling participant discussing the experience of journaling, in Eidse and Turner [2014], p. 246)

A key concern with regards to solicited journaling is the burden that it places on participants since journaling requires a prolonged commitment, unlike standard interviews. As shown in the above quote, this street vendor found it hard to find the time and energy to maintain her journal given her fatigue at the end of a long work day. She also had to negotiate a parsimonious landlord not wanting her to keep the light on at night to be able to write her entries. Other concerns include literacy, as mentioned earlier, and while alternative approaches such as audio diaries might be possible, they carry their own limitations since participants might not be comfortable with the technology needed (Kenten, 2010).

While journaling can offer a space to "download" emotional memories depending on the topic of study, this can be both empowering and emotionally draining or stressful (Morrison, 2012). It is therefore very important to recognize and reflect upon the vulnerabilities involved with journaling. Meth (2003) describes how many of the women in her study regarding violence in South Africa shared painful stories in their journals that they had never revealed before. Morrison (2012) notes that during this process the researcher is not able to provide any immediate emotional support or terminate the data collection, as might be done in a face-to-face interview. For sensitive topics, it thus becomes even more important for you to check in periodically with participants to review the process, and to remind participants that it is their right to withdraw from the project at any time. If working on a sensitive topic, you can also search for possible local resources for interviewees to connect with if the process becomes difficult for them; this may also be an institutional requirement for you to get permission to undertake your study, as noted earlier. It is critical that you weigh up the potentially very significant ethical considerations of undertaking such research before getting started, if the topic could result in anxiety or distressful memories for participants, and determine if the benefits of such research truly outweigh the risks and concerns it could raise (see Chapter 2). Another major concern is protecting participants' privacy and confidentiality since journals contain a written record of personal information. For example, participants in Meth's (2003) study of women and violence in South Africa were often at risk of abusive husbands discovering their diaries. Thus, adequate attention must be given to minimizing such risks. In such a situation, perhaps an option could be to have "meetings" at which participants write their diaries and you store them safely between meetings.

As a researcher deciding on whether to use this method and how, it is also important that you consider the fact that participants are producing their journal entries in isolation from the researcher and hence entries can vary substantially in the degree of personal reflections, coverage of events, and regularity of entries. This means that there might be noticeable differences in the quality of the journal data collected (O'Connell & Dyment, 2011). Similarly, participant efforts might be quite different and, due to the time commitment needed, it is likely that some participants will give up. Some—but not all—researchers utilizing this method have thus decided that journaling is best used in conjunction with other methods. This is your decision to make.

As has hopefully become clear in this discussion, careful preparation is vital for researchers to be able to use this method wisely and respectfully. However, the rewards of doing so can be immense, with powerful accounts of daily experiences, emotions, and personal histories of people–place relationships.

Narrative Mapping as a Means to Communicate Solicited Journaling and Other Results

Narrative mapping is a type of qualitative cartography that provides a visual representation of relationships between the experiences of individuals or groups and their socio-spatial environments. Narrative maps draw on a combination of spatial data and data produced from qualitative methods such as solicited journals, interviews, sketch mapping, **photo elicitation**, or other qualitative methods. The reasoning for combining these two forms of data is because qualitative data can often provide "a much richer representation of the lived experience of individuals as compared to traditional, quantitative GIS data" (Mennis, Mason, and Cao, 2013, p. 271). This can help to provide insights into individual or group access, restrictions, or usage of specific spaces, fear in certain environments, or historical or contemporary narratives produced about certain locales.

As with solicited journaling, narrative mapping has proved to be successful in highlighting the voices of marginalized communities, as well as demonstrating dissimilar experiences between different groups. Narrative maps can contribute greatly to geographers' understandings of how space and place are experienced and reproduced (Bell et al., 2015; Evans & Jones, 2011). In part, this approach has arisen due to calls from feminist and LGBTQ+ geographers to make GIS and cartography more critically engaged with each other (Kwan, 2002a), with some even calling narrative mapping approaches a form of "countermapping" that can highlight and denounce injustices (Hohenthal Minoia, and Pellikka, 2017). For example, Brown and Knopp (2008), aiming to destabilize the heteronormative logics that underpin map-making and GIS science, produced a map indicating areas of historical significance for the queer community in Seattle, Washington. Their map drew upon a collection of archival material including **oral histories**, community newspapers, and memorabilia to produce a rich spatial narrative that had previously been ignored. Kwan (2008) used narrative mapping to highlight the lived experience of a Muslim woman amidst the Islamophobic atmosphere in Columbus, Ohio, after the 9/11 attacks. After soliciting an oral history and accompanying the participant in her everyday activities, the author produced a visual narrative showcasing which urban spaces the participant frequented and the associated feelings of danger, before and after 9/11.

One key reason to represent data through narrative mapping can be to show a defined chronology of events, drawing on narratives to help express certain sequences (Kwan & Ding, 2008). For example, maps might show precise GPS tracked routes or perhaps outline routes that are drawn from **go-along interview** notes. Complementary data, such as participant quotes, excerpts from interviews, or stories and photographs may then be linked to specific locations on the map (Pearce, 2008: Watts, 2010). See Box 14.2 for an example of such a narrative map created from data collected by an undergraduate honours student who biked and walked alongside a street vendor on her daily route around Hanoi, Vietnam, and interviewed her during their trip.

As well as representing events chronologically, narrative maps can also pinpoint specific events in space and time, with data gathered from solicited journals, interviews, or oral histories. A notable example of this approach is Annette Kim's (2015) space-time map of a city block sidewalk in Ho Chi Minh City, Vietnam. She created a map in which coloured bands are plotted in a two-dimensional space, with different coloured bands representing specific spatial uses of the sidewalk, such as motorbike parking, or street food stalls. A z-axis was then added to integrate time into the map as well, representing the changing uses of the sidewalk over different periods of the day.

Another approach is to use narrative mapping to visualize the importance of different locations for specific individuals or groups and the reasoning behind this. Again, data collection can be through solicited journals, interviews, sketch maps, or a range of other methods including participatory approaches. For example, Margaret Wickens Pearce (2014) completed a three-year collaboration with the Penobscot Nation Cultural and Historic Preservation Department to map the traditional Indigenous place names of Penobscot territory in the state of

BOX 14.2 A Narrative Map Illustrating the Daily Mobilities and Concerns of a Young Woman Street Vendor in Hanoi, Vietnam

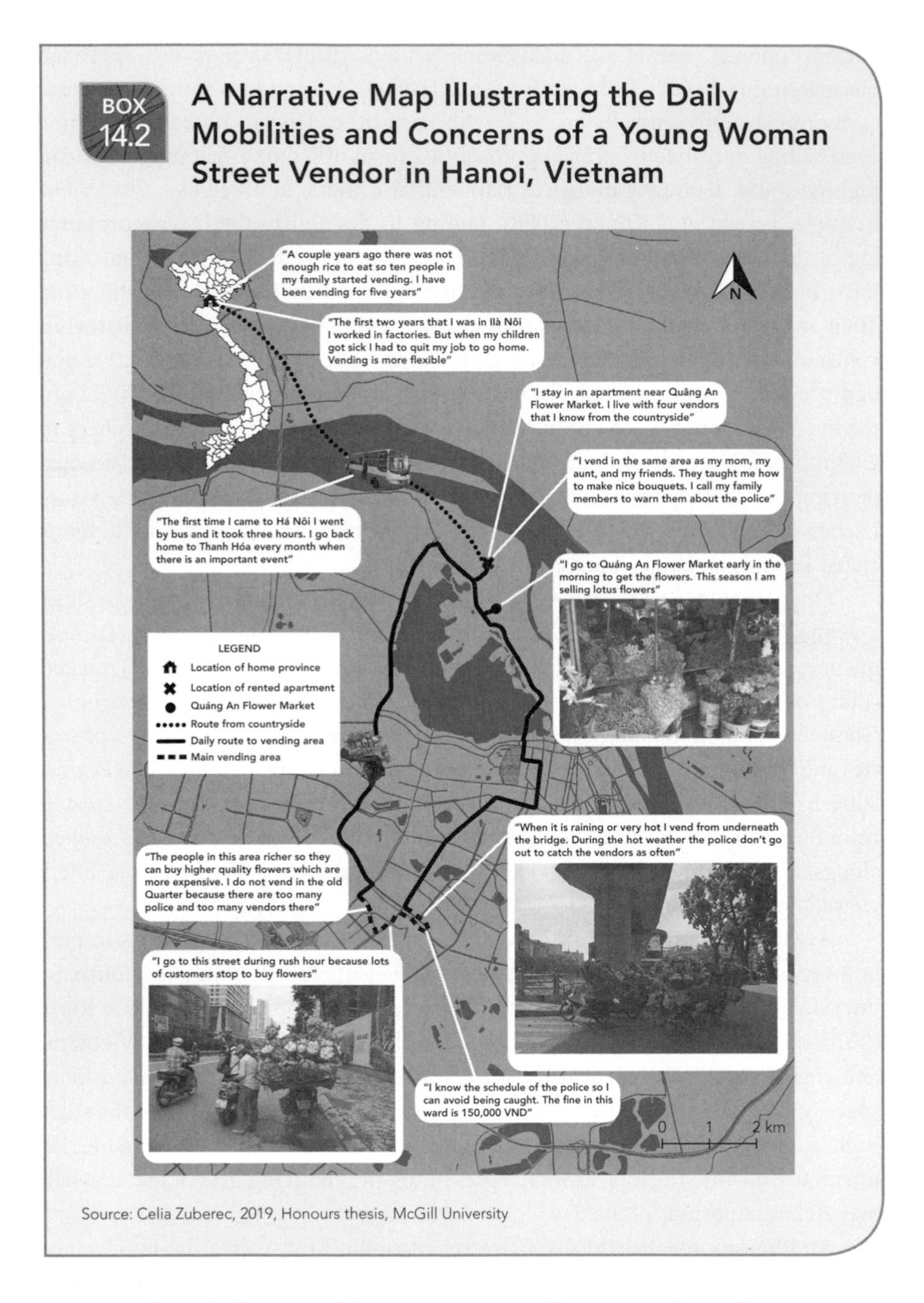

Source: Celia Zuberec, 2019, Honours thesis, McGill University

Maine in the US. The aim of this project was to illustrate traditional canoe routes and the surrounding landscapes, while highlighting the narratives that related to these. The maps that were produced had a base map on top of which traditional place names, canoe routes, and stories were mapped as well as community names. The routes and locales noted in stories about Gluskabe, a Penobscot hero, were also added. The resulting maps privileged succinct memories as efficient markers rather than relying on dominant western map-making conventions. As such, this approach could be considered a step towards helping **decolonize** cartography.

Conceptually, narrative mapping can help us to understand the meanings that we attach to specific spaces and how we infuse space with significance. This is what Yi-Fu Tuan (1977) has called "the transformation of space into place" (Knowles, Westerveld, and Strom, 2015, p. 237). Moreover, instead of advancing one uniform narrative, narrative maps are able to capture "multiple subjectivities, truths, and meanings" (Knigge & Cope, 2006, p. 2035). This provides the potential to challenge top-down and official narratives that tend to exclude the voices of marginalized communities, including queer communities (Brown & Knopp, 2008; Cieri, 2003), refugees (Caquard & Dimitrovas, 2017; Fawaz et al., 2018), women of colour (Bagheri, 2014; Kwan, 2008), Indigenous communities (Pearce, 2014), and low-income communities (Boschmann & Cubbon, 2013; McCray & Brais, 2007). Narrative maps are often characterized by their rejection of **positivist** principles of cartography. Some transgress the Cartesian confines of space by incorporating the dimension of temporality, and by mapping various forms of media, including excerpts from interviews, passages from stories or journals, photographs, sketches and paintings, and word clouds (Knowles, Westerveld, and Strom, 2015). Geographers have also undermined the norms of Euclidean geometry by depicting bodies and their movement at multiple scales (Kim, 2015; Pearce, 2008). These approaches mean that narrative mapping can "highlight the subjectivities that enter into virtually all representations of geographic space and human interaction" (Cieri, 2003, p. 149). In sum, narrative maps can be a positive means by which "visual images, words, and numbers are used *together* to compose contextualized cartographic narratives in geographical discourse" (Kwan, 2002b, p. 272, emphasis added).

Creating Narrative Maps

There are numerous ways in which narrative maps can be created and therefore the possible final representations take on a wide variety of forms. Most final narrative maps are two-dimensional renditions, but it is also possible to create three-dimensional models or interactive digital maps. There are many different software programs to create maps for this style of representation (however, map-drawing is beyond the scope of this chapter). Some basic approaches can be to work with Google Maps, free tools such as StoryMapJS (see Chapter 1 for example), or other map products found online. You can also often find shareware maps that can be a good starting point. Maps can be imported into PowerPoint as a very simple approach to editing and adding narrative materials, or there are more specialized programs such as Adobe Illustrator or ArcGIS that can be used. Caquard and Dimitrovas (2017) provide an overview of six different story-mapping applications, and detail the benefits and drawbacks of each. I recommend that you talk with your supervisor or cartography experts in your department to seek their recommendations as to the best way to proceed beyond these starting points, and also see Box 14.3 for some other initial ideas.

BOX 14.3

Reflections on Designing a Narrative Map

I experimented with creating narrative maps for my thesis that aimed to investigate the trading practices and livelihoods of Hanoi's young itinerant street vendors. During fieldwork in Hanoi, Vietnam, I completed semi-structured interviews with 35 young vendors who were all migrants to the city. Police regulations and tolerance of street vending fluctuates regularly in the city and is an ongoing cause for concern for these youth. I wanted to highlight their daily routes and routines around the city and the factors that supported or inhibited their mobility. Narrative maps seemed to me to be the perfect medium for doing so, since they allowed me to represent the vendors' relationships to their socio-spatial environments.

To create the narrative maps (including the one shown here as Box 14.2), I used data collected during "walking-while-talking" interviews with vendors, although sometimes I rode alongside them instead. This method provided me with the opportunity to ask vendors about their perceptions of the environment as we travelled through it. By shadowing the vendors along their entire route I became attuned to the spatiality and temporality of their movements. With their permission, I used a GPS tracking application to collect complementary data on their route through the city.

I generated a base-map of Hanoi city using QGIS, a free and open source Geographic Information System. I then imported this map as a PNG file into Photoshop where I added several layers of qualitative data including photos and quotes that I collected during the walking-while-talking interviews. I selected quotes that I felt best highlighted my overall research questions concerning the young migrants' reasons for taking up vending, how they became established, their relationship to public space, their treatment by local residents, and their strategies for coping with city vending regulations. Inspired by Kim's (2015) work in Ho Chi Minh

City, I added an insert map of Vietnam and depicted the province from which each vendor migrated, to illustrate the individual trajectories that underpinned their current work. By combining several forms of data I was able to clearly illustrate the vendors' experiences, perceptions, and everyday place-based practices.

Reflections from Honours student Celia Zuberec regarding her narrative mapping process (see Zuberec, 2019).

Narratives can be added via a range of different means such as those mentioned above, including inserting direct links to quotes in a digital project, summaries of interviewee statements, interviewee themes, photographs, sketches, or other art forms, or hyperlinks to similar media. Links are often made to interview excerpts and quotes, as shown in Box 14.2. Narrative maps are also an excellent way to represent data generated in a **PhotoVoice** project (see Chapter 16). Such mixed-method approaches of data representation allow patterns to be revealed that are often not possible when only one medium is used. They also allow understandings and interpretations gained from one method to be clarified and developed further through others (Bell et al., 2015; Knigge & Cope, 2006).

When designing narrative maps, it is important to explain in the accompanying text why particular narratives, in whichever form is chosen, are displayed on the maps (e.g., why specific quotes were used), and what sources they are from. It is also important to be clear as to what activities or events are being depicted in the maps. When designing the maps, also think through what degree of detail is required on each base map and how this will help the reader understand the representations you are creating. Should your base map be black and white or coloured? Will colours add necessary detail or be distracting? What degree of detail regarding buildings, roads, or political boundaries is relevant?

Some familiarity with map-making is required to create narrative maps, and not everyone will feel confident to take on this challenge. However, as noted earlier, it is often possible to start to develop such maps with fairly simple approaches such as a map displayed in PowerPoint and added narratives. GIS-based programs are very powerful but many other tools are available for those looking for user-friendly options.

Conclusion

Solicited journals require careful thought and preparation to make sure that everyone involved is clear regarding what is expected of them and that they are comfortable to proceed. Your overall research aim and questions or objectives will play a large part in how you work with this method, helping you to determine how you select participants, the length of time you will give them for writing their journals,

Selected Geographers and Some of their Narrative Mapping Projects

Geographers / Authors	Focus of narrative mapping projects
Nazgol Bagheri (2014)	Visually representing Iranian women's experiences, preferences, and usage of public space in Tehran
Eric Boschmann & Emily Cubbon (2013)	Spatial narratives of working poor individuals and the experiences of fear among lesbian, gay, bisexual, and transgender community members
Chris Brennan-Horley & Chris Gibson (2009)	Determining and mapping where unstructured forms of creativity are located in Darwin, Australia
Michael Brown & Larry Knopp (2008)	Creating maps to highlight spaces of historical significance to members of the lesbian and gay community in Seattle, Washington
Sébastien Caquard & Stefanie Dimitrovas (2017)	Mapping narratives regarding the life story of a Rwandan refugee
Julie Cidell (2010)	Mapping content clouds to compare how environmental issues are perceived and discussed according to location
Marie Cieri (2003)	Drawing upon qualitative mapping to challenge mainstream representations of lesbian and bisexual spaces
Sarah Elwood (2006)	Investigating the work of two community organizations to produce spatial narratives that advance different agendas for an inner-city Chicago neighbourhood
Jaes Evans and Phil Jones (2011)	Using GIS and walked interviews to capture the relationship of "what people say" to "where they say it"
Annette Kim (2015)	Investigating how the use of a sidewalk (pavement) changes over space and time in Ho Chi Minh City, Vietnam
LaDona Knigge and Meghan Cope (2006)	"Grounded Visualization"—developing an integrated method for analysing both qualitative and quantitative data through geographic information systems (GIS) and ethnography
Mei-Po Kwan (2008)	Representing the lived-experiences and narratives of Muslim women in the aftermath of the 9/11 attacks in the US
Anne Kelly Knowles, Levi Westerveld & Laura Strom (2015)	Creating GIS-based visualizations and inductive visualizations regarding the geographies of the Holocaust and survivor testimonies
Talia McCray & Nicole Brais (2007)	Exploring the relationships between transport and social exclusion for low-income women in Quebec City, Canada
Margaret Wickens Pearce (2008)	Mapping the historical geography of fur traders in the eighteenth century, in eastern Canada
Paul Watts (2010)	Mapping human actions and eyewitness accounts during the 1992 Los Angeles riots

and the style of journal approach (pen and paper or email, for instance). Your project will also direct the key theme(s) or question(s) you will want to ask. Start to think through how you want to analyze and write up your interpretations of results early on, so that you can adjust your research process as a whole if needed. In sum, the benefits from completing a journaling project can be immense, as participants can reveal important insights into your research topic including complexities and contradictions that you might not have even considered, as well as providing detailed understandings of a range of experiences, places, and events.

When you consider the ways in which you want to represent your data, either from solicited journals or other qualitative methods, narrative mapping can be an insightful approach. By simultaneously drawing on quantitative and qualitative data, narrative mapping allows for understandings gained from one method to be further clarified and developed through another (Knigge & Cope, 2006). Low-tech options and open source tools allow for quick and easy representations, while those with GIS skills have even more powerful resources to draw from. Narrative maps allow experiences and practices to be richly represented, and different points of view to be captured and displayed clearly. Pearce (2008, p. 17) sums up this approach nicely as helping advance geography's objective of "uncovering place."

The examples and suggestions that I have outlined here and in Box 14.4 are just a fraction of an exciting possible range of ways to both gain qualitative data via solicited journals, and to represent qualitative data in narrative mapping approaches. Both approaches allow you to create nuanced empirical and conceptual reflections regarding complex situations. While doing so, it is important to carefully reflect on the power relations and ethics involved and your own subjectivity as a researcher. If you are interested in gaining a deeper appreciation and knowledge of people's experiences of everyday life, and how these are shaped by places, social and political structures, and history, then these approaches can be extremely valuable.

Key Terms

diary
go-along interviews
narrative map(ping)
solicited journal

Review Questions

1. Journaling has been suggested to help researchers gain information on topics that might be particularly sensitive. Give some examples of how journaling could be a positive qualitative method option for potentially sensitive topics that you can think of.
2. What are some of the key concerns that researchers need to think about when planning a journaling exercise? Consider how these might vary in different parts of the world, or with different groups within a specific society.

3. What are some of the strategies that researchers can use to make sure that the journaling process goes smoothly for participants and researchers alike?
4. What are some of the main challenges associated with narrative mapping? Can you think of possible solutions or work-arounds to still utilize this approach?
5. Find an example of a narrative map. Did the authors manage to represent their data in a way that made it easier to understand and interpret their results than a written text? If yes, in what ways? If no, why not? Could they have improved their narrative map?

Review Exercises

1. Solicited journals

Find a scholarly geographical article that explains a research project that you think could be undertaken using a solicited journal approach. Consider how you would recruit participants, and who these participants would ideally be. Create the information sheet that you would provide for participants, and design the process of keeping in touch with them pre-, during, and after journaling. Think through ethical and logistical concerns as well.

2. Narrative mapping

Create a base map of your campus (for example a screen shot from google maps or the official campus map, then copied onto a PowerPoint slide). Create a brief interview schedule or journaling exercise that asks another student to reflect upon how the campus could be made more inviting and enjoyable—even fun—for users of diverse backgrounds. Add comment boxes in the appropriate locations on the map with the other student's key reflections of how these could be changed. Then go and find these sites and take photos of them to add to your narrative map too. Take your finished narrative map back to the interviewee and ask them whether they think the map is an accurate depiction of their reflections. If not, how could it be improved?

Useful resources

On solicited journals

Latham, A. (2003). Research, performance, and doing human geography: Some reflections on the diary-photograph, diary-interview method. *Environment and Planning A: Economy and Space, 35*(11), 1993–2017.

Meth, P. (2003). Entries and omissions: using solicited diaries in geographical research. *Area, 35*(2), 195–205.

Morrison, C.-A. (2012). Solicited diaries and the everyday geographies of heterosexual love and home: reflections on methodological process and practice. *Area, 44*(1), 68–75.

Thomas, F. (2007). Eliciting emotions in HIV/AIDS research: A diary based approach. *Area, 39*(1), 74–82.

On narrative mapping

Brown, M., & Knopp, L. (2008). Queering the map: The productive tensions of colliding epistemologies. *Annals of the Association of American Geographers, 98*(1), 40–58.

Caquard, S. (2011). Cartography I: Mapping narrative cartography. *Progress in Human Geography, 37*(1), 135–44.

Kim, A.M. (2015). Critical cartography 2.0: From "participatory mapping" to authored visualizations of power and people. *Landscape and Urban Planning, 142*, 215–25.

Kwan, M.-P. (2002a). Feminist visualization: Re-envisioning GIS as a method in feminist geographic research. *Annals of the Association of American Geographers, 92*(4), 645–61.

Emerging Digital Geographies

Jamie Winders

Chapter Overview

Digital geographies—the on-demand information and interactivity associated with the internet; the technologies, infrastructure, and devices used to access that information; and the codes and algorithms that enable this connectivity—increasingly shape and mediate many people's experiences with and in the world. How, though, does the digital era change qualitative research in human geography? This chapter examines emerging digital geographies and their impacts on the qualitative study of human geographies. In doing so, it explores two questions: first, what new research *topics* does a digital era/realm raise for geographers? And, second, how can geographers *use* digital technologies, devices, and software as research tools? Drawing from recent work by geographers, the chapter explores the methods used to study digital geographies; their impacts on cultural, social, and political identities and practices; their roles in social movements and migration; and digital technologies as research tools. As this discussion shows, thinking about the relationship between digital geographies and qualitative research means accounting for how digital devices and the connectivity they provide shape social lives and become the ways that we experience and encounter the world.

Introduction

> If digital devices mediate and are in considerable measure the stuff of social, cultural, economic and governmental lives in contemporary northern societies, then what does this mean for our methods for knowing those lives? (Ruppert, Law, and Savage, 2013, p. 24)

Digital technologies have profoundly changed the world and our interactions with social and physical environments. Survey data on our digital lives back up this claim, showing that around the world, most of us access the digital realm on computers or on our phones, with wide variation across countries as to which device is more common (e.g., phones in Sweden, computers in France) (Cole et al., 2018). In multiple ways, in most places, and for a growing number of people, the social world is increasingly "saturated" with digital devices and the on-demand information and interactivity they enable (Ruppert, Law, and Savage, 2013, p. 23).

Human geographers have been at the forefront of exploring the implications of these emerging digital practices. As has been well documented (e.g., Crampton et al., 2013; Graham, 2013a; Wilding 2006;), access to the on-demand information at the centre of **digital geographies** is itself geographically uneven across countries and from neighbourhood to neighbourhood, with rural areas in particular facing what Martin Dodge (2019, p. 37) describes as "not-spots" in internet connectivity and mobile phone coverage. That access is also socially differentiated along lines of race, income, gender, and age. While in some countries, like the UAE, France, and Taiwan, internet use by men and women is almost the same, in other countries, such as Egypt and Tunisia, there are significant differences, with men demonstrating much higher usage rates (Cole et al., 2018). Internet usage positively correlates with education and income but negatively correlates with age, with older residents in countries like Taiwan and Egypt having incredibly low usage rates (Cole et al., 2018). As Meghan Cope (2018, p. 102) notes, worldwide, young people have been "born digital," with "no recollection of a time before instantly and fully accessible information and communication." These social and spatial differences in access are important starting points in thinking about emerging digital geographies and qualitative methods. As nearly all scholars of digital geographies stress, there is no doubt that connectivity and the digital devices associated with it have created "a moment of radical shift from the past" (Wilding, 2006, p. 126). What is less clear is "the nature of this shift" (Wilding, 2006, p. 126) and, increasingly, how we collectively and individually should respond to it.

This chapter examines that shift by looking at the impacts of emerging digital geographies on the qualitative study of human geographies. As a number of geographers have suggested (Ash, Kitchin & Leszczynski, 2018a; Cope, 2018; Thatcher, 2019), digital geographies have changed the ways that human geographers think about qualitative research. Not only do we have new objects, practices, and spatial dynamics to study—virtual assistants; robotic vacuum cleaners that remember floorplans; and facial recognition software on doorbells, smartphones, and at international borders and airports—but we also have begun to articulate new ways to study these elements *and* to incorporate them into our research tools.

Despite a growing body of geographic scholarship on these digital developments (Ash, Kitchin, and Leszczynski, 2018b; Madge, 2007), many questions remain about the relationships among social media, the internet, and digital geographies and how interchangeable studies of these topics are. Within research on digital realms, there is debate over whether to focus on technologies, devices, social practices, or content, as well as on whether the technologies or the content they provide holds emancipatory or democratizing possibilities (see Kitchin et al., 2013). Although geographers acknowledge complex interactions between the material and virtual worlds (Graham, 2013a), there is no consensus about how human geographers should proceed vis-à-vis the digital realm or digital technologies themselves. Questions remain, for example, about how geographers can incorporate digital devices, practices, and software into the study of the **spatialities** of human life. What new research topics do emerging digital geographies raise? How might geographers use digital media and technologies as research tools?

This chapter is organized around these two questions.[1] The profound role of the digital in the everyday lives of many people means that we must consider how it can be used as part of qualitative research. As Evelyn Ruppert and her co-authors (2013, p. 24) note, "digital devices and the data they generate are both the *material* of social lives and form part of many of the apparatuses for *knowing* those lives." Thus, thinking about the relationship between digital geographies and qualitative research means accounting for how the digital realm, technologies, and practices shape social lives *and* become integrated in the ways that we experience and encounter the world. As this chapter will stress, these dynamics associated with the digital force us to rethink key assumptions about conducting qualitative research, at the same time that they reinforce key tenets in the qualitative study of the world "in real life" (IRL).

The following sections lay out different ways to think about the relationship between emerging digital geographies and qualitative research. The first explores definitions of the digital and how they compare with existing ideas in human geography, especially the link between social practices and material culture. The chapter then considers what human geographers can do with new digital technologies, highlighting innovative studies of the digital realm in geography and other fields. From there, I reflect on what digital geographies, as technologies, devices, and practices, can themselves *do*, particularly with regard to social movements and migration. In the chapter's last sections, I examine questions of identity and community wrapped up in digital geographies' emergence, before stepping back to identify themes in studying digital geographies and using them to conduct qualitative research.

What Is "the Digital"?

Before looking at digital geographies as a topic and tool of qualitative research in human geography, it is worth discussing what the digital is. As James Ash and his co-authors (2018a, p. 25) have noted, "Geography . . . is in the midst of a digital turn" in which the digital has become "both object and subject of geographical inquiry." They note further that the digital entails material *technologies* associated with binary computing architectures, the complicated *interactions* that result "from our spatial engagement with digital mediums," and the *logics* that structure both the technologies and our use of them (Ash, Kitchin, and Leszczynski, 2018a, p. 26). This definition, in many ways, echoes that of **new media** as both the digital devices that populate our daily lives and the connectivity that they enable, with an additional focus on the software, code, and algorithms (the logics) that "underwrite access to digital phenomena and mediations" (Ash, Kitchin, and Leszczynski, 2018b, p. 3).

Early work by foundational authors like Donna Haraway (1985), whose "cyborg manifesto" broke down perceived boundaries between humans, machines, and animals, has sparked geographic explorations of the social/digital constructions of place, space, and daily life, informed by critical perspectives on power relations, such as feminism and critical race theory (Bell & Reed, 2004; Whatmore, 1997). Recently emerging digital geographies build on this work, helping us understand the increasingly enmeshed nature of virtual spaces, social lives, and technological devices and "the hybrid and augmented ways in which the internet is embedded into our daily lives" (Graham, 2013a, p. 180). In practice, digital geographies encompass the on-demand access to information and interactivity provided by a range of digital devices, the devices themselves, and the code or software that links the two. Of course, what constitutes digital geographies today will change rapidly in the near future, but we can safely note a general trend toward more deeply enmeshed digital daily lives, to the extent that the boundary between what is digitally enabled and what solely inhabits a non-digital realm will be indistinct and, in effect, seem meaningless. Indeed, in much of the world, digital devices are "increasingly the very stuff of social life . . ., reworking, mediating, mobilizing, materializing and intensifying social and other relations" (Ruppert, Law, and Savage, 2013, p. 24). While the kinds of devices we use to access and mediate our digital lives and practices will change, the *idea* of digital devices as the way we do so will not. As a growing body of work clearly demonstrates (e.g., Geoghegan, 2019), the digital shapes the geographies many of us study, reconfigures the production of geographic knowledge, and itself is deeply geographical (Ash, Kitchin, and Leszczynski, 2018b).

This imbrication of digital and IRL spaces and practices raises complicated questions. When you watch a television show on your tablet or smart phone, are you using new digital media (smart phone and the on-demand information it provides) or more traditional media (television shows, film, etc.) or both? When you post a photograph on Instagram, are you using the traditional media of photography or the new digital media of image manipulation and shared images? At another level, how might either of these practices (streaming videos and taking photographs) become ways to conduct qualitative research? How can, and how *should*, geographers study the ways that digital geographies reconfigure understandings and experiences of space?

Thinking Spatially about and with Digitality: A Brief Review of the State of the Art

In 1997, Derek Alderman and Daniel Good pushed open the boundaries of existing digital geographies and raised some of the discipline's first questions about "electronic mass media" in a study of "the Virtual South—the production of the idea of a distinct American South through the cultural discourse and new electronic folklore of cyberspace" (p. 21).[2] What, they asked, does the Virtual American South look like? Where on the internet can it be found? By examining websites that promoted US southern cities and ideas about southern identity, they tried to identify "the possible differences and similarities between the cyber-representation of place and more traditional mediums of geographic portrayal" and, thus, to describe "the cultural geography of the Web" (p. 21). This early scholarship laid the groundwork for what has become an exciting line of scholarship on the relationships between the virtual and the material and the devices and technologies used to mediate these relationships.

Beginning in the mid to late 2000s, geographers showed an increasing interest in online practices and digital devices. Dydia DeLyser and her co-authors (2004) shared their experiences of buying historical materials on eBay and turning eBay itself into a research tool and archive. For human geographers more wedded to field-based research, digital geographies pose a number of thorny questions. When people are "on" their smart phones, where exactly are they? What are those spaces/places like, and how can geographers study them? How do we think geographically about social media like Instagram or Twitter that represent places, create a sense of community, and function as gathering sites for groups of people but are not "real" spaces in any obvious sense, aside from corporate headquarters, server farms, and the contract workers paid to flag inappropriate material? Can we study digital geographies from our laptops, tablets, or phones? Can we study human geographies *with* these devices? (Increasingly, we might ask, "Can we study human geographies *without* the digital?") Can geographers study digital engagements and dynamics without observing, or meeting, users in person—a question that the COVID-19 pandemic raised for many scholars in geography and beyond? In what ways can geographers help identify and understand the spatial constitution of digital lives?

More recent attention to digital geographies traces a longer genealogy for this emerging field. Ash and his colleagues (2018b), for example, situate digital geographies within geography's 1960s quantitative revolution, the popularization of GIS in the late 1980s, and the more recent growth of data science. From this perspective, digital geographies—with attention to everything from **big data** and algorithms to "smart cities" and video games—seems less like a new and unprecedented socio-technical development and more like a recent permutation of a much longer relationship between spatial analyses and digital data sources of various kinds.

Across the social sciences, scholars have asked such questions of and about the digital domain. Does distance matter in the age of the internet (Mok, Wellman, and Carrasco, 2010; Wilding, 2006)? (It does). How do digital geographies change face-to-face interactions (Green & Singleton, 2009; Mok, Wellman, and Carrasco, 2010)? Why are some parents fearful of what their children discover online (Valentine & Holloway, 2001), and how do other parents use technologies like Skype to manage long-distant relationships with their kids (Longhurst, 2013)? How quickly are new digital technologies and practices adopted and rendered passé (Wilding, 2006)? How important are devices to the individuals who own them (Green & Singleton, 2009)? How do digital geographies shape understandings of the multiple publics and audiences accessed and produced through interactive technologies like making memes or blogging (Kitchin et al., 2013)?

Attempts to address such questions have involved a range of methods. In many cases, scholars have used **interviews** to map with whom people regularly communicate and through what technologies, devices, and media (e.g., email, texting, social media, computers, mobile phones, face-to-face) (Mok, Wellman, and Carrasco, 2010). Others have used interviews to examine how online practices are interpreted by parents, teenagers, immigrants, and other groups (Arnado, 2010: Bonner-Thompson, 2017; Valentine & Holloway, 2001). Still others have found it "necessary to investigate what happens *in practice*" (Valentine & Holloway, 2001, p. 80, emphasis added) when people interact with digital geographies and practices. This attention to practice can focus on how people spend time online (e.g., Dean & Laidler, 2014), as well as how they interact and respond to technology like video games. Ash (2010), for example, **ethnographically** studied the testing of video games at a design company, interviewing game designers and keeping a **research diary** while he tested games for two years. Interested in the ways that **affect** could be manipulated for economic ends (in this case, more profit for game designers), Ash focused on how game designers tried to elicit particular responses and feelings from users and how users, especially game testers, became part of the process of manufacturing affective responses that made people play the game longer and more frequently.

Attention to practice in digital geographies can also highlight how information and communication technologies (ICT) that link us to digital spaces reshape the ways people use and design *material* spaces. Early research on this topic by Bjorn Nansen and his co-authors (2011), for example, examined "the relation between people, their media stuff, and their homes" (p. 697). How, they asked, do our uses *and design* of domestic space change as we embrace new digital technologies in work, study, and play? To answer these questions, Nansen and his co-authors followed the ICT use of four families in Melbourne, Australia, from 2004 to 2007. Each family was given cameras to photograph rooms in their house, scrapbooks to annotate and discuss those photographs, colour-coded stickers to tag forms of ICT, and diaries and maps to trace how ICT moved in and out of their house over time. Through this detailed look at the ways that families incorporated new technologies into their homes, Nansen et al. (2011) highlighted "the reciprocal relationship between ICT *stuff* and the structure they lived in" (p. 712). As they discovered, ICT does not just change how, where, or with whom we watch television or do homework. It also reshapes the very ways we use the spaces of our homes, leading to complete home redesigns in some cases.

In a related vein of work, several scholars have examined how technology, especially mobile phones, shape social practices and identities. In the 2000s, Eileen Green and Carrie Singleton (2007), for instance, wrote about the "mobile selves" that young Pakistani-British men and women in northern England associated with and experienced through their mobile phones, conducting **focus groups** to examine the gendered dynamics of phone use. Documenting gendered differences in how these Pakistani-British men and women used their mobile phones, Green and Singleton also found similarities—namely, that both groups communicated in multiple languages through their mobile phones and mixed local and global influences to create a specific youth culture experienced and expressed through digital practices and platforms.

More recent research on digital devices and practices has pushed beyond attention to use and toward more complicated questions about how the digital and the embodied become intertwined, as digital devices become more and more embedded in the daily lives of many people. Carl Bonner-Thompson (2017, pp. 1611–12), for example, has examined the ways that "digital spaces are deeply entangled with the fleshy corporeality of embodied experience" in a study of masculinities on the dating site, Grindr, the first gay geosocial app to formally launch through venues like iTunes. Geosocial apps collect users' location data, allowing people using the app not only to connect with other users but also to see which users are geographically close at any given time. Grindr, as Bonner-Thompson shows, is a key site in the performance and embodiment of sexualities, desire, and other facets of identity for gay men, even as some users expressed a strong desire to keep their "working and professional masculinities" entirely separate from their profiles and performances on the dating site. Here, rather than the blending or merging of different aspects of daily life and users' identities, we see digital technologies being used to mark bright boundaries between work and play, professional and personal.

Of course, digital technologies like mobile phones and smartwatches are not just objects that allow contact with friends and family or connect with new people. They are also objects that confer status and earn social capital (Green & Singleton, 2007). Phones can be personalized through cases, wallpaper, and ring tones and, thus, can be studied for the cultural and social messages they convey about their users. Smartwatches—the styles we wear and the ways (and places) we use them—signal something about us: we are health conscious and monitoring steps, we are busy and in high demand, or we are tech savvy and interested in the latest trends.

What is more, mobile phones and smartwatches can be traded, sold, given away, lost, or stolen, and tracked with GPS, becoming part of informal, formal, and even illicit exchanges bound up with and themselves shaping local cultures of commerce (Pfaff, 2010). These multiple ways that people perceive, use, and exchange digital devices like mobile phones highlight the complexity of studying digital geographies.

Although many of the studies discussed to this point mobilized common research approaches to study digital geographies (interviews, focus groups, study of material culture, etc.), Ruppert and her co-authors (2013) suggest that the digital era also calls key aspects of social-science research into question. Whereas traditional research methods, especially qualitative ones, are "deeply implicated in the formation of human subjects" (Ruppert, Law, and Savage, 2013, p. 33), the digital era complicates this focus on a fixed human *subject*. Methods like **questionnaires** or interviews presuppose identifiable research subjects who participate in them (although see Rose, 1997). When these methods shift into a digital context, the idea of a knowable, clear subject behind the screen becomes much hazier. In his study of the Occupy Wall Street (OWS) movement in Indonesia (2012), for example, Michael Oman-Reagan was never certain who his research subjects were. Indonesia's OWS movement took place online because activists faced harsh consequences if they protested in the streets. When Oman-Reagan (2012) began to interact with Indonesian activists through their online profiles, he wondered "if I was meeting the shadows, the puppets, or the puppet masters, and how I would know the difference" (p. 41). For Omar-Reagan, qualitative research conducted through digital media raised difficult questions about participants' identities, especially how to assess and understand the research subject accessed and encountered virtually. For more on personhood and **subjective** experience in "the digital," see Kinsley (2018).

Such questions create a quandary for researchers accustomed to seeing the identities of research participants as complex but at least "true" and legible in some sense. How, for example, do you obtain **informed consent** from participants in online research (Madge, 2007)? How do you know who is generating the content you examine in virtual correspondence? How do we study phenomena like fake Facebook or Twitter profiles, email services that allow anonymous communication, or catfishing (luring someone into a relationship via a fake online profile)? Such questions get even more complicated when we factor in generational differences in how people understand their digital lives and practices. Younger generations, especially those born in the twenty-first century, have come of age in the digital era in which, as Cope (2018, p. 102) points out, "not only are their everyday movements, interactions, and (Snap)chats fully logged and digitized, but also their own worldviews are already framed as digital ways of being." How that "digital way of being" influences young people's sense of identity and self remains an open question, but it has profound implications for scholarly understandings of digital geographies.

Of course, digital spaces can also be deeply politicized and marketized, and both practices merit critical attention from human geographers. As social media, from Twitter to TikTok, become key sites of political debate and contestation, users' opinions concerning how such sites should, or *should not*, be regulated by national governments vary from place to place. Whereas in the US and, to a lesser extent, Tunisia and the UAE, more surveyed respondents find it acceptable to express extreme ideas online, in many other countries, including Saudi Arabia, France, Cyprus, and Egypt, the pattern is reversed (Cole et al., 2018). In similar fashion, through phenomena like YouTube and Instagram influencers, digital performances and personas have become lucrative practices. In fields from gaming to beauty practices, comedy skits to tech demonstrations, individuals of all ages have built up enormous fan bases, allowing them to monetize their social media presence and amass significant cultural influence.

How might these efforts to politicize and monetize digital geographies reconfigure politics and economies IRL? What happens to understandings of identity—both self and other—in such contexts? Who is behind both sets of practices? As Ruppert, Law, and Savage (2013, p. 34) suggest, our examination of digital devices must consider how they "observe and follow activities and 'doings'—often, but not always or exclusively, those of people." In other words, "instead of tracking a subject that is reflexive and self-eliciting, they track the *doing subject*" (Ruppert, Law, and Savage, 2013, p. 35), which increasingly might not be human at all, as internet bots, autonomous systems designed to interact with digital users, pop up on sites ranging from Amazon to army recruitment pages. This idea of a "doing" subject that moves beyond humanist notions of the subject forces us to think carefully about with whom we are interacting when we move research into digital realms. At the same time, Sarah Elwood and Agnieszka Leszczynski (2018, p. 630) remind us that the digital itself "(re)produces power and extant sociospatial inequalities along lines of race, gender, class, sexuality, age, ability and more." In this way, even as understanding the digital means thinking beyond the human, it also means thinking along and across the multiple lines of difference and inequalities that contour the digital realm, its users, its representations, and its practices.

What Does the Digital *Do*?

The research described above pushes human geographers to consider what digital technologies and practices mean in and to people's lives, as well as to consider the connections and events they enable (i.e., how digital geographies change what people do). Jeffrey Juris (2012), for instance, looked at how social media like Twitter, and especially the ability to access social media through smartphones, played key roles in the Occupy Wall Street (OWS) movement. Through Twitter's shared temporality, OWS protesters in more than a thousand places around the world were able to exchange images and updates in October 2011 as part of a global protest produced through handheld phones. The "global" reach of OWS, however, was an overstatement because protesters in places like Egypt often lacked internet connections and because even in sites that were fully wired, face-to-face contact remained key to OWS's success. More recently, we have seen similar developments in the 2019 climate strikes, organized by young people, that involved an estimated four million people worldwide.[3] Although many websites went dark during the strike in solidarity with its efforts, significant portions of the climate organizing took place via Facebook, Twitter, and Instagram, even as the massive gatherings in city streets around the world were also necessary to sustain the movement. Thus, although digital technologies reconfigure how social movements function, it is important to remember that "places, bodies, face-to-face networks, social histories, and the messiness of offline politics continue to matter" (Juris, 2012, p. 260) and to shape the contours of digital geographies.

As this example shows, geographers interested in digital technologies can examine the kinds of practices they facilitate—in this case, new patterns of protest and new ways to link networks of global justice activists. At the same time, even as digital technologies—whether listservs in the 1990s or Twitter in the 2020s—create visibility for social movements beyond the local context and can quickly mobilize advocates and resources around the world (Froehling, 1999), they have their limitations. Twitter, for example, has allowed global movements to quickly blast out updates and images (Juris, 2012) and to connect activists around the world. At the same time, rapid, but brief, bursts of information via tweets cannot easily accommodate longer discussions and debates of these movements and their futures, in the ways that older technologies like email listservs can. This fact is an important reminder that while new media create new possibilities for social movements, they also create new challenges.

Rob Kitchin and his co-authors (2013) present yet another way that digital interactions—in their case, blogging—change how people interact with a wider world. As they showed, their collective blog about Ireland's financial crisis not only enabled them, as academics, to reach a wider and different audience than did their standard academic publications but also engaged in a different kind of knowledge through new means of **knowledge production**. Blogging as a digital practice changed how they understood and interrogated the Irish financial crisis. It brought them closer to their readership, through things like comment sections on blogs, but also weakened control over how their ideas were picked up, reworked, and reproduced elsewhere both online and in traditional media. In this way, blogging changed how these geographers worked with and engaged a wider public, altering their relationship with readers, with their own arguments, and ultimately, with the kind of writing they generated.

Digital geographies also enable new kinds of connections among people living apart. Particularly in research on long-distance migration, scholars have studied how families use digital media to stay in touch and, in the process, how they change family dynamics (Wilding, 2006). Geographer Robyn Longhurst (2013), for example, has examined whether the real-time communication offered through Skype is qualitatively different from written, text, and phone communication and, thus, whether it changes mother–child relationships. Working with a group of mothers in Hamilton, Aotearoa New Zealand, she found that Skype reoriented how mothers and their adult children experienced distance and how mothers assessed their children's well-being from afar. In some cases, Skype came to be seen as "part of the family," in the words of one mother, because it had "the capacity to bring [her] son and other children into focus" (Longhurst, 2013, p. 670). In other cases, mothers felt restricted in using Skype, which forced them to sit still in front of their computers, rather than to talk on the phone and move around. As such studies of family life show, digital technologies both enable families to feel closer and, thus, mediate a sense of separation *and* create new tensions around uneven adoptions of technologies and new expectations of constant availability through them. Further studies along these lines are clearly needed in the context of the global lockdowns and massive shift to digital interactions prompted by the COVID-19 pandemic.

It should be clear by this point that digital technologies and practices have the potential to reshape how things from social movements to families unfold, take place, and interact. In each case discussed, the digital realm impacted, but did not *determine*, the ways social, cultural, and political practices were "done" and created both new opportunities and new obstacles for users. One task for scholars interested in digital geographies, therefore, is to interrogate what the digital allows users to *do* and how that doing transforms social, cultural, and political geographies and practices. Another task is sorting out who we are and become, as well as whom we are with, when we engage digitally.

Who Are We in the Digital World? Who Is with Us?

Several scholars have explored the question of our digital identities—more specifically, how users of digital media understand and present a sense of self through these technologies. Koen Leurs and Sandra Ponzanesi (2011), for example, studied instant-messaging (IM) practices of Moroccan-Dutch teenage girls in the Netherlands. To examine "how gender, diaspora, youth culture and technologies intersect and influence each other" (p. 56), they surveyed over 1,500 teenagers in 2009 and 2010 on their use of digital media. They also interviewed Moroccan-Dutch girls about what they did and how they presented themselves online and analyzed the IM transcripts of six participants who agreed to save their IM conversations for two months. Through this work, Leurs and Ponzanesi found that Moroccan-Dutch teenage girls used IM as a key "communicative space of their own" and "a relatively safe playground" (p. 56) where they tried out different relationships and identities (see also Dean & Laidler, 2014).

Echoing findings from Green and Singleton's (2007, 2009) research on mobile-phone use, these girls used IM to create and manage public and private identities and relationships.[4] In the "*backstage* of IM" (Leurs & Ponzanesi, 2011, p. 58), young women engaged in private conversations, where they felt freer to talk about health, body issues, and discrimination with their friends, away from their parents' watchful eyes. "Onstage" in IM, through the selection of display names, photos, and their buddy lists, young women presented a different, more public version of *self.* Also echoing Green and Singleton, Leurs and Ponzanesi noted that IM created a way for Moroccan-Dutch teenagers to feel part of a Muslim global youth culture and to find a sense of belonging amidst hostile reception in the Netherlands. In this way, IM became a way of being in the world for these teenagers and "a space where they can negotiate several issues at the crossroads of national, ethnic, racial, age and linguistic specificities" (p. 56). Through IM and the digital devices that enabled it, Moroccan-Dutch teenage girls could experience being Muslim as "belonging to a particular Dutch as well as global youth-subculture grouping" (p. 63), all from their bedrooms and in ways that partially challenged the gendered restrictions they faced in everyday life as young Muslim women.

Putting It All Together

Mark Graham (2013a) argues that "Geographers should take the lead in employing alternate, nuanced and spatially grounded ways of envisioning the myriad ways in which the internet mediates social, economic and political experiences" (p. 177). The question this chapter has tried to address is, how should geographers do so in the context of digital technologies and practices? What do geographic concepts like space, place, and scale mean in digital geographies? If "Our 'place' is wherever our computer and phone are" (Mok, Wellman, and Carrasco, 2010), what does that location mean to and for geographic research?

The starting point for addressing this question is recognition that the virtual and material realm complement, rather than supplement, each other in practice and, thus, must do so in our research. Online and IRL aspects of social relations co-exist in complicated, recursive ways. As we explore different apps and platforms, we learn new ways to interact with people. These co-developing social relations and technologies, of course, are deeply gendered (Dean & Laidler, 2014; Green & Singleton, 2009), as well as raced, classed, and sexed in ways that geographers must explore. Given this recursive relationship between the virtual and the material, the social and the technological, geographers must examine both the *production* of meanings enabled by digital devices and the *materiality* of those devices, both *what* they enable and *how* they do so, whether in the context of online dating (Bonner-Thompson, 2017), rural livelihoods (Dodge, 2019), or national and global politics. Our study of digital geographies, in other words, must attend to how they create and rework social dynamics and relations, as people themselves engage in and perform "the digital."

A second point to keep in mind is that the *practices* of social media, from friending to tweeting to using Instagram filters, create "a distinct writing style with its own 'Internet-speak' norms" (Leurs & Ponzanesi, 2011, p. 58) and, thus, foster new cultural and social practices, giving rise to platform or app-specific "micro-cultures" (Ash, 2018, p. 147). Whether the Twitterspeak of 280 characters, emojis in texting, or pinned aspirational images in Pinterest, digital media shape how users communicate and, in the process, create new means of communication, which must also be incorporated into any study of digital geographies.

Third, the research discussed here has shown that it is important to learn as much as possible about whatever digital device, online practice, or virtual interactivity is under study. In their discussion of how historical geographers might use eBay, for example, DeLyser and her co-authors (2004) devoted much time to learning about how eBay works, how many users it has, how long it has been in existence, and how different groups of users interact with it (see also Gruzd et al., 2011). In this regard, studying digital realms is no different from studying the world IRL. Knowing as much as possible about one's topic is crucial to any study, and this knowing comes from both reading about one's research and spending time with it oneself.

Finally, as in qualitative research conducted IRL, research on and with digital geographies brings its own set of **ethical** issues. In a virtual context, many of the same ethical mandates apply, as Clare Madge (2007) has noted: informed consent, **confidentiality**, privacy, **debriefing**, and "netiquette" (p. 656). Figuring out how to meet those mandates, however, is more complicated, and this point continues to merit attention from geographers and other scholars (see Chapters 2, 3, and 4). As more geographers study digital realms and conduct research *with* and through them, more questions about research ethics will surface and demand consideration.

What, though, about using digital means to conduct qualitative geographic research itself? To conclude this chapter, let me discuss some convenient strategies. Technologies like mobile phones can facilitate qualitative fieldwork in all sorts of ways. Texting can be a way to unobtrusively take jottings of field observations that can later be the basis of extended **field notes** (Emerson, Fretz, and Shaw, 1995).[5] Fieldnotes taken on phones, tablets, or other digital devices can be automatically added to field diaries stored on the cloud, and backed up continuously, thus reducing the chance of lost fieldnotes that plague researchers dependent on hand-written hard copies.[6]

Second, phones can be used by researchers or research subjects to document and describe events and spaces that are central to a study—landscapes, social movements, monuments, neighbourhoods, and so on. With mobile devices and the ease of **geo-tagging** photographs (Jones & Evans, 2012), the task of asking research participants to document their daily lives or key events through images becomes nearly instantaneous and, through the ease of sending photographs or audio clips, can be done at great distances (see Chapter 14 for additional examples including journaling and narrative mapping). This practice allows research sites to be visually documented in multiple ways, and by multiple subjects. Crowdsourcing the meaning of a place through user-generated photographs or videos is just one of many possibilities here.

Third, **video call** tools like Skype, FaceTime, or Zoom can be used to conduct interviews or focus groups at a distance, "internationalizing research without adding costs to the funding body" (Madge, 2007, p. 656). Of course, as Kevin Dunn points out in Chapter 9 of this volume, interviewing online is not the same as interviewing in-person, and privacy concerns should be addressed, but videoconferencing technologies open new possibilities for qualitative researchers, particularly those who, for various reasons including pandemic restrictions on travel, cannot be in the field for extended periods of time or at great distances from home. Many of these tools can convene focus groups across places, bringing individuals with shared experiences but different locations together to discuss their experiences. Video chat technologies can also facilitate participatory research by enabling researchers to stay in closer contact and dialogue with research participants. Virtual "returns" to the field to present preliminary findings or conduct follow-up interviews become much easier, at least in places and among groups with access to these technologies. Finally, video chats can be recorded for future analysis and documentation, and many of them now provide auto-captioning, which, despite its many mistakes, can absorb some of the burden of **transcribing**.

Fourth, the popularity of apps like Twitter, Facebook, and Instagram makes them avenues for informal surveys or group dialogues on a given study, particularly for hard-to-reach groups. For example, Facebook networks were essential to the success of the *Roots Migration* project, exploring why residents of rural American states left or stayed in their places of origin (Morse & Mudgett, 2018). Obviously, selection biases limit how representative any one person's Facebook friends or Twitter followers will be. Additionally, as Madge (2007) discusses, there are serious ethical considerations for online research of this kind, ranging from whether online conversations are public or private to how informed consent online is obtained and confidentiality ensured (see also Chapter 13). Even with these issues, though, the social networks and communities of interest that digital communications enable create exciting possibilities for rethinking the "where"—and, thus, the how—of qualitative geographic research.

Finally, the GPS-enabled phones that many of us carry at nearly all times hold much potential for examining daily experiences and lives, as well as innovative uses of **volunteered geographic information** in our research. Software programs like Google Earth can be used to explore research sites before in-person visits, allowing researchers to become familiar with material places before leaving home.[7] As Taylor Shelton and his co-authors (2014) showed in a study of tweeting patterns associated with Hurricane Sandy in 2012, many digital devices create "data shadows," or "imperfect representations of the world derived from the digital mediation of everyday life" (p. 167). These geographic data (which some call the **geoweb**) can be useful for both qualitative and quantitative study in human geography. Geo-tagged tweets, photos, social media posts, and so on can be the basis of geographic analyses of, among other things, how places are represented, how those representations are reproduced and contested by different groups, and how places are linked or separated through social relations among residents. In the process, these digital geographies—the geographic relations produced through digital technologies and media—can help us as scholars visualize and interrogate different social geographies that might not be evident through other methods, such as **landscape** analysis, interviews, or **discourse analysis** of non-virtual texts and images.

Conclusion

Digital geographies have fundamentally changed political, social, cultural, and economic geographies for many peoples and places. The platforms, technologies, and devices discussed in this chapter capture and materially shape different kinds of relationships and geographies. They also shape "the pathways that guide how we use information" (Graham, 2013b, p. 78) and, thus, are also important as tools to conduct those studies. How much digital geographies reshape the doing of qualitative research depends on how willing we are to think creatively about the practice of qualitative research and to think critically about the questions of ethics, **power**, **subjectivity**, and practice that digital geographies provide for us.

Key Terms

big data
digital geographies
geo-tagging
information and Communication Technologies (ICT)
social media
Volunteered Geographic Information (VGI)

Review Questions

1. What new possibilities and new challenges do digital geographies raise for qualitative research in human geography?
2. How do digital geographies change, or not change, how human geographers think about key elements of research (fieldwork, data collection/analysis, etc.)?

3. How might a qualitative study in geography incorporate all the different components of digital geographies (devices, software, on-demand information, interactivity, etc.)? What might such a study look like?
4. How do digital geographies change the way we think about research subjects (their identities and subjectivities) in qualitative research in human geography?

Review Exercises

1. Select a group of digital users that you want to study (youth, immigrant group, men, college students, etc.) and an aspect of digital technology (a particular device, app, or online practice). Create five questions to use in interviews or focus groups that examine how the aspect of the technology you selected *shapes social and spatial dynamics and relations* for your group and how those digital practices/devices are themselves made meaningful by group members.
2. Pick a method discussed in this book (participatory action research, interviewing, questionnaires, etc.). List three ways that you could use digital technologies or devices to conduct this kind of qualitative research. What would they add to this method? What might be challenging about using digital technology in this capacity?

Useful Resources

Ash, J., Kitchin, R., & Leszczynski, A. (2018). Digital turn, digital geographies? *Progress in Human Geography, 42*(1), 25–43.

Ash, J., Kitchin, R., & Leszczynski, A. (Eds). (2018). *Digital geographies.* Thousand Oaks: SAGE. (See especially Cope, M., *Qualitative methods and geohumanities*, pp. 95–105.)

Green, E., & Singleton, C. (2009). Mobile connections: An exploration of the place of mobile phones in friendship relations. *Sociological Review, 57*(1), 125–44.

Madge, C. (2007). Developing a geographers' agenda for online research ethics. *Progress in Human Geography, 31*(5), 654–74.

Nansen, B., Arnold, M., Gibbs, M., & Davis, H. (2011). Dwelling with media stuff: Latencies and logics of materiality in four Australian homes. *Environment and Planning D: Society and Space, 29*(4), 693–715.

Ruppert, E., Law, J., & Savage, M. (2013). Reassembling social science methods: The challenge of digital devices. *Theory, Culture and Society, 30*(4), 22–46.

Notes

1. For a discussion of the implications of digital devices and data for social science methods in general, see Ruppert, Law, and Savage (2013).
2. For a critique of the metaphor of cyberspace, see Graham (2013a).
3. www.nytimes.com/2019/09/20/climate/global-climate-strike.html. Accessed 6 June 2020 These tools were also key to the 2020 protests against racial injustice, which were similarly global.
4. This finding echoes a wider characteristic of the internet: that "there is no clear agreement about what is public and what is private" in its content or experience (Madge, 2007, p, 661).
5. I thank Ricardo Millhouse for sharing this practice.
6. I thank Jesse Swann-Quinn for bringing this use to my attention.
7. I thank Jessie Speers for bringing this use to my attention.

Participatory Action Research: Collaboration and Empowerment

Sara Kindon

Chapter Overview[1]

Participatory action research (PAR) is increasingly common in human geography. It involves academic researchers in research, education, and socio-political action with members of community groups as co-researchers and decision-makers in their own right (Thomas-Slayter, 1995). As such, it is quite different from many other approaches to research and demands different types of attitudes and behaviours from a researcher. In this chapter, I present the overall process of PAR and its cycles of action and reflection. I also discuss different types of relationships and various strategies and techniques that can be used to enable collaboration with research participants at all stages of the process. These strategies and techniques aim to establish a more democratic research process, which respects and builds co-researchers' capacity and generates more rich, diverse, and appropriate knowledge. If done well, PAR has many benefits for human geographers, particularly for those committed to challenging unequal power relationships and increasing social justice. PAR can also be challenging to carry out: the emphasis on power, relationships, and change is a potent mix. Finally, I consider how to present PAR-generated research information to a range of audiences in effective and ethical ways.

What Is Participatory Action Research?

Participatory action research seeks to:

> engage people in a learning process that provides knowledge about the social injustices negatively influencing their life circumstances. The knowledge about social injustice includes understanding methods for change and thus organizing skills necessary to remedy the injustice. (Cammarota & Fine, 2008, pp. 4–5)

Participatory action research (**PAR**) has been evolving since at least the 1970s and as the above quote states, it is fundamentally different from many other approaches to social science research because its goal is not just to describe or analyze social reality but to help change it (Pratt, 2000). This change occurs through active collaboration between researchers and participants in the focus and direction of the research (Kindon, Pain, & Kesby, 2007a; mrs c kinpaisby-hill, 2019). To clarify, as an academic researcher thinking about embarking on PAR, you do not usually determine a research agenda independently. Rather, you work with a social group to define the issues facing the members (see also Chapters 2, 3, and 4). Together, you then generate and analyze information, which hopefully leads to action and, ultimately, positive change for those involved (for excellent examples, see Cahill et al., 2008; Cameron & Gibson, 2005; Gibson-Graham, 2005; Pain & Askins, 2011). In short, a PAR researcher does not conduct research *on* a group but works *with* that group to achieve change that *those involved* desire.

PAR has its origins in **action research** (Lewin, 1946) in the US, which sought to inform change by testing theory through practical interventions and action, and **participatory research** (Hall, 2005), which emerged out of Africa, India, and Latin America where educators and others involved in community development devised a new **epistemology** grounded in people's struggles and local knowledges. When combined, participatory action research sought to "develop new alternative institutions and procedures for research that could be **emancipatory** and foster radical social change" (Kindon, Pain, & Kesby, 2007b, p. 10).

PAR has many similarities with **community-based research (CBR)**, which has evolved from public health as medical and health-care practitioners seek ways to better involve community members in research to improve health outcomes. Many universities, particularly in North America, have adopted CBR to reconnect academic interests with the communities in which they are located. Where geographic information systems (GIS) and spatial mapping are also prominent in university service-learning courses, such an approach has been identified as **community geographies** (Robinson, 2010; Robinson et al., 2017).

PAR also shares many aspects of feminist and Indigenous research (see Chapter 4) through its commitment to the destabilization of dominant ways of knowing, principles of collaboration, and negotiation, and its recognition of the importance of place, embodiment, and our senses in the **co-constitution of knowledge** and change-based action (Askins, 2018). However, PAR may not always be **decolonizing** in its design or implementation, and may not engage **more-than-human** participants effectively. The tensions that persist between participatory and Indigenous approaches require careful attention and negotiation to find mutually-suitable ways of "being-in-the-world" (Coombes et al., 2014).

These variations and tensions withstanding, PAR generally involves a number of stages through which academic and social group members work together to define, address, and reconsider the issues facing them (Kindon, Pain, &

Kesby, 2007a; Parkes & Panelli, 2001). Such issues commonly include the lack of access to information or resources, the threat of removal of services or subsidies, or the need to respond to and mitigate unanticipated events. The emphasis on an iterative cycle of **action–reflection** is one of the key distinguishing features of PAR. It can also enable multiple perspectives of different **stakeholders** to be taken into account throughout the research, which can lead to more informed decision-making and more equitable and potentially sustainable outcomes (Halvorsen, 2019).

PAR attends to **power** relations (Kesby, Kindon, & Pain, 2007) and as such can be challenging, particularly in the context of an undergraduate research project. It often involves the researcher in a facilitative rather than **extractive** role and demands considerable attention to **ethics** and issues of **representation**. That said, PAR can be very rewarding, and even if it is not possible to involve research participants deeply in every step of a research project, it may be possible to make your research more participatory by adopting some of the ideas discussed in this chapter.

Conducting "Good" Participatory Action Research

PAR is an approach that ideally grows out of the needs of a specific context or place, a research question or problem, and the relationships between researcher and research participants. It is more about the value orientation of the work and its approach (epistemology) than about the specific techniques used, although participatory techniques are certainly important (see Kesby, Kindon, and Pain (2005) for discussion of deep versus other forms of **participation**). It is also an approach that values the process as much as the product so that the "success" of a PAR project rests not only on the quality of information generated but also on the extent to which skills, knowledge, and participants' capacities are developed (Chatterton, Fuller, & Routledge, 2007; Cornwall & Jewkes, 1995; Kesby, Kindon, & Pain, 2005; Maguire, 1987).

According to some practitioners (see Chambers, 1994), the most important aspects of participatory work are the attitudes and behaviours of "outside" researchers (usually us, as academics or practitioners). These attitudes and behaviours affect the relationships formed with research participants and the outcomes achieved. Whether you, as a researcher, respect people's knowledge or perpetuate unequal power relations and extract information largely for your own benefit depends to an extent on your attitudes and behaviours and the nature of the research relationships you establish. To illustrate this point further, Box 16.1 shows common connections between the attitudes of the researcher (illustrated here by things a researcher may say to a researched group), the kind of research relationships they form, the resultant mode of participation possible, and the relationship between the researched group and the research itself.

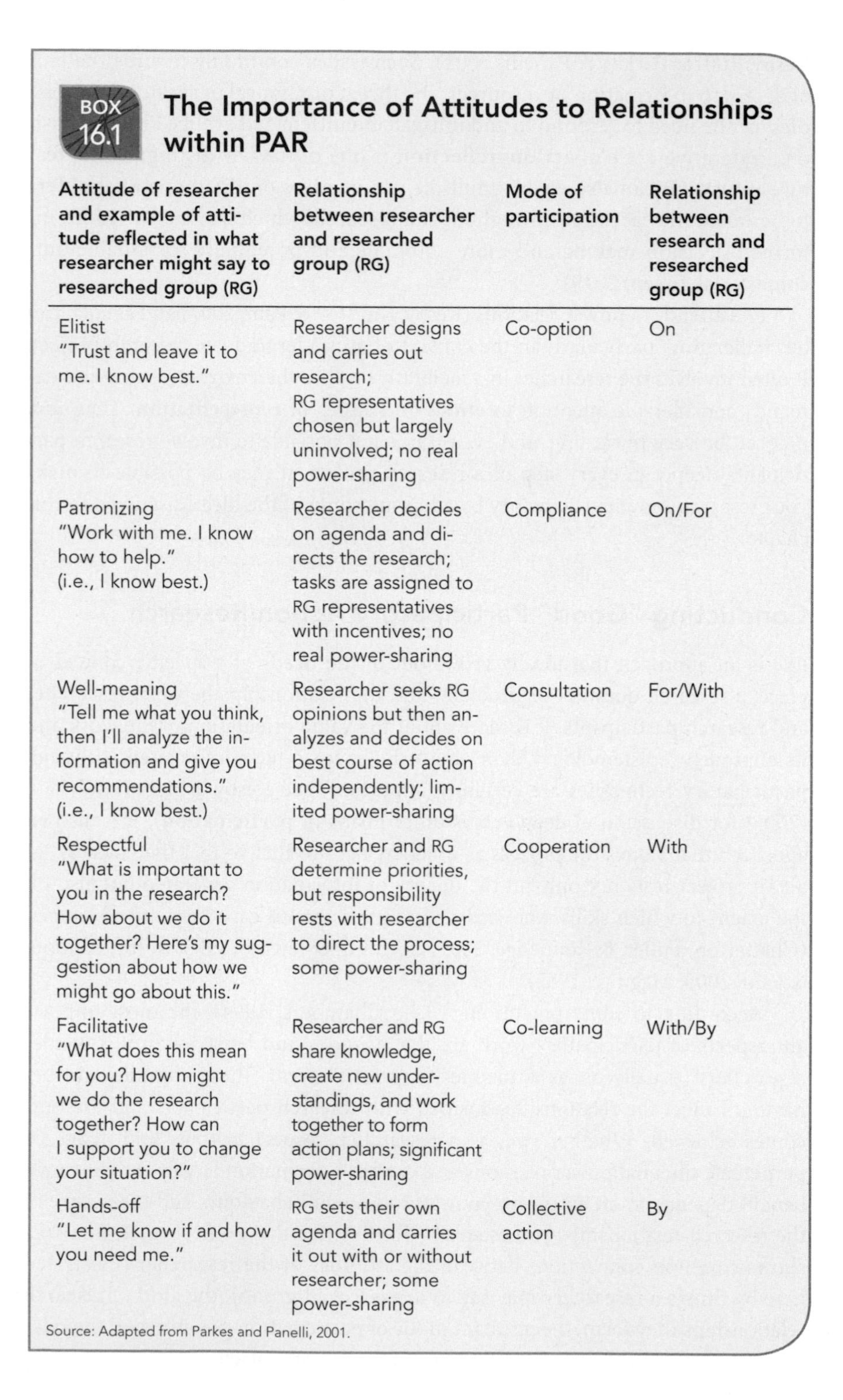

BOX 16.1

The Importance of Attitudes to Relationships within PAR

Attitude of researcher and example of attitude reflected in what researcher might say to researched group (RG)	Relationship between researcher and researched group (RG)	Mode of participation	Relationship between research and researched group (RG)
Elitist "Trust and leave it to me. I know best."	Researcher designs and carries out research; RG representatives chosen but largely uninvolved; no real power-sharing	Co-option	On
Patronizing "Work with me. I know how to help." (i.e., I know best.)	Researcher decides on agenda and directs the research; tasks are assigned to RG representatives with incentives; no real power-sharing	Compliance	On/For
Well-meaning "Tell me what you think, then I'll analyze the information and give you recommendations." (i.e., I know best.)	Researcher seeks RG opinions but then analyzes and decides on best course of action independently; limited power-sharing	Consultation	For/With
Respectful "What is important to you in the research? How about we do it together? Here's my suggestion about how we might go about this."	Researcher and RG determine priorities, but responsibility rests with researcher to direct the process; some power-sharing	Cooperation	With
Facilitative "What does this mean for you? How might we do the research together? How can I support you to change your situation?"	Researcher and RG share knowledge, create new understandings, and work together to form action plans; significant power-sharing	Co-learning	With/By
Hands-off "Let me know if and how you need me."	RG sets their own agenda and carries it out with or without researcher; some power-sharing	Collective action	By

Source: Adapted from Parkes and Panelli, 2001.

Researchers using PAR generally strive to adopt and practise the attitudes and behaviours that result in people's **co-learning** and **collective action**. They also generally follow an iterative process of action–reflection (see Box 16.2), although the specifics of what actually happens, how, and when vary depending on the particular context and circumstances of those involved.

The cycles of action–reflection outlined in Box 16.2 ideally involve researchers and collaborators in each stage. However, this may not be possible within the confines of an undergraduate research project (see Pain et al. [2013] for a helpful discussion here). Do not be put off, since this "ideal" process can and should be adapted collectively to suit the particular needs and constraints of the situation.

BOX 16.2

Key Stages in a Typical PAR Process

Phase	Activities
Getting started	• Assess information sources. • Scope problems and issues. • Initiate contact with researched group (RG) and other stakeholders. • Seek common understanding about perceived problems and issues. • Establish a mutually agreeable and realistic time frame. • Establish a **memorandum of understanding** (MoU) if appropriate.
Reflection	**On problem formulation, power relations, knowledge construction process**
Building partnerships	• Build relationships and negotiate ethics, roles, and representation with RG and other stakeholders. • Establish team of co-researchers from members of RG. • Gain access to relevant data and information using appropriate techniques (see Box 16.3). • Develop shared understanding about problems and issues. • Design shared plans for research and action.
Reflection	**Reformulation, reassessment of problems, issues, information requirements**
Working together	• Implement specific **collaborative research** projects. • Establish ethically-appropriate ways of involving others and disseminating information (see Box 16.3).
Reflection	**Evaluation, feedback, re-participation, re-planning for future iterations**
Looking ahead	Options for further cycles of participation, research, and action with or without researcher involvement.

Source: Adapted from Parkes and Panelli, 2001, p. 98.

histories of particular crops, animals, or trees, and changes in land use, population, migration, fuel uses, education, health, credit, and so forth may enable analysis of cause and effect factors over time.
- *Seasonal calendars.* Focusing on seasonal variation of particular factors (for example, rain, crop yields, workload, travel) can enable insights into matters such as climatic variation, labour patterns, migration, diet, and local decision-making processes.
- *Daily time-use analysis.* Indicating the relative amount of time, degree of drudgery, and level of status associated with various activities may reveal local power relations and identify the best times for research activities (see also Chapter 14 on journals)
- *Institutional or Venn diagramming.* Drawing out the relationships between individuals and institutions using overlapping circles signifying the importance or closeness of the relationships enhances understanding of power relations, local and surrounding contexts, and the identification of where there are blocks to or possibilities for change.
- *Well-being grouping (or wealth ranking).* Grouping or ranking households according to local criteria, including those considered poorest and worst off, can be a helpful lead into discussions about the livelihoods of the poor and how they cope depending on the particular cultural context.
- *Matrix scoring and ranking.* Drawing matrices of resources, such as different types of trees, soils, or methods of health provision, then using seeds to score or rank how they compare according to different criteria (such as productivity, fertility, or accessibility) can reveal local preferences (what scores highly) and the aspects that inform decision-making strategies.

Many of these techniques can generate information that can subsequently be digitized to create infographics or visual representations that present data in appealing and accessible ways.

Use audiovisual technologies and the internet to engage people and disseminate information:

- Mobile phones and their various apps (Facebook, Messenger, WhatsApp, Instagram) can be helpful tools for communicating and involving people in research, particularly if this is their usual mode of communication (see Chapters 9 and 15).
- Mobile phones, or small digital audio or video cameras, are helpful to engage people in mobile interviews and analysis of local issues. Short clips can be shared with others to stimulate discussion and further analysis.

- Establish closed Facebook, Messenger, WhatsApp, and Viber groups or use GoogleDocs to create spaces in which co-researchers and participants can share data, ideas, and responses to emerging analysis. As with all group spaces, ethical considerations about identifiability and storage of people's information need to be carefully managed and moderated (see Chapters 6 and 20 for more ideas).
- Design and promote project websites as a productive means of engaging participation and establishing a core identity for a group, as well as disseminating the process, findings, and outcomes of participatory action research.
- Sites like YouTube or Vimeo can be useful to share information widely, but take care with settings if you do not want images and videos to be used and appropriated by others for their own means.

It is vital to discuss the various technologies associated with access, storage, and sharing of data during and after the research as part of your original ethical agreements or MoUs negotiated between collaborating parties, and to revisit these processes and aspects of ethical consent regularly throughout your work together.

Engage people in joint analysis, reflection, and future planning

- *Shared presentations and analysis.* Involving people in the presentation and analysis of maps, diagrams, and other information generated throughout the research shares power and allows information to be checked, corrected, and discussed.
- *Contrast comparisons.* Asking group A to analyze the findings of group B and vice versa can be a useful strategy for raising awareness and establishing dialogue, particularly between different groups. This technique has been used for gender awareness, asking men, for instance, to analyze how women spend their time.

Further information on these and other strategies and techniques, with examples of their use, can be found in Pretty et al. (1995) and Pain et al. (2017).

While the above list of techniques may seem exhaustive, it is not prescriptive. Integrating one or several of them, where appropriate, will enhance your research. However, a few participatory techniques will not, in and of themselves, make your project PAR. For this to occur, the open negotiation of the research design and methodology with the people with whom you are working is critical, as is an emphasis on supporting people's capacity to do their own research and analysis.

For more ideas about what this might mean in practice, the work of Caitlin Cahill and the Fed Up Honeys is instructive. Over several weeks, a US academic researcher (Cahill) met with six young women (also known as the Fed Up Honeys) to carry out a PAR project, Makes Me Mad: Stereotypes of Young Urban Womyn of Color, in New York City's Lower East Side. The process involved a period of time deciding first on the research focus and then on the approach they would adopt. They decided that each of them would engage in reflective journal writing and analysis of their own thoughts, plus the reading and discussion of each others' writing, to explore how they had internalized racist and sexist stereotypes and applied the same representations to others. It was an intense and at times frustrating process for participants because of the project's close attention to detail and the challenge it presented to their personal beliefs. The project required dedicated and **reflexive** facilitation by Cahill and commitment from each participant to enable them to work through their differences. It was worthwhile, however, for the depth of friendships it produced, the degree of increased awareness it provoked, and the level of political action that the participants engaged in through their development of a website and their implementation of a sticker campaign (see Cahill [2004]). Box 16.4 summarizes some ideas about how to approach doing PAR. The points integrate and reinforce ideas about the attitudes, behaviours, relationships, research design, process, and techniques discussed above.

BOX 16.4 Some Ways to Promote Participation in Geographic Research

- Involve/be involved with the group with whom you are working as equal decision-makers to define the research questions, goals, and methods and as co-researchers and analysts of information generated.
- Show awareness that you are an outsider to the group you are researching, even if you are working together as co-researchers.
- Be clear about the potential impacts people's involvement may have and what will happen to the information generated especially if it is to be shared digitally or via online sites (ideally through an MoU).
- Take care not to promise too much or inflate people's expectations of what might happen as a result of the research.
- Develop facilitation skills, which can stimulate initiative and sensitively challenge the status quo without imposing your own agenda.
- Work at fostering participatory processes and research techniques, which will release creative ideas and enthusiasm but not take too much time for those involved (see Box 16.3).

- Seek out the perspectives and participation of the most vulnerable and marginal people.
- Find ways of limiting the dominance of interest groups and powerful people (including yourself, where appropriate).
- Find ways to build capacity of others to generate, analyze, and present information.
- Acknowledge that process is as important as product (and sometimes more important), and factor in enough time to involve people appropriately at various stages of the research, including time for reflection.
- Support the group with whom you are working to share the benefits of their involvement with others and to take initiative to address their concerns.
- Involve the group with whom you are working in the writing and dissemination of relevant information; at the very least, acknowledge their contributions to any sole-authored work.
- Practice honesty, integrity, compassion, and respect at all times.
- Keep a sense of humour!

Source: Adapted from Botes and van Rensburg (2000, pp. 53–4) and Kesby, Kindon, and Pain (2005), and the author's own experiences.

The Value and Rewards of Participatory Action Research

Many uses of PAR are oriented at social change because it offers a tangible way of putting the aims and principles of **critical geography** into practice (Kesby, 2000). Often, this means specifically addressing issues of racism, ableism, sexism, heterosexism, and imperialism (Ruddick, 2004, p. 239) and how they are manifested through people's unequal access to and control over resources or their positions within inequitable social relationships. A helpful example of this can be found in the work of feminist geographer Geraldine Pratt who has worked for more than 20 years with different generations of Filipina migrants to Vancouver, British Columbia, using PAR (Pratt & Philippine Women Centre, 1999, 2009). Together, they have explored different dimensions of women's lived and work-life experiences. Storytelling and testimony have been central methods that have led to collective mobilization and action to challenge working conditions, racism, and to raise awareness about mental health challenges of living in transnational families. Together with Caleb Johnson, they developed a testimonial play that has been performed in Vancouver, Berlin, Manila, and Whitehorse, Yukon, since 2009.

In my own work with Te Iwi o Ngaati Hauiti in the central North Island of Aotearoa New Zealand, several *iwi* (tribe) members established themselves as a community video research team to explore the relationships between place, cultural identity, and social cohesion (Hume-Cook et al., 2007; Kindon, 2012).

of clarifying expectations, establishing greater collaboration, and specifying roles and responsibilities (Kindon & Latham, 2002; Manzo & Brightbill, 2007).

Participatory techniques can generate information quickly, but they are not a substitute for more in-depth social research methods (Kesby, 2000) such as those discussed elsewhere in this book. Understanding the contexts within which information is generated is critical to our ability to rigorously analyze it. For this reason, you might wish to consider first undertaking PAR within a community or location already familiar to you. For an undergraduate dissertation, this could provide you with some of the necessary contextual information, freeing you to focus more energy on the process.

Sometimes our desire to avoid exploitation or extractive research relationships can mean that we become so involved with our co-researchers that we are unable to work effectively for change. Establishing outside support networks (see Bingley, 2002) can help to prevent this situation and sustain our endeavours.

A related point is that long-term relationships, even friendships, with participants and co-researchers commonly develop through PAR, and while some studies may become lifetime projects, we typically have to leave the group with whom we have been working. Investing time into a sensitive and appropriate leaving strategy at the beginning of the research can help to avoid raising expectations and assist in navigating the changing status of relationships (Kindon & Cupples, 2016). Formal meetings, celebrations, feedback sessions, visits, and the exchange of gifts may all be appropriate mechanisms to assist with closure.

In other cases, it may be academically and professionally important to maintain a sustained engagement long after the official research project is over—particularly to ensure the spread and impacts of any empowering change (Kesby, 2005) and/or to practice what Alissa Starodub (2019) has named "horizontal participatory action research." Your engagement may be in the capacity of support person, community board member, fund writer, publicist, or campaigner. There may be ethical challenges if your status changes from co-researcher to "friend" or "resisting other" (Chatterton, Fuller, & Routledge, 2007), and being clear about what you can commit to in any of these relationships is vital. Overall, a key way to manage this and other challenges associated with PAR is to be realistic with yourself, your co-researchers, and other stakeholders about what is possible within the time and resources available to you.

Sharing Results

As a student or academic, the results of your work need to meet the requirements of the academy. It can be challenging to present the "results" of an iterative and participatory research process within the context of a typical thesis or dissertation but certainly not impossible (Kale, 2017; Klocker, 2012).[3] Within your dissertation, thesis, or research report including direct quotations from a range of people, which tease out common or disparate threads, can illustrate the multiple perspectives in circulation. Citing a disagreement or exchange between people can highlight where there are tensions or differences of perspective. However, take care to contextualize and analyze them adequately or they could be overwhelming to the reader. In addition, discussing aspects of methodology—so important to the participatory process—can honour people's involvement, acknowledge that process is as important as product, and enrich the analysis of the "results" produced.

Engaging in PAR provides you with the opportunity to negotiate explicitly how you will use information and how you will represent others' experiences and/or views (Cahill, Sultana, and Pain, 2007; Kindon & Latham, 2002). Although sometimes time-consuming, such steps can temper your powerful position as the sole author in what has been, until now, a collaborative process. Sharing your choices and discussing how you intend to construct your argument continues the participatory process and goes some way toward ensuring that your final product respects the people and diversity of issues involved. An MoU item about this at the beginning of the research can prevent misunderstandings when you later want to quote people or include maps or diagrams produced through the research process.

It may be appropriate and courteous to include co-researchers as co-authors on any papers that may emerge from PAR (see for example, Box 16.5; Hume-Cook et al., 2007). We may also work behind the scenes to enable our co-researchers to publish or disseminate their understandings independently of us. Or we can establish a collective name under which we write and publish. These strategies can demonstrate to others a collaborative and collective approach to research and writing (for example, see mrs kinpaisby [2008] and mrs c kinpaisby-hill [2008, 2011, 2013]). However, with increasing pressures on academics to demonstrate their scholarship through publications and citation indexes, we need to weigh up the benefits and potential limitations of not having our names identified as an author.

An Example of Participatory Publishing

The following text is from American geographer Sarah Elwood, who worked for many years with urban-based community groups and municipal councils in Chicago, Illinois, to support research, education, and community capacity-building within processes of neighbourhood redevelopment. It is a collaboratively authored chapter and represents one way in which multiple research partners' voices can be respected and shared.

> In this chapter, we provide a collectively authored discussion of our use of GIS as a negotiating tool, and the lessons we have learned about sustaining community-university PGIS (**participatory**

(continued)

geographical information systems) projects. Our author group includes a university-based researcher (Sarah), the executive directors of the Near Northwest Neighborhood Network (NNNN) and the West Humboldt Park Family and Community Development Council (the "Development Council") (Bill and Eliud), a university-based research assistant (Nandhini), and past and present staff members of NNNN and the Development Council (Kate, Reid, Niuris, Lily and Ruben). . . .

From the perspective of the community organisation staff, PGIS partnerships are likely to be sustainable and effective if they produce immediately applicable results, and if university partners do not dictate results. One staff participant writes:

> There has to be a clear understanding that the action and the research produced are usable for the community, not just something to publish.

Both directors argue that leaders of organisations have a special role to play in sustaining GIS resources and skills. One writes:

> Part of the job of executive director is recognizing the importance of the project, but also finding a way to keep it for the organisation, and not let it go.

In sum, for community organisations a great deal of the power and impact of PGIS stems from the diversity of ways it can serve as a process for community change.

Source: Elwood et al., 2007, pp. 170–8.

Finally, perhaps after you have submitted your dissertation, thesis, or research report, it is worth investigating how to share the findings of your work together. Universities are increasingly being asked to demonstrate the impact and value of their research for participating communities (Pain and Askins, 2011), and **public scholarship** is often promoted by university media (Mitchell, 2008). Project or policy reports are powerful advocacy tools to advance action plans developed during the research. Clear, simple presentations work best, providing policy-makers with a sense of the process and how it generated reliable, meaningful "data" upon which practicable policy can be developed. Presenting the results of PAR at public meetings, conferences, or other gatherings with co-researchers can be appropriate and often enjoyable. If their direct participation is not possible, then discussing what they would like you to emphasize in a presentation can go some way towards addressing the power imbalance and lend you some authority to speak on their behalf.

Sharing reports, photographs, diagrams, maps, videos, and other products generated throughout a PAR process can be valuable in public spaces like libraries, community halls, and schools (see Chapter 20; Cahill & Torre, 2007; mrs c kinpaisby-hill, 2011). It can be relatively easy (though not necessarily quick) to develop on-line materials too, such as the beautiful ebook sharing quilts and stories of migrant women living in Johannesburg who participated in the Mwangaza Mama project (https://issuu.com/move.methods.visual.explore/docs/mwangaza_mama_ebook). If you do decide to disseminate your work in creative ways like these, however, take additional care to renegotiate ethical consent and to consider how to prepare co-researchers for potentially unpredictable responses from diverse audiences (Luchs & Miller, 2016).

Overall, having multiple research products produced by different combinations of people can be a helpful strategy. It can enrich the knowledge produced and be critically important if ongoing action is to be sustained, while also meeting institutional requirements and agendas.

Conclusion

A key question of PAR is often "to whom is the research relevant?" (Pain, 2003, p. 651). For us as researchers, if we accept that we have an opportunity and an obligation to co-construct responsible geographies (McLean, Berg, and Roche, 1997; Williams et al., 2003), then PAR offers us an exciting means of undertaking relevant, change-oriented research. While academia does not usually reward such **activism**, the central role of space in many people's oppression (Ruddick, 2004) means that human geographers are uniquely positioned, and morally beholden, to adopt ways of researching that build collaborative communities of inquiry (Reason, 1998, cited in Hiebert and Swan, 1999, p. 239) and challenge oppression. Moreover, with the rise of the **impact agenda**, there are growing calls for the co-production of knowledge with research participants and its creative dissemination (as discussed above). This agenda provides scope for more critically engaged research along the lines of examples in this chapter, but also requires vigilance on the part of researchers not to fall victim to utilitarian applications of participatory methods at the expense of more demanding shifts in power relations (Pain, Askins & Kesby, 2011; Pain et al., 2016).

Fortunately, certain parts of human geography, such as social geography, have a rich tradition of activism and participatory action research. PAR is not without its challenges, particularly within the confines of student research projects, but it is possible to adopt many of the principles discussed in this chapter to enable a rigorous research approach, which also results in tangible benefits for those involved. Perhaps the greatest challenge of all is for academics, including undergraduate researchers, to "cross boundaries of privilege and confront their personal stake in an

- If you make more aspects of your project participatory, are there resource or time costs that you need to consider?

After 20 minutes, share your insights and/or methodological adjustments with a classmate or other group, and discuss their ethical and operational implications.

Useful Resources

Breitbart, M. (2016). Participatory action research. In N. Clifford, M. Cope, S. Gillespie, & S. French (Eds), *Key methods in human geography* (pp. 198–216). Thousand Oaks, London, New Delhi: SAGE Publications.

Cameron, J., & Gibson, K. (2005). Participatory action research in a post-structuralist vein. *Geoforum, 36*(3), 315–31.

Halvorsen, S. (2019). Participatory action research. In B. Warf (Ed.), *Oxford bibliographies in geography*. New York: Oxford University Press.

Kesby, M., Kindon, S., & Pain, R. (2005). "Participatory" diagramming and approaches. In R. Flowerdew & D. Martin (Eds.), *Methods in human geography* (2nd edition) (pp. 144–66). London: Pearson.

Kindon, S. (2010). Participation. In S. Smith, R. Pain, S. Marston, & J.P. Jones III (Eds), *The handbook of social geography*. London: SAGE.

Kindon, S., & Cupples, J. (2016). "Nothing to declare": Leaving the field. In R. Scheyvens (Ed.), *Development fieldwork: A practical guide*. London: SAGE.

Kindon, S., Pain, R. & Kesby, M. (Eds). (2007). *Participatory action research approaches and methods: Connecting people, participation and place*. London: Routledge.

Klocker, N. (2012). Doing participatory action research and doing a PhD: Words of encouragement for prospective students. *Journal of Geography in Higher Education, 36*(1), 149–63.

Mountz A., Miyares, I., Wright, R., & Bailey, A. (2003). Methodologically becoming: Power, knowledge and team research. *Gender, Place & Culture, 10*(1), 29–46.

mrs c kinpaisby-hill. (2011). Participatory praxis and social justice: Towards more fully social geographies. In V. Del Casino, M. Thomas, P. Cloke & R. Panelli (Eds), *A companion to social geography* (pp. 214–34). London: Blackwells.

mrs c kinpaisby-hill. (2013). Participatory approaches to authorship in the academy. In A. Blunt (Ed.), *Publishing and getting read: A guide for new researchers in geography* (Section 4.2, p. 24). London: Wiley-Blackwell.

Pain, R., Finn, M., Bouveng, R., & Ngobe, G. (2013). Productive tensions: Engaging geography students in participatory action research with communities. *Journal of Geography in Higher Education, 37*(1), 28–43.

Pain, R., & Askins, K. (2011). Contact zones: Participation, materiality and the messiness of interaction. *Environment and Planning D: Society and Space, 29*(5), 803–21.

Helpful Online Resources

Pain, R., et al. (2016). Mapping Alternative Impact: Alternative approaches to impact from co-produced research. Centre for Social Justice and Community Action, Durham University. www.dur.ac.uk/resources/beacon/MappingAlternativeImpactFinalReport.pdf

Pain, R., Whitman, G., Milledge, D., & Lune Rivers Trust. (2017). Participatory Action Research Toolkit: An Introduction to Using PAR as an Approach to Learning, Research and Action. Department of Geography, University of Durham. http://communitylearningpartnership.org/wp-content/uploads/2017/01/PARtoolkit.pdf

Notes

1. Readers of this chapter are strongly encouraged to also read Chapters 2, 3, and 4, which provide useful, complementary overviews of key issues also needing careful attention in PAR.
2. The research team may consist of researchers only, researched people only, or both researched people and researchers.
3. If you are using PAR in a thesis, for example, and because universities typically expect a thesis or dissertation to be the "original work" of the student alone, you should discuss matters of authorship with your supervisor when preparing your project.

PART

Making Sense of Your Data: Co-producing Geographic Knowledge and Sharing with the World

Revealing the Construction of Social Realities: Foucauldian Discourse Analysis

Gordon Waitt

Chapter Overview

Foucauldian discourse analysis is a well-established interpretive approach in geography to identify the sets of ideas, or discourses, used to make sense of the world within particular social and temporal contexts. The challenging, yet, insightful ideas of the French philosopher Michel Foucault underpin this interpretive approach. Following Foucault, discourse is a mediating lens that brings the world into focus by enabling people to differentiate the validity of statements about the world(s). The goals of this chapter are twofold. The first is to outline why Foucauldian discourse analysis is a fundamental component of geographers' methodological repertoire. The second goal is to provide a methodological template. The chapter begins by introducing Foucault's concept of discourse. His interest in discourse was to explain how some lines of thinking/being/doing are generally accepted as "true," while other possibilities are marginalized or dismissed. Foucault's notions of discourse point towards the production of categories of knowledge, thereby governing what it is possible to talk about, and what is taken as "common sense." For Foucault, discourse simultaneously produces and reproduces knowledge and power (power/knowledge) through what it is possible to think/be/do/experience. Next, the chapter outlines a methodological template for the conduct of discourse analysis. Examples illustrate the benefits of discourse analysis for geographical research, particularly projects committed to addressing uneven social relationships and environmental injustice.

Introducing Discourse Analysis

What is discourse? This chapter relies upon Foucault's concept of **discourse**, which coheres around the cultural production and circulation of knowledge. Foucault does not provide us with a clear-cut dictionary-like definition of discourse but employs the concept to interpret what, and *how*, we know about the world about us. In Foucault's work at least three overlain explanations of the concept of discourse can be identified:

1. all meaningful statements or texts that have effects on the world;
2. a group of statements that appear to have a common theme that provides them with a unified effect; and
3. the rules and structures that underpin and govern the unified, coherent, and forceful statements that are produced.

Crucially, Foucault's (1972) use of discourse is different from conventional linguistic definitions, which understand discourse as passages of connected writing or speech. Foucault is interested in how particular knowledge systems convince people about what exists in the world as well as shaping what they say, do, experience, and become. Given Foucault's interest in the role of **knowledge production** in making and remaking the co-constitutive relationships between people and place his conception of discourse is inherently geographical.

Following Foucault, the concept of discourse allows us to understand better what everyone would like to believe as true about people, animals, plants, things, events, and places. This a **constructionist** approach that demands asking questions about the ways in which distinct social "realities" or categories become normative ways to think/be/do/experience. Thus, the aim of **discourse analysis** is to investigate why particular lines of argument (constructed in and as discourse) become taken for granted as truths, while dismissing others. For example, why are individual households rather than corporations in Western societies positioned as accountable for reducing carbon emissions? Alternatively, why in Western societies are tropical islands often categorized as earthly paradises; or national parks as place for nature? Likewise, why are certain plants categorized as native while others are labelled weeds or invasive? Simply put, a Foucauldian discourse analysis seeks to uncover the cultural and social mechanisms that maintain or rupture the structures and rules of validity embedded in statements about the world. The productive effect of **power/knowledge** between actors is always uncertain; on the one hand, discourses may reinforce social norms, on the other hand, discourses may mobilize possibilities for change. In political terms, discourse analysis allows geographers to expose the production and reproduction of inequalities and injustices by the resilience of certain underlying normative categories.

At the start, please note that there are many different types of discourse analysis. A website entitled How to Use DAOL hosted at Sheffield Hallam University, UK, provides a helpful discussion on the range of different discourse analysis techniques and their application in various disciplines (http://extra.shu.ac.uk/daol/howto). Furthermore, Foucauldian discourse analysis is distinct from other forms of qualitative analysis including **semiology** (which considers qualitative source materials in terms of inherent meanings) and **manifest content analysis** (which quantifies the number of occurrences of particular themes or words, see Chapter 9). So, how to begin conducting a Foucauldian discourse analysis?

Interestingly, Foucault struggled with describing how to conduct discourse analysis. He feared that a methodological template would become too formulaic and reductionist. This absence is perhaps what has often led others to describe Foucault's methodological statements as "vague" (Barret, 1991, p. 127). Several qualitative methods handbooks in the social sciences are equally hesitant to give formal guidelines (Phillips & Hardy, 2002; Potter, 1996). Some scholars argue that guidelines undermine the potential of discourse analysis as a "craft skill" (Potter, 1996, p. 140); by limiting the researcher's ability "to customize" (Phillips & Hardy, 2002, p. 78); or by inhibiting demands for "rigorous scholarship" (Gill, 1996, p. 144) and "human intellect" (Duncan, 1987, p. 473). For these scholars, the maxim is "learning by doing." This intuitive approach to discourse analysis requires a combination of scholarly passion and practice. Thus, the methodology is often left implicit rather than made explicit. For scholars who defend Foucauldian discourse analysis as an art, any methodological template would be understood as too systematic, mechanical, and formulaic (Burman & Parker, 1993). In practical terms, such counsel is not especially helpful for those seeking advice on how to do discourse analysis! To help grapple with the methodological implications of doing a Foucauldian discourse analysis, this chapter draws on important lessons offered by geographer Gillian Rose (1996, 2001) and linguist Norman Fairclough (2003).

Doing Foucauldian Discourse Analysis

To help novices conduct a discourse analysis, Rose (2001) provides seven stages through which the technique moves (see Box 17.1). These axioms should not stifle your own interpretations. Instead, they offer starting points in a conversation for your own thoughtful analysis.

In what follows, examples from tourism geographies and geographies of sexuality provide a means to review each stage. This choice of examples is not accidental. They illustrate a larger effort of the application of discourse analysis in geography to challenge what can be said, what is worthy of study, and what is regarded as factual. For example, until the 1990s, the discipline often shunned research on both tourism and sexuality as unworthy of geographical analysis (Binnie & Valentine, 1999). At that time, professional geographers would obtain greater academic kudos and authority by quantitatively analyzing the economic geographies of men at work in the industries of steel, textiles, coal, and automobiles, spatialized by the metaphor of "industrial heartlands." There is nothing intrinsic to sexuality and tourism that makes them more, or less, relevant for geographers to study than, say, ethnicity or car manufacturing. Instead, dominant sets of ideas (discourses) in geography eschewed the topics of sexuality and tourism. The very idea of spaces and bodies being gendered and sexed, let alone mutually constituted

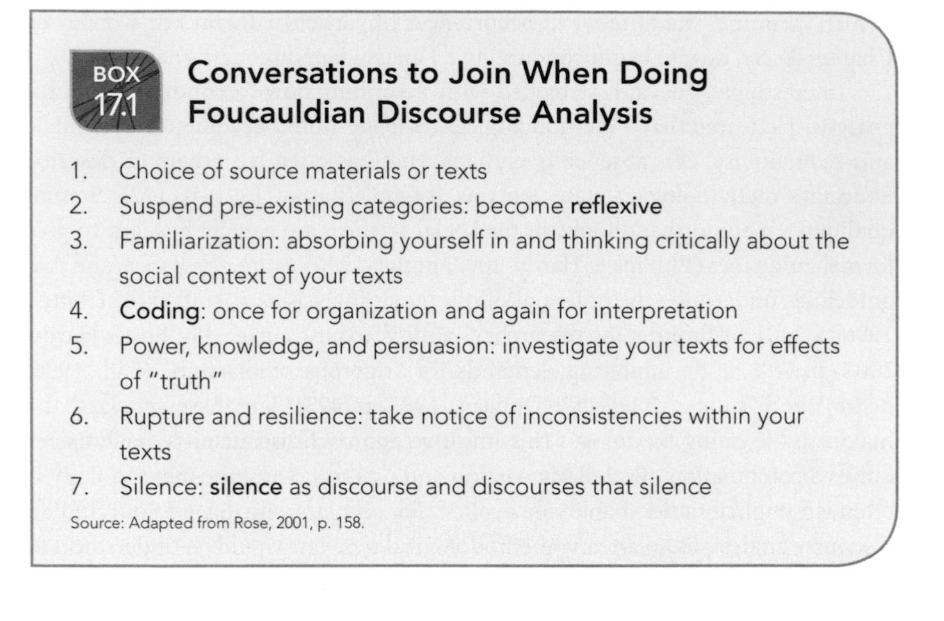

BOX 17.1

Conversations to Join When Doing Foucauldian Discourse Analysis

1. Choice of source materials or texts
2. Suspend pre-existing categories: become **reflexive**
3. Familiarization: absorbing yourself in and thinking critically about the social context of your texts
4. **Coding**: once for organization and again for interpretation
5. Power, knowledge, and persuasion: investigate your texts for effects of "truth"
6. Rupture and resilience: take notice of inconsistencies within your texts
7. Silence: **silence** as discourse and discourses that silence

Source: Adapted from Rose, 2001, p. 158.

through space, was considered trivial. Equally, tourism involved people having fun. Thus, within masculinist academic geography it too was marginalized until the 1990s, irrespective of its value as a topic of study.

Choice of Source Materials or Texts

What kinds of source materials or texts are required for discourse analysis? Source materials or texts may include advertisements, brochures, maps, novels, statistics, memoranda, official reports, interview transcripts, diaries, paintings, sketches, postcards, photographs, websites and social media, and the spoken word. The research goals will inform the choice of source materials for discourse analysis. Here four examples will be helpful to illustrate the diversity of applications of discourse analysis and relevant sources materials. First, John Urry (2002) investigated the role of the tourism industry in the transformation of people and places into tourist destinations. Urry envisaged the tourist experience as an exercise in Foucauldian "gazing." Urry selected brochures and guidebooks of the tourism industry to illustrate how tourist sights/sites are constructed by drawing on different sets of ideas—myth, science, cultural histories—that enable those sites/sights to be fashioned as extraordinary and then to be "consumed" by visitors. Drawing on the example of Niagara Falls, he argued that because of the vast number of tourism industry images instructing visitors on how to "consume" or "gaze" at the Falls as extraordinary/spectacle, it is no longer possible for tourists to "see" the Falls themselves. In the second example, Kevin Markwell and Gordon Waitt (2009) explored the role that pride festivals play in processes of social change surrounding dominant ideas of heterosexuality in Australia. Source materials for this work included articles published in LGBTQ+ media. In Kylie Budge's (2019) work, the "creative" industries provide the third example. A discourse analysis of public policy documents and social media demonstrated divergent understandings of the creative economy in Sydney. Creative economy discourses circulating within the policy domain reproduce the categories of "artist," "art," and "culture" but silence the importance of the role of design and the categories of "craft" and "manufacturer" evidenced in maker movements. In the final example, Andrew Gorman-Murray (2007) employed discourse analysis to examine how people who understand themselves as lesbian or gay challenge taken-for-granted ideas around heterosexuality that help stabilize notions of the Australian home. Source materials for this work were transcribed interviews conducted with participants in their homes. As these examples illustrate, the expression of discourses is evident in a wide variety of source materials.

Genre refers to the subdivision of source materials into different categories. For example, there are different genres of written texts, including tweets, blogs, travel diaries, **solicited diaries**, biographies, autobiographies, fiction, travel writing, government policy papers, newspapers, scientific reports, interview **transcripts,** and so on. Being mindful of the genre of text is important, because the producer of each genre or category of text (e.g., journalist, traveller, blogger, academic, medical practitioner, diarist) is addressing a specific **audience** and may have to conform to a particular writing style. Furthermore, specific technologies are integral to the production, circulation, and display of specific texts (such as audio-recorder, printing, painting, photography, handwriting, and email).

Your research project may dictate the collection of one genre of text—say, solicited diaries, semi-structured interview transcripts, tourist maps, music album covers, tourist brochures, holiday photographs (tourist selfies), sketches/doodles/mind maps, or picture postcards. In each case, to inform the interpretation, you will need to remain alert to the social circumstance of *production*, *text*, and the *author* in question. If you are using transcripts from interviews or **focus groups**, such **critical reflection** is an integral part of the method (see Chapters 9 and 11). Questions to help guide you through this process of critical reflections are outlined in Box 17.3.

Alternatively, your project may require that you to establish a research archive by collecting and interpreting several different types of text, including photographs, advertisements, letters, newspaper articles, and so on. A research archive requires that you conduct a good deal of background research to answer the questions outlined in Box 17.3. Consider, for example, Lynda Johnston's (2005) research on the practice of sun tanning on the beaches of Aotearoa New Zealand. Her analysis drew on semi-structured interviews and young women's magazines (*She*, *Dolly*, *New Idea*, and *Cosmopolitan*), which advertise sun-tanning products. Her analysis required working across these source materials to explore how people navigate the dominant discourse that shapes the practice of changing the colour of one's ski—sun-tanning. Johnston highlights how the practice of sun-tanning is fashioned by Western ideas of femininity associated with leisure, youth, and beauty and contests sets of ideas that locate the skin as a "true" biological marker of the body's gender and race. Robyn Longhurst's (2005) analysis of pregnant bodies in Hamilton, Aotearoa New Zealand, is another excellent example of the application of **intertextuality**, the assumption that meanings are produced as a series of relationships *between* texts, rather than residing within the text itself. To explore how pregnant women use clothing to constitute their subjectivity and social meaning in different places, Longhurst's discourse analysis relied on a research archive comprising interview transcripts, maternity wear brochures, advertisements for maternity wear, internet sites marketing maternity wear, newspaper clippings, and magazine photographs and reports (*New Zealand Women's Weekly*, *Woman's Day*, and *New Idea*). Intertextuality acknowledges the co-creation of meanings with an active audience, and, as with other methods employing multiple data sources, it stimulates productive exploration of where sources agree with/support each other and where they might contradict or rub up against each other.

Your initial problem will most likely be having *too many* potential source materials. To begin your analysis, identify the key source materials to address your research aim. What, then, might inform your choice of texts? Use the following questions to help you to both identify and justify your preliminary or opening texts:

- *Which source materials to sample?* In making decisions about selecting or excluding source materials, student researchers are often advised to select "rich" or "in-depth" texts. Attention to richness of source materials allows the researcher to interpret the effects of discourse in normalizing understandings. However, what qualities make a source material qualitatively rich? For example, for an interview transcript to be categorized as a rich text, it requires more than "yes" or "no" answers. A qualitatively rich interview transcript may therefore require longer-term involvement and consideration of particular techniques to elicit thoughtful and in-depth responses (see Chapters 9 and 10 and also Riley and Harvey [2007] for further discussion on how to enhance the richness of interviews).
- *How to select source materials?* To help identify source materials that are likely to provide insight, deploy a **purposeful sampling** strategy (see Chapter 6). For purposeful sampling to work most effectively requires early and ongoing secondary research. Yet, because the research process is a continuing building process, you are likely to find that, as the project

unfolds, other, unexpected source materials may become relevant and important to its success. As a reflexive researcher, you must remain open to such unexpected possibilities. Being able to justify why a particular source material is meaningful to your project is an integral step in establishing **rigour**. Such justification is important in writing **transparent** research that is open to scrutiny by both the **participants** and **interpretive communities**. (See Chapter 6 for additional discussion.)

- *How many source materials?* As other chapters of this book outline, there are no pre-determined rules for sample size in the context of qualitative research. Whereas in inferential statistics sample size can be prescribed by demands of **representative samples** and **validity**, in discourse analysis the number of source materials depends on what will be useful and *meaningful* in your project context. Background research may lead you to justify a sample size as small as one or extend into hundreds of texts. Again, explaining the number of texts you selected for inclusion in your project becomes an integral step in establishing qualitative rigour.

What, then, might be a meaningful choice of source materials for investigating research questions that aim to better understand how discourses create and shape social expectations and experiences? Take for example the discourses that constitute the re-emergence of tiny house living in Western societies. What source materials may be helpful to understand better the discourses that shape tiny houses? What might your initial starting point be? What sort of background research of related works might be of assistance? What other source materials might you wish to consult to create a research archive that enables comparison with discourses that constitute "conventional" house designs? Who might you wish to interview? (See Box 17.2.)

Suspend Pre-existing Categories: Become Reflexive

Foucault (1972) regarded the starting point of discourse analysis as reading, listening, or looking at your texts with "fresh" eyes and ears. The imperative of shelving preconceptions is underpinned by the objective of discourse analysis—to disclose the created "naturalness" of constructed categories, **subjectivities**, particularities, accountability, and responsibility. Foucault (1972, p. 25) pointed out that all preconceptions

> must be held in suspense. They must not be rejected definitively, of course, but the tranquility with which they are accepted must be disturbed; we must show that they do not come about by themselves, but are always the result of a construction the rules of which must be known and the justifications of which must be scrutinized.

BOX 17.2

Tiny Houses, Sustainability, and Affordability: Choosing Texts for Discourse Analysis

The tiny house is an emerging Western cultural phenomenon supported by **social media** and lifestyle television platforms. The growth of smaller housing options is likely to continue in the Western world against a backdrop of housing affordability, tightened monetary lending, energy price hikes, and urban sustainability.

Imagine that your project goal is to understand better whether the tiny house offers a solution to housing affordability and/or sustainability. First: which initial source materials to select? In the first instance, you may find census statistics of dwelling size. These statistics will tell you about trends in housing size and average floor space per person. Moreover, they may provide a helpful starting point to think about the politics of housing. Does the census even count tiny houses? The numbers alone tell us nothing about the ideas that may motivate people to live in a tiny house. Instead, source materials that may be helpful include social media, print media, advertising, and businesses. This research can build upon the discourse analysis conducted by April Anson (2014) and Hilton Penfold, Gordon Waitt, and Pauline McGuirk (2018), who point towards sets of ideas that fashion tiny houses in ways that reproduce rather than trouble environmental romanticism and cultures of consumption. The next step of the research may be to interview people living in tiny houses to learn more of their motivations for choosing this housing type.

Foucault acknowledges that this request to defer pre-existing categories is an impossible task, given the socially constituted qualities of all knowledge. There is no independent position. Instead, he calls for researchers to become self-critically aware of the ideas that inform their understandings of a topic. One strategy by which to implement Foucault's call is to deploy the techniques of **reflexivity** outlined by John Paul Catungal and Robyn Dowling, earlier in this volume (see Chapter 2). A key starting point for a critically reflexive researcher is to acknowledge why you selected a research topic (context) and your initial ideas about the subject (partiality). The term **positionality statement** refers to a researcher explicitly locating themselves in the project context, and may include sets of ideas alongside their lived experiences (**embodied knowledge**). In acknowledging your positionality, remain alert to Gillian Rose's (1997) warning that it may be impossible to fully locate oneself in research. The ideas we carry into the work may change while conducting the research. Hence, a vital part of a positionality statement is noting and reflecting upon these changes as the project unfolds. One crucial question to ask is: how has your self-understanding changed while conducting the research? Finally, as discussed by Elaine Stratford and Matt Bradshaw (see Chapter 6), keep careful and transparent documentation of your interpretation process. Document your search for regularities and the ways you developed hunches. The strategies outlined in the following section will help prevent you from imposing your own taken-for-granted understandings upon the source materials and will boost the **trustworthiness** of your work.

Familiarization: Absorbing Yourself in and Thinking Critically about the Social Context of Your Source Materials

Familiarization with your source materials is essential. A helpful starting point is to begin thinking critically about their "social production," in becoming alert to their *authorship*, *technology*, and intended *audience*. These social dimensions of sources are a good starting point for a critical interpretation because discourses operate as a process, restricting what can be said about the world and who can speak with authority (Wood & Kroger, 2000). Foucault understood discourses as grounded within social networks in which certain groups are empowered and disempowered relative to one another. He saw discourse as subtle forms of social control and power. One effect of discourse is the privileging of relatively powerful social groups. That is, certain voices and technologies are favoured over others, often counted as sources of "truthful" or "factual" knowledge, while other voices may be excluded and silenced, perhaps by becoming positioned as untrustworthy, anecdotal, hearsay, or folklore. Take, for example, the distinction in Australian national parks authorities between Indigenous and "scientific" approaches to environmental management. Until relatively recently, so-called **objective** scientific knowledge was valued by most national park authorities at the expense of Aboriginal Australian knowledge. Scientific methods were equated with research, objectivity, and environmental management solutions. In contrast, oral Aboriginal Australian knowledge was linked with negatively valued concepts, including the irrational and subjective (Lawrence & Adams, 2005). Increased recognition of Indigenous communities' knowledge has led to a critical re-examination of "scientific" perspectives globally (see Chapter 4).

All texts are the outcome of an uneven power-laden process, fashioned within a particular social context. Hence, an integral part of the familiarization process is to conduct background research to help anchor your texts within their historical and geographical context. Box 17.3 provides a series of questions to help you interpret all texts as expressions of knowledge production and a subtle form of social power that constitutes particular social realities.

First, consider *authorship* as the outcome of a highly social process. Regardless of whether your source material is a transcript, photograph, painting, song, or novel, you must investigate the author's relationship with the intended audience. Reflect carefully on the embedding of social dynamics into the production of the source material that may operate as a subtle form of social control. Box 17.3 provides a starting point to interpreting the relationship between discourse, knowledge, and power. *Who* produced your selected materials? Give attention to the positioning of the author within historically and geographically specific understandings of gender, class, sexuality, and ethnicity. Remember that these are not "natural" categories but are instead relations of power constituted through discourse. *When* and *where* was your selected source material produced? *Why* was it produced? Are the social identities of the author or audience portrayed in your selected sources: that is, does the text help to naturalize, as selfies and family snapshots do, belonging to a social group in society?

Second, ask questions that are more specific about the social conventions linked to the source material production. The key point is that you must examine the *how* of your source materials—in particular, the implications of technologies. More subtle questions then follow, including thinking carefully about how different categories of material have their own social histories and geographies.

Michael Brown and Paul Boyle provide a helpful discussion of how the technology underpinning the census operates to sustain an objective understanding of the population of a nation state (Brown, 2000). Yet, they remind us that what is taken-for-granted as factual knowledge of the nation-state is always a reflection of what those in authority deem to be important about the attributes of a nation. The census, through statistical techniques of data collection and dissemination, is one way that production of citizenship and the nation occurs through the state. The results of the census are always contingent upon the questions asked by the state, the categories a variable is given, the ways in which they are analyzed, and how questions are interpreted by respondents.

Similarly, Lisa Law (2000) provides an interesting discussion of how map-making of the Health Department of Cebu, the Philippines, helps to maintain gendered sets of ideas about Filipina women. She noted how the appearance of the so-called red-light district as an empirical "reality" on the map made apparent the Catholic-influenced moral distinction between marital and commercial sex. This example illustrates how mapping technology is a power-laden process. In this case, the Cebu Health Department map may have helped convince many health officials that all women living in Cebu could be categorized along a simplistic spatialized division of good/bad. An integral part of discourse analysis is remaining alert to the role that different technologies play in **strategies of conviction** by producing, or mobilizing, certain "truths."

Finally, Box 17.3 offers strategies to investigate the role of audiences. Audiences are not pre-given, but produced through on-going exchange in a highly social process. The intended audience shapes the initial production of all texts and becomes—in a sense—a co-author. In other words, an author will draw on certain discourses, mindful of the intended audiences' needs, demands, and fantasies.

For example, writing in a pre-digital context, both Crang (1997b) and Markwell (1997) illustrate the active role of prospective audiences as co-authors of vacation snapshots. They demonstrate how the genre of tourist photograph is always about framing, making, and circulating social realities for a particular audience. Vacation snapshots are the outcome of a highly selective social process influencing the choice of subjects to photograph. At one level, this choice reveals how a person reproduces, or challenges, particular common-sense understandings of, for example, tourism discourses. Making vacation snapshots is not value-free but reflects specific ideologies and sets of ideas about appropriate leisure practices for a specific place, produced and circulated in tourist brochures, guidebooks, and internet sites. At another level, each photographer frames a place in a particular way to meet audience needs, which could be themselves, family, friends, or strangers. The vital point, as Crang (1997b, p. 362) emphasizes, is that: "It is not a case of pictures showing what is 'out there' . . . but rather how objects are made to appear for us." How vacation snapshots are framed through the tourist gaze is an example of how individuals make sense of and then, often with an audience in mind, communicate particular versions as "truths" about places that are made "real" by the photograph.

With the advent of the web and social media platforms, tourist photography is an integral part of network travel that enables travellers to stay in touch with distant others and larger social networks. Jonas Larsen and John Urry (2011) argue that an audience is at the forefront of selfie tourist photography in the age of the social media tourist, who wishes to convey travel experiences through photographs as-they-happen. Consequently, Anja Dinhopl and Ulrike Gretzel (2016) argue that the selfie is one example of tourist photography that illustrates how audiences of social media sites both enable and constrain tourist experiences. An integral part of the digital tourist's experience is their relationship to smartphones and social media audiences. The genre of selfie tourist photograph prioritizes the extraordinary within themselves over that of the destination in the framing of the photography.

Of course, Foucault did not envisage audiences as passive recipients. As emphasized earlier, people are always socially positioned in relation to the world. For example, Josh Whittaker and David Mercer (2004) illustrated how different social groups understood the increased intensity and occurrence of wildfires in Victoria, Australia.[1] Their discourse analysis of the reporting of the wildfires of 2002–3 in print media provides an excellent example of how different audiences made sense of, and understood, this event. They identified two dominant positions. First, some people drew on the storyline of the wildfires as natural and inevitable, including members of government departments and environmental groups. Rather than configured as hazard, wildfires could be regarded as a productive resource that enabled regenerative ecological functions. Management solutions to wildfires thus become configured in terms

BOX 17.3

Strategies for Investigating the Circumstances in Which Discourse is Produced

For each set of questions, pose the following: why is this answer important in the context of establishing or maintaining particular social realities?

Social Circumstances	Questions about Authorship	Questions about Source Material	Questions about Audience
Social	• Who made the source material? • Who commissioned the source material? • Who owns the source material? • What are the relationships between the maker, the owner, and the subject of the source material? • When was the source material made? • Where was the source material made? • Why was the source material made?	• Does your selected source material re-circulate texts found elsewhere? For example, can the same source materials be found elsewhere—say, on the internet, in tourist brochures, or on postcards?	• Who is/are the original audience(s)? • What are the conventions of how an audience engages with this source material? (e.g., What are the social norms of attending a pop music dance party? What are the social conventions for engaging with the music at this event?) What are the social conventions for engaging with other people at the event? (e.g., How do the social norms of a music festival differ from attending a classical music concert, or a Christmas Eve service in a church?) • How actively does an audience engage with your selected source material?
Technological	• What technologies does the author rely upon to produce the source material—for example the internet, a paper diary, or a camera?	• How has technology affected the production of the source material? For example, what use is made of print, colour enhancement, photography, digital technologies (airbrushing)? • What are the distribution networks of the source material?	• How is your selected source material displayed? • How is your selected source material stored? • Do the technologies of display/storage affect the audiences' interpretation? • What are the social norms of reading, viewing, or listening to your selected source material?
Content/ aesthetic	• Does the subject matter of the source material address the social identities and relationships of its maker, owner, or intended audience?	• What is the subject matter? • What is/are the genre(s) of your selected source material? For example, if a photograph, is it a family snapshot or advertisement? If your source material is in an electronic form, how was it generated (sound, images, cultural references) and what platform was it intended for? • Is your source material one of a series? • What are the conventional characteristics of your selected genre of source material? • Is your selected source material contradictory, critical, or in some way different from those circulated elsewhere? • What is the "vantage point" of the viewer/reader in relationship to the source material? That is, how is the viewer/reader positioned in relationship to the source material?	• Where is the reader/viewer/listener positioned in relationship to your selected source material? • What sort of relationship is produced between the reader/viewer/listener and your selected source material?

of fuel reduction burning. Second, others shaped a counter storyline that the wildfires were unnatural and avoidable. Farmers, community groups, non-government organizations and rural landholders circulated this set of ideas that configure wildfire as a disaster. An integral part of this storyline was positioning environmentalists as to blame for wildfires for having locked up public land in National Parks. The management solution thus became configured in terms of increased involvement of rural community groups in land management decisions deploying "local knowledges." This discourse analysis of wildfires illustrates two key points. First, the meanings of a text are never singular or uni-directional. Each audience segment or social group will bring different meanings and power dynamics linked to their personal affiliations. Second, critical reflexivity may help you identify the ideas you bring to a project, and help prevent you from imposing your own taken-for-granted understandings on to the source materials (see Chapter 2).

Coding: Once for Organization and Again for Interpretation

Coding is an interpretation process. When doing discourse analysis, coding serves two primary functions: organization and analysis of source materials. One possible way to start coding is to draw on two different types of codes: descriptive and analytical. In Chapter 18, Meghan Cope provides a helpful discussion of the difference between these two coding structures. **Descriptive codes** offer one way of organizing your data initially using obvious category labels. For example, Waitt and Warren (2008) employed four category labels when developing a list of descriptive codes to interpret the transcripts of diaries kept by surfers:

1. Context: *where*, *when*, and *who* participated in surfing
2. Practices: the *events* (what happened while surfing), *interconnections* (who influenced the surfers' style of surfing and how they interact with other surfers and beach-goers), and *actions* (what type of maneuvers they performed while surfing)
3. Attitudes: statements of judgment about other surfers, beach-goers, the ocean, or sharks
4. Experiences/emotions: statements of emotions about surfing or interactions with other surfers or beach-goers

For each category label, start a list of codes. However, remember that coding is an iterative process so your initial descriptive codes will change. In some cases, you will need to divide initial codes into finer detail. In other cases, you will have to amalgamate descriptive codes into broader categories. Following each reading of your source material, your coding structure will become more refined.

Alternatively, Rose (2001) suggests content analysis as the starting point for discourse analysis. Content analysis is essentially a quantitative descriptive coding technique involving coding key **themes** and then quantifying (counting) the instances.

However, what is a key theme? Again, it may be helpful to start with the headings of context, practices, attitudes, and experiences. Thomas McFarlane and Iain Hay (2003) provide a helpful example that illustrates the application of content analysis. They used word clusters to code articles published in *The Australian* that discussed protests to the World Trade Organization (WTO) ministerial conference in Seattle during December 1999. They demonstrated how articles in *The Australian* effectively demonized and marginalized anti-WTO protestors. The free software Wordle One is one useful way to generate graphic word clouds or clusters from text (www.wordle.net). Those words that appear most frequently in your text will appear most prominent in the Wordle cluster. Remember, when doing discourse analysis, descriptive codes, or content analysis, are envisaged only as an initial starting point. The frequency of words may help identify shared understandings but provide little insight into the particular sets of ideas that maintain statements as taken-for-granted or "true."

Coding for a Foucauldian discourse analysis also involves devising a list of **analytical codes** to offer an interpretation. Analytical codes involve some form of abstraction or reduction. In discourse analysis, analytical codes typically provide insights into why an individual, or collective, holds sets of ideas by which they make sense of places, themselves, and others. Take for example Goss's (1993) discourse analysis of the Hawai'i Visitors Bureau's (HVB) portrayal of the Hawai'ian Islands for the North American tourism market. For his analysis of the 34 advertisements commissioned by HVB and published between 1972 and 1992, he initially coded the recurring images and words employed to portray the island, including Indigenous Hawai'ians, plants, volcanoes, location, and climate. This coding then suggested five analytical themes deployed by the HVB to "invent," or pitch, the Hawai'ian Islands outside the "normal" construction of North American geography: earthly paradise, marginality, liminality, femininity, and aloha. Goss went on to interpret each of these themes, demonstrating that the tourism industry reproduced long-standing colonial interpretations about Hawai'i. To provoke travel desire among potential tourists, the advertisers' preferred image of Hawai'i remained with the socially constructed colonial fiction of a permissive, Indigenous, female sexuality and verdant tropical nature. The portrayal of Hawai'i is as a timeless location, a portal to the past, inhabited by "naturally" friendly people. In this (socially constructed) timeless and luxuriant tropical island paradise, the promises made to visitors are possibilities to become themselves, free from the repressive regulations of mainland North America. The lesson for doing discourse analysis is how each analytical theme deployed by Goss is comprised of a set of ideas from which he defined a particular condition of existence.

Power, Knowledge, and Persuasion: Investigate Your Source Materials for "Effects of Truth"

Foucault conceptualizes *persuasion* as a form of disciplinary power that operates through knowledge. Persuasion entails establishing and maintaining sets of ideas, practices, and attitudes as both common sense and legitimate. Foucault positions the mutually interdependent relationship between power and knowledge as indistinguishable, arguing that: "Truth isn't outside power. . . . Truth is a thing of this world; it is produced only by virtue of multiple forms of constraint. And it induces regular effects of power" (1980, p. 141). Hence, questions about the "truth" of knowledge are fruitless, for truth is unattainable. Instead, Foucault focuses on the stabilizing or sustaining of particular knowledge as truth (truth effects). According to Foucault, underpinning the mutual relationship between power and knowledge are **discursive structures**.

Discursive structures are the relatively rule-bound sets of statements that impose limits on what gives meaning to concepts, objects, places, plants, and animals (Phillips & Jørgensen, 2002). Foucault uses this term to refer to sets of ideas that typically inform dominant or common-sense understandings of and interconnections between people, places, plants, animals, and things. In Western thought, for example, a set of binary rules characterizes "rational" thinking underpinned by Descartes's mind/body separation. Today, this dualistic thinking is still evident in a whole series of hierarchically valued dichotomies, including rationality/irrationality, man/woman, mind/body, straight/gay, masculine/feminine, and humanity/nature. Therefore, while Foucault understands discourses to be inherently unstable, discursive structures operate to "fix" ideas of the world within particular social groups at specific historical and spatial junctures. For both individuals and collectives, discursive structures establish limits to, or operate as constraints on, the possible ways of being and becoming in the world by establishing normative meanings, attitudes, and practices. Simply put, discursive structures are a subtle form of social power that fix, give apparent unity to, constrain, and/or naturalize as common-sense particular ideas, attitudes, and practices. Foucault refers to this form of social control as the "effects of truth."

Take, for example, the concept of sexuality. There are many different discourses about sexuality in the world. However, within a particular time and place, a specific set of ideas will come to define socially acceptable practices of sexuality. For example, in nineteenth-century Europe, North America, and Russia, religious, scientific, and medical institutional narratives inspired the privileged knowledge about sexuality. Among most scientists of this time, theories of gender inversion were taken for granted. Mistakenly, these ideas constituted homosexuals as effeminate men. When combined with a lethal mix of fundamental Christian morals, same-sex-attracted men were "invented" through rhetorical strategies that grounded meaning in a binary opposition to heterosexual men as the "natural" and "healthy" standard of masculinity. This constituted the cultural identity of the homosexual man as lacking. Homosexuality became associated with negatively valued concepts: the primitive, the irrational, the feminine, the diseased, and the sinful. In this way, it seemed self-evident that the homosexual person was only worthy of study in efforts to find a medical cure. Simultaneously, the scientific knowledge of the time made same-sex-attracted women "invisible" by constituting sex as a penetrative act. Thus, understandings about same-sex-attracted women and men produced social limits. Indeed, the casting of homosexuality in the nineteenth century with so many negatively valued concepts often resulted in laws criminalizing sodomy in the West.

This example illustrates several key implications for doing discourse analysis. First, persuasion, or the effect of truth, is where an institution or individual deploys knowledge as a mechanism of social control. In this case, social power operates through the way that male–female binary distinctions came to define the appropriate and dominant conception of gender. Drawing on scientific knowledge, relatively powerful groups in society were able to naturalize meanings, attitudes, and practices towards another social group constituted as diseased and mad. When doing discourse analysis, it is imperative that you remain alert to institutional dynamics and the social context of source material. Second, while discourses are always inherently unstable, multiple, and contradictory, discursive structures operate to give a sense of fixity, bringing a common-sense order to the world. Sets of ideas become accepted and repeated by most people as "common sense," unproblematic, unquestionable, and apparently "natural." Hence, an essential part of doing discourse analysis is becoming aware of the ways in which particular kinds of knowledge become understood as valid, legitimate, trustworthy, or authoritative. Appropriate knowledge on a topic at one level may use particular technologies in the production of texts (for example, computers, maps, and photographs). At another level appropriate knowledge may encompass the way that sets of ideas are legitimized by the subtle deployment of different knowledge-making practices (statistics, medicine, policies, anecdotes) or categories of spokesperson (politician, scientist, academic, lawyer, judge, priest, eyewitness).

Resilience and Rupture: Take Note of Inconsistencies within Your Sources

While one rationale for doing discourse analysis is to identify the limits on how a particular social group talks and behaves (that is the discursive structures), another is to explore inconsistencies within your source materials. Taken-for-granted sets of ideas about who and what exist in the world help to impose bounds beyond which it is often very hard to reason and behave, while common-sense relationships set limits to the cultural know-how of a particular social group. As we saw earlier, Foucault understood dominant or common-sense understandings as discursive structures (see Phillips & Jørgensen, 2002), which may appear eternal, fixed, and natural. Yet, embedded within different social networks those structures are fragile and continually ruptured, meaning there are always possibilities for the challenge and change of meanings, attitudes, and practices. Therefore, an essential part of doing discourse analysis is to be alert to possible contradictions and ambiguities in texts.

For example, take again Gorman-Murray's (2007) research on the Australian home. He demonstrated the resilience of social norms within Australian building design and government policy that shape and reshape the detached ("single-family") suburban dwelling as "natural" for the heterosexual nuclear family. Consequently, the parental home may simultaneously be, for some LGBTQ+ people, a site of belonging *and* alienation as they reconcile their sexuality. Gorman-Murray illustrated how the homemaking practice of people living their adult lives as lesbians and gay men in suburbia can rupture normative heterosexual domestic ideals. Inconsistencies in the discourse of suburbia then become evident. The domestic realm is a site of affirmation of self and family. Yet, not all people living in suburban homes are involved in reproducing heterosexuality and nuclear families through their home-making practices. Consequently, domestic spaces may affirm, legitimize, and nourish sexual difference. To generalize, doing discourse analysis involves remaining alert to contradictions and ambiguities within your texts.

Lesley Head and Pat Muir's (2006) work on suburban gardens in Wollongong, New South Wales, also illustrates the concept of resilience and rupture. Interviews with suburban gardeners provided insights into their understandings, attitudes, and practices towards nature. Head and Muir discuss the resilience of the idea among many European-Australian gardeners of cities as places devoid of nature. Some of these gardeners locate nature beyond the spatial limits of the city in, say, national parks or the Australian bush. Yet others talked about cultivating nature in their backyards, thereby rupturing conventional ideas about where nature is found and repositioning humans as nurturing rather than destroying nature. Remaining alert to such ambiguities and contradictions is a fundamental challenge of doing discourse analysis.

Silences: Silence as Discourse and Discourses that Silence

Finally, becoming attuned to silences in your texts is as important as being aware of what is present. Gillian Rose (2001, p. 157) reminds us that "silences are as productive as explicit naming; invisibility can have just as powerful effects as visibility." Similarly, Elizabeth Edwards (2003), while discussing her approach to discourse analysis as "dense context," draws attention to the importance of silences:

> "Dense context" is not necessarily linked to the reality effect of the photograph in a direct way, indeed to the extent that it is not necessarily *apparent* what the photograph is "of." Often it is what photographs are *not* "of" in forensic terms which is suggestive of a counterpoint. (Edwards, 2003, pp. 262–3, emphasis in original)

The key point is that identifying silences produced by texts is an integral part of discourse analysis. However, becoming alert to silences is always challenging.

Clearly, to be able to interpret omissions within your texts requires conducting background research into the broader social context of your project and texts. Only then will you become aware of the existence of various social structures that inhibit what is present in your texts.

According to Foucault (1972), silences operate on at least two levels. First, silence as discourse is a reminder of how speakers create subjectivities within discourses. Who has the right, or authority, to speak is itself constituted through discourse. The social circumstances of authorship lead us to consider how silence occurs through the sets of ideas that establish the authority to speak, and give attention to how silence may become evident through the way that sets of ideas configure the voice of a person at the intersections of gender, age, class, ability, sexuality, and race. Being mindful of whose voices are silenced within your texts is an integral part of discourse analysis. For example, think about an advertisement for McDonald's restaurants. These advertisements often contain the "voices" of children, while the voices of parents who are resolving the conflicting sets of ideas about fast food and parenting are silenced. You can speak to these silences.

Second, Foucault's ideas alert us to how a **privileged discourse**—or dominant discourse—operates to silence different understandings of the world. Here, of vital importance are Foucault's arguments concerning the intersection between power, knowledge, and persuasion. According to Foucault, silence surrounding a particular topic is itself a mechanism of social power within established structures. For example, Sarah Holloway, Mark Jayne, and Gill Valentine (2008) demonstrate how policy debates, media attention, and epidemiological research about alcohol in England have centred on questions of drinking in public spaces, particularly by younger people participating in the city centre nighttime economy of metropolitan areas. As they go on to argue: "Like many geographical imaginaries, this vision is a highly partial one and is as much of interest for what it excludes as that which it includes" (p. 533). The city centre focus silenced questions about alcohol consumption in the domestic realm, yet that is a significant part of the English market. Hence, on the one hand, the focus on the nighttime economy served the interests of municipal, law-enforcement, and health authorities seeking to curb binge-drinking. On the other hand, the focus of the night economy and the youthful binge-drinker also served the interests of older alcohol consumers drinking in the domestic realm, supermarkets, and alcohol companies supplying the retail sector.

Another helpful example of a discourse that silences the voices and interests of some is the elaborate colonial Western fiction of the *frontier*. In the nineteenth century, frontier discourses among European politicians, scientists, priests, ministers, social scientists, and many settlers helped to naturalize and justify colonial settlement and nation-making in North America (see Goss, 1993; Turner, 1920) and Australia (see McGregor, 1994; Schaffer, 1988; and Ward, 1958). Frontier discourses relied upon maintaining the Western understanding of colonized places as "uncivilized," "empty," "natural," "wild," and "timeless" (see also Chapter 4). In frontier places, colonial sets of ideas configure European history as yet to begin. As Deborah Bird Rose (1997) argues, settlers' discursive strategies towards the frontier erased the presence of Indigenous people or, at best, cast them as "primitive" people whose natural fate according to social Darwinism was extinction. In Australia, frontier discourses that portrayed Australia as a place waiting for history to begin became law, dictating that before the arrival of Europeans, the continent belonged to nobody (*terra nullius*). It was only in 1993, following the Australian High Court's Mabo decision, that the "truth" of *terra nullius* was overturned. In decisions over land ownership Aboriginal Australians' knowledge could no longer be ignored.

Nevertheless, in their discursive analysis of texts from tourists on the Thelon River in Arctic Canada, Grimwood, Yudina, Muldoon, and Qiu (2015) illustrate how Western sets of ideas of nature as wilderness inform understandings of responsibility around an ethic of "leaving no trace." In turn, these sets of ideas operate to silence understanding of the Thelon River area as Indigenous homeland. Likewise, Waitt and Head (2002) note how the colonial frontier discourse still enjoys widespread currency within the contemporary Australian tourism industry. The offer to potential tourists is a portal to a timeless land. Imagined as timeless, the Kimberley—a northwest region of Australia—can then fulfill the specific market demands of primarily metropolitan Australian and international visitors, including: to discover the "real" Australia, imagined as the outback; to experience places imagined as wilderness; to gaze upon places portrayed as offering sublime "natural" beauty; or to explore the Kimberley as an adventure setting. However, the Miriuwung-Gajerrong's (the Indigenous people) knowledge constituting this place as "home"—named, known, and cared for over tens of thousands of years—is silenced by the colonial frontier discursive structures. To summarize, the work of revealing silences requires appropriate background research to identify who or what is missing from source materials.

Conclusion

The goal of Foucauldian discourse analysis is to reveal the process through which particular ideas that forge social and spatial realities become dominant. In political terms, discourse analysis allows insights into processes of social and environmental injustice. Conducting Foucauldian discourse analysis requires both understanding the concept of discourse and familiarity with a methodological template. Foucault's concept of discourse alerts us to the social constitution of *all* knowledge. Further, Foucault warns us that within many competing knowledge systems, particular sets of ideas emerge as dominant in the definition of appropriate forms of knowledge. According to Foucault, the constitution of certain knowledge as "truth" is not an accidental or haphazard process. Particular forms of knowledge become ascendant in ways that serve the interests of particular social groups. To unravel how, and why, this happens requires careful tracing of ideas expressed in the authorship, technological production, and circulation of diverse texts including government reports, policy documents, paintings, diaries, photographs, selfies, websites, social media, and interviews. When conducting discourse analysis these texts become your source materials.

Foucault did not provide a formal set of guidelines for the analysis of discourse. Instead, geographers, along with other social scientists, have contributed to designing approaches to analysis (Box 17.1). Are there disadvantages to relying on a methodological template? One objection, perhaps, is that the effect of writing geography that employs these criteria is inevitably both selective and prescriptive. To implement the checklist criteria uncritically would mask their potential to both silence and demand particular responses. In other words, do not equate the checklist with objectivity and rationality at the expense of subjectivity and creativity.

Nevertheless, with this caveat in mind, the checklist has at least four crucial advantages for the researcher new to Foucauldian discourse analysis. The first is to keep in mind the iterative relationship between the researcher and their work: the researcher shapes the analysis as much as the analysis shapes the researcher. The researcher is integral to rather than separate from the discourse analysis. Therefore, reflexivity, or writing one's self into a project, is a vital part of discourse analysis (for a more detailed discussion of related ideas, see Chapter 19). The second is that the checklist encourages researchers to remain alert to a key aim of discourse analysis: discourse analysis is not about determining the "truth" or "falsity" of statements but instead seeks to understand the geographical and historical circumstances that privileged and fixed particular discourses within discursive structures. Thus, in addition to selecting, familiarizing, and coding, discourse analysis requires background research into a source material or text's social circumstances, including authorship, production, and circulation. Third, the checklist operates as a reminder of the **situated** qualities of knowledge production. Knowing the world is a highly political process that operates to privilege and silence sets of ideas and voices, discourse analysis involves being alert to different strategies of conviction deployed by authors to help persuade audiences that a particular form of knowledge is intrinsically better than another. Finally, it is crucial to understand that while discourses may manifest themselves in ways that bring order to social life as rules, maxims, common sense, or the norm, they are always unstable and open to change. At the heart of discourse analysis is remaining alert to such instability, ambiguity, and inconsistency. Well-conducted and thoughtful Foucauldian discourse analysis enables insights into the resilience and rupture of multiple and sometimes conflicting discourses that produce and reproduce meaning of our always spatially located everyday lives.

Key Terms

audience
constructionist approach
discourse analysis
discursive structure
embodied knowledge
genre
intertextuality
manifest content analysis
metaphor
positionality statement
power/knowledge
privileged discourse
rigour
semiology
silence
text

Review Questions

1. Why should both background research, and the researcher, be regarded as integral parts of a Foucauldian approach to discourse analysis? To answer this question, think about how discourse constructs knowledge that produces generally accepted ways of thinking/being/doing.
2. According to Foucault's notion of discourse, in what ways are the terms *real* and *truth* misleading?
3. What are the differences between descriptive and analytic codes? How do these different codes relate to one another in a Foucauldian approach to discourse analysis?
4. Why are there tensions surrounding presenting a checklist or methodological plan for "doing" Foucauldian discourse analysis?

Review Exercises

1. *Nature Talk.* Foucault argues that discourse governs what is possible to talk about through the production of categories of knowledge. This process is not accidental. At the same time discourses reproduce power by governing what is generally accepted as true. To explore this concept, think about your own "nature talk."

   First, make a list of all the things/places/people that you generally talk/think about as belonging to nature.

   Next, using your nature list, explore the following questions:

   a. What is disallowed?
   b. What is normalized?
   c. Whose interests are being mobilized and served by this list?

   To further explore ideas associated with this exercise, see Castree, N. (2013). *Making sense of nature.* London: Routledge.
2. *Sweaty Geographies.* Foucault argues that discourse defines categories of thinking or lines of argument that frame ways of thinking/being/doing

as generally accepted truths, at the same time as other ways of thinking/being/doing are dismissed. To explore this concept, write down the places where you think it is both appropriate and inappropriate to sweat. Start with the places where sweating is appropriate, or even expected. What sets of ideas makes it possible and or desirable to sweat here? Why is sweat normalized in these places? How is sweat understood in these contexts? How does the appearance of sweat in these locations help constitute a particular subject? What interests are being mobilized and served by thinking/talking about sweat in this context? Now turn to the places where you think sweating is inappropriate. What makes sweat a problem in these locations? Why is sweat disallowed? What identities are made impossible? Explore the differences and similarities between class members' answers to these questions. What are the social norms governing "truths" about where sweaty bodies are welcomed or marginalized?

To further explore ideas associated with this exercise, see Waitt, G. (2014). Bodies that sweat: The affective responses of young women in Wollongong, New South Wales, Australia. *Gender, Place & Culture, 21*(6), pp. 666–82. DOI: 10.1080/0966369X.2013.802668

Useful Resources

Linda Graham, Queensland University of Technology, and Helen McLaren, University of South Australia provide helpful papers about the application of Foucauldian discourse analysis in education and feminist research. They are available at:

- Graham, L. (2005). http://eprints.qut.edu.au/archive/00002689/01/2689.pdf ; and
- McLaren, H. (2009). www.unisa.edu.au/Documents/EASS/HRI/foucault-conference/mclaren.pdf

Helpful starting points for more comprehensive reviews of Foucault's concepts include Cramption and Elden (2016), Hall (1997), Hook (2001), McNay (1994), Mills (1997), and Thiesmeyer (2003).

Alternative discussions of "doing" discourse analysis are provided by Dryzek (2005), Jones, Chik, and Hafner (2015), Phillips and Jørgensen (2002), Rose (2001), and Shurmer-Smith (2002).

Note

1. Clearly a project that would benefit from being repeated: as this chapter was submitted to the publisher in early 2020, Australian fires burned with unprecedented ferocity and arguments about climate change made the front page around the world.

Organizing, Coding, and Analyzing Qualitative Data

Meghan Cope

Chapter Overview

This chapter discusses some ways qualitative data can be organized and analyzed systematically and rigorously to produce trustworthy new knowledge. I begin with some comments on how to make sense of your data, and explore practices of "memoing," concept mapping, and coding. Much of the subsequent discussion revolves around coding as a process of distilling data and identifying themes. The chapter reviews different types of codes and their uses, as well as several ways to get started with coding in a qualitative project. Specifically, a distinction is drawn between *descriptive* codes, which are category labels, and *analytic* codes, which are thematic, theoretical, or in some way emerge from the analysis. An example from a project on children's urban geographies is used to illustrate coding. The building of a "codebook" is also discussed, stressing the importance of looking critically at the codes themselves, identifying ways in which they relate, minimizing overlap between codes, and strengthening the analytical potential of the coding structure. Finally, several related issues are covered, such as coding with others, the use of computer-aided qualitative data analysis software (CAQDAS), and integrating coding and mapping.

Introduction: How Can We Make Sense of Our Data?

It is an exciting time to be doing qualitative geographic research—as discussed in Chapter 1, growth in digital tools and creative methods is generating more diverse engagements with geographic inquiry. Appropriately, the standards of practice are high, requiring practitioners to be well-informed and attentive to principles of **transparent** data collection, trustworthy methods, rigorous scholarship, and well-communicated results. As is evidenced by this volume and the notable expansion, over the past two decades, of publications on qualitative methods generally in the field of geography, scholars and students are increasingly engaged not only in *doing* qualitative research but also in thinking and writing critically about methodologies, including the ways that we evaluate, organize, and make sense of our data (see also Clifford, Cope, Gillespie, & French, 2016; DeLyser et al., 2010; Gomez & Jones, 2010; Tuck & McKenzie, 2015). As the qualitative turn in geography and related fields has matured, the techniques for gathering and producing data have expanded in scope and depth; this, in combination with critical reflection on issues of methods, context, and researcher **positionality**, has produced ever more thoughtful, reflexive, and creative research projects and, ultimately, more robust **knowledge production**.

Toward these goals, the practices of organizing and analyzing one's data deserve special attention. The job of the researcher is, in effect, one of synthesis and translation; we are tasked with observing and engaging with the world, making sense of the resulting data, and representing (re-*presenting*) the facts, stories, ideas, and events that were shared with us, all in a coherent manner. The focus for this chapter will be on the middle piece of this—making sense of data—while recognizing that it cannot be wholly separated from the other parts of the process (for which see other chapters in this book).

Qualitative data are taking more diverse forms, well beyond the classical oral, text-based, and observational sources (though these are all still very important), into more varied visual data sources (images, video, maps, sketches, memes); emerging digital forms (tweets, social media scrapings, geo-tagged routes); sources resulting from **participatory** engagement with community partners (collaborative maps, exhibits, activism); and creative approaches such as artwork, **geopoetics**, theatrical productions, **PhotoVoice**, self-directed video, **narrative mapping**, material culture analysis, and other methods that foster more collaborative interaction with participants. In the context of such an expansion of research techniques, particularly those based less on texts, it is natural that we should revisit some ideas about "making sense of data." I propose, however, that many of the same principles that have appeared in previous editions of this volume remain valuable: organizing and reducing data to manageable chunks, identifying themes, and paying attention to rigorous interpretation. Toward that end, I review some common techniques for the purposes of description, classification, and connection (Dey, 1993) of qualitative data; these techniques are: **memoing**, **concept mapping**, and **coding**.

Making Meaning I: Memos

A memo (short for memorandum) is usually a short note to oneself or research collaborators, jotted to capture a quick insight, to serve as a reminder of a future task, or to draw connections between multiple referenced items. In **ethnography**, researchers typically rely on heavy use of field notes to record the events, dialogues, and **observations** made while engaging with a community, yet ethnographers often use memos as a sort of intermediate-level mechanism to remind themselves of something, to reflect on patterns or connections, to contextualize events, and to forge new links between emerging themes. For example, Annette

Watson and Karen Till (2010) provide a particularly rich example of the use of memos as a first stab at interpretation and reflection on their experiences and the material in their field notes:

> Both of us understand our writing as a safe space to explore, think through, and represent knowledges, desires, and fears. In memos, we question our experiences and assumptions, pay attention to processes, respond to our embodied and emotional presences, consider the material and visual cultures that constitute what is being studied, scrutinize various relationships with research fields and partners, and elaborate upon our insights. We also make connections to other studies of previous work in memos, and raise critical questions that inform future theoretical readings. (Watson & Till, 2010, p. 128)

The crucial elements of memos, as demonstrated by Watson and Till, can be summarized as follows. First, memos are quick and informal. They are jottings that by their very nature allow the researcher space and freedom to explore possible connections and viewpoints, to put an idea down in its earliest undeveloped form, and to remind oneself to pay attention to this or that in the future. Second (and despite its informal nature), memoing constitutes a valuable *interpretive* practice toward making meaning of the data. Along with margin comments, sticky notes, sketches, and other annotations researchers commonly make on and about their data, memos are useful for sorting out ideas, identifying patterns and similarities, recording "Aha!" moments, and generally beginning the process of organizing and analyzing. Finally, memos are **reflexive**. That is, they foster critical review and contemplation, including self-reflection; memos represent a chance to step back and consider alternate interpretations, to critique one's own role in the research process, and identify—however tentatively—linkages between events and discussions that might not have been seen while in the midst of data collection mode.

The benefits of memoing are not limited to participant-observation research practices, of course. Memos can be useful for any research project across the spectrum of qualitative, quantitative, and mixed-methods scholarship because they serve the important functions of reminding us of ideas to return to, as well as connecting, processing, and critiquing the process and early findings of the research endeavour. Reflecting the value of memoing, most computer-aided qualitative data analysis software (CAQDAS) programs include memo functions, including some that allow one to make memos about memos!

Making Meaning II: Concept Mapping

Despite the word *mapping* in the name, there is little formal discussion of concept mapping in the qualitative geography literature, yet it is an important organizational and analytic strategy that most researchers probably do without really thinking too much about, perhaps because there is a certain intuitiveness to the process. At its most basic level, concept mapping involves visualizing data and their relationships. This could be performed through the process of categorizing, that is, sorting data into piles (literal or digital) that have some internal cohesion. For example, in studying teen mobility my research partner and I used open-ended survey questions to query parents' perceptions of their teens having drivers' licences; we immediately saw that "safety" and "convenience" were major themes and, in one analytic exercise, began to sort survey responses along those lines, with subordinate factors contributing to the main concepts. By sorting and re-grouping responses on a large white board, using statements from parents who mentioned the convenience of having teens drive, we were able to quickly identify some clusters: "working parent(s)" signified that both parents were employed full time (or that a single parent was employed full time); "rural/remote" indicated that these families tended to live in more rural areas and at farther distances away from school, home, and workplaces; and "younger children" indicated the households in which teens had younger siblings who also needed chauffeuring around to activities and events (Cope & Lee, 2016). Discerning these kinds of connections through the relatively quick process of concept mapping—really a visual brainstorming exercise—allowed us to form some initial insights that were then followed up with more systematic examinations of the connections between "convenience" and the household situations of those parents and teenaged drivers. While sorting and piling are common components of concept mapping, other methods are also frequently employed, such as word clouds, flow charts, relationship trees, sketch maps, and non-hierarchical graphical plotting that begin to drill down into key associations within data.

Although the example mentioned here was intended as a relatively quick, exploratory, and coarse-level analysis, concept mapping has evolved into a tool that is much more formalized and the results can in themselves be part of the final representation of the research; this approach has gained traction in such fields as planning, organizational research, and evaluation studies, but could lend itself to diverse disciplinary fields. Kane and Trochim (2006) acknowledge both the informal notion of concept mapping (as employed above) and the much more structured, systematized methodology in the very first sentences of their book on the topic:

> *Concept mapping* is a generic term that describes any process for representing ideas in pictures or maps. [Here] however, we use the term only to refer to one specific form of concept mapping, an integrated approach whose steps include brainstorming, statement analysis and synthesis, unstructured sorting of statements, multidimensional scaling and cluster analysis, and the generation of numerous interpretable maps and data displays. (Kane and Trochim 2006, p. 1, emphasis in original)

In this method, which combines qualitative and quantitative analysis, researchers begin their analysis of, say, open-ended survey questions by isolating unique "statements" (single-topic fragments of longer responses), then they sort those statements into groups based on their substantive relationships. This sort/pile part of the method could be done by hand using slips of paper and actual piles or using CAQDAS functions digitally (for example using the network functions in Atlas.ti, or the sort and filter functions in NVivo). This in and of itself may be very helpful to researchers and, indeed, in the above example, my research partner and I stopped at this point having identified some valuable conceptual–empirical linkages that warranted further exploration. Kane and Trochim, however, go on to perform analyses of the clusters that result from the sort/pile method, including multidimensional scaling (which assesses clusters depending on the strength of their relationship or similarity) and cluster analysis (to assess connectivity, conceptual distance, and so forth). For those who are interested in these latter, more statistical, functions, the work of Kane and Trochim (2006) and their various research partners is recommended.

Ultimately, concept mapping can be as informal or formal as is suitable for the goals of making sense of data. For many geographers and other visual thinkers, it can be a helpful way of seeing data and relationships in different dimensions, it works well as a group analysis exercise with multiple researchers and research assistants, and, most importantly, concept mapping leads to new insights and connections being made.

Making Meaning III: Coding

Coding social data such as text, images, talk, or interactions, is intriguing and very valuable to the research process, though its operations are often derided as tedious. The purposes of coding are partly **data reduction** (to help the researcher get a handle on large amounts of data by distilling key **themes**), partly organization (to act as a **finding aid** for researchers sorting through data), and partly a substantive process of data exploration, analysis, and theory-building. Further, different researchers use coding for different reasons depending on their goals and epistemologies; sometimes coding is used in an exploratory, inductive way such as in **grounded theory** in which the purpose is to generate theories from empirical data, while other times coding is used to support a theory or hypothesis in a more deductive manner. Several approaches are discussed here, with pointers on how to organize and begin the coding aspect of a research project.

A short caveat is necessary: it is important to recognize that coding is not always the best or only way to rigorously understand qualitative data. Even in his coding manual, Saldaña points out, "No one, including myself, can claim final authority on coding's utility or the 'best' way to analyze qualitative data . . . *there are times when coding the data is absolutely necessary, and times when it is most inappropriate for the study at hand*." (2012, p. 2, emphasis in original). **Narrative analysis** and **discourse analysis** are two methods that are more appropriate for many projects (see Chapter 17; also Cope & Kurtz, 2016; Dixon, 2010; Doel, 2010).

Types of Codes and Coding

One common type of coding is **content analysis**, which is essentially a *quantitative* technique and by no means represents the full extent of coding for qualitative research. Content analysis can be done by hand or by computer (see discussion of CAQDAS programs below), but either way it is a system of identifying terms, phrases, or actions that appear in a text document, audio recording, or video and then counting how many times they appear and in what context. For example, a researcher might be interested in how many times the word *democracy* is used in newspaper articles from a particular country or they might be interested in how a particular place is portrayed in a visual media. Frequently in content analysis, **sampling** is used in similar ways to quantitative analysis of populations; perhaps only front-page newspaper stories are included in the analysis, or a television program is sampled for five minutes out of each hour. Similarly, researchers using content analysis typically subject their coded findings to standard statistical analysis to determine frequencies, correlations, variations, and so on. There are many good guidebooks and instructions for conducting content analysis, including some available on the web and broader methods texts (see also Chapter 9, regarding interviews). While content analysis has its place, the practices of coding qualitative data go well beyond merely *quantifying* them, which is what we consider next.

There are many ways to approach coding (see Crang [2005] and Saldaña [2012] for excellent examples), but to simplify, I discuss two main types of codes—descriptive and analytic. **Descriptive codes** reflect themes or patterns that are obvious on the surface or are stated directly by research subjects. Some descriptive codes can be thought of as category labels because they often answer who, what, where, when, and how types of question. Examples of descriptive codes that might interest geographic researchers include demographic categories (male, female, young, elderly), site categories (home, school, work, public space), or even scale identifiers (local, regional, national, global).

One special type of descriptive code is called ***in vivo*** codes; these are descriptive codes that come directly from the statements of subjects or are common phrases found in the texts being examined (Strauss & Corbin, 1990). For example, Jacquie Housel did interviews with elderly women about their daily routines and spaces in Buffalo, NY, and they repeatedly mentioned concern with crime in their neighbourhoods (see Housel, 2009). In this instance, "East Side response" became an *in vivo* descriptive code to indicate a type of slow and ineffective police protection (as is common in the majority-Black East Side of Buffalo), because the term is used by and describes something important to the subjects. *In vivo* codes are a good way to get started in coding, particularly in projects that are designed to be **inductive** (moving from data to theory) or exploratory. Other descriptive codes are generated by the researchers' interactions with the data as they sort through the sources, but they tend to—as their name implies—be fairly superficial.

Qualitative researchers also develop **analytic codes** in order to reflect a concept or theme the researcher is interested in or that has already become important in the project. Analytic codes typically dig deeper into the processes and into the context of phrases or actions. For example, in Housel's work it became apparent that the elderly white women mentioned above were experiencing a change of status in the neighbourhood as its demographics shifted to younger Black families: while in the past the women had been seen as the matriarchs of the local community, they increasingly experienced what they perceived as being ignored, disrespected, or even threatened by newcomers marked by race and cultural differences. Based on this shift, "erosion of white privilege" served as an analytic code (Housel, 2009, p. 134). This code could then be applied to the rest of the data to identify other responses based on fear and loss, categorizations of Others (Black people) as violent, perceptions of particular spaces as heightening their vulnerability (parks, streets and sidewalks, front porches), and loss of control in the disinvested and deteriorating public spaces of the neighbourhood.

Often, descriptive codes bring about analytic codes by revealing some important theme or pattern in the data or by allowing a connection to be made (for example, crime, fear of young men in public, shifting meanings of "race"), while other times analytic codes are in place from the beginning of the coding process because they are embedded in the research questions. For instance, Housel was also interested from the start in how elderly women navigate urban spaces, so their personal mobility, daily routines, and perception of their neighbourhoods were themes reflected in the analytic codes from the project's very beginning. The recursive strength of coding lies in its being open to new and unexpected connections, which can sometimes generate the most important insights.

The Purposes of Coding

Three main purposes for coding qualitative material can be readily identified: data reduction, organization and the creation of searching aids, and analysis. As the prolific French theorist Henri Lefebvre noted, "Reduction is a scientific procedure designed to deal with the complexity and chaos of brute observations" (Lefebvre, 1991, p. 105). Qualitative research often produces masses of data in forms that are difficult to interpret or digest all at once, whether the data are in the form of interview **transcripts**, hours of video, boxes of **archival** documents, or pages of observation notes. Therefore, some form of reduction is desirable to facilitate familiarity, understanding, and analysis. Coding helps to reduce data by putting them into smaller "packages." These packages could be arranged by topic, such as "instances in which environmental degradation was mentioned," or by characteristics of the participants such as "interviews with women working part-time," or by some other feature of the research context or subjects such as "observations in public spaces." By reducing the "chaos of brute observations," data reduction helps us get a handle on what we have and allows us to start paying special attention to the contents of our data.

The second purpose of coding is to create an organizational structure and finding aid that will help you make the most of qualitative data. Similar to data reduction, the organizational process mitigates the overwhelming aspects of minutiae and allows analysis to proceed by arranging the data along lines of similarity or relationship. An important early step is setting up and maintaining a complete record of sources, dates (of participant observation, interviews, or focus groups, for example), participants' contact information, and other relevant information, often in a **research diary** or logbook. While this information is not part of the coding process per se, it constitutes important record-keeping and allows the researcher to find specific data more easily. For example, interview transcripts might be coded not only for their content but also by their circumstances—Was the interview conducted in the participant's home? Were others present? Did the subject seem nervous?—which can help organize information. With better computer-aided qualitative data analysis software available now (see below), organizing and searching within electronic documents is greatly simplified. Additionally, coding itself is also an important aspect of organizing and searching because it is essentially a process of categorizing and qualifying data. "Codes are primarily, but not exclusively, used to retrieve and categorize similar data chunks (words, phrases, images) so the researcher can quickly find, pull out, and cluster the segments relating to a particular research question, hypothesis, construct, or theme." (Miles, Huberman, & Saldaña, 2020, p. 75). While the development of the **coding structure** is by no means a simple process, it is one that—if done well—enables the data to be organized in such a way that patterns, commonalities, relationships, correspondences, and even disjunctures are identified and brought out for scrutiny. Good organization also means the process will be more rigorous, an important consideration for defending one's work as **reliable**, transparent, and trustworthy (Baxter & Eyles, 1997; see also online resources of Pacheco-Vega, 2019).

The final, and principal, purpose of coding is analysis. While strategies for analytical coding will be examined in greater detail below, at this point it is sufficient to note that the *process* of coding is an integral part of analysis. Rather than imagining that analysis of the data is something that begins after the coding is finished, we should recognize that coding *is* analysis (and is probably never truly "finished"!). Coding is in many ways a recursive juggling act of starting with codes that come from the research questions, background literature, and categories inherent in the project and progressing through codes that are more interpretive as patterns, relationships, and differences arise.

Coding, as with memoing, also opens the opportunity for **reflexivity**, that critical self-evaluation of the research process (see Chapter 2). By recursively reviewing data and the connections between codes, researchers can also come to see elements of their own research practice, subjects' representations, and broader strategies of knowledge construction that had not previously been apparent. Even if it is difficult to be self-critical in the midst of fieldwork or data collection, the process of coding is inherently more contemplative and analytical and thus offers a moment ripe for reflection.

Getting Started with Coding

The above discussion of types of codes addressed descriptive and analytic codes, although other terms are also used in the literature (for example, **initial codes** and **interpretive codes** or first-round and second-round). To recap, the key distinction is that one type of code is fairly obvious and superficial and is often what the researcher begins with, such as simple category labels. The other type of code is interpretive, analytic, and has more connections to the theoretical framework of the study; it tends to come later in the coding process after some initial patterns have been identified. When coding is done manually, researchers develop a **codebook**—a list of codes that are categorized and organized repeatedly. Although current qualitative software packages typically do not use the term codebook, it is a useful concept that has relevance whether the codebook is actually a tangible item in manual coding or merely an abstraction in electronic coding. To start, begin with the most obvious qualities, conditions, actions, and categories seen in your data and use them as descriptive codes. These elements will emerge quite rapidly from background literature, your own proposal or other research-planning documents, and the themes that stick out for you from gathering qualitative data (for example, memorable statements in interviews, notable actions seen while doing participant observation, or key words that jump out in first readings of historical documents).

Thus, the first step is to make a list of what *you* think are the most important themes upfront, with the understanding that some of them will be split into finer specifics while others will remain largely unused. But how do you know what is important? Anselm Strauss, one of the founders of grounded theory, had a helpful system for beginning this daunting task (best represented in Strauss and Corbin [1990]). Here I introduce Strauss's approach with examples from several excellent studies in geography, and then I extend his system with one I developed and use my own fieldnotes as an example.

Strauss suggested paying attention to four types of themes:

- conditions
- interactions among actors
- strategies and tactics
- consequences

First, "conditions" might include geographical context (both social and physical), the circumstances of individual participants, or specific life situations that are mentioned or observed (for example, losing a job, moving to a new place, changing schools). By thinking along the lines of conditions and beginning by coding only for them, the coding process is easily started, and you may learn a lot about your data in a short time.

Secondly, look for interactions among actors—if you focus on relationships, encounters, conflicts, accords, and other types of interactions, a series of powerful codes will emerge that will be helpful throughout the research. For example, in their PhotoVoice project with Indigenous youth in urban Canada, Goodman et al. found that, for the young people involved in their project, social relationships were compromised by frequent mobility due to evictions, being placed in state care, or other events: "moving [to a different residence] made it difficult to maintain relationships with friends and family" (2018, p. 319). By coding their data along lines of mobility and social relationships, Goodman and her co-researchers were able to identify respondents' related concerns about isolation, instability, and "broken homes."

Strauss's "strategies and tactics"[2] dimension is a little more complicated than the first two types of themes because it suggests a certain level of purposeful intent among the participants that may demand additional inquiry on the part of researchers. Goodman et al. (2018) found that some of the Indigenous young people who moved frequently developed negative coping mechanisms (e.g., drug use) and destructive behaviours (e.g., dropping out of school) as tactics to deal with stress, gain control, or take care of themselves. These strategies built up over time until, as one young participant shared, "I didn't want to make friends because like I knew I wouldn't stay in the area for long. . . . Most of the time I was scared to trust them because a lot of people that I've trusted have just walked out of my life, or passed on, or moved away and it hurt a lot and I just chose not to trust nobody" (Participant 2, Goodman et al., 2018, p. 320). Broadly, some types of strategies or tactics to look for include emotional self-preservation, education and career decisions, housing choices, family negotiations, health-related decisions, political activism, or even subversion.

Other times, coding for strategies and tactics may be more subtle—and more analytical—as when respondents do not explicitly state their reasons for certain actions but a connection emerges through observation, review of interview text, or other data. Many geographers have paid attention to ways that people engage in *resistance* against diverse forms of oppression, which may be seen as strategies for empowerment, rights, or merely survival. For example, Orna Blumen (2002, p. 133) took "dissatisfaction articulated in subtle terms" by ultra-orthodox Jewish women as small but significant indicators of the women's resistance to their families' economic circumstances and, more broadly, to the status and roles of women in that community. For the women in Blumen's study, referring to fatigue, hoping their husbands would soon find paying work, and "minor, personal, nonconformist remarks suggestive of ambivalence" (2002, p. 140) could all be coded as tactics of resistance, in part because Blumen—through careful qualitative work—had sufficiently analyzed the broader context of the women's lives and goals to recognize them as such.

Similar to the above, "consequences" is a slightly more complicated category of codes. On the surface, there are descriptive indicators for consequences, including terms such as *so*, *because*, *as a result of*, and *due to* that may be present in subjects' statements and can be good clues to consequences. For example, a young Indigenous participant in Goodman et al.'s study said, "When you leave a geographical spot, for young people, you are leaving behind your friends that are there a lot of the time. So that affects your relationships, stability, schools, just everything" (Participant 1, in Goodman et al., 2018, p. 319). Note the word *so* in this statement, which is a good tip-off that a consequence is embedded in the text. Beyond the obvious connector words, though, there are also more analytically subtle ways of discovering and coding consequences that are dependent on the unique empirical settings and events of each study. Some consequences will be matters of time passing and actions taking place that result in a particular outcome—the passage of a law, a change in rules or practices, and so on. However, other consequences are more nuanced and personal, or they are not the result of changes over time and therefore may be trickier to identify and code as such. For example, going back to Housel's work with elderly white women, and noting that they felt uncomfortable in some areas of their long-time neighbourhoods, she coded her interview transcripts for the consequences of feeling "out of place in my own space" due to the women's sense of being left behind as the city's demographics and investment patterns changed. Coding for consequences of this kind requires sensitivity to both the subjects and their community context but is potentially a rich source of analysis and insight if done with care.

As an example of what a sample of coded material looks like, Box 18.1 demonstrates a small selection of field notes from my project on children's urban geographies in Buffalo, New York, along with codes, themes, and notes. In the example below, I use what I call the CRAFT method, which is derived from and an extension of Strauss's system, identified above, but somewhat simplified. In this method, I look for text related to Conditions, Relationships, Actions, Feelings, and Themes and then generate codes that best reflect those. By providing a bit of structure and an easily remembered anagram, but also opening up a variety of considerations, the CRAFT method is multi-dimensional and flexible, and thus serves as a good place to start. For example, one of my original interests in this project was how children in the 8- to 12-year-old age group identified community spaces and how they perceived boundaries between public and private space. These themes were obvious starting points for my codebook, but the CRAFT approach also revealed other interesting dynamics. The fieldnotes are from a neighbourhood mapping project, which explored what kinds of things children would like to see in their neighbourhoods in place of vacant lots. Subsequently, from this, in combination with other findings from my Children's Urban Geographies (ChUG) Project, I generated a theory of how children define and ascribe meaning to the idea of "neighbourhood" (Cope, 2008). Theory-building, after all, is an important goal of most qualitative research!

BOX 18.1

Coded Field Notes: Summer Kids' Program, Buffalo, New York (NSF #BCS 99-84876)

Field note text	Descriptive Codes	Analytic Themes	Notes/Reflections
Recorded by Meghan Cope *Friday July 22, 2005; 10:00 am*	CRAFT*		
It is our second week of the Lots of Opportunity Project (LOOP). As I walk in I'm greeted by several children, and the staff say hello. Soleil was the first to come hug me and asked if I remembered. I said yes, the wagon was in my car. She gave me a big smile and seemed excited. Walking through the main room I saw Marianna and went to say hello. I asked if she wanted to join us and she asked if it required walking outside. When I said yes she said she didn't want to because it was too hot. After a bit of organization we gathered the LOOP kids. Rochelle commented that "a lot of the kids don't want to walk today." It was overcast and threatening rain, but not nearly as hot as last week. While we got going in the art room Rochelle noticed that it was raining. My three grad students and I looked up at each other and groaned. We hadn't prepared for this possibility! We scrambled for a minute thinking we could do the "visioning" part of the project but then the rain largely stopped and I suggested to the children that we live in Buffalo and aren't scared of a little rain, and besides it's not hot like last week, so let's just go out. There was a little grumbling but I also sensed that the kids welcome a chance to get out of the club. Someone had found some balloons in the art room and the kids were blowing these up with various results.	*Relations*: Ms B and Ms L are friendly to us (unlike some staff) *Feelings*: excitement, affection, affirmation *Theme*: conditional participation (Marianna is self-assured and knows what she likes!) *Conditions*: warm, muggy summer day, rain forecast *Feelings*: unhappy/happy	*ChUG Effect* (Note: The code "ChUG effect" refers to the ways that our presence in the club makes ripples, affects rhythms, or in some ways shifts people's behaviour in the moment *Child–Adult Translator* (Rochelle is very observant and communicative and informs us of group sentiments)	Soleil had been so reluctant to participate and walk last week but wanted to join us that I had suggested bringing my wagon so she could ride. Poor planning on our part

After some more preparation, handing out folders and maps, sharpening coloured pencils, etc., we finally set out. We had 8 children. L.K. and I went with Soleil, Tayana, Shelby, and Danisa. We arranged that my group would go south on Balister St and the other group would go south on Shallow and we'd meet and cross paths on Ulster.			Logistics
I got the wagon out of my car and grabbed my big red and white umbrella, both of which Soleil immediately appropriated. I put some of the children's balloons in my car because they were very distracting. Tayana was pulling Soleil in the wagon and Danisa and Shelby were walking along. As we walked past the two houses just east of the parking lot, Shelby went up to the porch (as she did last week) to say hello to the people there and hugged her little sister, who is 4. Everyone on the porch was African-American: several men in their 20s and 30s and one woman, along with a few smaller children. As Shelby started up the steps Soleil said "you can't go up there" and I said it was ok because these were Shelby's cousins and little sister, which I had learned last week.	*Action*: taking control of material objects *Relationships*: family, sister *Conditions*: racially diverse neighborhood *Action*: calling out others' behaviors	*Researcher/Adult Promises*: stakes were high for Soleil, testing me *Boundaries* between public/private space Children "*policing*" each other spatially	I have met Shelby's mother, who is white with blond hair. I wonder about Shelby's position as a mixed-race child in this racially divided city and how she makes sense of that.
We all crossed Mansfield Ave (Tayana bumping Soleil over the curb) to Balister and stopped at the eastern side of Balister. As Tayana pulled Soleil up the curb I suggested she get out of the wagon because she was too big and she immediately said "you think I'm fat!" and crumpled into near-tears. [I felt terrible that she had interpreted my comment that way (and am especially aware of girls' fragile body images)] but my efforts to explain that I only meant that the wagon is meant for toddlers did nothing to make her raise her head from her hands.] Finally, giving in entirely to her tantrum, I offered her the camera to take some photos. [While I felt like I was violating every rule of parenting toddlers at this moment, I did not feel like having this outburst affect the entire day's activities.]	*Feelings:* hurt *Actions*: provocative statement; crying (or pretending to) *Action*: use technology	*Body image* – girls	Wow, I blew that. Soleil (age 10) has been emotionally delicate and volatile the whole time I've known her. Her deliberate misunderstanding was certainly a dramatic way to get the attention she craves. Makes me wonder about her home life. As a mom of toddler girls, I knew distraction would break Soleil out of this mode.

(continued)

Field note text *(continued)*	**Descriptive Codes**	**Analytic Themes**	**Notes/Reflections**
We crossed Balister to go see the community garden, which the girls had pointed to immediately. As we approached the garden, Soleil (from the wagon) started saying we couldn't go in there, that it belonged to the lady who lives in the house next to it (north), and that this lady would yell at us if we went in. Danisa said no, it's for everybody. Soleil repeated her claims and was quite adamant. I glanced at L.K. [her dissertation is on community gardens] who shook her head "no" regarding the garden being the property of that resident. I assured Soleil that it was ok, that it was community space and that Danisa was right, it's for everyone. She refused to budge and gave up on us saying something like "y'all gonna get yelled at." L.K., Danisa, Tayana, Shelby, and I went into the garden and noted the raised beds, we talked about the plants there, some of which the children could identify (lettuce) and others that they read off the little signs (okra, collard greens). We walked deeper into the garden (me keeping half an eye on Soleil, who was still on the sidewalk) and Shelby wanted to take pictures. Everyone took turns with the camera a few times, then went back to Soleil, where we took more photos of all the girls on the wagon with the umbrella.	*Theme*: community spaces *Feeling*: superior knowledge? *Feeling*: fear of consequences of trespassing *Theme*: children's knowledge of urban space *Theme*: children's knowledge of foods, plants *Conditions*: technology soothes upset kids	*Boundaries* between public/private space Children *"policing"* each other spatially Children's understandings of the *"rules of place"*	Did Soleil want to be the "expert" on the expectations of this space? Was she especially peeved when we agreed with Danisa that is was public? It's hard to interact with Soleil in this mode I missed an opportunity to ask the girls if their families ever have okra and collard greens, given their association with African-American cooking and soul food. Perhaps next time.
We paused at the next parcel, which was vacant and covered in rubble, gravel, and weeds. There was one car parked on it. Danisa said there was a house there but it burned down. I asked her what could go there and she said a house, a garden, or "anything you want, just imagine." Tayana joined in and started making suggestions, saying "I would put my swing-set, my pool, and my trampoline here." Someone – a young boy – was watching us from just inside the house next to the lot, so we moved along.	*Conditions*: vacant lots *Theme*: children's knowledge of urban space *Theme*: neighbourhood desires *Theme*: neighbourhood desires	*Eyes on the street*	Tayana's phrasing was kind of sweet – as if she owned these things and had just needed a spot to keep them.

All children's names and streets have been changed

Sometimes codes can be too general and become cumbersome. Because much of my children's urban geography research was centred around issues of neighbourhood and community, I found that I needed to break these into more specific codes, such as codes for specific community spaces, a sense of Hispanic "West Side pride," children's views on trash and broken infrastructure on their streets, known "dangerous" spots in the area, etc. This is a frequent characteristic of coding: an initial category turns out to be overly broad and must be refined and partitioned into multiple codes.

The opposite also occurs—some codes die a natural death through lack of use. For instance, in my project I had naively expected the children, who were for the most part in low-income families, to talk about a lack of money or not being able to afford something they wanted. However, after four years in the project, I found little evidence of children discussing their own poverty (though that absence is itself an interesting research question). So, while demographic data in the project necessitated that I keep "low income" as a code, it was rarely used in analyzing direct quotes from children. As Miles and Huberman acknowledged: "Some codes do not work; others decay. No field material fits them, or the way they slice up the phenomenon is not the way the phenomenon appears empirically. This issue calls for doing away with the code or changing its level" (2013, p. 82).

Developing the Coding Structure

Using Strauss's themes and/or the CRAFT method will take you a long way towards constructing a codebook, and you may find other types of themes that are helpful to you, such as "meanings," "processes," or "definitions." But how many is too many? Using the combination of descriptive and analytic codes, you may find yourself with well over 100 codes at some point, which is unwieldy at best and counterproductive at worst. Lists of codes that have not been categorized, grouped, and connected will be hard to remember, have too much overlap and/or leave uneven gaps, and will not enable productive analysis. Therefore, the next step is to develop a coding structure whereby codes themselves are grouped together according to their similarities, substantive relationships, and conceptual links. This process requires some amount of work but is well worth the effort, both for ease of coding your material and for discerning significant results from your findings.

Developing the coding structure can proceed in various ways, and there are many resources available that demonstrate different approaches (see Denzin & Lincoln, 2018; Miles, Huberman, & Saldaña, 2020; Silverman, 2001), but the main purpose is to organize the codes—and therefore the data and the analysis process. Some codes will automatically cluster; for example, codes relating to the setting of interviews (for example, home, office, public space, clinic), characteristics of subjects (for example, age, gender), or other categories (for example, occupations, leisure activities, life events). Other codes seem to fit together because of their common issues; for example, you might have a group of codes related to people's goals or intentions or a group of codes related to people's experiences of oppression. Finally, codes based on the substantive content of text or actions—and most likely related to the analytic themes you are developing—will create another cluster of codes: for example, perceptions, meanings, places, identities, memories, difference, representations, and associations.

Once the codebook is relatively comfortable (I hesitate to say "complete") and the coding structure is devised, you will want to go through much of your data again to capture connections that may have been missed the first (or second or third) time around. Remember that coding is an iterative process that feeds back on itself—only you can decide when it is time to move on. As Miles, Huberman, and Saldaña (2020) point out, it is sometimes simpler when time or money pressures put some finality on projects that otherwise could always benefit from one more case study or endless additional tweaking of the coding structure.

Coding with Others

Depending on the size and resources of the research project, there may be a case for using multiple coders for the data, which can add considerable complexity to the process, but also can bring in different perspectives and thus be more rigorous. There is an inherent tension in using multiple coders on a project: is the goal to make everyone code as consistently as possible, or is the goal to allow each coder to interpret data in their own way within the bounds of the coding structure in order to capture many diverse meanings? The answer will depend on the project and the epistemological leanings of the lead researcher(s), but in fact both of these goals are important. In the first instance, **reliability** of the data is undoubtedly enhanced when several coders independently code a piece of data the same way—a common interpretation of data means that there is agreement on its meaning and less of a chance of **confirmation bias**. One position suggests that for the sake of time and data reduction, having multiple coders can certainly be helpful, but only if they are truly consistent in their coding, which is rare but could be accomplished by achieving conformity on the meanings of codes and providing thorough definitions for each code. On the other hand, text and video—as social data sources—are inherently subject to multiple interpretations and understandings, many of which may be correct or "true." While there may be some interpretations that are farfetched or extreme, in general we as social researchers will be interested in capturing diverse understandings, and having multiple coders can be a great benefit for the project, making deeper and broader connections from the data.

Computer-Aided Qualitative Data Analysis Software (CAQDAS) and QualGIS

Anyone who has coded research material by hand with notecards and coloured pens can understand the attraction of computer assistance in this endeavour—the idea that one's codes could be organized and employed using software holds many potential benefits in terms of time-savings, consistency, and data security, particularly when compared to stacks of handwritten materials. However, there are also some cautions that should be kept in mind before embarking on the investment of time and resources to use CAQDAS. This brief overview is not intended as instructional or comprehensive, but rather, identifies some key considerations and points the reader toward further resources.

There are numerous CAQDAS packages on the market today, ranging in price, sophistication, functions, and utility. However, the basic premise of all of them is the same: to bring coding and analysis into a computerized operation. In fact, some of the most basic functions of coding can be completed in word processors or using spreadsheets, and for smaller or fairly basic analyses, these might be preferable. For example, a first-run content analysis is easily accomplished using the find function in word processors, and pieces of text can be highlighted and commented upon using review functions such as track changes in electronic documents. Lists of codes and memos, as well as notes about when to use which codes, are easily stored and managed in spreadsheet programs or table options in word processors. For more complex projects with more codes, more data, and/or multiple researchers learning how to use a CAQDAS (and, sometimes, investing in purchasing a licence) can be well worth it. I highly recommend reading the review and self-critique Bettina van Hoven (2010) wrote about her experiences learning and using CAQDAS in her dissertation project. She reveals several "advantages" of using CAQDAS, including managing large quantities of data, convenient coding and retrieving, and quick identification of deviant cases, as well as "concerns" that include obsessions with volume of data, mechanistic data analysis and a "taken-for-granted" mode of data handling, alienation from one's data, and an over-emphasis on grounded theory (van Hoven, 2010, p. 462). No matter how much packages change, these advantages and concerns are likely to be timeless.

Some examples of this type of software being used by geographers include NVivo, Atlas.ti, MaxQDA, and Dedoose. All of these have demo versions and online manuals in PDF on their respective websites, and most have lower-cost student versions; therefore, one can easily test-drive a package before committing fully to it. Many universities now have site licences for these packages and a lucky few even have in-house support for qualitative researchers. Because packages are constantly being updated and changing in their functionality, an assessment of the relative strengths and merits of these would be quickly outdated, and therefore they will not be reviewed here *per se*; I do recommend visiting the University of Surrey's CAQDAS support site for reviews of the latest versions and consulting Silver and Lewins (2014), which is an e-resource that is frequently updated. Miles, Huberman, and Saldaña (2020) also provide an in-depth comparison of features and uses of CAQDAS. These resources can guide you on what might suit a given project, the capacities of the researchers, and the available time and budget. Still, there are some features that are worth mentioning here that are relevant to consider, regardless of what new offerings emerge.

First, CAQDAS has three primary functions: *text retrieval* (especially helpful for the "constant comparison" approach to analysis), *coding*, and *theory-building*. It is, of course, essential to realize that the program does not *do* the coding for you, except in the most rudimentary way based on a set of instructions you provide; the researcher still must engage in thoughtful and thorough building of codes and applying them to text segments, photo elements, or video footage clips. However, a CAQDAS is an excellent organizational tool for storage, retrieval, and interpretation of data, and all packages have at least some memoing capacity. Targeted searches or browsing by coded material can reveal connections that would have been difficult to see with paper index cards, and the **theory-building** functions (such as hierarchies or concept maps that show relations between data, or even between codes) are useful for analysis and the generation of original interpretations and conceptual understanding.

The second feature to consider is the degree to which a CAQDAS can be geographically linked, if this is something that is relevant for your research. First, virtually all programs allow maps to treated as static images and coded, just as a photo might be—this is useful if you are working with sketch maps or other geographic annotations provided by research participants. Integrating geographical maps into analysis is more tricky, but some programs are moving in this direction: Atlas.ti allows the importing of KML files and Google maps can be opened from within the program and MaxQDA allows hyperlinks to Google Earth images. In what Fielding (2009) identifies as a "technological convergence" these developments should be promising for many geography scholars and students. Further, with the rapid growth of online mapping tools (both vernacular and professional), volunteered geographic information, crowdsourcing, and spatially referenced social media—what we might more generally refer to as qualitative **digital geographies** (see Chapter 15)—the possibilities for blending mapping with qualitative research are growing fast. These raise important questions of how the massive saturation of spatially referenced digital data will affect qualitative research in geography and beyond (Ash, Kitchin, & Leszczynski, 2019).

New work is emerging that pushes the boundaries of coding and leads to creative analyses and representations of qualitative geographic data. With the rise of participatory research in geography (see Chapter 16 in this volume), qualitative GIS (Cope & Elwood, 2009; Garnett & Kanaroglou, 2016), participatory mapping, mixed methods, and other integrative practices, we need to stay attuned to how coding can keep up with new research processes and technologies. Emerging approaches seem to come from multiple directions, suggesting that diverse skills and creativity will help scholars forge their own paths: CAQDAS programs are increasingly incorporating geographical mapping functions, GIS programs and spatial data tools are increasingly accommodating non-numerical (that is, qualitative) data, and in the area of **GeoHumanities** we see the foregrounding of artistic experimentation that blends analysis and representation in creative ways. One early example of this that is still worth reviewing is Jin-Kyu Jung's (2009) experiments with using codes as a bridge between qualitative analysis using CAQDAS and spatial data analysis using GIS. In his work, the code literally serves as a software-level link between databases, allowing analysis programs to "speak" to each other in a platform he calls CAQ-GIS (computer-aided qualitative geographical information systems). At a conceptual level in Jung's work, the code also serves as an analytical connection between social contextual data and spatially referenced data, allowing researchers to develop new understandings of social–spatial relations. With the many possibilities for mixed-method and qualitative research facilitated by better and faster digital technologies, the necessity for researchers to be well-trained, rigorous, reflexive, and ethical has never been greater.

Conclusion

Being in the World, Coding the World

By way of conclusion, let me reiterate the point we made in this volume's introduction: interpreting and analyzing the social world is not a mysterious process that must be learned from scratch but rather is one that we are all already actively practising in our everyday lives. The recognition that we are all constantly "coding" and making sense of the world around us may be a helpful realization for getting started in a research project and can also assist us in critiquing our own practices of data reduction, organization, and analysis. As Silverman (1991, p. 293) pointed out, there are many ways of "seeing" and interpreting the world, and—as social beings—we never really shut those lenses off, so why not embrace diverse perspectives and turn our gaze to the process of interpretation?

> How we code or transcribe our data is a crucial matter for qualitative researchers. Often, however, such researchers simply replicate the positivist model routinely used in quantitative research. According to this model, coders of data are usually trained in procedures with the aim of ensuring a uniform approach. . . . However, ethnomethodology reminds us that "coding" is not the [sole] preserve of research scientists. In some sense, researchers, like all of us, "code" what they hear and see in the world around them [all the time]. . . . The ethnomethodological response is to make this everyday "coding" (or "interpretive practice") the object of inquiry. (Silverman, 1991, p. 293)

Being in the world requires us to categorize, sort, prioritize, and interpret social data in all of our interactions. Analyzing qualitative data is merely a formalization of this process in order to apply it to research and to provide some structure as a way of conveying our interpretations to others.

Constructing Meaningful Geographical Knowledges, Writing Qualitative Geographies

Juliana Mansvelt and Lawrence D. Berg

Chapter Overview

In this chapter, we examine the process of writing-up the results of qualitative research in human geography. Our aim, however, is to *contest* simplistic understandings of the relationship between research, writing, and knowledge production that arise from describing the process in this manner. The very phrase "writing-up" implies that we are somehow able to unproblematically reproduce the simple truth(s) of our research in our writing. In this chapter we challenge a view of writing that suggests it is a mirror that innocently reflects the reality of research "findings." Instead, we argue that our writing and the conventions of presentation that we draw on when writing are socially constructed, that is created by human activity and imagination. Often writing styles and conventions may seem straightforward—"academic" conventions of writing can appear to be guidelines that we simply learn to apply. Rather, recognizing that writing styles are socially constructed, made through repeated acts of writing and expressed (codified) in such forms as theses, reports, and journal articles, enables us to consider how and why these conventions arise. Understanding that the act of writing does not simply *reflect* meaning, but *produces* meaning also enables researchers to consider the potential effects of our writing for the audiences who read it. Consequently, in this chapter we draw on post-structuralist approaches to argue that writing is not merely a mechanical process that reflects the "reality" of qualitative research findings but rather shapes, in part, how and what we know about our research. Thus, the chapter begins by tracing key philosophical and epistemological developments that engage with the question of how knowledge is co-constituted through action and reflection. We then demonstrate that writing is not so much a process of writing-*up* as one of writing-*in*, a perspective that has significant implications for how our research is made meaningful for those who read it.[1]

Styles of Presentation

We begin this chapter by considering some taken-for-granted aspects of academic writing, considering how different styles of presentation draw on different understandings of the relationship between the researcher, the research, and the audiences for whom we are writing. Much of our argument in this chapter revolves around the idea that a number of powerful **dichotomies**—such as subject/object, researcher/researched, data/conclusions, and research/writing—currently exist and that these dichotomies structure our understanding of research (see also Chapter 17). These dichotomies have developed historically as part of a **neo-positivist** framework that, we suggest, limits our understanding of the writing process. Understanding something of the neo-positivist framework that historically underpinned much research writing in geography is helpful to identifying these dichotomies in our writing, and to understanding the potential impact of them in the presentation of research for a range of audiences.

Positivist and Neo-positivist Approaches: Universal Objectivity

The philosophical approach to scientific knowledge known as **positivism** was founded by Auguste Comte (1798–1857), a French philosopher and sociologist (see Gregory, 1978; Kolakowski, 1972). Comte argued that scientific knowledge of the world arises from **observation** only. The notion of a singular central truth that provides explanations for how the world is structured and understood was associated with this. The logical positivists, whose work developed during the 1920s as a critique of strict positivism, differed from the Comtean positivists in their conception of scientific knowledge. While Comte allowed only statements arising from **empirically** verifiable knowledge (that is, information available to the senses) as the basis of factual knowledge (or truth), the logical positivists accepted the validity of more than just empirical observation as the basis for truth. Logical positivists argued that the truth of the social and natural worlds could be constructed from analytical statements that comprised theories or models, with the truth of these theories (and the hypotheses or assertions that underpin them) being confirmed through empirical observation. Karl Popper (1959) developed an alternative form of neo-positivist thought he termed *critical rationalism*. Unlike the logical positivists, who focused upon *verification* as the basis of knowledge, Popper argued that **falsification** should be the basis for making decisions about hypotheses, based on the idea that we can never know for sure whether a particular hypothesis is true or not, but we do have the ability to ascertain whether such statements are false.

Although few geographers have ever fully taken up the ideas of any single school of positivist thought, positivism of one sort or another has played a foundational role in the way that many geographers have come to understand the world. This became most explicit during the so-called quantitative revolution of the 1960s (Gregory, 1978; Guelke, 1978), which saw geography develop as what we might broadly term a neo-positivist "science." Geographers maintained a strict distinction between facts and values (Gregory, 1978), giving emphasis to observational statements over theoretical ones (Berg, 1994a) and often universalizing their findings across all contexts (Barnes, 1989). They also made a strict distinction between objective and subjective knowledge.

Objective knowledge is seen as "scientific," rigorous and detached, and consequently valid. It is constituted in opposition to **subjective** knowledge, which is seen as personal, value-based, non-scientific, and non-academic (and therefore unacceptable as a basis for establishing "the" truth). Objective knowledge is founded on interrelated and highly gendered notions of rationality, disembodied reason, and universality (Berg, 1994a, 1997; Bondi, 1997). Trevor Barnes and Derek Gregory (1997, p. 15) suggest that "scientific geographers" thus imagined themselves:

> as a person—significantly, almost always a man—who had been elevated above the rest of the population, and who occupied a position from which he could survey the world with a detachment and clarity that was denied to those closer to the ground (whose vision was supposed to be necessarily limited by their involvement in the mundane tasks of ordinary life).

Knowledge from the vantage point of the objective researcher looks the same from any perspective—it is monolithic, universal, and totalizing. This concept of un-located and disembodied rational knowledge draws on powerful metaphors of mobility (the researcher can move to any and all perspectives) and transcendence (the researcher is not part of the social relations they are examining but instead can rise above them to see everything) for its rhetorical power to convince readers of its claims to the truth (for powerful critiques, see Barnes & Gregory, 1997; Haraway, 1991).

In adopting a broadly neo-positivist model for their work, geographers at the time also developed a specific approach to writing their research. They developed commonly used approaches or forms of writing (or what we call **tropes**) in their academic studies that attempted to erase the authorial self from their written work. Similarly, they tried to create, through their writing, distance between themselves as researcher/author and their research objects. These tropes are most evident in the practice of writing in the third person—a practice that is still prevalent in much academic writing today. Many geography undergraduate students, for example, are still required to use the formal **third-person narrative** form in their essay assignments. In other words, third-person narratives are needed to maintain the appearance of impartiality and objectivity, two cornerstones of the neo-positivist model. The third-person narrative constructs an **objective modality** (Fairclough, 1992), which effectively removes the author from their writing while at the same time implying the author's full agreement with the statements being made. In so doing, it does the work of transforming interpretative statements into factual statements (see Box 19.1).

Writing in the objective mode is often accompanied by **nominalization**, a process that further removes the writer from the text (see Box 19.1). Nominalization involves the transformation of adjectives and verbs into nouns. It occurs, for example, when in the process of researching, researchers and research subjects are rhetorically transformed into things such as "the research" or when actions and processes are given particular kinds of argumentative power as nouns, such as can be found in statements such as "the analysis suggests" (Fowler, 1991). Nominalization deletes a great deal of helpful and important information from sentences. For example, it removes information about the participants—normally the agent or researcher (the person "doing" something), and the affected participant (or the person having something "done to" them). It also changes the **modality**—that is, the implicit indication of the writer's degree of agreement with the statements being made in a text. Nominalization creates **mystification** because it permits concealment, hiding the participants, details of time and place, and the stance of the participants involved. It also results in **reification**, whereby complex, uncertain, and often contradictory processes, such as are involved in human geographic research, assume the much more certain status of "things." Thus, the complex processes of research are smoothed over into a one-dimensional and simple thing: "the research" (see, for example, Fowler, 1991, p. 80). Ironically, while positivist **epistemology** is founded on the idea that any statement can be contested through recourse to empirical evidence and logic, the language used to communicate research findings within this framework implies an unquestionable accuracy that is a very poor approximation of the messiness of social life.

Interestingly, writers drawing on neo-positivist approaches have shown ambivalence with regard to their own writing practice by framing it in contradictory ways. On the one hand, they have explicitly acted as if language has no impact on meaning, yet ironically, they have implicitly acknowledged—through their insistence on writing in the third person—the significant role that language plays in constructing knowledge and meaning.

Post-positivist Approaches: Situated Knowledges

Although the positivist-oriented science model dominated during the quantitative theoretical period of the 1960s and 1970s, its **hegemony** (or conceptual dominance) began to be contested in geography during the 1980s and early 1990s by post-positivist approaches such as humanism, Marxism, political economy, and post-structuralism (for example, Barnes, 1993; Berg, 1993; Dixon & Jones, 1996;

BOX 19.1 Removing the Writer—Third Person and Nominalization

Writing in third person and a process called nominalization help to create a more formal written text. This has the effect of distancing the author from the narrative being constructed and creating a sense of neutrality and objectivity.

As you read the sentences below, consider how the use of third person, by removing terms that describe oneself (such as *I*) and substituting descriptors of others (for example, *the investigator*) creates a sense of distance, from the research and the research subjects, and makes the statement appear more factual.

- The investigator examined five types of place-based knowledge.
- The researcher demonstrated that understandings of home are complex.
- One tries not to influence the responses gained from interview questions.

Nominalization is the process of turning verbs into nouns. It involves taking actions and events and changing them into objects, concepts, or things—making sentences seem more formal and abstract. Like writing in third person, nominalization contributes to a sense of detachment from the text (and the research), removing both personal reflections and the context of the subject being discussed.

For example, the verb developed is nominalized to the noun development. A **first-person narrative** such as:

- I *developed* a framework for analyzing interviews.

might become:

- The researcher commenced the *development* of a framework for analyzing interviews.

Below are two more examples of text. In the informal text, few verbs have been nominalized. The second paragraph is an example of more formal research writing in which verbs and verb phrases have been nominalized. As you read these two texts, think how the agency of the researcher appears to be removed in the second, concealing the researcher's emotions and experiences. Note, for example, how the first sentence in the formal text suggests that there is much to be learned from the research (now assumed to have the status of a thing) rather than from the researcher-led processes that constituted it, as implied by the more informal excerpt.

Informal text:

> Designing and conducting the research was a learning process for me. Because I did not explain the purpose of my study clearly, the people I approached for interviews often refused to participate. My structuring of the interview questions was also too complex, and my participants struggled to answer them. Consequently I had difficulty in analyzing the information I had obtained from my interviews and my insights were limited. I learnt a lot through the process but concluded that my research project was ill-conceived.

Formal text:

> Much was learned from the design and conduct of the research. A higher than expected level of refusal to participate was a consequence of a lack of clarity in the explanation of the research. The interview questions were too complex, and negatively influenced participant responses. The research analysis produced limited insights. The research was an ill-conceived learning exercise.

The formality achieved through the use of third person in the second excerpt has the effect of producing the author as an all-seeing and all-knowing (seemingly objective and blameless) surveyor of the research rather than an active participant, complicit in the production of both text and research. Such distancing is an illusion, negating the ways in which power and subjectivity are constituted through the research process.

England, 1994; Massey, 1993). Perhaps the strongest critiques from within geography arose in response to the works of feminist **post-structuralist** writers in the wider social sciences (Anzaldúa, 1987; Frankenberg & Mani, 1993; Haraway, 1991; Mohanty, 1991) who were keen to confront the universalism, mastery, and disembodiment inherent in positivist notions of objectivity, criticizing masculinist and Eurocentric concepts of universal knowledge. This has led to a rich vein of work in geography that examines the relationship between geographers as situated knowledge workers and their relationship to the objects of knowledge that they "produce" in their research (see Berg, 2004, 2012; Clough & Blumberg, 2012; de Leeuw, Cameron, & Greenwood, 2012; Pain, 2014; Radice, 2013, Sotoudehnia, 2016). All of these authors would agree that simplistic notions of "objectivity" play a significant role in marginalizing those who do not fit into dominant conceptions of social life.

Donna Haraway's (1991) evocative metaphor of **situated knowledges** provides perhaps the most useful approach to contest universalist forms of knowledge. She argues that within dominant ideologies of scientific knowledge, objectivity must be seen as a "God Trick" of seeing everything from nowhere. She proposes a different concept of "objectivity," one that attempts to situate knowledge by making the knower accountable to their *position* (see also Chapter 2). All knowledge is the product of specific embodied knowers, located in particular places and spaces: "there is no independent position from which one can freely and fully observe the world in all its complex particulars" (Barnes & Gregory, 1997, p. 20). Research that draws upon situated knowledges—including much of the work cited in this volume—is thus based on a notion of objectivity much different from that posed by positivists, and this conception of objectivity also requires a different form of writing practice.

Post-structuralists and feminists also contest approaches to inquiry that conceptualize writing and language as simple reflections of "reality." They argue that "language lies at the heart of all knowledge" (Dear, 1988, p. 266). It should be made clear, however, that such arguments do not assume that language and ideas are the same as "real" phenomena, objects, and material things. Instead, arguments about the centrality of language express the fact that all processes, objects, and things are understood by humans through the medium of language. Thus, while we might experience the very material process of hitting our "funny bone" on a table in ways that do not necessarily entail language (for example, as a very visceral sensation we know as "pain"), we come to understand the process and the objects involved through language (with categories such as table, funny bone, pain, and so on). Accordingly, language must be seen as not merely reflective but instead as *constitutive* of social life (for example, Barnes & Duncan, 1992; Bondi, 1997; Dear, 1988).

The post-structuralist critiques of language as a direct representation of the "real" world and of disembodied concepts of universal objectivity have significant implications for writing practices. If language is **co-constitutive of knowledge** and meaning, then it would seem to matter *how* we write our knowledges of the world. Likewise, if we are to *locate* our knowledge, then we must locate ourselves as researchers and writers within our own writing. Accordingly, post-structuralist writers reject the ostensibly "objective" modality of writing their work in the third person. Instead, they opt for locating their knowledge-defining objectivity as something to be found not through distance, impartiality, and universality but through contextuality, partiality, and **positionality** and consideration of one's **subjectivity** (Pacheco-Vega & Parizeau, 2018). However, as Gillian Rose (1997) has argued, given the difficulty in completely understanding the "self," it may be virtually impossible for authors to fully situate themselves in their research. Notwithstanding such difficulties, it is possible for authors to go some way towards locating themselves within their work. Risa Whitson (2017) brilliantly argues that reflexivity must involve more than reflecting on one's positionality, that it is vital also to scrutinize the researcher (or writer's) own subjectivity—the self we wish to portray to our readers. This might involve reflecting on how our aspirations, intentions, and goals might influence our research and considering how we might use our writing to "paint a picture of ourselves" (the title of Whitson's article) in relation to our readers, and recognizing that our social identities can shift and change through the research and writing process.

Certainly, the first step to acknowledging our subjectivity and positionality is to reject the third-person narrative, replacing it with a first-person narration of our essays.[2] At the same time, it is not enough to merely adopt the first-person narrative form. Instead, it is important to both reflect upon and analyze how one's position in relation to the processes, people, and phenomena we are researching actually affects both those phenomena and our understanding of them. Thus, rather than writing ourselves *out* of our research, we write ourselves back *in*. This is, perhaps, one of the most important distinctions to be made between what we will refer to as the writing-up (distanced, universal, and impartial) and the **writing-in** (located, partial, and situated knowledge) models. Another significant difference arises from the ways that post-positivists conceptualize the relationship between observation and theory.

Balancing Description and Interpretation—Observation and Theory

The Role of "Theory" and the Constitution of "Truth"

We argue in this section that there currently exist a number of powerful dichotomies—observation/theory, subject/object, researcher/researched, data/conclusions, and research/writing—that structure our understanding of research. It is important to remember that these dichotomies are not recent developments in Western thought. Instead, they arose within the long history of dualistic thinking in Western philosophy (Berg, 1994a; Bordo, 1986; Derrida, 1981; Foucault, 1977b; Jay, 1981; Le Doeff, 1987; Lloyd, 1984; Nietzsche, 1969). These dichotomies became racialized and gendered through a long historical process of developing a singular Eurocentric and masculine concept of rational thought. For example, through a process that Susan Bordo (1986) terms the "Cartesian masculinisation of thought," Descartes's mind–body distinction came to define appropriate forms of knowledge. The mind was conceptualized as rational, and it came to be a property of European men. The body was seen as irrational (have you ever heard the phrase "mind over matter"?) and was associated with everything that was not European or masculine: women, racial minorities, and sexual dissidents, for example. In other words, the mind was unmarked, but the body was a mark of difference. Further, Descartes's dualistic philosophy of knowledge formed the basis for the dominant present-day conception of objectivity as impartial, distanced, and disembodied knowledge. Accordingly, Cartesian dualistic thinking forms the foundation of positivist thought (Karl Popper, of whom we spoke earlier, for example, was a well-known adherent of the mind–body dualism).

As we have already discussed, positivist-inspired geographers make a rigid distinction between objective and subjective knowledge and between theory and observation. As with the observation/theory binary, the so-called "objective" is valued at the expense of the subjective. Such hierarchically valued dichotomies form parts of a whole series of other binary concepts—including (but not limited to) mind/body, masculine/feminine, rationality/irrationality, and research/writing—that are interlinked through complex processes of signification (Derrida, 1981; Jay, 1981; Le Doeff, 1987; Lloyd, 1984). In the observation/theory binary observation is equated with objectivity, mind, masculinity, rationality, and research. Theory is constituted as lacking, and it is associated with all those other negatively valued concepts: the subjective, the body, the feminine, and the irrational.

Conceptualizing one side of the binary as a *lack* of the other leads to a devaluation of the subordinate term. Thus, in the case of the positivists who conceive of theory as a lack of empirical observation, the ways in which theory constitutes our understanding of empirical "reality" (in addition to explaining it) are underestimated. Indeed, positivist constructions of factual knowledge as phenomena that are available to the senses (empirically observable) tend to efface the very theoretical nature of positivist thinking itself. As we have already suggested, positivism has both a history and a geography associated with Europe. It is an epistemology—a theory of knowledge—that has developed relatively recently and has come to dominate contemporary intellectual life in the West. Nonetheless, it is not the only theory of knowledge; rather, it is one of many competing theories of knowledge. However, because it is dominant, or hegemonic, it rarely has to account for its own epistemological frameworks. With this in mind, we argue for "recognition that we cannot insert ourselves into the world free of theory, and neither can such theory ever be unaffected by our experiences in the world" (Berg, 1994a, p. 256). Observations are thus *always already* theoretical, just as theory is always touched by our empirical experience (see also Chapter 17 on Foucauldian **power/knowledge**). Recognition of this relationship has important consequences for the way we write-*in* our work (and *work*-in our writing).

Writing and Researching as Mutually Constitutive Practices

Metaphors that allude to research as "exploration" or "discovery" are hard to avoid, because their meanings are so taken for granted (how often have you heard lecturers speak of their research in terms of "exploring," "examining," "discovering," and "uncovering"?). This is particularly the case with the so-called "writing-up" of research. The term *writing-up* powerfully articulates the written aspect of the research process in a way that renders it somehow less significant and/or less problematic than other aspects of the research process. Writing-up is usually seen as a phase that occurs at the end of a research program; indeed, many textbooks about conducting research (this one included) include the section on writing-up at the end of the book (for example, Bryman, 2012; Flowerdew & Martin, 2005; Kitchin & Tate, 2000). Writing is also often seen as a neutral activity, although ironically, scientific and positivist modes of writing and thinking are subject to rhetoric, creativity, and intuition (Creme, 2003).

The writing-up phase may be discussed in such a way that it appears to be merely a matter of presenting the results and conclusions in an appropriate format at the *end* of a research program. We argue here that in writing research, the researcher is not so much presenting their findings as *re-presenting* the research through a particular medium. Rather than reflecting the outcome of a particular research endeavour, we believe the act of writing is a means by which the research is constituted—or given form—and that this process occurs throughout the research process. Writing involves a process of selecting categories and language to describe complex phenomena and relationships. But we are getting at much more than that here. Research and writing are *iterative* processes, and writing helps to shape the research as much as it reflects it. This is why as academic supervisors we encourage our students to write from the very beginning of their thesis process.

In addition, the way we conceptualize the author (as distant and impartial or as involved and partial, for example) has significant implications for the ways that the very processes of research itself can be understood. Perhaps just as important, writing involves very clear decisions to include some narratives and to exclude others. It thus brings some ideas into existence while (implicitly) denying the existence of others (Ely, 2007). For us, then, writing is not so much a matter of writing-*up* as of writing-*in*, a perspective that has considerable implications for how qualitative research is conceptualized and undertaken.

Writing is not devoid of the political, personal, and moral issues that are a feature of undertaking research, nor is writing devoid of our embodied emotions as we sense and feel the narratives we construct. As a creative process, writing can be enjoyable for authors and readers (Bradford, 2003), opening up new possibilities for representation, narrative, and engagement with those who may participate in and reflect on the research. Further, the separation between fieldwork and writing is artificial (Denzin, 2008). Whatever the qualitative research technique utilized, some means of recording the researcher's interpretations, impressions, and analysis must be used, and although such accounts may be recorded as digital audio or video files the words with which they are constructed are an integral part of the research, not simply a result, recollection, or recording of it. The research cannot be separated from the labels, terms, or categories used to describe it and interpret it, because it is through them that the research is made meaningful.

For example, after I (Juliana) had conducted several qualitative **interviews** with local authority economic development officers for my PhD research, I realized that using the word *traditional* as a label for a certain form of local economic initiative was problematic. This was because definitions of "traditional economic initiatives" were contested by officers and because the term appeared to position local authorities who undertook this type of initiative as "old-fashioned" or "not progressive." I learned a valuable lesson about the power embodied in words I had simply drawn from the literature on local economic development and about the kind of assumptions embedded in my own research agenda. This understanding enabled me to construct my questions (and consequently my entire research project) in a different way.

I (Lawrence) had a similar experience of reorienting research agendas in a **participatory action research** project I undertook with a number of colleagues (Berg et al., 2007). This project involved the fusion of **Indigenous methodologies**, participatory action research, and white studies (Evans et al., 2009) to understand the exclusion of urban Indigenous people from the so-called "universal health care system" in Canada. However, before we even started our formal research, we had to come to grips with the hegemonic ideology among white Canadians that Indigenous people do not live in cities in Canada but instead are to be found out in the countryside, on rural reserves, disengaged from modernity. Such categories are especially problematic given that, as of the 2006 census, more than 51 per cent of all Indigenous people now live in urban areas. What it means to be "Indigenous" in Canada is thus contested (Berg et al., 2007) and so was our own understanding of our research project as a "participatory" partnership with urban Indigenous people (Berg et al., 2007; Evans et al., 2009). It should be clear that any attempts to write about such work involve significant politics in the construction of our categories and in the decisions about what to say (and not to say) about any specific events. Accordingly, the writing-up of the research is only the initial phase of an iterative process of negotiating the production of knowledge between the researchers and the participants. I am not suggesting here, however, that negotiated knowledge is intrinsically better than other forms of knowledge. Instead, what I want to point out is that the process of making explicit the act of negotiation helps to make the research *accountable* in ways that are appropriate given the specificity of positions and power relations. All research is caught up with power relations, and to deny this is to deny an important aspect of **knowledge production** (see virtually all other chapters in this volume). Taking this process seriously has enabled me to rethink the role of writing in my intellectual endeavours.

Thoughts, observations, emotions, and interpretations that occur during the research become important components of any research endeavour, not because they record events or ideas but because they are **signifiers** of them (in this sense, they act to "define" complex constellations of ideas and thoughts about the research in more simplified categories of knowledge). Whether interpretations are noted by way of a personal research **diary**, log, **field notes**, video, or audio recording, they can provide insight into the researcher's own speaking position and how it is articulated, challenged, and modified through the research journey.

Writing-in is not a matter of "telling"—it is about knowing the world in a certain way. The process of writing constructs what we know about our research, but it also speaks powerfully about who we are and where we speak from. As we suggested in a previous section, the detached third-person writing style so common in academic journals and reports implies that the researcher is omniscient—that they have a perspective that is all-seeing and all-knowing. However, what may appear to be the truth spoken from "everywhere" is actually a partial perspective spoken from some*where* and by some*one*. Knowledge does not, according to a post-structuralist perspective, exist independently of the people who created it; knowledges are partial and geographically and temporally located. Because all knowledges are situated, how we study, understand, and perceive the social world is always situated, influenced by where we are, and our subjectivity—who we are and the beliefs, biases, and blind-spots that emerge from this (Proudfoot, 2019; Simandan, 2019). As the researcher writes and inscribes meaning in the qualitative text, they are actually constructing a particular and partial story. Richardson and St Pierre (2008) suggest that writing creates a particular view not only of what we are talking about (and what we do not say) but also of ourselves. Power is connected with a speaking position through text, so a qualitative researcher should consider not only their standpoint in constructing a research account but also the implications of their interpretations for those who may have been involved in the research.

Because the practice of writing is not neutral, the voices of qualitative researchers do not need to hide behind the detached "scientific" modes of writing. **Reflexivity** is the term often used for writing self into the text. Kim England (1994, p. 82) defines this as "self-critical sympathetic introspection and the self conscious analytical scrutiny of self as researcher." A reflexive approach can make researchers more aware of their necessary connection to the research and their effects on it, and also of asymmetrical or exploitative relationships (i.e., where a researcher has more social power and influence than their participants), but reflexivity cannot remove inequities (England, 1994, p. 86; also see Rose, 1997, and Chapters 2, 3, and 4 of this volume). One way in which reflexivity can be encouraged in the writing-in of qualitative research is by the use of personal pronouns (for example, *I*, *we*, *my*, *our*).

However, it is important that the use of personal pronouns does not become merely an emotive tool. Alison Jones (1992) suggests that in academic spheres, the use of *I* may result in the insertion of "emotion" as a replacement for "reason," thereby creating a work of "fiction" hiding power relations as much as it might make them explicit. Employing reflexivity through use of first person should instead make explicit the politics associated with the personal voice and draw attention to assumptions embedded in research texts. Reflexivity is also concerned with constructing research texts in a way that gives consideration to the voices of those who may have participated in the research. Reflexivity is about writing critically, in a way that reflects the researcher's understanding of their position in time and place, their particular standpoint, and the consequent partiality of their perspective. Writing situated accounts may involve acknowledging the role of embodied emotions in research (Bondi, 2014; Cahill, 2010; Whitson, 2017; Wright, 2010), thinking about how one's positionality is "mutually constituted through the relational context of the research process" (Valentine, 2003, p. 377) and considering the ways in which our texts might constitute performances, by enacting and valorizing certain metaphors and meanings and silencing others in the spaces where they might circulate (Denzin, 2003). This understanding of the **dialogic nature of research and writing** (in the sense of a dialogue between various aspects of the research process) enables qualitative researchers to acknowledge in a meaningful way how their assumptions, values, and identities constitute the geographies they create. It also provides an opportunity to play and experiment with writing as a way of knowing, representing, and practice (DeLyser & Hawkins, 2014).

Just as there are many ways of knowing, there are also numerous ways in which qualitative researchers may construct their research narratives as "re-presentations"—constructions that evoke what we as researchers have lived and learned (Ely, 2007). Richardson and St Pierre (2008) believe that to write "mechanically" shuts down the creativity and sensibilities of the researcher. They encourage researchers to explore text and genre in the (re)presentation of qualitative research through a variety of media, including oral and visual and to experiment with diverse forms of the written word (prose, poetry, play, autobiography). See, for example, Kaya Barry's (2018) use of interviews, personal observations, and time-lapse photos to study the relational intensities and spatial encounters involved in backpackers' experiences of travelling and "Packing a Bag." We support Richardson and St Pierre's (2008) metaphorical construction of writing as **staging a text** and encourage geographical researchers to consider how they are (re)presenting the research "actors," creating the plot, action, and dialogue of a research "tale," how they are constructing the stage, the setting of the "research" play; and at whom (i.e., the audience) the "production" is aimed.

Of course, most undergraduate geography students will be required to write their research within a given format—the essay or report (for some advice on the conventions associated with these forms of writing, see Bradford, 2003; Hay, 2012; Hay, Allington, & Bochner, 2021; Kitchin & Tate, 2000; and Chapter 20 of this book). Nevertheless, it may be possible to persuade your professor to let you produce another form of geographic representation: a play, a video, a poem, a short story, a poster-board, to name a few options. For example, Lawrence's third-year students in his Culture, Space, and Politics course create two- to three-minute videos involving critical analyses of "landscapes" as part of their major assignment for the class and then upload these videos to YouTube for public viewing, and Juliana's Consumption and Place postgraduate students are asked to create an **auto-ethnography** (Butz, 2010) of their learning in the course during the year, presented through any medium they choose. Despite writing constraints and "staging" conventions imposed by self and audience (such as for an academic publication or a thesis), we argue that there is no single correct way to "stage" a text (see Box 19.2). By exploring the varied ways in which the text can be staged and how in such staging different stories may be emphasized and other voices may come to the fore, the researcher has the potential to create dynamic and interesting research pieces that engage and challenge both writer and reader.

BOX 19.2

Writing-in: Alternative Representations

Geographers' use of poetry in their writing remains unconventional, yet according to Clare Madge (2014), poetry can be a powerful means of expressing "an embodied, **affective** Geopolitics" of the world. (See also Chapter 1.)

Syrian Not Knowing
You went to buy vegetables today
I heard, but never returned. . . .
Bundled into a white van,
screaming; squashed and squelched
as a bruised aubergine
As the van disappeared through a
grotesque worm hole
you evaporated too. And in this moment,
from such a lofty distance,
the air sucked out of me, punctured,
Imagine that. . . .
Un-knowing!
Gaping Emptiness!
As your child stares at her empty plate
Suspended in a petrifying, pitiful, fissure
lingering, longingly for your return
Just imagine that. . . .

Madge suggests creative works such as this can be a catalyst for creative thinking and invoking debate. She acknowledges both the power and limitations of such writing for opening up multiple interpretations across different times and places, depending on the positionality of the reader. Readers of this poem might relate to horror expressed by Claire of the abduction, connected as it is to the everyday, the banal, the repetitive act of food provisioning, a geopolitics that reverberates and is made real by the child's empty plate. She suggests her poem is not so much a view of the world, but a point of view in it—a form of writing that may challenge those reading it to consider "not only how emotions can 'travel' but also how geopolitical events can 'speak through the body'." (Madge, 2014, p. 181)

The truth and validity of knowledge arising from geographical and social science research has been the subject of discussion for four decades (see for example Harrison & Livingston, 1980). Post-structuralist thinking has challenged the assumption of a singular truth and the privileging of certain claims to knowledge. Associated with this is what has been termed the **crisis of representation** (Marcus & Fisher, 1986). That is, doubts have arisen over researchers' authority to speak for others in the conduct and communication of research (Alvermann, O'Brien, & Dillon, 1996). In recent years, an increased sensitivity to power and control on the part of some qualitative researchers has encouraged a rethinking of research design and implementation (Evans et al., 2009). It has also meant a growing concern over how researchers appropriate and assume participants' voices in the writing of research accounts (Opie, 1992). Moreover, it has become important to acknowledge that one's research writing might have an active role in constructing and reconstructing research relations in unintended ways and that research accounts are in themselves a kind of creative invention (Freeman, 2007).

Post-structuralist thinking casts doubt on foundational arguments that seek to anchor a text's authority in terms such as **reliability**, **validity**, and **generalizability**. How the validity and authenticity of qualitative research accounts might be assessed are well articulated elsewhere in this book (see Chapters 1 and 6) but suffice to say the reflexive writing-in of research experiences and assumptions is not a licence for sloppy research or monographs based solely on personal opinion.

Clifford Geertz (1973) has argued that good qualitative research comprises **thick description**. Such descriptions take the reader to the centre of an experience, event, or action, providing an in-depth study of the context and the reasons, intentions, understandings, and motivations that surround that experience or occurrence. While it may not be possible to assess the authenticity of such partial descriptions, the interpretations upon which they are constructed can be articulated. For many years Pamela Moss had written about the illness experiences of others, but it was her own condition that caused her to reflect on the challenges of writing her own illness, a challenge that enabled her to consider how illness derives from discourse, material things, and social relationships. She notes that attention to her autobiographical writing of her own illness became a means of "accentuating precarious, ambiguous and diffident aspects of the generation of 'I' as an academic subject" (2016, p. 85). Moss's autobiographical writing in drawing on thick descriptions of moments and places in which tiredness, pain, and illness were experienced, allowed her to reflect critically on how her illness experiences formed and the contexts in which they emerged. Writing autobiographically in this way also demonstrated the porous boundaries between health and illness, and how these informed her becoming as an academic.

Communicating qualitative research is about choices—for example, about how we show the workings of our research, how we present and convey others' voices, how we locate our own subject position, and what effects our writing may have for those who read it and who want to engage with it (Denzin, 2008; Holliday, 2007). Although such choices are not always conscious and not necessarily made in circumstances of our own choosing, researchers can attempt to take responsibility for their perspective and position in research through reflexively acknowledging and making explicit those choices that have influenced the creation, conduct, interpretation, and writing-in of the research, which is sometimes referred to as **transparency.** Such choices are likely to be guided by principles of **ethics** and truthfulness. Transparency may make researchers, and the audiences for whom they write, more aware of the constraints on interpretation, of the limitations imposed by the "textual staging," and of the implications of the former for research participants. For example, the choices surrounding the use of research participants' quotes in a written text comprise far more than a simple matter of how, where, how many, and in what form participants' voices are to be included. The inclusion of quotations raises issues of representation, authority, appropriation, power, and participation. David Bissell (2018), for example, provides a particularly compelling description of the intensity of community life juxtaposing quotations from his participants with his own observations and theoretical interpretations. Acknowledging the situatedness of his accounts, and the ways in which commuting experiences are themselves a product of individual and collective experiences produced in and across the times and places of the commute, Bissell has produced a piece of writing that enables the reader to see how the smallest encounters matter, taking his audience beyond the usual depictions of the banality of journeys to and from work (see Box 19.3).

Jamie Baxter and John Eyles (1997) suggest that the criteria of **credibility**, **transferability**, **dependability**, and **confirmability** are useful general principles for guiding an evaluation of the rigour (**trustworthiness**) of a piece of qualitative research. They see these categories as broadly equivalent to the concepts of validity, generalizability, reliability, and objectivity that have been used to evaluate the quality of quantitative research endeavours. It is important to note, however, that we avoid using the term *rigour* in our approach to constructing pieces of qualitative research, since we believe the term is too closely tied to quantitative approaches that are predicated on systematic exposure of pre-existing truth. Instead, we prefer to use the term *trustworthiness*, which speaks much more directly to qualitative geographical research as a reflexive practice that constitutes and understands meanings in place: one that actively recognizes that knowledge is constructed, open-ended, and fluid (but see Chapter 6, for example, for a different interpretation of rigour). The characteristics of the particular audience to whom writing is intended should be kept in mind, and in writing qualitatively it is important for researchers to explore and make explicit their own research agendas and assumptions and to elaborate on how they believe their research text constitutes the "truth" about a particular subject. Communicating qualitative research can be as much about *how* we come to know as it is about *what* we know, as examples of auto-ethnographic writing demonstrate (see Box 19.4).

Impression Zones: Understanding the Skills Required in Commuting

In his research on commuting, David Bissell's (2018) in-depth fieldwork on the journey Australians have to and from work involved interviews with communities, journalists, policy-makers, and transport advocates. David also spent many hours travelling himself in and out of Sydney making detailed observations of the social and spatial interactions he witnessed. This research project forms the basis of his book *Transit Life*, which highlights the intensities, skills, powers, and politics created through seemingly mundane journeys to work. In these selected excerpts on "Impression Zones" from his book (2018, pp. 12–13), rather than occupying a distant and objective subjectivity and positioning as a researcher David acknowledges the vulnerability, uncertainty, and interpretive questioning that accompanied his research. Here we see the juxtaposition of David's fears with Rob's (his "instructor") confidence and control in negotiating Sydney's streets via bicycle.

> For me [David], the very idea of cycling in Sydney seems, frankly, terrifying. Rob laughs like a man who has heard this too many times before. "It's a bit of a Sydney problem," he admits. "It seems like, in Sydney, a lot of people are really in a bit of a rush to get where they're going and have a very—I don't know if *selfish* is the right world, but it seems like there's a very self-centred entitled view about, you know, 'I have a right to get to where I am going and I am going to get there as quickly possible'." Then his tone changes, "A lot of people will talk about a war on the roads. . . ."

David narrates how Rob goes on to explain how his cycling course teaches people to cycle "graciously"—a disposition that creates a calmness and implies making good road decisions. Rob also notes how shifting one's perspective away from the city streets as zones of conflict means one is able to control more of the cycling experience. In writing this excerpt David continues his account to explain how his street ride with Rob not only improved his skill level but began to shift his own disposition towards both his cycling and his research. David writes:

> As we do something again and again, things are changing, however slightly. Every time I performed this drill, it felt a little different. This logic could be extended to whole journeys. Did you

> really take "exactly" the same journey today as you did yesterday? Could this ever be possible? What exactly has changed?

David's acknowledgement of his lack of expertise in city cycling positions him in this narrative as the hesitant learner rather than an all-knowing researcher—a positioning that changes as he gains in confidence cycling with Rob. David's vulnerability, acknowledged in his writing, also becomes a part of narrating the process of his learning as a researcher, leading him to question his own assumptions of the homogeneity of the daily commute. Such writing provides productive openings for his research audience too. In closing this excerpt David notes that these questions also fascinated philosopher Félix Ravaisson, and then explains how these ideas become encapsulated in Ravaisson's notion of *habit*. David's skill in incorporating his insights as a cyclist, researcher, and learner with those of others (theorists and his participants) in his writing demonstrates that academic writers need not distance their subjects and themselves from research accounts in order to produce "honest" accounts of the social world that are theoretically sophisticated.

BOX 19.4 Auto-ethnography: A Method for Reflexive Writing

Ellis (2004, p. xix) defines auto-ethnography as

> research, writing, story, and method that connect the autobiographical and personal to the cultural, social, and political. Autoethnographic forms feature concrete action, emotion, embodiment, self-consciousness, and introspection portrayed in dialogue, scenes, characterization, and plot. Thus autoethnography claims the conventions of literary writing.

Ian Cook has encouraged his students to use auto-ethnography as a means of writing reflexively and critically, bringing the insights from their reading, visual, and aural material research and their felt experiences, emotions, and knowledges into their assessment for a course on geographies of material culture (Cook et al., 2007). Similarly, David Butz and Katharine Besio

(*continued*)

(2004) have used auto-ethnography to better understand the implications (for research and knowledge production) of their own positionality as white, Western researchers working in the Karakoram. Auto-ethnographies provide one way by which the emotions, experiences, contradictions, and inconsistencies in our research journeys (and those whose lives and experiences intersect with them) can be re-presented. The use of literary rather than objective modalities and the active inclusion of self means that one's writing,

> far from being a dispassionate process of producing what was, is instead a product of the present, and the interests, needs, and wishes that attend it. This present, however—along with the self whose present it is—is itself transformed in and through the process at hand. (Freeman, 2007, pp. 137–8)

Writing thus becomes a method of knowing. Examine this piece of auto-ethnographic writing with footnotes, by Sarah, one of Ian Cook's students:

> It's evening. To go out-on-the town I need an image refinement. Understatement! My cybernetic mask thickens with the make-up and lip-gloss, hair spray. . . . A reach for the wardrobe and a selection of T-shirts. Removing my glasses I read the small print—a third "gaze": no glasses—close-up-focus but not *understanding*. Black tops from Hong Kong and EEC. Or should I go to Greece? Or with the red tops—more classy—to Thailand, England (no?!), Indonesia? I must emphasize never having looked at the origin of my clothes before. Think I will wear the silver one which has no label. Safer to not know where it comes from. Would "made in Syria" really make a difference? I still wouldn't know how, where, when it was made and by whom. And it goes with my chain—connecting my outfit as well as me with the world.
>
> Again, the concept of "commodity fetishism" is important. As a cybernetic being I am embedded in global networks and yet I am unaware of those that I connect.
>
> Through the clothes that I wear I am connected not only to the people who made them but to the machines themselves that manufactured my outfit. As a cyborg, are those machines a part of me? The amalgam of man [*sic*] and machine, to which Haraway's (1991) "cybernetic organism" alludes, where does it end? What is the extent of my hybridity?

What effects might such writing have? Sarah's writing, like Madge's (2014) earlier "staging" (see Box 19.2), might be seen as opaque, diffuse, and uncertain. While auto-ethnographic pieces might seem disconcerting, even confusing, relative to conventional objective modalities of writing research, such narratives can provoke an awareness of the situated problematics and politics of knowing, representing, and transforming knowledge. Ian Cook believes this approach can have radical effect—an effect that arises out of a less straightforward or didactic connection between what is known and how it is interpreted. What do you think?

Conclusion

We have discussed the importance of language in the social construction of knowledge, how power is articulated through dichotomies, and how meaning is inscribed in language. We believe models of writing that construct the writer as a disembodied narrator are inappropriate for communicating qualitative research. Our focus has been on written rather than visual texts, since writing remains the predominant means of communicating qualitative research. Breaking down dichotomies—through an interpretative understanding of writing and researching as mutually constitutive processes—and an understanding of which principles might guide valid qualitative research are critical to writing "good" qualitative research. It is also crucial to understand how power and meaning are inscribed in the words that we use (and those we choose not to use) to constitute the research process, to recognize our subjectivities, standpoint, and locatedness (shifting and partial though they might be), and to acknowledge the voices of those with whom we undertake research. We believe that doing this enables qualitative researchers to have confidence in the "validity" and truthfulness of their interpretations and to write confidently for particular audiences. Consequently, this chapter has not been a how-to guide but a means of raising important issues that are inherent in the writing-*in* process.

Key Terms

auto-ethnography
dialogic nature of research and writing
dichotomy
falsification
first-person narrative
hegemony
modality

mystification
neo-positivism
nominalization
objective
objective modality
positionality
positivism
post-structuralism
quantitative revolution
reflexivity
reification
signified/signifier
situated knowledge
staging a text
subjective
thick description
third-person narrative
trope
validity
writing-in

Review Questions

1. What implications do post-structuralist perspectives have for writing qualitative research?
2. In what ways is the term *writing-up* misleading?
3. Why should writing be seen as an integral part of the entire research process?
4. How can a researcher endeavour to produce "trustworthy" research?
5. Why can writing be seen as a method of knowing?

Review Exercises

1. *Exploring different writing modalities*: Using the first-person narrative write a paragraph describing your experiences of eating your evening meal last night. Swap your account with a friend and re-write the paragraph they wrote using third-person narrative. Together discuss your paragraphs: which version seems more "truthful," and why?
2. *Dismantling dichotomies*: With a partner or in a small group, identify a dichotomy that you would like to work on (e.g., masculine/feminine, public/private, rational/emotional, etc.). On a sheet of paper or a whiteboard write each element on one side with lots of space in between, then brainstorm to develop ideas, evidence, and experiences that fill the space between and challenge the binary. Share your results and discuss how other dichotomies can be dismantled.

Useful Resources

Bondi, L. (1997). In whose words? On gender identities, knowledge and writing practices. *Transactions of the Institute of British Geographers*, 22, 245–58.

Chandler, D. (n.d.) Semiotics for beginners. www.aber.ac.uk/media/Documents/S4B/semiotic.html. This site is a comprehensive introduction to some post-structuralist understandings of the power of language.

Cook, I. (2009). Geographies of food—afters. http://food-afters.blogspot.com. This is Ian Cook's blog for co-authoring an article for the journal *Progress in Human Geography*. Ian is an associate professor of geography at the University of Exeter and has been active in various forms of participatory qualitative research. The blog gives a wonderful insight into many of the issues discussed in this chapter.

DeLyser, D., & Hawkins, H. (2014). Introduction: Writing creatively—process, practice, and product. *Cultural Geographies,* 21, 131–4.

Ellis, C., Adams, T.E., & Bochner, A.P. (2010). Autoethnography: An Overview. *Forum Qualitative Sozialforschung / Forum: Qualitative Social Research,* [S.l.], *12*(1). ISSN 1438–5627. Available at: www.qualitative-research.net/index.php/fqs/article/view/1589/3095 Presents a useful summary of the origins and practice of auto-ethnography.

Jones, A. (1992). Writing feminist educational research: Am "I" in the text?. In S. Middleton & A. Jones (Eds), *Women and education in Aotearoa.* Wellington: Bridget Williams Books.

McNeill, D. (1998). Writing the new Barcelona. In T. Hall & P. Hubbard (Eds), *The entrepreneurial city: Geographies of politics, regime and representation.* London: John Wiley.

Richardson, L. (2003–7). Laurel Richardson. www.sociology.ohio-state.edu/lwr. A key advocate for creative writing in research, Richardson refers to a substantial list of publications.

Rose, G. (1997). Situating knowledges: Positionality, reflexivities and other tactics. *Progress in Human Geography,* 21, 305–20.

Notes

1. Readers seeking specific how-to advice on stylistic conventions associated with the presentation of research are advised to consult Chapter 20 of this volume, Hay (2012); Hay, Allington, and Bochner (2021); Kneale (1999); or Stanton (1996) in conjunction with the conceptual material of this chapter.
2. It is appropriate to note, however, that it can be difficult institutionally to present some forms of research this way. For instance, social or environmental impact statements prepared by government departments or consulting firms will frequently not list authors, in which case the "we" would be poorly defined. Moreover, neither group would want individuals to be sued for their opinions.

Small Stories, Big Impact: Communicating Qualitative Research to Wider Audiences

Dydia DeLyser and Eric Pawson

Chapter Overview

This chapter aims to help beginning qualitative researchers effectively communicate their research to wider audiences, as part of a broader public scholarship. To do so, it explores different modes of communication, including written work and public presentations, websites and social media, and considers audiences, including supervisors and examiners, research participants and respondents, along with the public at large. It asks who the audience is, as well as how to reach the audience, and how to convince the audience. It considers options for organizing and reporting the kinds of information and insights that come from qualitative research. To do so, it describes ways to focus creativity, along with techniques that can help in the imaginative expression of ideas. Finally, it addresses the power and credibility of qualitative research, and suggests ways to harness the results of such research in or for public discussion and debate.

Introduction

How can we communicate the findings of qualitative research to an audience? This is a central and often challenging issue for any researcher. Yet with the exception of a short chapter on "writing up," it is a topic about which most books say very little. Even the term *writing up* is itself misleading, implying that writing is something that occurs straightforwardly at the very end of the research process. As Chapter 19 of this book has argued, writing and research are mutually constitutive practices, with writing—or **writing-in**—being continuously woven throughout the research process, and research-driving ideas occurring throughout the writing process as well. Even so, writing—a fruitful means for communicating research—is today far from our only avenue for sharing our research with broader **audiences**. Public presentations, group talks, webpages, podcasts, and social media, are just some of the other ways that can be employed to present and represent the richness of qualitative work. All of those can serve as means of **public scholarship**, of mobilizing research to reach audiences in ways that seek to make a positive and often collaborative contribution to communities. Public scholarship is part of choices about who and what our research can be for.[1]

Many of the most vexing contemporary issues lend themselves particularly well to qualitative methods, for the rich and focused insights of qualitative research can give voice to social difference and convey meaningful insights into events and experiences (Braun & Clarke, 2013)—insights that even the most compelling statistics cannot convey. Huge issues can be rendered not just intelligible but palpable by engaging with the kinds of data-based stories that qualitative research is built from. In this chapter we use examples from environmental disasters to show how stories grounded in qualitative data can hook an audience and make overwhelming phenomena really meaningful. We begin by showing the power of one person's story—what quantitative researchers call "an 'N' of one" (see also Chapter 7)—in understanding individual human agency in the midst of tragedy. We then build on this to illustrate how qualitative research can mobilize large amounts of data and many stories in compelling ways, ways that, we show, have been useful in research, and also in the very communities affected.

Begin with a Story

Each year the sitting US president gives a presentation before Congress known as the State of the Union Address, which reaches an audience of many millions. Here a specific form has evolved over decades that starts with statistics highlighting political accomplishments, citing, for example, numbers of new jobs created or figures about an administration's response to a tragedy. Each State of the Union speech is, in this way, jammed with statistics. But though such lists can be impressive, they can also be dull, and they mask the real lives of the people involved. So presidents then turn to specific, individual examples by inviting Americans who have lived the experiences the statistics describe to sit in the first lady's box and have their stories told as part of the address.

President Barack Obama gave his second State of the Union in 2013, when the US was still reeling from Hurricane (Superstorm) Sandy. The monster storm had devastated much of the US eastern seaboard and Upper Midwest, particularly parts of the states of New York and New Jersey. At least 186 people lost their lives, and damage estimates exceeded US$65 billion (Ovink & Boeijenga, 2018). Obama alluded to global climate change and the widespread predictions of worsening storms as a possible cause. But to make the example more real, and to urge Americans to reflect on ways to contribute in such circumstances, he also drew from the experience of one woman, a registered nurse in a New York City neonatal unit. Menchu Sanchez, a Filipina immigrant living in Secaucus, New Jersey, knew her *home* must be flooding from the storm, but when her *hospital* began flooding she kept her focus on the twenty premature babies in her care. As night fell the hospital lost power, leaving the babies on the ninth floor with just hours to reach another hospital before back-up batteries in their ventilators and other medical devices would fail. Administrators searched for options, but it was nurse Sanchez who devised a human plan to safely evacuate these most fragile infants. With non-medical personnel lighting the dark stairwells by cell phone and flash-light, and with medical staff in trail holding the bottles and machines serving as each child's lifeline, Menchu and other nurses and doctors carefully carried the babies in their arms down to waiting ambulances. With Menchu making multiple trips herself, all twenty reached safety and survived. Menchu, Obama said, was one whose example we should all follow (Tengco, 2013). One person can make a difference.

So too can one story, drawn from qualitative data, and activated to pull into focus a human and potentially inspiring side of often-otherwise dark national and international issues. In ways such as these, qualitative data can reveal the complex and changing truths of human lives and social worlds, conveying them through real struggles and situations to bring meaning even to those far outside the original story's location. Thus, the power of qualitative research often lies in how it can lend the insight of human experience to challenging social issues and vexing ordeals—even, or perhaps especially, when the sample size is just *one*.

This chapter uses a range of examples, and suggests a number of techniques, to show how qualitative research, which can speak so powerfully to the lives of individuals and communities, can be mobilized to communicate effectively, and to reach a broader public. It draws on specific experiences to underline its main points. It also identifies and discusses issues like how to persuade audiences of the power and **credibility** of qualitative research, how to use qualitative research to contribute to or advance public debate, and considers ways to engage qualitative research and research practice to contribute to public scholarship. To do so, it is necessary to first consider the nature of the audiences that we as researchers wish to reach.

Understanding Audiences

To reach an audience effectively we must first know something about who the audience is or might be. Just as a personal conversation or online message is shaped by knowledge of the person we interact with, so must the audience for a piece of research—which is likely to have involved far more preparation and effort than most conversations, texts, or emails—be considered. But "who is the audience?" is not a straightforward question, and the answer is different for every project, and even for different expressions of each project.

The initial answer may be surprising—we first communicate our research with ourselves. We write first for and to ourselves, because the very act of writing reveals what it is that we think and know and what we do not (Chapter 19; Richardson & St Pierre, 2008). In other words, writing is an iterative process, with reading, writing, and research being interlinked phases and each activity clarifying the direction that needs to be taken in the others. Reading makes clearer what to write and vice versa. Writing helps to elucidate what we think, and being clearer on what we think enables more effective communication of those thoughts to others. Understood in this way, it becomes evident that the writing process itself is *formative*; writing is an essential part of clarifying thinking (Becker, 2007; DeLyser, 2010) because "Writing *is* thinking" (Wolcott, 2008, p. 18).

Since academic life takes place within social networks and involves social responsibilities, in order to fully participate in the academic world, many qualitative researchers contribute back in different ways both to academic life and, in expressions of public commitment, to the communities that have provided the research opportunities (see Becker, 2007), and here *audience* becomes a critical issue. Students as scholars have a wide public, being accountable within the academy (to meet course or degree requirements, or to peers), as well as having obligations beyond the university (for example to those who gave access to networks or responded to requests to participate). Depending on which audience or which part of the audience we wish to communicate with, the ground rules vary. Universities, departments, or instructors make known their expectations for class presentations, term papers, dissertations, or theses, and such rules are best adhered to. Here practice and feedback from peers will go a long way, such as to ensure that a talk does not run over time, or that written drafts are clear and polished.

More challenging is the issue of obligations to wider audiences, particularly those who have helped in the research, and those one may wish to make a public contribution to (see also Chapters 3, 4, and 11). Qualitative inquiry in general seeks to reduce **power** differences and encourages sharing of meaning-making between researchers and **participants**. In this it differs from the traditional conception of quantitative research, where the researcher, as the ultimate source of authority, does not encourage participants' active contribution to the research process (Karnieli-Miller et al., 2009). Giving back to those who have helped with your research recognizes this reciprocity and may facilitate additional visits or co-operation in the future (Cupples & Kindon, 2014; and Chapter 16). One way in which this can be done is by returning to respondents their interview **transcripts**, or those parts of a written account that use their words. Such strategies need careful consideration, however; although often motivated by ideals of participant co-ownership of the research and participant empowerment, they can as likely result in "surprise and embarrassment" (Forbat & Henderson, 2005, p. 1118). Because it is people's realities, feelings, and lives that are represented in qualitative research, this must be done sensitively.

Such sensitivity begins before the research even starts, for some populations feel heavily (even too heavily) researched by people from outside. This can easily arise in small towns that are popular for university field trips, or with particular neighbourhoods or social groups (e.g., refugees) that attract frequent attention.

In post-disaster areas, well-meaning researchers acting without coordination run the risk of asking similar, emotionally fraught questions time and again. The result can be "research fatigue." In 2005, after Hurricane Katrina flooded much of New Orleans, many who stayed behind or returned to their city faced not only the devastation of the storm but also an inundation by earnest experts, who "got parachuted in very quickly . . . the communities felt both excluded and unheard" (Ovink & Boeijenga, 2018, p. 95). Who benefits in these situations is a question that has also long been asked in Indigenous societies, where the alternative term **helicopter researchers** is well-known (see Chapter 4). Linda Tuhiwai Smith characterizes this sort of behaviour as a reproduction of **colonialist** attitudes and argues that "sharing is a responsibility of research" (Smith, 2012, p. 162). Local sensitivities and needs should be recognized with use of more participatory methods that build trust and co-operation (Chapter 11; and Willyard, Scudellari, & Nordling, 2018).

Engaging (with) Communities

In September 2010 what was to become a more than three-year sequence of thousands of earthquakes began to strike the city of Christchurch, Aotearoa New Zealand. Several particularly violent events destroyed buildings and caused ground liquefaction (where water-saturated ground acts like a liquid, rather than a solid, failing to support the weight of structures built upon it) across much of the urban area. Quantitative studies revealed the profound extent of the devastation in numbers of buildings, numbers of homeless, and hectares of land rendered useless. The real devastation, though, was not just to houses but to *homes*, and not only to a city, but to *people's lives and livelihoods*. While many different kinds of research are needed to understand and recover from complex tragedies like this, in Christchurch qualitative researchers—both students and faculty—have engaged in meaning-filled qualitative research that has sought to trace how people are creating new and unfolding *ways of being* in post-disaster situations (Cloke & Conradson, 2018).

As we have observed, such situations demand that the conduct of research—and teaching—is actively reshaped to respect context. This is a fundamental obligation of public scholarship. An example of this has been the development of **community-based** learning courses, in which, for example, groups of senior geography students begin by negotiating the topics for research *with* community representatives, and then take responsibility for ensuring that the findings are both shaped by and communicated between the parties in multiple ways (Pawson, 2016). One method has been through an off-campus conference where each student group presents a 15-minute illustrated talk in the community. Such talks are an effective way of involving an audience, and giving them the opportunity to participate through listening, questions, and discussion. Box 20.1 outlines the key approaches that will engage an audience, as well as those that will have the opposite effect.

BOX 20.1

For Successful Public Presentations

Consider:

- Preparation: if you know your material, you will be less nervous about presenting it, and that will give your audience more confidence in what you are saying.
- Content: your audience has come to hear what you have to say and only secondarily how you say it.
- Visuals: use them to illustrate your points, ensuring that they add value to the talk.
- Practice: even in an empty room, practice builds confidence and helps ensure the presentation is the right length.
- Interaction: engage the audience by including them. It is often easiest to look at two or three people in particular whilst presenting; then leave time for questions.

Avoid:

- Rambling: both you and your audience will understand what is being said if you have a clear structure and keep to it.
- Poor time-keeping: make sure you can see a clock or some means of telling the time, and keep to the allotted length.
- Mumbling: ask the audience if they can hear you. Voice clarity for the audience comes from speaking clearly, loudly enough, and not too quickly. Don't be shy to use a microphone, particularly as many venues employ means of sound-deadening such as carpeting.

Organizing and Representing Qualitative Findings

Whether we choose to present our qualitative research through verbal, visual, or written means, the basic academic fundamentals remain critical. These include cogent thinking, good preparation, and transparent structure. Developing a transparent structure, one that is clear to both presenters and listeners/readers, is worth both extra effort, and the extra time that effort demands. It cannot be achieved if research questions, aims, and goals are not lucid and convincing to the researcher or research team first. A well-planned outline must signpost or map the contents of the presentation from the outset. There are various ways of doing this, but many find going analog with a piece of paper or whiteboard the most effective. The whiteboard is great for brainstorming amongst members of a group, and for working out a collective and agreed-upon plan. A well-planned outline that is clear to us as authors will more likely to be clear to the audience as well.

This heavy emphasis on structure may seem unconvincing, or even annoying. Writers facing projects like term papers or theses might feel that any externally imposed form or structure hinders creativity, forcing the work into "boring" moulds or "stifling" formats. But the opposite proves true. Highly restrictive forms like the *haiku* or the sonnet reveal that such structures are stimulants to creativity—after all, no writer today would deny the creativity in Shakespeare's work. Contemporary examples of equally restrictive structures bear this out as well. Twitter, whose original 140-character restriction put a premium on identifying the point of the message, has become a potent tool in the hands of politicians and citizen journalists (Corson, 2015). PechaKucha, a method of presentation named for the Japanese word for "chit chat", restricts (in its purest form) storytelling to a sequence of 20 PowerPoint slides, each used for 20 seconds, for a total of 6 minutes and 40 seconds (20 x 20 = 6:40). Here the objective is to keep presentations brief and focused, at the same time as stimulating discussion.

More productive, then, is an understanding that structure harnesses creativity. Consider the now-well-known TED (Technology, Entertainment, Design [www.ted.com/talks]) talks, growing in popularity in recent years with exposure through YouTube. Their maximum length of eighteen minutes again puts a premium on clarity and purpose—though they may seem spontaneous, invariably they have been carefully planned and practiced. Successful academic writers and speakers also use structure in carefully considered ways. Box 20.2 highlights some of the key techniques that enable the presentation of the richness of qualitative research and the expression of originality within clear structures. In specific ways, the pointers in the box apply to different options for organizing and reporting qualitative findings, such as written reports, public talks, webpages, or conference papers. Clarity of communication comes first through attention to the basics of structure.

Today digital presentation tools (such as PowerPoint, Prezie, and GoogleSlides) are often the staple of an academic presentation to a class, at a conference, or in a public meeting. But the standard sequence of slides does not in itself provide structure and meaning. Rather, relying on slides makes the focus, clarity, and storyline of a presentation even more important. The alternative is the kind of "death by PowerPoint" too often seen in lecture halls: overloaded slides heaving with bullet points that no-one remembers by the half-way mark, let alone at the end. Excellent resources like *Beyond Bullet Points* (Atkinson, 2018) and *presentationzen* (Reynolds, 2019) can take straightforward simplicity to the level of art form. But "simplicity" in this context does not mean oversimplification, "rather, it comes from an intelligent desire for clarity that gets to the essence of an issue" (Reynolds, 2011, p. 115).

BOX 20.2

Using Structure to Communicate

- Begin with clear research questions and aims: they focus attention for both authors and speakers by sharpening creative abilities, and they focus the audience on the key elements of what is to be said.
- Develop an outline: essential to clear organization, outlining ensures that the main points are made, and in the most appropriate order. The outline can change and grow as preparation proceeds, providing it does not become detached from the research questions or aims.
- Use headings and subheadings: when ideas need to shift, call attention to this with a new heading on the slide, or a subheading in the text. Headings and subheadings are also invaluable tools for organizing and clarifying the argument in early drafts.
- Develop clear ways to order the argument into sections or slides: in written work, this is the purpose of the paragraph. Locate the thesis statement (the "point" of the paragraph) at the beginning or end, and use the middle to build that idea.

This sort of simplicity is not easy to achieve. It requires the kind of hard work in advance that combines planning, focus, and structure. To illustrate the true challenge of this kind of simplicity Reynolds asks, "What is my absolutely central point?" "If the audience will remember only one thing (and you'll be lucky if they do), what do you want it to be?" (2011, p. 63). Identifying that central point comes readily from an open process of discussion with oneself, collaborators, or members of the community, allowing input and focusing ideas. Again, the sketchpad or the whiteboard can be an excellent tool for this. The more effort put into preparation, the better the end result, and the more likely that we will carry the audience with us.

A further means of communicating research results is via a website; setting up webpages while doing research is another aspect of the formative process of writing (see also Chapter 15). Blogs (web logs) and websites that encourage public comment are useful as a means of enabling research respondents to give feedback on the initial results, assisting in their verification (James & Rashed, 2006). They have proved valuable in Indigenous communities in Canada (e.g., northern Manitoba and Nunavut) where community-based research, consultation, and feedback are now expected (Stewart & Draper, 2009; Chapter 4). This is also the case in post-disaster situations, where increasingly it is recognized that communities must be involved for recovery to be in any way effective (Ovink &

BOX 20.3

Presenting Effectively

For Success with Slides, Use them Creatively

- Minimize the number of slides and bullets: simplicity is the ally of the audience.
- What is the central point? If the audience is to remember just one thing, what do you want it to be?
- Use the slides to reinforce your words, not to repeat or anticipate them. The slides should convey your message in fewer words than you speak.
- Choose images carefully and consider **representational** issues including permissions, sensitive cultural questions, and *who* is or is not depicted.
- Avoid dizzying dissolves, spins, or other amped-up transitions, and anything that distracts from your message.

Death by PowerPoint

Think back to all the bad presentations that you've witnessed:

- with far too many slides and bullet points, ones that are at odds with or overload the message;
- with the speaker repeating the slides word for word and adding nothing to them;
- with complex background schemes or silly colour choices that distract the audience;
- with illustrations that detract from the point of the presentation; and
- that confuse quantity of information with clarity of message.

To present well, we must do things differently, avoiding these pitfalls.

Boeijenga, 2018). Online invitations provide an opportunity for **respondents** to feel that they are valued for their roles and are part of the research process rather than merely being "the researched." In other words, such outreach can demystify the research process and contribute to the maintenance of meaningful research relationships in particular places.

Some choose to describe their research not through individually designed websites but through a social networking utility like Facebook or Instagram. Others elect to make information about their research topics available through publicly editable online wiki pages. Still others update followers of their progress or provide links to digital resources via Twitter. Carefully and attractively done, such sites and postings may draw an audience—they can give you contacts working in similar fields or potential respondents; they may also help you to make your research available to those outside your department or discipline and to give something back to those you worked with in the field—they can build towards public scholarship. A web presence is also a way of giving something back to your department—it may attract others to think about studying there in the future.

Writing for the web requires design as careful as any other means of communication. Balance words with images, but be aware of the politics of visual **representation**—it is only possible to use images for which you have permission to do so, and research subjects should not be misrepresented by picturing them insensitively (e.g., portraying them in ways devoid of context or placing them in inappropriate juxtapositions) (Cupples & Kindon, 2014). Whether the chosen medium is words, sounds, or images, questions of representation have to be considered. Chapter 19 argues that writing research re-presents findings, thereby underlining the mediated character of communication. One aspect of this is the use of photographs, the staple of many forms of geographical communication, be they field reports, theses, webpages, social media, or slide presentations. It is vital to think carefully and critically about the purpose for which photographs were taken, how a specific selection of images reflects or silences particular points of view, and even whether there are not other forms of representation that might extend what can be conveyed by visual means (Phillips, 2015).

The Power of Qualitative Research

Some wonder how qualitative research can be used effectively to address important issues when it cannot be quickly summarized in tables or statistics and must be presented and engaged in full (Richardson & St Pierre, 2008). The power of qualitative research lies precisely in this rich interaction with human stories that matter, and in the fact that it cannot be reduced to a row of numbers but must retain the lives of its participants. So, although some qualitative researchers worry that their research needs to be "defended" against those of a quantitative persuasion, the value of a qualitative approach should instead be actively promoted (see Chapter 1, and Marshall & Rossmann, 2015).

Box 20.4 outlines the valuable attributes of qualitative research and shows how, if good academic procedures are employed to ensure its credibility, it will readily convince your audience. With its focus on experience, and giving voice to those who otherwise might not be heard, it yields invaluable insights into lived social worlds.

For example, in the Christchurch earthquakes, it was widely assumed that Aotearoa New Zealand's official Earthquake Commission (EQC), which provides

BOX 20.4

The Power and Credibility of Qualitative Research

What is valuable about qualitative research?

- It provides insight into and renders meaning from human experiences.
- It focuses on how complex social worlds are produced, experienced, and interpreted.
- It is flexible and sensitive to the social context in which the data are created.
- It is based on methods of analysis and explanation that seek understandings of complexity, detail, and context in our dynamic social worlds.
- It can help build strong connections with communities and publics.

Promoting its value to your audience

- Qualitative research allows respondents to speak for themselves.
- In moving beyond mere description, qualitative research can provide explanations to intellectual puzzles.
- When data have been gathered according to principles of qualitative research and analyzed in ways consistent with a body of **theory** or **conceptual building blocks** (as argued in the previous chapters), it is clear that work was conducted systematically and **rigorously**.
- **Triangulation**, that is drawing data from different sources to **corroborate**, elaborate, or further illuminate the argument you seek to forward, is convincing.
- When these points are met, the findings will have wider resonance, **transferability**, and may be applicable to other circumstances and other places (quantitative researchers call this **generalizability**).
- When your research subjects are able to recognize themselves and their experiences in your work, they will receive your work as **credible.**
- With rich community connections, qualitative research can inform public scholarship, exchanging the knowledge of academics with the expertise of communities towards a greater good.

natural-disaster insurance for residential-property owners, would take care of people in the wake of the disaster. But the sheer extent of property damage overwhelmed the EQC, which had to increase its staff nationally from 26 before the earthquakes to around 1,600 to handle the 467,000 claims that it received

(Marsh, 2014). Many of these claims exceeded the maximum coverage that EQC was permitted to provide, resulting in tens of thousands of householders becoming mired in lengthy assessments and disputes with more than one insurer, at the same time as having to deal with the physical and emotional stresses of broken houses, neighbourhoods, and even lives.

None of this human anguish is conveyed by the widely available statistics of the extent of damage. The three biggest Christchurch temblors, on 4 September 2010, 22 February 2011, and 13 June 2011, were respectively the fifth, the third, and the tenth costliest "insured earthquakes" in the world to date (Marsh, 2014). Yet the challenges people have faced in recovery have been variable and complex, ranging far beyond financial cost. Qualitative research has been able to reveal some of the human realities behind the numbers, for example those associated with the loss of home and community in the "red zoned" part of the city that had to be abandoned due to weakened ground conditions (Ellison-Collins et al., 2017). But in general we know far more about disaster preparedness and emergency response than we do about the traumas associated with disaster recovery and what it means (is there, for example, an end point?) (Medd et al., 2015). These are all contributions that qualitative research has the opportunity to make, and their impact can be directly felt in future events by members of the public, thanks to researchers' engaged scholarship and purposeful communication.

This means that qualitative research can also be used effectively *in conjunction with* quantitative research, and the two should not necessarily be construed as in opposition to each other (see Chapter 1)—both can be a part of public scholarship and both can make positive contributions to community. In the sort of post-disaster situations that have been a backdrop to this chapter, it is also often the case that qualitative research that reveals challenging individual experiences, when convincingly conveyed by academic or media reporters, can lead to social action.

Conclusion

To conclude we offer another case to show the lasting power of stories from qualitative data, and how these speak to people's hearts and lives. Here we show also some of the ways that qualitative research can be important, both as research about and *for* communities. Profound issues like environmental disasters reach across days, months, and years of people's lives; while the disaster may begin in one instant, its precursors of vulnerability and its after-effects of destruction and dislocation stretch much farther. Sensitive qualitative research can attend to multiple dimensions of community recovery, resourcefulness (Mackinnon & Derickson, 2013), and liberation.

When disaster strikes, the human cost can be high, in terms of lives lost, workplaces disrupted, homes broken or abandoned, and the edgy anxiety that comes with uncertainty about what happens next. As we have indicated in this chapter, the qualitative researcher can contribute to individual and community healing in such times and places. Giving voice to people's experiences, seeing them in their contexts, allowing them to tell their stories, can be an important part of the recovery process. Earlier, we made reference to the extensive lands that were "red zoned" as a result of the Christchurch earthquakes. This was partly due to liquefaction (a condition where saturated ground loses strength under stress) and partly due to land subsidence, greatly increasing the risk and incidence of flooding. Whole communities near the coast were forced to move, through a process of government-imposed "voluntary buyouts" and managed retreat. But to listen to the stories of those 7,000 people is to realize that this process was experienced as neither voluntary nor managed. And the result was a 600-hectare vacated space, in which all that was left was stories.

These stories have been collected informally by student researchers and community groups over the years, and are now being coordinated in a Red Zone Stories app and website (www.redzonestories.nz). It is not merely a repository, but a living resource to which former residents, citizens, and visitors can contribute their narratives and memories of the red zone. Of how places that now regularly flood were once home to people who treasured living close to nature, or how places that seem empty but for the residual street pattern were where richly textured lives, feelings, and journeys were interwoven. We have given examples in this chapter of how student researchers can work to tell such stories. In turn, a resource like Red Zone Stories is but one instance of how whole communities can come together around public scholarship, create and share qualitative data—like text, video, images, and photographs—for audiences including themselves and extending to all of those who otherwise will not know or cannot remember.

Key Terms

audience
communication
creativity
credibility
social media
stories
public scholarship

Review Questions

1. Why does the audience matter?
2. How is writing also thinking? Give an example of an insight you came to through a writing process.

3. In what ways can your data be shared?
4. Why does structure help to harness creativity?
5. How would you promote the value of qualitative research as public scholarship?
6. Think of a single, exemplary story based on your research data. How can you use that to drive your point home?

Review Exercises

1. Using a current class assignment, list the five or six key points that could form the core of a slide presentation about it.
2. Many journalists' reports on disasters and other major events use a story of one person as a hook for their readers or as an exemplary case. Browse a news source and find an example of this and identify the strategies the author uses to convince or inform readers of the issue.

Useful Resources

A wide and constantly changing array of academic and commercial websites is available to help with writing and presentation skills. An internet search will yield many. Some sites that we found helpful are listed below, together with useful published resources.

Atkinson, C. (2018). *Beyond bullet points: Using PowerPoint to tell a compelling story that gets results* (4th edition). New York: Pearson Education, Inc.

Becker, H.S. (2007). *Writing for social scientists: How to start and finish your thesis, book, or article* (2nd edition). Chicago: University of Chicago Press.

Braun, V., & Clarke, V. (2013). *Successful qualitative research: A practical guide for beginners*. Los Angeles: SAGE.

DeLyser, D. (2010). Writing qualitative geography. In D. DeLyser et al. (Eds), *Handbook of qualitative geography* (pp. 341–58). London: SAGE.

Hay, I. (2012). *Communicating in geography and the environmental sciences* (4th edition). South Melbourne: Oxford University Press.

Presentation Magazine. (2019). www.presentationmagazine.com/ Offers guidance and resources for effective speeches and presentations, from voice-improvement techniques to thousands of PowerPoint templates.

Reynolds, G. (2019). *presentationzen: Simple ideas on presentation, design and delivery* (3rd edition). Berkeley, CA: New Riders.

Smith, L.T. (2012). *Decolonizing methodologies: Research and Indigenous peoples* (2nd edition). London: Zed Books.

The Writing Center at UNC Chapel Hill. (2019). https://writingcenter.unc.edu/tips-and-tools Offers comprehensive writing advice on topics ranging from audience to argument, from commas to conclusions.

Ulibarri, N., Cravens, A.E., Nabergoj, A.S., Kernbach, S., & Royalty, A. (2019). *Creativity in research: Cultivate clarity, be innovative, and make progress in your research journey*. Cambridge: Cambridge University Press.

University of Toronto. (2019). Writing advice. https://advice.writing.utoronto.ca/general A helpful and comprehensive review of academic writing, geared to students.

Note

1. Many universities support public scholarship with diverse on-campus initiatives. Carleton College's Center for Community and Civic Engagement (in the US) offers a positive example: https://apps.carleton.edu/ccce/scholarship/what_is/. See also Mitchell (2008) and Robinson and Hawthorne (2018).

Glossary

abstraction Process of making broad statements about how the world works based on connections between specific processes and contexts.

accession (accessioning) Term used in the management of archival repositories referring to the formal inclusion of materials into a collection and assignment of appropriate reference numbers to make the materials available to other researchers.

accidental sampling See *convenience sampling.*

action research Term coined by sociologist Kurt Lewin in the 1940s to talk about the idea of an iterative (or repeated) cycle of action and reflection in research, oriented towards solving a problem or improving a situation.

action–reflection Periods of action followed by times when participants reflect on what they have done and what can be learned. The learning informs the next phase of action, creating an iterative cycle of action and reflection. This process enables change to occur throughout the research process. (See also *reflexivity.*)

activism Political and practical action usually intended to bring about social, economic, or other change. (See also *applied people's geography* and *critical geography.*)

Actor Network Theory (ANT) Approach to social theory and research in which objects (e.g., non-humans) are treated as part of social networks. (See also *more-than-human.*)

affect (affective) The visceral forces beneath, alongside, or generally other than conscious knowing; see Pile (2010) on "affectual geography" and its relations to emotion.

analytic code Code that is developed through analysis and is theoretically informed; a code based on themes that emerge from relevant literature and/or the data. (See also *interpretive code*; compare with *descriptive code.*)

analytical generalization Strategy for creating in-depth, rich, and credible concepts/theory. Rather than achieving generalization through large probability samples, analytical generalization is focused on the qualitative notion of transferability, specifically (a) the careful selection of informative cases and (b) the creation of theory that is neither too abstract nor too case-specific. Readers of research narratives must be able to see how the concept might apply to other phenomena or in other contexts.

analytical log Critical reflection on substantive issues arising in an interview. Links are made between emergent themes and the established literature or theory. (See also *personal log.*)

anecdote A story, often personalized to the author or presenter and directly related to the point of the paper or presentation, that captures the attention of an audience and persuades them of the importance, relevance, and/or interest of what they are reading or hearing.

animating archives Refers to various strategies to make the contents of archival collections and stories behind archival collections more visible to the wider public. This often involves, but is not limited to, digitization projects and may involve the integrated use of text, cartographic, and pictorial records as well as sometimes material objects in presenting new, forgotten, and marginalized accounts from the past.

anonymity Assurance given to participants that their names and identifying information are disassociated with responses entirely; distinguished from *confidentiality*, in which responses will be held securely by the research team but names are still connected to data.

applied people's geography Term coined by David Harvey to refer to geographical research that is consciously "part of that complex of conflictual social processes which give birth to new geographical landscapes" (Harvey, 1984, p. 7). (See also *activism* and *critical geography.*)

archival research Research based on historical documentary sources (for example, public records, photographs, newspapers, personal diaries and correspondence, etc.).

archives Narrowly defined as the non-current records of government agencies but also includes company and private papers. Typically managed by a specialist in a government agency dedicated to the records' long-term use and preservation. A distinction can be made between archives as surviving records and archives as the institution dedicated to their preservation.

archivist Professional curator of non-current records who has expertise in the accession, arrangement, and preservation of such records (in contrast to the current files of a central or local government agency that are controlled by records managers).

arts-based methods A broad set of techniques in which artistic production is woven into other research methods or takes central stage as the primary method; common in *participatory research*.

asymmetrical (power) relations Research situation characterized by an imbalance in power or influence between researcher and participant. Sometimes used to refer specifically to relationships in which informants are in positions of influence relative to the researcher. (See also *studying up*.)

asynchronous interviewing When the answers in an interview do not occur immediately after the questions—that is, there is some extended delay between the putting of a question and the receipt of answer. It is most common in interviews undertaken using email in which there is a delay between the sending of a question and the making of a reply, since the researcher and informant are unlikely to be undertaking the interview at the same time. (Compare with *synchronous interviewing*.)

attribute database Set of information compiled from measurable characteristics, such as the census.

audience People with whom you wish to communicate the results of your research. Identifying who constitutes the audience is fundamental to the success of communicating with them. In discourse analysis, the term audience refers to the social process through which different collective social categories are forged, say, "American" or "tourist." Audiences are conceived to emerge through how texts are produced, circulated, intersected, and interpreted. Hence, discourse analysis sees an audience "taking shape" rather than being a pre-given category. For example, a presidential inauguration speech may help to create an audience fashioned by shared values of the collective "we" of a nation. Alternatively, infomercials selling beauty products may rely on cultural norms of attractiveness and femininity to generate an audience (market) interested in purchasing their products.

auto-ethnography Qualitative method involving the explicit writing of an embodied and situated self into the research. It often entails writing in literary fashion, involving autobiographical narratives in which the author/researcher actively reflects on their choices, emotions, and knowledges as a vital part of the construction of the research.

axis/axes of difference Refers to the different and intersecting modes of power that produce distinctions between and groupings of people through classification, identification, and enforcement. Axes of difference also manifest materially in inequalities between different people. Examples of intersecting axes of difference include race, gender, sexuality, class, nationality, religion, etc. Usually used in the plural, *axes*, to heighten awareness that these dimensions of difference and marginalization intersect and compound each other. (See also *intersectionality*.)

bias Systematic error or distortion in a data set that might emerge as a result of researcher prejudices or methodological characteristics (for example, case selection, non-response, question wording, interviewer attitude).

big data Very large data sets analyzed computationally to identify trends and patterns; often includes *geo-tagged* data scraped from social media and communication, and/or *volunteered geographic information*.

blogging Short for *web log*. Blogging is the act of creating discrete posts or entries published online. Blogs can have individual or multiple authors and typically include space for reader comments and interactions.

born digital records Materials that originate in digital form as opposed to textual or other pictorial or cartographic material that has been digitized from its original form.

canon Body of work, such as texts, held by some critics to be the most important of their kind and therefore worthy of serious study by all interested in the field.

CAQDAS See *computer-assisted qualitative data analysis software.*

CAQ-GIS (computer-aided qualitative geographical information systems) The integration of computer-assisted qualitative data analysis software (CAQDAS) with geographic information systems (GIS) using techniques that bridge the two software systems and move towards the goal of creating complementary analysis of qualitative and spatially referenced data.

case Example of a more general process or structure that can be theorized. (See also *case study.*)

case study Study of a single instance or small number of instances of a phenomenon in order to explore in-depth nuances of the phenomenon and the contextual influences on and explanations of that phenomenon. Research may take the form of within-case temporal comparisons or spatial (place-oriented) cross-case comparison.

chain sampling See *snowball sampling.*

Chicago School of Sociology Both a body of work and a group of University of Chicago researchers from the 1920s and 1930s involved in pioneering work in urban ethnography. The Chicago School helped to establish the in-depth case study as a legitimate and powerful means for conducting relevant social science. Notable researchers from this era are William Thomas (1863–1947), Robert Park (1864–1944), Ernest Burgess (1886–1966), and Louis Wirth (1897–1952).

cis(gender) Refers to the assumed natural match between one's gender identification and one's gender assignment by society.

cis-heteronormativity The institutionally enforced normalization of cisgender identity and heterosexuality in tandem, recognizing that the latter only makes sense as normative in the presence of a binary categorization of gender into the essentialist categories "male" and "female."

closed questions Questions for which respondents are offered a limited series of alternative answers from which to select. Respondents may be asked, for example, to select one or more categories, to rank items in order of importance, or to select a point on a scale measuring the intensity of an opinion. (Compare with *open questions* and *combination questions.*)

CMC interviews See *computer-mediated communications (CMC) interviewing.*

co-constitution of knowledge The idea, informed by post-structural theory, that knowledge does not exist "out there" ready to be discovered by objective researchers using discerning research methods but instead is built intersubjectively through interaction between the researcher and the research subjects. Qualitative research encounters are thus seen as social relationships (however fleeting) in which the researcher and the researched are closely involved in the process of construction of knowledge. (See also *co-learning* and *knowledge production*)

codebook Organizational tool for keeping track of the codes in a project, including their meanings and applications, as well as notes regarding the coding process.

coding (code) Processes of assigning qualitative or quantitative values to chunks of data or categorizing data into groups based on commonality or along thematic lines for the purposes of describing, analyzing, and organizing data. A simple version typically involves marking the transcript margin with a colour, number, letter, or symbol code to represent key themes or categories.

coding structure Organization of codes into meaningful clusters, hierarchies, or categories.

co-learning Philosophy of teaching and learning that positions researchers and participants as equals in a process of mutual inquiry and education. Co-learning challenges traditional assumptions that the researcher has more knowledge and is dominant in the research relationship. It requires considerable self-reflexivity on the part of the researcher and a genuine desire to facilitate a process in which researcher and participants collectively create a learning community. (See also *co-constitution of knowledge.*)

collaborative research Research designed, conducted, interpreted, and disseminated by a team of local and non-local members, with local members directing the process or sharing equally in decision-making. (See also *participatory action research* and *participatory research*.)

collective action Process in which a group of individuals takes action together to affect social or political change. This action may be the outcome of research and analysis in which individuals have come to understand the ways in which they are each implicated or affected by wider inequalities. The desire to take collective action reflects an interpersonal commitment and what some authors have called a "we intention."

colonial (colonialist) research Imposed, often exploitative research in both imperial and non-imperial contexts that maintains distance from, and domination of, the marginalized Others that it seeks to study and which denies the validity of their knowledge, ways of knowing, experience, and concerns. (Compare with *decolonizing research*, and *post-colonial research*.)

combination questions Questions made up of both *closed* and *open* components. Their closed component offers respondents a series of alternative answers to choose between, while their open component allows respondents to suggest an additional answer not listed in the closed component or to elaborate on the reason why a specific option was selected in the closed component.

common questions Asked of each participant in oral history interviews. They build up varying views and information about certain themes. (Compare with *orientation questions*, *specific questions*, and *follow-up questions*.)

community geography Practices and organizations that enable academic geographers to build university–community partnerships that facilitate access to research expertise and tools (e.g., spatial technology, data, and analysis) for the benefit of both the community organizations and students/academics involved in a project. (See also *participatory GIS*.)

community-based research (community-based teaching) Participatory approach in which members from a community engage directly in research and/or teaching as equal partners in the discovery and sharing of new knowledge. (See also *participatory action research*.)

comparative analysis Form of analysis used in case study research that compares similarities and differences across multiple instances of a phenomenon to enhance theoretical/conceptual depth. It is also known as *comparative case study* or *parallel case study*.

comparative case study See *comparative analysis*.

complete observation Situation in which observation is overwhelmingly one-way and the researcher's presence is masked such that they are shielded from participation.

complete participation Situation in which the researcher's immersion in a social context is such that they are first and foremost a participant. As a result of this level of immersion, the researcher may need to adopt critical distance to achieve an observational stance. That critical distance might be gained by reflection out-of-hours in the field or through short-term exits from the field.

computer-aided qualitative data analysis software (CAQDAS) Both a general acronym and the specific acronym for the CAQDAS network based in Surrey, UK. These are software programs that enable the storage, retrieval, sorting, coding, and analysis of text and visual data; common programs include Atlas.ti, NVivo, MaxQDA. For a review of programs see Silver and Lewins (2014): http://methods.sagepub.com/book/using-software-in-qualitative-research-2e

computer-assisted telephone interviewing (CATI) Questionnaire/interview conducted by telephone with questions read directly from a computer file and responses recorded directly onto a computer file.

conceptual building blocks Concepts that can be operationalized or acted upon within the context of a research project; it sometimes helps to think of these as the *connectors* between empirical data and the abstractions of the theoretical framework.

conceptual framework Intellectual structure underlying a research project that emerges from an integration of previous literature, theories, and other relevant information. The conceptual framework provides the basis for framing, situating, and operationalizing research questions.

concept mapping Related to *concept building* but refers to the specific ability to visually represent data in some form. For example, QSR NVivo™ software uses *hierarchical tree* structures, and ATLAS/ti uses network diagrams for concept mapping. Inspiration and Decision Explorer are purpose-built concept mapping programs for qualitative research.

conditions of use form Sometimes known as an *informed consent form*, a form outlining what will happen to the material research participants share with you—what their rights are, who will own copyright, where recordings will be stored and for how long, what they will be used for, and so on. (See also *memorandum of understanding*.)

confidentiality Assurance given to participants that their responses will be held securely by the research team (and, in the case of *focus groups*, by other participants); distinguished from *anonymity*, in which names are disassociated with data entirely.

confirmability Extent to which results are shaped by respondents and the structure of the research, and not by researchers and their biases

confirmation bias Tendency of researchers to search for and find what they expected and (re) affirm their beliefs in a predetermined manner.

consent (also **informed consent**) Process of transparently informing potential research participants of the risks and benefits associated with the research; can also include negotiating expectations for knowledge co-production, as well as obtaining administrative permissions; a part of research protections for human subjects, consent is a formal process involving communication and documentation, usually required by ethical review boards at universities and other research institutions.

constructionist approach Approach for challenging assumptions of coherence and truth within positivist knowledge (sometimes referred to as either rationalist, objectivist, or Cartesian knowledge—see *positivism*). Draws attention to social practices in the production of all knowledge, including scientific knowledge.

content analysis Quantitative method for analyzing texts based on identifying the frequency and relationship of particular words or phrases.

convenience sampling Involves selecting cases or participants based on expedience. While the approach may appear to save time, money, and effort, it is unlikely to yield useful information. Not recommended as a *purposive sampling* strategy.

copperplate script Writing style used in the nineteenth century (and earlier) produced using a sharp metal-nibbed pen. It is characterized by an elegant looping script in which the lettering is thicker on the heavy downward stroke and thinner on the upward loops. Typewriters typically replaced copperplate handwritten records in most government records from the 1880s and 1890s onward. (See also *modern hand* and *secretary hand*.)

co-researcher In cross-cultural research, an individual of different, "Other" social/cultural identity and positionality with whom one carries out research collaboratively with equal participation in decision-making. (See also *Other*, *positionality*, and *collaborative research*.)

corroboration Strategy for guarding against threats to the credibility of a theory or concept. Corroboration often involves the process of checking that a concept/theory makes sense to the participants in a case study (see *participant checking*; *member checking*). Longitudinal studies provide a useful context for corroboration whereby concepts developed through an intensive case study are checked for enduring relevance in later time periods.

credibility (credible) The plausibility of an interpretation or account of experience; the trustworthiness of a source or method.

crisis of representation In geography, a set of debates amplified by feminist scholars (e.g., England, 1994) about who has the right to represent a group and the group's interests.

criterion sampling Choosing all cases that satisfy some predetermined standard.

critical consciousness Process by which members of a group become aware of their cultural oppression, and colonized mentality, and by doing so discover that they have a popular culture, identity, and societal role.

critical geography Ideas, theories, and practices that are committed to challenging unequal power relationships, developing and applying critical theories to geographical problems, and working for political change and social justice. (See also *activism* and *applied people's geography*.)

critical inner dialogue Constant attention to what an informant is saying, including *in situ* analysis of the themes being raised and a continual assessment of whether the researcher fully understands what is being said.

critical reflexivity See *reflexivity*.

cross-case comparison Strategy for comparative analysis that compares different case studies. Such comparisons help to develop richly detailed conceptual explanations of phenomena.

cross-cultural research Research requiring the translation of values and concepts with contextual meanings and associations

cross-sectional case study A case study conducted at one point in time. Contrast with *longitudinal case study*.

cross-sectional data Information derived from many subjects (e.g., individuals, businesses, groups) at a single point in time. (Contrast with *longitudinal data*.)

cultural protocols Local-, community-, or group-defined codes of appropriate behaviour, interaction, and communication to which outsider researchers are expected to adhere.

cultural–historical geographer Historical geographer who adopts a broadly defined social theory approach in their research, which gives them a more acute sensitivity to issues of power, gender, class, and race; in contrast to those of an earlier generation who worked in synchronic and diachronic traditions and were more concerned with reconstructing geographies of the past or geographical change over time.

data cleaning Identifying and correcting errors in a data set.

data management software Generic term for any software that facilitates the entry, organization, retrieval, and/or coding and mapping of input data.

data reduction Using categorization and qualification to consolidate data into larger packages, thereby reducing the complexity and number of data points but increasing the level of understanding of trends, processes, or other insights. (See also *abstracting*.)

data retrieval Refers to the process of getting access to data that has already been entered into a computer system.

data storage Refers to the process of introducing data into a computer system so that it may be archived in some way (for example, as a document, spreadsheet, or graphics image).

debriefing Procedure by which information about a research project (some of which may have been withheld or misrepresented) is made known to participants once the research is complete.

deception Practice in some types of research in which participants are not fully informed of the nature or practice of the research; used when knowledge of the research project's details would preclude participation or when a researcher's safety is at issue. (See *ethics*.)

decolonizing research (decolonize) Research whose goals, methodology, and use of research findings contest imperialism and other oppression of peoples, groups, and classes by challenging the cross-cultural discourses, *asymmetrical power relationships*, and institutions on which they are based. (See also *applied people's geography*, and *post-colonial research*.)

deconstruction Method for challenging assumptions of coherence and truth within a text by revealing inconsistencies, contradictions, and inadequacies (for example, when matters that are problematical have been naturalized).

deduction (deductive) Reasoning from principles to facts. (Compare with *induction*.)

dependability Consistency of matching conceptual constructs with empirical data over space and time; minimizing variability in interpretations of information gathered through research; usually checked by researchers' intellectual and community partners.

dependent variable Study item with characteristics that are influenced by an *independent variable*.

For example, flooding is heavily dependent on rainfall.

descriptive code A *code* describing some aspect of the social data, typically aspects that are obvious. (See also *manifest message* and *initial code*. Compare with *analytic code*.)

deviant case sampling Selection of extraordinary cases (for example, outstanding successes, notable failures) to illuminate an issue or process of interest.

dialogic (nature of research and writing) Research and writing are dialogic in that they are relational. In this sense, research is always informed and constructed by writing, just as writing is always already informed and constructed by research. The dialogic nature of research and writing extends beyond their relationality to each other and alludes to their relationality to the wider set of social and spatial relations within which academic researcher-writers work.

diary This can refer to a number of different formats and types of texts, including (1) personal diaries/journals (often historical) that reflect moments, actions, places, and the inner thoughts of a person's life; (2) calendar diaries that record daily observations (e.g., weather), appointments, and tasks; and (3) solicited journals that are kept specifically for research purposes, such as food diaries, travel journals, activity diaries, etc. Any of these can be in hardcopy or digital forms. The terms *diary* and *journal* are often used interchangeably and care should be taken to specify which use of these terms is in play.

diaspora Groups or individuals with cultural roots tied to a specific place, who maintain those cultural identities in their new homes across generations.

dichotomy Division or binary classification in which one part of the dichotomy exists in opposition to the other (for example, light/dark, rich/poor). In most dichotomous thinking, one part of the binary is also more positively valued than the other.

digital geographies Refers to "geographies produced *through*, produced *by*, and *of* the digital" (Ash, Kitchin, & Leszczynski, 2016, p. 1; emphasis added). Digital geographies include the digital devices we use, the connectivity and interaction those devices enable, and the code, software, and algorithms that create the connectivity in the first place.

digital interviewing Mode in which there is no direct access to the informant, which includes techniques such as email exchanges and video call interviews. These interviews can be asynchronous, taking the form of either email exchanges or postings to a web-based platform, or synchronous, by using video conferencing, chat-rooms, or instant messaging boards. (See also *asynchronous interviewing* and *synchronous interviewing*; and *video call*.)

digitization Process of making hardcopy materials (maps, photos, archives, etc.) *digital* by scanning, transcribing, indexing, or other means, and—often—sharing them via online archives or other digital projects.

disclosure When a researcher reveals information about themself or the research project or when research participants reveal information about themselves.

disconfirming case Example that contradicts or calls into question researchers' interpretations and portrayals of an issue or process.

discourse There is no fixed meaning for discourse. The term has accrued a number of meanings that are in circulation in both academic and popular cultures. Even among cultural theorists, whose ideas human geographers draw upon, discourse is used in differing ways (for example, Mikhail Bakhtin's double-voiced discourse). In this book, discourse is generally understood as it was used by Michel Foucault. To make matters more complex, Foucault employed the term in at least three different ways: (1) as written/visual texts or statements that have meaning and effect, (2) an individual system or group of texts or statements that have meaning, and (3) a regulated practice of rules and structures that govern specific texts or statements. Particular attention in this book is given to his third definition because it is this rule-governed quality of discourse that is of primary importance to geographers. This definition of discourse evokes how it shapes social practices, influencing our actions, attitudes, and perceptions. (See also *discourse analysis*.)

discourse analysis Method of investigating rules and structures that govern and maintain the production of specific written, oral, or visual *texts*. (See also *discourse, narrative analysis*.)

discursive structures or **formations** A key concept of Foucauldian *discourse analysis*. The rules and structures governing the production of *discourse* that affect the way individuals think, act, and express themselves—for example, through travel, comportment, clothes, make-up.

document In archival research, an individual archived item such as a memorandum or letter, handwritten or typed, that constitutes a single item or part of a larger file.

emancipatory Term used to describe a process by which someone is freed from political or other restrictions.

embodied knowledge Refers to how bodies are experienced. The ways people make sense of their experiences, and themselves, cannot be separated from competing and contradictory discourses through which bodies are given meaning. For example, in the context of the uneven gendered social relationships of a public bar, there are normative assumptions about what straight women should do and wear to become "attractive" to men. Hence, because of the possibilities of feeling self-confident, attractive, and feminine, some women may make a deliberate decision to make their bodies more visible through their choices about how they dress. Others may make deliberate decisions to make their bodies less visible to avoid being marked as sexual objects, experiencing inappropriate sexist behaviour, and being disrespected.

emic/etic Parallel terms to insider/outsider; positionalities to be reflected upon in the research process.

empirical Broadly, this refers to "real-world" data (as compared to abstractions, theories, or the conceptual realm).

empiricism Refers to philosophies of science (such as Comte's, see Chapter 19) that argue that the only true knowledge is that derived from empirical facts (that is, they are observable directly through the human senses). (See *positivism* and *objectivity*.)

empowerment Process of increasing the social, political, spiritual, economic, and/or psychological potential of individuals and communities.

episteme In the writings of Michel Foucault, episteme refers to the whole sets of discursive structures/formations within which a culture thinks. An episteme refers to the social processes by which certain statements about the world are considered as knowledge and others are dismissed. Episteme therefore requires critically addressing the range of methodologies that a culture employs at a specific time as "common sense" to allow certain statements to become knowledge about particular people, events, and places. (See also *discursive structures/formations*.)

epistemology (epistemological) Ways of knowing the world and justifying belief. A critical assessment of how knowledge is produced, by whom, and for what purposes. Summed up by the question: "How do we know what we know?" (See also *ontology*.)

essentialism The idea that words (language) have some clear, apparent, and fundamental/intrinsic meaning rather than being given a meaning by users (readers, writers). This idea can be extended to (false) assumptions about social groups, such as that there is assumed to be an "essence" of Blackness, whiteness, femininity, masculinity, and so on.

ethics (ethical) Refers to the moral conduct of researchers and their responsibilities and obligations to those involved in the research.

ethnographic research See *ethnography*.

ethnography Research method dependent on direct field observation in which the researcher is involved closely with a social group or neighbourhood. Also, a mode of academic writing focused on an account of events that occur within the life of a group, paying special attention to social structures, behaviour, and the meaning(s) of them for the group.

extensive research Research typically involving large scale questionnaires or other standardized or semi-standardized methods to identify regularities, patterns, and distinguishing features of a population and to yield descriptive generalization. (Compare with *intensive research*.)

external validity See *generalizability.*

extractive Process of using the research process for one's own goals without regard for the community whose knowledge has been extracted from one place or community and used in another, without deep engagement or compensation.

facilitator Person who encourages or moderates the discussion in a *focus group*. In *participatory action research*, a person who helps others to learn by guiding an appropriate process rather than imposing their own agenda, using techniques that enable people's self-reflection and analysis.

falsification Derived from the writings of Karl Popper (1902–94), the concept of falsification suggests that it is possible to demarcate scientific theory from non-scientific theory on the basis that scientific theories could be falsified or proven untrue by empirical observation and testing. Popper noted that it was not possible to prove a scientific theory through recourse to empirical evidence, since scientific knowledge is always contingent (that is, since scientific observations cannot include all aspects of a phenomenon, we can never be sure that a scientific theory covers the total population of a phenomenon). Accordingly, Popper developed his concept of falsification, which allowed for the contingency of scientific knowledge and allowed scientists to state that a specific theory had yet to be proven false and thus that theory could still be considered valid. Popper's famous example is that it only takes one black swan to falsify a theory that "all swans are white." Although this term is used largely by quantitative researchers, in the context of qualitative studies falsification may help to develop more robust concepts and/or open up new areas of inquiry by exploring negative cases.

field notes Accumulated written record of the fieldwork experience. May include personal reflections, but primarily focused on recording notes from participant observation in a faithful record. Field notes are usually written soon after the period of observation either digitally or by hand, often using notes jotted down while doing fieldwork. (Compare with *research diary.*)

fieldwork diary (See *field notes* and *research diary.*)

files In *archives*, a set of papers and related documents—typically held together in a box or folder—created by an official agency, organization, or individual's *records* and relating to a common topic or theme. Usually organized in reverse chronological order with the newer material overlaying older documents. (See also *provenance.*)

finding aids In the area of *archives*, this is equivalent to a library catalogue, typically taking the form of a list of accession of *files* by name as originally organized by creating agencies. Some archives now have computer-based systems that allow material to be located by use of keywords. In a more general sense, the term is used to indicate organizational tools to help researchers structure their analyses.

first-person narrative In the context of a research narrative (e.g., research report, journal publication, public talk), the presentation of the researcher(s) as "I" or "we." In first-person narrative, researchers or "narrators" can insert their stance, beliefs, emotions, and assumptions explicitly into the text. In so doing, they become more accountable, and their knowledge can be better "situated." (Compare with *third-person narrative.*)

focus group Research method involving a small group of between 6 and 10 people discussing a topic or issue defined by a researcher, with the researcher facilitating the discussion.

follow-up questions Sometimes known as *prompts*, these are questions that permit the interviewer to ask the participant to elaborate on certain elements of an earlier response. (See *primary* and *secondary questions.*)

fonds Overarching term in archival science to mean all the material created by a single administrative entity; maintenance of the relationship between materials as originally and organically created within a fonds is an essential task for archivists.

funnel structure Interview question ordering such that the topics covered move from general issues to specific or personal matters. (Compare with *pyramid structure.*)

gatekeeper Person (or institution) who controls, facilitates, or denies researcher access to a particular community or institution. They may influence matters such as whether individuals choose to participate in a research project, how quickly they can be recruited, and the credibility and cultural skills of the researcher within the community.

genealogy Refers to the ideas of Michel Foucault that question histories of human societies, especially how certain knowledge excludes certain individuals from civil society through, for example, specific portrayals of sexuality, insanity, and illness. To conduct this critique, Foucault (1980, p. 50) refers to the term genealogy as an interpretation that can account for the knowledges, discourses, and objects of human societies without reference to the notion of the Cartesian subject, an individual, unified self with agency and control over itself through thought and reason. Instead of a unified self, Foucault's writing focuses on the processes that he thought were crucial in the constitution of a person's subjectivity.

generalizability (generalization) Degree to which research results can be extrapolated to a wider population group than that studied. A general statement made, or concept proposed, on the basis of inference from specific cases. (See also *transferability*.)

genre In discourse analysis, a subcategory of source forms. For example, oral sources can be subdivided into genres including oral histories, semi-structured interviews, focus groups, conversations, and life narratives. Remaining alert to the genre of the source form is crucial because specific accounts of the world are articulated through differences between the intended audience, social relationships, and technologies of production.

geographic information system (GIS) Generic title for several integrated computer tools for the processing and analyzing of geographical data, including specialized software for input (digitizing) and output (printing or plotting) of mappable data.

GeoHumanities A creative (re)engagement between geography and humanities (arts, philosophy, literature, history, performance) resulting from a combination of the "spatial turn" in humanities in which place, space, and geographic characteristics have gained new attention, and the "cultural turn" in the field of geography that embraces creative practices, collaborations with artists, and experimental material and digital projects.

geopoetics A transaesthetic approach that juxtaposes contemporary poetics, particularly in the realm of ecopoetics, with critical human geography.

geo-tagging Association of data with a GPS-derived location; particularly useful in digital geographies engaging with social media and other web-based research.

geovisualization Combined form of the terms *geographic* and *visualization*. Geovisualization refers to the analysis of geospatial data using interactive visualization. By depicting geospatial data visually, its interpretation may be enhanced and extended.

geoweb Combined realm of digital geographic software, location-enabled devices, and geographic content.

GIS See *geographic information systems.*

go-along interviews Also called "walk-along interviews," as the name suggests, the go-along interview involves the researcher accompanying the respondent within the "field" and engaging in a direct discussion of spatial engagement. This technique combines aspects of well-established techniques such as *interviewing*, *oral history*, and *participant observation*.

grounded theory Systematic inductive (data-led) approach to building theory from empirical work in a recursive and reflexive fashion. That is, using a method of identifying themes or trends from the data, then checking through the data (or collecting more), then refining the themes using repeated checks with the data to build theory that is thoroughly "grounded" in the real world. Initiated by Glaser and Strauss (1967) and reinterpreted and refined greatly since then by them and other authors. For a geography-based application, see *grounded visualization*,

grounded visualization Analytical technique for generating new insights from mixed methods research that includes spatial data; based on iterative, recursive rounds of evaluating findings from multiple sources in relation to each other, rooted in a combination of grounded theory and exploratory spatial data analysis (see Knigge & Cope, 2006, 2009).

guide See *interview guide.*

hegemony Social condition in which people from all sorts of social backgrounds and classes come to interpret their own interests and consciousness in terms of the *discourse* of the dominant or ruling group. The hegemony of the dominant group is thus based, in part at least, on the (unwitting) consent of the subordinate groups. Such consent is created and reconstituted through the web of social relations, institutions, and public ideas in a society.

helicopter research (researcher) Figurative term describing a researcher who drops in to study a place or community and conducts their study with minimal engagement with local people or community groups After the researcher's departure, they may never be seen again, their work is published to their own professional advantage, and the local community experiences no improvement in their situation as a result of the study.

hermeneutic circle Circle (or more broadly, *process*) of interpretation of qualitative information, which accounts for the point that no such interpretation is free from the values, experiences, attitudes, and ideas of the observer or researcher. Implicit in this realization is a need for the researcher to be clear about their position and to ensure that interpretation is participatory and iterative—that is, involves participants and is done in one or more collaborative rounds.

hermeneutics Study of the interpretation of meaning in *texts*, whether there is assumed to be a single dominant meaning or a multiplicity of meanings.

heteronormative Assumption that heterosexuality is the only normal and natural form of desire and social organization. (See also *cis-heteronormativity*.)

identity Set of characteristics, memories, and qualities that we draw on to make a sense of self. (See also *situated knowledge, positionality, reflexivity*.)

idiographic Approach to knowledge that highlights the particular, subjective, and contingent aspects of the social world. Cases are understood holistically. Credible and authentic descriptions are emphasized instead of statistical *generalizability*. Contrast with *nomothetic*.

impact agenda Refers to policy makers' demands that scholarly research not only serve to advance knowledge but that it offers value, typically measurable in terms of productivity and well-being, to society more broadly.

independent variable Study item with characteristics that are considered to cause change in a dependent variable. For example, the independent variable rainfall may promote flooding (the *dependent variable*).

Indigeneity Describes the evolving pan-Indigenous movement and corresponding identity among peoples who, despite often considerable cultural divergence, share similarities founded in an ancestral birth right in the land, a common core of collective interests, and the shared experience of dispossession precipitated by on-going colonialism.

Indigenous methodologies In many Indigenous settings, the term *research* is deeply enmeshed in colonizing processes and experiences. In developing approaches to contemporary research, many Indigenous groups are exploring ways of conceptualizing, executing, disseminating, and evaluating research against their own explicit criteria. These methodologies affirm Indigenous peoples' ways of knowing, research purposes, and protocols as well as critiquing and adapting "Western" research paradigms. Adoption of Indigenous methodologies shifts the balance of power, responsibility for ethical oversight, and judgment about the value, meaning, and utility of research away from traditional research institutions such as universities and funding agencies and towards contemporary structures of Indigenous governance, decision-making, and accountability—including new structures and processes within some universities and agencies.

induction (inductive) Process of generalization involving the application of specific information to a general situation or to future events. (Compare with *deduction*.)

informant Person interviewed by a researcher. Some refer to those who are interviewed as "subjects" or "respondents." Others argue that someone who is interviewed, as opposed to simply observed or surveyed, is more appropriately referred to as an informant. That is because an interview informant is likely to have a more active and informed role in the research encounter.

informed consent Informant/subject agreement to participate in a study having been fully apprised of the conditions associated with that study (for example, time involved, methods of investigation, likely inconveniences, and possible consequences).

initial codes Codes that are pre-determined in some way, usually because they are a prominent theme in the research questions or are inherent in the topic of the research. (See also *descriptive codes.*)

insider Research position in which the researcher is socially accepted as being "inside" or a part of the social groups or places involved in the study. (Compare with *outsider.*)

intensive research Research typically of individual agents or small groups, involving semi-standardized or unstructured methods (e.g., interviews, oral histories, ethnography). Focuses on causal processes and mechanisms underpinning events and specific cases. (Compare with *extensive research.*)

interpretive codes Codes that emerge from analysis and interpretation of the data along themes and toward theory generation. (See also *analytic codes.*)

interpretive community Involves established disciplines with relatively defined and stable areas of interest, theory, and research methods and techniques. Influences researchers' choice of topics and approaches to and conduct of study.

intersectionality Recognition and scrutiny of how multiple dimensions of difference (gender, race, class, age, sexuality, etc.) intersect and compound each other to affect people's identities, social context, and experiences. A theoretical framework that understands processes of social differentiation not as discrete from each other, but as thoroughly overlapping, co-constitutive, and reinforcing of each other. The Black critical legal scholar Kimberle Crenshaw is often credited with coining the term in 1989, although other critical race feminists have similarly theorized how systems of power overlap to shape marginalized people's experiences of oppression as complex and multifaceted. (See also *positionality* and *axes of difference.*)

intersubjectivity Meanings and interpretations of the world created, confirmed, or disconfirmed as a result of interactions (language and action) with other people within specific contexts. (See also *subjectivity* and *objectivity.*)

intertextuality Necessary interdependence of a *text* with other texts that have preceded it. Any text is built upon and made meaningful by its associations with others.

interview Means of data collection involving an oral exchange of information between the researcher and one or more other people.

interviewer effects When a person is interviewed, they are not doing their normal everyday activities. An interview is a formal data-gathering process. This formality, and the unusual discursive style of an interview, can have an influence on what an informant says and how they say it. This is one example of an interviewer effect. There could also be effects that flow from the demeanour, dress, accent, and physiology of an interviewer. More broadly, in ethnographic work, they are referred to as researcher effects.

interview guide List of topics to be covered in an interview. May contain some clearly worded questions or key concepts intended to guide the interviewer. (Compare with *interview schedule.*)

interview schedule Ordered list of questions that the researcher intends to ask informants. Questions are worded similarly and are asked in the same order for each informant. In its most rigid form, an interview schedule is a questionnaire delivered in face-to-face format. (Compare with *interview guide.*)

***in vivo* codes** Codes that emerge from the body of the work being examined; phrases and terms used by respondents in the course of ethnographic research or already appearing in examined texts that suggest a theme worthy of analysis. (See also *coding.*)

journals See *diaries.*

knowledge (co-)production Creation of new knowledge through diverse social and cultural practices, including intentional research; represents the understanding that knowledge is *pro-*

duced in social and geographic contexts by certain agents and that—due to power relations—some groups' knowledge is valued more than others'. (See also *co-constitution of knowledge.*)

lacuna Gap or hole in the scholarly literature; something missing from a body of work.

landscape Term used broadly to mean a built, cultural, or physical environment (and even the human body), which can be "read" and interpreted.

latent content analysis Assessment of implicit themes within a text. Latent content may include ideologies, beliefs, or stereotypes. (Compare with *manifest content analysis.*)

legitimacy Approval and respect accorded researchers, research projects, and research methodologies and methods that are considered appropriate and are valued and welcomed by the people(s) and communities with whom researchers work.

life history Interview in which data on the experiences and events of a person's life are collected. The aim is to gain insights into how a person's life may have been affected by institutions, social structures, relations, rites of passage, or other significant events. (Compare with *oral history.*)

literature review Comprehensive critical summary and interpretation of resources (for example, publications, reports) and their relationship to a specific area of research.

longitudinal case study Type of case study conducted over time; typically involves a revisit whereby the researcher returns to the case after an intervening time period during which no appreciable research is done.

longitudinal data Information relating to the same sample (of individuals or groups or other entities) at different points in time. (Contrast with *cross-sectional data.*)

longitudinal interviews Interviews that are planned to repeat over a specified stretch of time, for instance, one interview per month, per year, or per five years.

manifest content analysis Assessment of the surface or visible content of text. Visible content may include specific words, phrases, or the physical space dedicated to a theme (for example, column centimetres in a newspaper or time in a video). (Compare with *latent content analysis.*)

masculine (masculinist) gaze This term speaks to the ways in which a viewer looks upon the people either present or represented (e.g., via photography, painting). Feminist theory has added to understanding by speaking of the masculine gaze to express an asymmetric (or unequal) power relationship between viewer and the person or population viewed.

maximum variation sampling Form of *sampling* based on high diversity aiming to uncover systematic variations and common patterns within those variations.

member checking See *participant checking.*

memoing In qualitative software systems, a process whereby the researcher may write memos or reflections on the research process as she or he works and then incorporate these memos as electronic data for further investigation.

memorandum of understanding (MoU) Document specifying the aims, process, roles, responsibilities, and rights of parties involved in a research project. (See also *conditions of use form.*)

metaphor Expression applied to something to which it is not literally applicable in order to highlight an essential characteristic.

method Means by which data are collected and analyzed (e.g., in-depth interviewing).

methodological nationalism Refers to particular tendencies in social scientific research to take the nation-state and its boundaries for granted as natural and ahistorical. It occurs when researchers treat the nation-state as a stable unit of analysis that research participants base their primary social and political identification with.

methodology Philosophical and theoretical basis for conducting research that is much broader and sometimes more politically charged than method alone (e.g., feminist methodology).

mixed methods Combination of techniques for tackling a research problem, often used specifically to mean a combination of quantitative and qualitative methods; distinguished from *multi-method research* through its purposeful and iterative integration of data collection and analysis of the data.

modality See *objective modality* and *subjective modality*.

moderator See *facilitator*.

modern hand Way of referring to the array of more simplified ways of forming linked letters constituting writing, as opposed to printing, from the later nineteenth century onwards. Initially produced by metal-nibbed pens dipped in ink and later with fountain pens. From an archival researcher's perspective, this is often a sort of "dark age" when handwriting standards slip and file material can be very difficult to read and may be badly smudged. It is, however, important not to ignore handwritten notes and comments in favour of typewritten records because the former can often provide vital clues as to the concerns underpinning decisions. (See also *copperplate script* and *secretary hand*.)

more-than-human Since its inception more than 20 years ago, the sub-discipline of more-than-human geographies centres on new ways of approaching relationships between the bio (life) and the geo (earth). Geographers working in this way explore how the social, cultural, and political dimensions of power and space are forged in relation to a range of non-human subjects and forces including animals, weather and earth systems, technologies and scientific models. Through attention to species- and neuro-diversity, relationality, mobilities, and sensibilities, and using diverse theorists and methods, a common aim has been to disrupt taken-for-granted ways of knowing, challenge hierarchical power relations and empower marginalized voices to achieve greater spatial justice and sustainability. (See also *Actor-Network Theory*)

multi-method (multiple method) research Refers to research employing diverse data collection and analysis techniques (qualitative and quantitative) that generally proceed parallel to each other until the final stages of the project; compare with *mixed methods* in which data collection and analysis are integrated and shed light on each other *throughout* the research process.

multiple ontologies The appreciation that different societies (or groups and individuals in society) might have different views of the world, or have diverse ways of being in the world. (See also *ontology*).

multiple voices Reference to the need to listen to alternative literatures, texts, expressions, or opinions and therefore avoid the assumption that there is only one view of merit.

mystification In a research narrative (for example, research report, journal publication, public talk), the concealment of particular details in ways that give other details legitimacy and/or coherence. For example, scientific third-person writing, and grammatical structure may sometimes involve obscuring details of thought, feeling, emotion, time, place, embodiment, and stance in ways that mystify the social agents and participants being described. Third-person writing is particularly useful for mystifying the social location, partiality, and embeddedness of authors (researchers). Haraway refers to this as the "god trick" of being everywhere and nowhere simultaneously.

narrative analysis Refers to an analytical technique applied to data that are presented as a story; using common strategies to identify protagonists, antagonists, conflicts, story arcs, and resolution, this method is well-suited to some kinds of texts and oral data (compare with *discourse analysis*).

narrative mapping Type of qualitative cartography that provides a visual representation of relationships between the experiences of individuals or groups and their socio-spatial environments.

neo-positivism Variants of positivism as they have evolved from the original axiom of logical positivism developed by the Vienna Circle in the 1920s and 1930s. Neo-positivists have responded to various critiques (especially Popper's critique that logical positivism's reliance on verifiability was too strong a criterion for science and instead we should rely on falsifiability as a primary criterion for scientific knowledge) and incorporated these critiques into their work. Neo-positivism was most strongly represented in geography during the quantitative revolution of the 1960s and 1970s.

new media On-demand information and interactivity associated with the internet and the technologies and devices used to access that information. New media include anything that enables digital interactivity—the devices, the software or applications, and the online sites they access. (See also *digital geographies.*)

nominalization Transformation of verbs and adjectives into nouns. Nominalization reduces information available to readers, and it mystifies social processes by hiding actions and the identity of actors.

nomothetic Approach to knowledge that emphasizes generalizability for understanding the social world. Social phenomena are reduced to variables for the purposes of generating statistically generalizable findings. The credibility, authenticity, and holistic understanding of each sub-unit studied is of lesser importance. (Contrast with *idiographic.*)

NVivo™ Software package to help organize and analyze qualitative data. This is a specific form of computer-assisted qualitative data analysis software (CAQDAS). Allows importing and coding of textual data, text editing, coded data retrieval and review, word and coding pattern searches, and data import/export to quantitative analysis software. (See also *computer-assisted qualitative data analysis software.*)

objective/objectivity Unaffected by feelings, opinions, or personal characteristics. A (contested) assumption that social research and its findings can be stripped of power relations, researcher positionalities and world-views, and that knowledge can somehow be "pure," free from bias or context. Often contrasted with *subjectivity.* (See also *intersubjectivity.*)

objective modality Form of writing that implicitly hides the writer's presence in the text (for example, third-person narrative form) but which clearly signals agreement with the statement being made.

observation Most literally, purposefully watching worldly phenomena. Increasingly broadened beyond seeing to include apprehending the environment through all our senses (for example, sound, smell) for research purposes.

observer-as-participant Research situation in which the researcher is primarily able to observe but in so doing is also participating in a social situation. (Compare with *participant-as-observer.*)

online focus group Focus groups that are conducted online using real-time technology such as chat-rooms or asynchronous technology such as bulletin boards.

ontology Beliefs about the world. Understanding about the kinds of things that exist in the universe and the relations between them. (See also *epistemology* and *multiple ontologies.*)

open questions Questions in which respondents can formulate their own answers, unrestricted by having to choose between pre-determined categories. (Compare with *closed questions* and *combination questions*).

opportunistic sampling Impromptu decision to involve cases or participants in a study based on leads uncovered during fieldwork.

oral history Prepared *interview* conducted in question-and-answer format with a person who has first-hand knowledge of a subject of interest. (Compare with *life history.*)

oral methods Verbal techniques, such as *interviews* or *focus groups*, as opposed to written methods for seeking information.

oral testimony Probably the oldest kind of evidence in human communication. Spoken research information gained by means of informal, semi-structured, or structured interviews, and/or key informant interviews.

Orientalism As used by Edward Said (1978), a key post-colonial term (see *post-colonialism*) referring to Western (mis)representations and construction of an imagined Orient in discourses that serve to produce and legitimize imperialism.

Other The non-Self, groups and peoples perceived as fundamentally different from one's self and against which a person might compare themselves and establish their own social position, meaning, and identity. Also taken to mean that which is oppositional to the mainstream—marginal or outside the dominant ideology. Initially developed by Simone de Beauvoir in her 1949 book *The Second Sex* to characterize patriarchal representations and subjugation of women, the term was extended by Franz Fanon (1967) and Edward Said (1978) to the cross-cultural representations and relationships that underlie colonialism.

outsider Research position in which the researcher is rendered "outside" a social circle or feels "out of place" on account of differences such as visible appearance, unfamiliarity, or inability to speak the language or vernacular used. (Compare with *insider*.)

over-disclosure Can occur in an interview or focus group when research participants reveal personal, sensitive, or confidential information that goes beyond the scope of the research or that they may regret having mentioned after the interview or focus group. (Contrast with *under-disclosure*.)

panopticon Circular prison with cells surrounding a central guards' station. In the panopticon, inmates may be observed at any time, but they cannot see the observers and thus internalize the discipline, resulting in self-surveillance. (See also *power/knowledge* and Foucault.)

paradigm Set of values, beliefs, and practices shared by a community (e.g., members of an academic discipline) that provides a way of understanding the world.

paralinguistic clues Tacit signs perceptible in face-to-face interviewing. The tone of speech used by an informant is an important indicator of their emotional disposition when answering a question. It can also indicate an informant's level of comfort and the degree of rapport between the researcher and the informant. Other non-spoken clues include eye contact, fidgeting, furtive glances, and aggressive or defensive postures.

parallel case study Study of multiple cases at the same time for the purposes of *comparative analysis*.

participant Person taking part in a research project. Usually the *informant* rather than a member of the research team.

participant checking Providing research participants the chance to review the transcript of their contribution to an *interview* or *focus group* for accuracy and meaning. May also involve the participant or the project's community partners review of the overall research output (for example, thesis, report). Also serves as a means of continuing the involvement of participants in the research process.

participant community One's research participants; the community may be known as such to one another—members of a formal or informal grouping—or may be understood as a community by the researcher.

participant observation Fieldwork method in which the researcher studies a social group while being a part of that group.

participant-as-observer Research situation in which the researcher is primarily a participant in a social situation or gathering place but in so doing can maintain enough critical distance to observe social dynamics and interactions.

participation Process in which people play active roles in decision-making and other activities affecting their lives.

participatory action research (PAR) Umbrella term covering a range of participatory approaches to action-oriented research involving researchers and participants working together to examine a situation and change it for the better. (See also *action research*, *participatory research*, *collaborative research*, and *decolonizing research*.)

participatory diagramming Technique whereby a group of people, with support from a *facilitator*, collectively produces a visual representation (for example, drawing, diagram, chart, mind-map, sketch) for subsequent analysis using locally appropriate materials (for example, stones, leaves, chalk, ground, pens, paper, whiteboards), criteria, and symbols. Diagrams usually convey relationships between key stakeholders, institutions, or resources, sometimes over different time periods. (See also *participatory mapping*.)

participatory geographic information systems (PGIS) Process that adapts GIS software to incorporate local expertise and knowledge, usually within a participatory action research framework to enable mapping and data analysis by non-professionals. PGIS emerged in response to criticisms that the high financial, time, and training requirements of GIS can discourage grassroots groups from using it to inform their own research and development. (See also *participatory mapping* and *community geography*.)

participatory mapping Technique whereby a group of people, with support from a *facilitator*, collectively produces a map for subsequent analysis using locally appropriate materials, criteria, and symbols. Maps often focus on material aspects of life such as a watershed, a village, a body, or the distribution of specific resources within a particular area. (See also *participatory diagramming* and *participatory geographic information systems*.)

participatory research Community-based approach to research involving local people and their knowledges as a foundation for social change. Participatory research was developed by educators like Paulo Freire in Brazil and others in the global South to support consciousness-raising and political action. It now informs participatory action research and processes of participatory development within the global South.

pastiche A *text* that is a medley or mash-up drawn from various sources.

patriarchal Relating to a system of oppression in which male people (literally "the father") are dominant; reinforced by cultural, economic, legal, and political practices.

patriarchy Refers to the ways that particular forms of masculinity come to occupy privileged positionings in social life, including and especially in institutions, knowledge, and other systems of power. It is thus accompanied by the categorization of femininity and non-normative forms of masculinity as inferior. Patriarchal processes and logics are those that uphold the supposed supremacy of masculinity in social life.

performativity Employed in human geography to denote the manner in and through which ideas and concepts are performed or acted-out. In doing so, roles and expectations are given physical expression. The acts people perform, are expected to perform, and/or struggle against performing are not natural but are complex engagements with power.

personal log Recorded reflections on the practice of an interview. Includes discussions of the appropriateness of the order and phrasing of questions and of the informant selection. Also contains assessments of matters such as research design and ethical issues. (See also *analytical log*.)

photo-elicitation Technique related to diary photographs and *diary interviews*. However, according to Harper (2002), the principal objective of this technique is to elicit alternate ways of seeing and understanding the same image among respondents. The outcome of this is to invoke a deep discussion of values and meaning.

PhotoVoice Methodology that was developed within health education and has now been adopted and adapted within the social sciences, including geography. Participants are asked to represent their perspective by taking photographs that are meaningful and then add verbal or written descriptions of those meanings, which subsequently offer a window into how they conceptualize their circumstances. Photography "gives voice" to otherwise marginalized groups and has been associated with social action.

pilot study Abbreviated version of a research project in which the researcher practises or tests procedures to be used in a subsequent full-scale project.

pluriversal world The idea that the world is made up of manifold, heterogeneous, dynamic ways of being and knowing.

polycentric epistemologies Different groups may have diverse ways of knowing, asking different types of questions about the world and transmitting them in varied ways. In other words, there are many ways of conceptualizing knowledge. (See also *epistemology*).

population The larger group from which a *sample* has been selected for inclusion in a study. In quantitative research, based on *probability* (random) *sampling*, it is assumed that the sample has been selected such that the mathematical probability of sample characteristics being reproduced in this broader population can be calculated. In qualitative research where *purposive sampling* is used, no such assumption is made.

positionality A person's social, locational, and ideological placement relative to the research project or to other participants in it. May be influenced by biographical characteristics, such as class, race, and gender, as well as various formative experiences.

positionality statement Record of how a researcher is situated in the production of knowledge over time. This requires careful reflection on the researcher's points of connection and disconnection with a project. Positionality statements are not stand-alone testimonials. Instead, they document the co-constitution of researcher and project. (See also *co-constitution of knowledge*.)

positivism (positivist) Approach to scientific knowledge based around foundational statements about what constitutes truth and legitimate ways of knowing. There are a number of variants of positivist thought, but central to all is the belief in an objective reality that lies at the heart of a singular, universal, and value-free knowledge, which in turn is based on empirical observation and the scientific method. This position is in contrast to critical perspectives (e.g., feminism, postmodernism, post-structuralism, queer theory, etc.) that embrace an understanding of the social construction of knowledge, which is imbued with multiple truths, subjectivity, positionality, and the rejection of the possibility of "value-free" objectivity.

post-colonial research Research that rejects imperialism and the goals, attitudes, representations, and methods of imposed, colonial research and instead seeks to conduct research that is welcomed and that fosters egalitarian relationships and openness, values local knowledge and ways of knowing, and contributes to self-determination and locally defined welfare. (See also *decolonizing research*, and contrast with *colonial research*.)

post-colonialism (post-colonial) Approach to knowledge that seeks to represent voices of the *Other*, especially colonized peoples and women, and to recognize knowledge that has been ignored through processes of colonization and patriarchy.

postmodernism Movement in the humanities and social sciences that includes *post-colonialism* and embraces the pluralism of multiple perspectives, knowledges, and voices rather than the grand theories of modernism. Individual interpretation is considered partial because it is to some degree socially contingent and constituted.

post-structuralism School of thought that endeavours to link language, *subjectivity*, social organization, and *power*.

power In Foucauldian discourse analysis, power is central to thinking about *discourse* as something that has an impact. It is through power that the elements of discourse have effects on what people do and think and how they express themselves. Yet power is not conceptualized in terms of acting upon people in an oppressive way. Rather, the individual is seen as an effect of power. That is, power makes things possible as well as restricting possible actions and attitudes. These possibilities are instances of power/knowledge relationships. (See also *power/knowledge*.)

power/knowledge A key concept of Foucauldian discourse analysis. Foucault argues that the relationship between power and knowledge is essential to thinking about the effects of *discourse*. He argues that statements that are accepted as knowledge are themselves the outcome of power struggles. For example, what has constituted geographical knowledge in universities has been a constant struggle over different versions of what constituted space/place.

praxis Use of research findings by researchers to make constructive social and environmental change; sometimes thought of as the combination of theory + action.

preguntando caminamos Translates to "asking we walk"—a commitment to a relationship of reciprocity and mutuality where one learns by walking, listening, talking, and doing.

preservation The durable, secure maintenance of research artefacts, including recordings, photos, maps, and other products for future review and research.

pre-testing See *pilot study*.

primary question Interview question used to initiate discussion of a new topic or theme. (Compare with *probe question*.)

primary sources From a historical geography perspective, primary sources in the narrower sense are public and private records created in an earlier period of time that is of interest to the researcher. They include letters, diaries, and journals, original census returns, minutes of organizations, and files of government departments. In addition, original maps, survey plans, and photographs would be included. In some circumstances, period newspapers and published official documents might also be regarded as primary sources. (Compare with *secondary literature*.)

privileged discourse While there are always competing and sometimes contradictory cultural discourses to make sense of the world, privileged discourse is one that takes priority over others in shaping social, cultural, and political meanings. For example, there exists a range of conflicting and competing discourses about Hawai'i. Yet in Western society, that which is privileged portrays Hawai'i as an earthly paradise.

probability sampling Sampling technique intended to ensure a random and statistically representative *sample* that will allow confident generalization to the larger *population* from which the sample was drawn. (Compare with *purposive sampling*.)

probe (question) Gesture or follow-up question used in an interview to explore further a theme or topic already being discussed. (See also *prompt*. Compare with *primary question*.)

prompt Follow-up question in an *interview* designed to deepen a response (for example, "why do you say that?", "what do you mean?"). (See also *probe question*. Compare with *primary question*.)

provenance Generally, the background of something, its history of coming into being (as for a document or institution). More specifically refers to the organizational principle for *archives* that stresses the importance of the original internal arrangement of a collection of *files* and the order of information in files as devised by their creating agencies as a means of understanding past events.

public scholarship Term given to the varied efforts of academics to use their resources, expertise, and standing to engage with broader society, whether through community-based research, scholar-activism, media outlets (op-eds, websites, etc.), or other community-university partnerships.

purposeful sampling See *purposive sampling*.

purposive sampling Sampling procedure intended to obtain a specific group for study on the basis of the specific characteristics they possess. (Compare with *probability sampling*.) Aims to uncover information-rich phenomena/participants that can shed light on issues of central importance to the study.

pyramid structure Order of interview questions in which easy-to-answer questions are posed at the beginning of the interview while deeper or more philosophical questions/issues are raised at the end. (Compare with *funnelling*.)

quantitative methods Statistical and mathematical modelling approaches used to understand social and physical relationships using primarily numerical data.

quantitative revolution Period in the mid-twentieth century (particularly from the late 1950s) during which transformations in information and computer technologies, developments in mathematical modelling, and sophisticated quantitative techniques influenced the form and nature of research being conducted. The quantitative revolution led to an increased use of statistical techniques for collating and analyzing large amounts of data, with these techniques being linked to empirical testing of models, theory, and hypotheses.

questionnaire Often used interchangeably with *survey*; a set of questions that have been carefully designed to elicit information from respondents, which can be delivered orally by a researcher or in written form (on paper or digitally).

radical vulnerability A collective ethos, "in which the individual ego must surrender to a politics of co-travelling and co-authorship." (Nagar, 2019, p. 7)

random sampling See *probability sampling*.

rapport Productive interpersonal climate between informant and researcher. A relationship that allows the informant to feel comfortable or confident enough to offer comprehensive answers to questions.

reciprocal power relation Research situation in which researcher and *informant* are in comparable social positions and experience relatively equal costs and benefits of participating in the research. (See also *asymmetrical power relation* and *studying up*.)

records Generic term for files, maps, plans, and other documents held in an *archive*.

recruitment Process of finding people willing to participate in a research project. Recruitment strategies can range from asking people on the street (perhaps to fill in a questionnaire) to inviting key individuals to participate (in a focus group, for example).

reflexivity (reflexive) Self-critical introspection and a self-conscious scrutiny of oneself as a researcher. Reflexivity refers to an intentional pause and reflection on one's own positionality and identity in relation to one's work, and often a published or unpublished discussion of how that positionality might have shaped experience and fieldwork (from safety to who might share what in interviews). (See also *action–reflection*.)

reification When the complexity of social life is reduced to concrete and simplified "things" in the construction of texts. For example, reference might be made to "the research" or "the participant," thereby obscuring the way in which these nouns come to be constituted and expressed through a variety of social and spatial relations.

relativism Approach to knowledge in which it is held that there is no means of significantly differentiating between the merits of arguments. Suggests that there are no absolute, unequivocal standards of true/false or right/wrong.

reliability (reliable) Extent to which a method of data collection yields consistent and reproducible results when used in similar circumstances by different researchers or at different times. (See also *validity*.)

replicability (replication) Ability to be repeated or tested to see how general the specific findings of a study are in the wider *population*.

representation Way in which something (the world, human behaviour, a city, the landscape) is depicted, recognizing that it cannot be an exact depiction. An important insight from post-structuralist thinkers is that representations not only describe the social world but also help to shape or constitute it.

representative sample Term indicating that a group that closely matches the characteristics of its population as a whole, meaning that the sample is a fairly accurate reflection of the population from which the sample is drawn.

research design Framework, encompassing question, theory, method, and procedures that are used to conduct research.

research diary This term can be used in several ways. The most common are, first, a place for recording observations in the process of being reflexive. Contains thoughts and ideas about the research process, its social context, and the researcher's role in it, similar to *memoing*. Alternatively, a research diary can be more of a logbook of a project's important dates, sources, participants' contact information, site descriptions, etc.

respondent Participant in a survey, interview, or other method involving a "response." (See *informant*.)

rhizomatic Emerging from the work of Gilles Deleuze and Felix Guattari and borrowing from the biological world, the idea that research and knowledge emerge from multiple, non-hierarchical, but still-connected sprouts; proposed in contrast to "arborescent" (tree-like, hierarchical) thinking in which an idea or method emerges from and depends on its trunk and branch structure.

rigour Trustworthiness of data collection and analysis; in qualitative research this might be established through practices including triangulation, member-checking, reflexivity, etc. Baxter and Eyles (1997) suggest rigour has credibility, transferability, dependability, and confirmability.

sample Phenomena or participants selected from a larger set of phenomena or a larger *population* for inclusion in a study.

sampling Means of selecting phenomena or participants for inclusion in a study. A key difference between qualitative and quantitative inquiry is in the logic underpinning their use of *purposive* and *probability* (random) *sampling* respectively.

sampling frame List or register (for example, electoral roll, phone directory) from which respondents for a questionnaire are drawn.

satisficing behaviour Conduct in which the decision-maker or agent acts in ways that yield satisfactory outcomes rather than optimal or "maximizing" outcomes.

saturation Point in the data-gathering process when no new information or insights are being generated. This is one method used by researchers to determine when to stop gathering data.

secondary data Information collected by people/agencies and stored for purposes other than for the research project for which they are being used (for example, census data being used in an analysis of socio-economic status and water consumption).

secondary literature From a historical geography perspective, an existing set of published material in the form of books, essays, and articles produced after the events that they discuss and based on primary sources and/or oral testimony and personal observations. In terms of a research exercise, it also refers to the existing academic writing on a specific topic. This literature may be obviously divided into groups that offer contrasting interpretations of the same events. (Compare with *primary sources.*)

secondary question Interview prompts that encourage the informant to follow up or expand on an issue already discussed. (See also *follow-up questions.*)

secretary hand Style of writing used by professional clerks that became increasingly widespread in the UK in the sixteenth and seventeenth centuries. Twenty-first century researchers require some special skills to be able to decipher the calligraphy and the now archaic English prose. (See also *copperplate script* and *modern hand.*)

selection bias Skewing the participants in a research endeavour; while this is usually seen as a negative factor (and indeed, should not be used to produce certain results), there are often reasons to employ *purposive sampling* in order to explore issues that are unique to a particular sub-group of the population.

semiology See *semiotics.*

semiotics System or language of signs (sometimes referred to as semiology). (See also *signifier* and *signified.*)

semi-structured interview Interview with some predetermined order but that nonetheless has flexibility with regard to the position/timing of questions. Some questions, particularly sensitive or complex ones, may have a standard wording for each *informant.* (Compare with *structured interview* and *unstructured interview.*)

series lists List of individual files from a specific organization held in an archive. Typically, they replicate the referencing system used by the creating organization; they are usually numerical or alphanumeric and can include sub-series as well as large files that are broken into several individually recorded parts.

signifier Images such as written marks or features of the landscape with which meaning is associated.

silence In discourse analysis, *silence* refers to how frameworks of understanding always conceal as much as they reveal about the world. Concealment inevitably occurs when all frameworks of understanding are conceived of as the outcome of a highly social process. For example, life narratives often frame events surrounding discriminatory legislation around personal accounts, often silencing the role of political organizations. In contrast, a spokesperson for a political organization may choose to frame the same events through ideas about the collective nation. Such official accounts may erase differences in lived experiences based on various intersections of gender, sexuality, class, race, and so on.

situated knowledge Metaphor that evokes recognition of the positionality (or contextual nature) of knowledges. The inscription and creation of knowledge is always partial and "located" somewhere. See Haraway (1988).

situated solidarity Solidarity emerging from shared places and struggles.

snowball sampling Sampling technique that involves finding participants for a research project by asking existing informants to recommend others who might be interested. From one or two participants, the number of people involved in the project snowballs. Also known as *chain sampling.*

social construction Perspective that considers human cultures, institutions, and even notions of humanity and nature as produced (and reproduced) through social interactions, norms, and agreed-upon meanings; geographers make the point that spatiality (the mutual constitution of society and space) is inherent in social construction and its expressions in material landscapes and lived experiences of place.

social justice Situation in which there is an equitable and respectful negotiation of social and spatial difference and a fair distribution of resources; linked to research and activism with the goal of identifying and dismantling oppression in its many interlocking forms (racial, class, gender, etc.).

solicited journal See *diary*.

social media Social dynamics and interactions among people in virtual contexts, typically mediated through websites and mobile apps. Common examples (which fluctuate in popularity) include Facebook, Instagram, SnapChat, WhatsApp, Pinterest, YouTube, TikTok, and Twitter. Social media are defined by interactivity among users and user-generated content. Clicks, likes, followers, and other expressions of interest by users drive the commercial side of the platforms as competition for consumers' attention, data, and digital profiles becomes increasingly sharp.

soundscape Compound word joining *sound* and *landscape* to create a term that captures the way a sound or combination of sounds can contribute to the sensory environment. Sounds can be "natural" (for example, water) or contrived (for example, music), and their apprehension can be understood as a non-visual form of observation.

spatial Relating to the *geographical* arrangement of phenomena, human processes, and environments; can suggest absolute locations, distance, etc., or more relative or metaphorical locations. Commonly seen in human geography as both contingent upon and constitutive of social relations and interactions.

spatiality Relates to the social production of space and the spatial influences on society, which are usually seen as dialectical, recursive forces.

specific questions Relate to respondents' individual experiences and are developed through follow-up work. (Compare with *orientation questions*, *common questions*, and *follow-up questions*.)

staging a text Theatrical metaphor for *writing-in* research that encourages geographical researchers to consider how the construction of a research text is actually a form of cultural production. The author of the text is a creator, director, and performer in the particular narrative that they are constructing.

stakeholder Any individual or group that has an interest in a project because of the way it may benefit, harm, or exclude them.

standardized questions Uniform set of questions that are repeated for all *interviews* or *focus groups* in a research project, in contrast to spontaneous questions that develop out of the conversational flow of an interview or focus group.

strategy of conviction Refers to the processes by which particular social realities or ways of making order of the world become accepted within specific spatial and historical contexts as common sense or as truth. The term is derived from the work of Michel Foucault. In particular, the term strategy of conviction is derived from his argument that knowledge and power are inseparable. According to Foucault, the process by which particular social realities are produced, circulated, and maintained requires thoughtful consideration of the intersection between authorship, technologies, and the type of text.

story-telling (story) Narrative retelling of one or multiple experiences; a valuable method for eliciting oral traditions, cultural legends, and personal experiences

structured interview Interview that follows a strict order of topics. Usually the order is set out in an *interview schedule*. The wording of questions for each interview may also be predetermined. (Compare with *unstructured interview* and *semi-structured interview*.)

studying up Asymmetrical power relationship in which the research participant is in a position of power relative to the researcher. (See also *asymmetrical power relation*.)

subaltern Oppressed, exploited, marginalized minority peoples and groups. The term derives from the work of Antonio Gramsci and from the "subaltern studies" project undertaken by South Asian historians since the early 1980s that endeavoured to write history from the perspectives of colonized peoples, but is often used in a wider sense.

subject A research participant. The term is often avoided in contemporary critical human geography for its disempowering positioning and dehumanization of participants. (See *informant*, *respondent*, *participant*.)

subjective/subjectivity Refers to the insertion of the personal resources, opinions, and characteristics of a person into a research project. Based on the notion that social phenomena are interpreted in ways that are relative to and contingent upon people's background, positionality, locational context, and life experiences. Often contrasted with objectivity. (See also *intersubjectivity*.)

surveillance Involves monitoring the behaviour of people or objects to observe (and sometimes record) deviation from conformity. In French, surveillance means "watching over," and in one sense this covers all types of all observation. Contemporary technology such as closed-circuit television epitomizes surveillance as social control. (See also *panopticon.*)

survey Systematic set of questions organized to elicit both open-ended (more qualitative) and closed responses (more quantitative). (See *questionnaires.*)

synchronous interviewing Interview in which the informant's answers immediately follow the questions. Face-to-face and telephone interviews are common examples. Some forms of internet-based interviewing can also be synchronous, such as conversations that occur within video conferencing, online video (Zoom, etc.) and other digital platforms. (Compare with *asynchronous interviewing.*)

synergistic effect A key feature of focus groups that occurs when participants are prompted and provoked by the things others say. Generally results in a lively discussion as participants respond to each other and add new thoughts and ideas. Best summed up by the phrase "the whole is greater than the sum of its parts."

talk-to-text Computer software that converts spoken words into text. It has three principal uses: (1) to support dictation use; (2) for accessibility purposes (e.g., so that people who are vision-impaired or who cannot use keyboards can use internet communications); and (3) to aid the transcription of research interviews into text. (See also *transcript.*)

text Traditionally synonymous with the written page but now used more broadly to refer to a range of source forms such as oral texts (including *semi-structured interviews* and *oral histories*) and images (including painting, photographs, and maps) as well as written and printed texts (including newspapers, letters, and brochures).

textual analysis Reading and constant reinterpretation of texts as a set of *signs* or signifying practices. (See also *discourse analysis, narrative analysis.*)

theme In *coding*, an important process, commonality, characteristic, or theory that emerges from the data and can be used to analyze and abstract the data.

theory (theory generation) The *inductive* process of identifying concepts to explain a phenomenon. One of the major advantages of qualitative research is that it facilitates the generation of new or revised theory. The theory may be as simple as a few loosely connected concepts or as complex as many tightly integrated concepts (for example, Marx's theory of capitalism).

theory building software Computer programs that deal with relationships between data categories to aid the researcher in developing higher-order conceptual insights and to formulate and assess propositions or assertions. Examples include ATLAS/ti, and NVivo.

thick description Term made common by anthropologist Clifford Geertz, involves not just describing an event, occurrence, or practice but the detailed context in which it occurs. This allows researchers to interpret the situated nature of the "event," reflecting on the ways in which the subject or object of study is constructed symbolically in relation to broader cultural and social relations and discourses.

third-person narrative In the context of a research narrative (e.g., research report, journal publication, public talk), involves constructing the narrative without reference to the researcher's thoughts, opinions, or feelings. This consequently conveys a distanced and seemingly neutral, omniscient, and "objective" point of view. Third-person narratives have dominated in the presentation of "scientific" research. (Compare with *first-person narrative.*)

transcript (transcription) Written record of speech (for example, interview, focus group proceedings, film dialogue). May also include textual description of informant gestures and tone.

transferability Extent to which the results of a study might apply to contexts other than that of the research study. (See also *generalizability.*)

transformative reflexivity Process through which a researcher and researched group reflect on their (mis)understandings and negotiate the meanings of the information generated together. The shared process has the potential to transform each person's own understandings.

transparency Clearly sharing details of data collection and analysis methods, including how participants were recruited/selected, choice of key analytical themes, etc.; a practice to ensure rigorous and trustworthy research, transparency also includes identifying potential weaknesses or gaps in research.

triangulation (triangulate) Use of multiple or mixed methods; drawing from and checking with multiple respondents, researchers, and data sources to confirm or corroborate results.

trope Figure of speech that allows writers or producers of other forms of *text* to say one thing but mean something else. May involve use of *metaphor* or metonymy. More broadly, refers to shared cultural meanings conveyed in words or images, such as motifs, clichés, or even memes.

trust Belief in the honesty of a research relationship, and mutual respect between parties of that relationship.

trustworthiness Principle in ethics and description applying to those who are given trust by others and who do not violate that faith; qualitative data and insights gained from them need to be trustworthy in order to be reliable and generalizable.

typical case sampling Selection of samples that illustrate or highlight what is considered typical or normal.

under-disclosure When participants in an interview or in focus groups provide very little information. (Contrast with *over-disclosure*.)

unstructured interview Interview in which there is no predetermined order to the issues addressed. The researcher phrases and raises questions in a manner appropriate to the informants' previous comment. The direction and vernacular of the interview is informant-driven. (Compare with *structured interview* and *semi-structured interview*.)

validity Truthfulness or accuracy of data compared with acceptable criteria; measures reflecting strong, transparent, and reliable research design and data collection practices. (See also *reliability*.)

vernacular Occurring in the location where it originated. Vernacular language is the language of a place.

video call Technologies for the reception and transmission of audio-video signals by users at different locations, for communication between people in real time (Skype, Zoom, FaceTime, etc). (See also *digital interviewing*.)

voice capture Voice capture software can be used in questionnaire survey work conducted by telephone or face-to-face mode and is particularly useful for recording responses to open-ended questions. Software packages are available to digitally record responses in the respondents' actual voice, capturing information about, for instance, the intensity, emotion, or intonation of the response. The responses can then be played back for electronic coding to produce numeric tables and charts for further analysis.

volunteered geographic information (VGI) Refers to personal data (location, photos, reviews, posts, and other interactions with social media) that is geo-tagged and thus provides geographical information, often in vast amounts and often not entirely voluntary because it is collected by default and in the background of many programs.

voyeurism Approach to data collection and the conduct of research wherein the researcher directs their analytical gaze at research subjects without the latter's explicit consent and/or without recognition of the power imbalances between the former and the latter. Voyeurism emplaces the researcher as the sole source of knowledge about research subjects, which constructs the former as passive objects to be studied rather than interlocutors in knowledge production.

warm-up Set of pre-interview techniques intended to enhance rapport between interviewer and *informant*. May include small talk, sharing food, or relaxed discussion of the research.

wicked problem Term deployed by Rittel and Webber (1973, p. 160) to mean challenges that were not "themselves ethically deplorable" but that were or could be malignant, viciously circuitous, or aggressive in effect. Rittel and Webber argued that wicked problems are simultaneously unique and the symptom of other problems. They have no "stopping rule," and solutions to them have neither immediate nor ultimate tests of success, there being no capacity to learn by trial-and-error "because every attempt counts."

writing-in Active and situated process of writing in which the author engages with the ways in which meanings are constructed through the creation of their text.

References

Chapter 1

Ash, J., Kitchin, R., Leszczynksi, A. (2016). Digital turn, digital geographies? *Progress in Human Geography, 42*(1), 25–43.

Baxter, J., & Eyles, J. (1997). Evaluating qualitative research in social geography: Establishing "rigour" in interview analysis. *Transactions of the Institute of British Geographers, 22*(4), 505–25.

Browne, K., & Nash, C. (2010). *Queer Methods and Methodologies, Intersecting Queer Theories and Social Science Research.* New York: Routledge.

Burt, E. & Atkinson, J. (2012). The relationship between quilting and wellbeing. *Journal of Public Health, 34*(1), 54–49.

Carr, C. & Gibson, C. (2017). Animating geographies of making: Embodied slow scholarship for participant-researchers of maker cultures and material work, *Geography Compass, 11*(6), 1–10.

de Leeuw, S. & Hawkins, H. (2017). Critical geographies and geography's creative re/turn: Poetics and practices for new disciplinary spaces. *Gender, Place & Culture, 24*(3), 303–24.

Deleuze, G. & Guattari, F. (1987). *A thousand plateaus* (Brian Massumi, Trans.). Minneapolis: University of Minnesota Press.

Eaves, L. (2017). Black geographic possibilities: On a queer black south. *Southeastern Geographer,* 57(1), 80–95.

Eshun, G., & Madge, C. (forthcoming). Poetic methods as creative writing. In von Benzon, N., Holton, M., Wilkinson, C., & Wilkinson, S. (Eds), *Creative methods for human geographers*. London: SAGE.

Goodman, A., Snyder, M., & Wilson, K. (2018). Exploring Indigenous youth perspectives of mobility and social relationships: A Photovoice approach. *The Canadian Geographer / Le Géographe canadien, 62*(3), 314–25.

Haraway, D. (1988). Situated knowledges: The science question in feminism and the privilege of partial perspective. *Feminist Studies, 14*(3), 575–99.

Hawkins, H. (2019). Geography's creative (re) turn: Toward a critical framework. *Progress in Human Geography, 43*(6), 963–84.

Hay, I. (2020) Qualitative data. In D. Richardson, N. Castree, M.F. Goodchild, A.L. Kobayashi, W. Liu, & R. Marston (Eds), *The international encyclopedia of geography: People, the Earth, environment and technology*. London: Elsevier.

Hofverberg, H., & Kronid, D. (2018). Human-material relationships in environmental sustainability education – an empirical study of a school embroidery project. *Environmental Education Research, 24*(7).

Johnson, J., & Larsen, S. (2013). (Eds) *A deeper sense of place: Stories and journeys of collaboration in Indigenous research.* Corvallis, OR: Oregon State University Press.

Leszczynski, A., & Elwood, S. (2015). Feminist geographies of new spatial media. *The Canadian Geographer / Le Géographe canadien, 59*(1), 12–28.

McKittrick, K., & Woods, C. (Eds) (2007). *Black geographies and the politics of place.* Cambridge, MA: South End Press.

McQuoid, J., Thrul, J., Ozer, E., Ramo, D., & Ling, P. (2019). Tobacco use in the sexual borderlands: The smoking contexts and practices of bisexual young adults. *Health & Place,* 58, 1–10.

Magrane, E. (2015). Situating geopoetics. *GeoHumanities, 1*(1), 86–102.

Moss, P. (Ed.). (2002). *Feminist geography in practice: Research and methods.* Oxford: Blackwell.

Palomino-Schalscha, M., Kindon, S., & Guiloff, K. (2017). Latin American women stitching memories and lives in Wellington, Aotearoa New Zealand. In A. Carrascosca, I. Momoitio Astorkia, and M. Oianguren Idigoras (Eds). *La embarcada artivista: Arterapia y activismo* (pp. 225–9). Bilbao: Fundacion Museo de la Paz de Guernika.

Patchett, M. (2014). Witnessing craft: Employing video ethnography to attend to the more-than-human craft practices of taxidermy. In C. Bates (Ed.), *Video methods: Social science research in motion.* New York: Routledge.

——— (2016). Taxidermy workshops: differently figuring the working of bodies and bodies at work in the past. *Transactions of the Institute of British Geographers, 42*(3), 390–404.

Poorthuis, A., Zook, M., Shelton, T., Graham, M., & Stephens, M. (2016). Using geotagged digital social data in geographic research. In N. Clifford, M. Cope, S. Gillespie, and S. French (Eds), *Key methods in human geography* (pp. 248–69). London: SAGE.

Price, L., & Hawkins, H. (2018). *Geographies of making, craft and creativity*. Oxon: Routledge.

Riley, J., Corkhill, B., & Morris, C. (2013). The benefits of knitting for personal and social wellbeing in adulthood: Findings from an international survey. *British Journal of Occupational Therapy*.

Sayer, A. (1992). *Method in social science: A realist approach* (2nd edition). London: Routledge.

______ (2010). *Method in social science: A realist approach* (revised 2nd edition). Milton Park and New York: Routledge.

Scoffham, S. (2016). *Teaching geography creatively*. Routledge, London.

Smith, L.T. (2013). *Decolonizing methodologies: Research and indigenous peoples* (2nd edition). London: Zed Books.

Stephens, M. (2013). Gender and the GeoWeb: Divisions in the production of user-generated cartographic information. *GeoJournal, 78*(6), 981–96.

Stephens, M., Poon, J.P.H., & Tan, G.K.S. (2021). *Misinformation in the digital age*. Cheltenham, UK: Edward Elgar.

Sui, D., Elwood, S., & Goodchild, M. (2013). *Crowdsourcing geographic knowledge: Volunteered geographic information (VGI) in theory and practice*. Dordrecht: Springer.

Tuck, E., & McKenzie, M. (2015). *Place in research: Theory, methodology, and methods*. New York: Routledge.

Valentine, G. (1996). Children should be seen and not heard: The production and transgression of adults' public space. *Urban Geography, 17*(3), 205–20.

Watson, A. & Till, K. (2010). Ethnography and participant observation. In D. DeLyser, S. Herbert, S. Aitken, M. Crang, and L. McDowell (Eds) *The SAGE handbook of qualitative geography* (pp. 121–37). London: SAGE.

Wilson, M. (2017). *New lines: Critical GIS and the trouble of the map*. Minneapolis: University of Minnesota Press.

Chapter 2

AAG Indigenous Peoples Specialty Group's Declaration of Key Questions about Research Ethics with Indigenous Communities. (n.d.). Available online: www.indigenousgeography.net/IPSG/pdf/IPSGResearchEthicsFinal.pdf

American Association of Geographers. (2009). Statement on Professional Ethics. www.aag.org/cs/about_aag/governance/statement_of_professional_ethics

Arnold, R., Burke, B., James, C., Martin, D., & Thomas, B. (1991). *Educating for a change*. Toronto: Between the Lines Press and the Doris Marshall Institute for Education and Action.

Bailey, C. (2001). Geographers doing household research: Intrusive research and moral accountability. *Area, 33*(1), 107–10.

Bain, A.L., & Nash, C.J. (2006). Undressing the researcher: Feminism, embodiment and sexuality at a queer bathhouse event. *Area, 38*(1), 99–106.

Butz, D. (2008). Sidelined by the guidelines: Reflections on the limitations of standard informed consent procedures for the conduct of ethical research. *ACME: An International E-Journal for Critical Geographies, 7*(2), 239–59.

Catungal, J.P. (2017). Feeling bodies of knowledge: Situating knowledge production through felt embeddedness. *Tijdschrift voor economische en sociale geografie, 108*(3), 289–301.

Clark, T. (2008). "We're over-researched here!" Exploring accounts of research fatigue within qualitative research engagements. *Sociology, 42*(5), 953–70.

Coram, S. (2011). Rethinking indigenous research approval—The perspective of a "stranger." *Qualitative Research Journal, 11*(2), 38–47.

Dawson, L., & Kass, N.E. (2005). Views of US researchers about informed consent in international collaborative research. *Social Science and Medicine, 61*(6), 1211–222.

Desmond, M. (2004). Methodological challenges posed in studying an elite in the field. *Area, 36*(3), 262–69.

Driver, F. (2001). *Geography militant: Cultures of exploration*. Oxford: Blackwell.

England, K.V.L. (1994). Getting personal: Reflexivity, positionality, and feminist research. *Professional Geographer 46*, 80–89.

Farrales, M. (2019). Repurposing beauty pageants: The colonial geographies of Filipina pageants in Canada. *Environment and Planning D: Society and Space, 37*(1), 46–64.

First Nations Information Governance Centre. (2019). https://fnigc.ca/

Gordon, E.J. (2000). When oral consent will do. *Field Methods, 12*(3), 235–8.

Gould, K.A. (2010). Anxiety, epistemology, and policy research "behind enemy lines". *Geoforum, 41*(1), 15–18.

Han, J.H.J. (2010). Neither friends nor foes: Thoughts on ethnographic distance. *Geoforum, 41*(1), 11–14.

Haraway, D. (1988). Situated knowledges: The science question in feminism and the privilege of partial perspective *Feminist Studies, 14*(3), 575–99.

Hay, I. (1998). Making moral imaginations: Research ethics, pedagogy, and professional human geography. *Ethics, Place & Environment, 1*(1), 55–75.

Hay, I. (2013). (Ed.). *Geographies of the super rich.* Cheltenham, UK: Edward Elgar.

Houston, S.D., Hyndman, J., McLean, J., & Jamal, A. (2010b). The methods and meanings of collaborative team research. *Qualitative Inquiry, 16*(4), 285–97.

Houston, S.D., McLean, D.J., Hyndman, J., & Jamal, A. (2010a). Still methodologically becoming: collaboration, feminist politics and "Team Ismaili". *Gender, Place & Culture, 17*(1), 61–79.

Israel, M., & Hay, I. (2006). *Research ethics for social scientists: Between ethical conduct and regulatory compliance.* London: SAGE.

Laliberté, N., & Schurr, C.L. (2016). Introduction: The stickiness of emotions in the field–complicating feminist methodologies. *Gender, Place & Culture, 23*(1), 72–78.

Ley, D. (2010). *Millionaire migrants: Trans-Pacific life lines.* Oxford: Wiley-Blackwell.

Longhurst, R., Ho, E., & Johnston, L. (2008). Using "the body" as an "instrument of research": Kimch'i and pavlova. *Area, 40*(2), 208–217.

McDowell. (1998). Illusions of power: Interviewing local elites. *Environment and Planning A,* 30, 2121–32.

Maguire, H., McCartan, A., Nash, C.J., & Browne, K. (2019). The enduring field: Exploring researcher emotions in covert research with antagonistic organisations. *Area, 51*(2), 299–306.

Moss, P. (1995). Reflections on the 'gap' as part of the politics of research design', *Antipode, 27*(1), 82–90.

Neufeld, S.D., Chapman, J., Crier, N., Marsh, S., McLeod, J., & Deane, L.A. (2019). Research 101: A process for developing local guidelines for ethical research in heavily researched communities. *Harm Reduction Journal, 16*, 41. https://harmreductionjournal.biomedcentral.com/articles/10.1186/s12954-019-0315-5

Nowicka, M., & Cieslik, A. (2014). Beyond methodological nationalism in insider research with migrants. *Migration Studies, 2*(1), 1–15.

Nuu-chah-nulth Tribal Council Research Ethics Committee (2008). Protocols and principles for conducting research in a Nuu-chah-nulth context. https://icwrn.uvic.ca/wp-content/uploads/2013/08/NTC-Protocols-and-Principles.pdf

O'Connell-Davidson, J., & Layder, D. (1994). *Methods, sex and madness.* London: Routledge.

O'Neill, P.M. (2001). Financial narratives of the modern corporation. *Journal of Economic Geography 1*, 181–99.

Oswin, N. (2019). Enough. In *Keywords in Radical Geography: Antipode at 50* (pp. 113–17). Wiley.

Pratt, G. (2004). *Working feminism.* Philadelphia: Temple University Press.

______ (2012). Families apart: Migrant mothers and the conflicts of labor and love. Minneapolis: University of Minnesota Press.

Pratt, G., in collaboration with Philippines-Canada Task Force on Human Rights. (2008). International accompaniment and witnessing state violence in the Philippines. *Antipode, 40*(5), 751–79.

Pratt, G., in collaboration with the Philippine Women Centre of BC. (2009). Circulating sadness: Witnessing Filipino mothers' stories of family separation. *Gender, Place & Culture, 16*, 3–22.

Pulido, L. (2002). Reflections on a white discipline. *The Professional Geographer, 54*(1), 42–49.

Rose, G. (1997). Situating knowledges: Positionality, reflexivities and other tactics. *Progress in Human Geography 21*, 305–20.

Routledge, P. (2002). Travelling east as Walter Kurtz: Identity, performance and collaboration in Goa, India. *Environment and Planning D: Society and Space, 20*(4), 477–96.

Sangtin Writers' Collective and Nagar, R. (2006) *Playing with Fire. Feminist Thought and Activism through Seven Lives in India.* Minneapolis: Minnesota University Press.

Smith, L.T. (2013). *Decolonizing methodologies: Research and indigenous peoples* (2nd edition). London: Zed Books.

Sundberg, J. (2003). Masculinist epistemologies and the politics of fieldwork in Latin Americanist geography. *Professional Geographer,* 55, 181–91.

Tauri, J.M. (2014). Resisting condescending research ethics in Aotearoa New Zealand. *AlterNative: an international Journal of Indigenous Peoples*, *10*(2), 134–50.

Tri-Council Policy Statement: Ethical Conduct for Research Involving Humans (2018). www.pre.ethics.gc.ca/eng/policy-politique_tcps2-eptc2_2018.html

Valentine, G. (1993). (Hetero)sexing space: Lesbian perceptions and experiences of everyday spaces. *Environment and Planning D: Society and Space, 11*(4), 395–413.

Wilton, R. (2000). Sometimes it's OK to be a spy: Ethics and politics in geographies of disability. *Philosophy and Geography*, 1: 91–97.

Chapter 3

Arondekar, Anjali. (2005). Without a trace: Sexuality and the colonial archive. *Journal of the History of Sexuality, 14*(1), 10–27.

Berry, Maya J., Chávez Argüelles, C., Cordis, S., Ihmoud, S., & Velásquez Estrada, E. (2017). Toward a fugitive anthropology: Gender, race, and violence in the field. *Cultural Anthropology, 32*(4), 537–65.

Bujra, J. (2006). Lost in translation? The use of interpreters in fieldwork. In V. Desai, & R. Potter (Eds). *Doing development research* (pp. 172–79). London: SAGE.

Cahill, C. (2007). Doing research with young people: Participatory research and the rituals of collective work. *Children's Geographies, 5*(3), 297–312.

Carpiano, R. (2009). Come take a walk with me: The "Go-Along" interview as a novel method for studying the implications of place for health and well-being. *Health & Place, 15*, 263–72.

Castillo, R., & Hernández, A. (2016). *Multiple InJustices: Indigenous women, law, and political struggle in Latin America*. University of Arizona Press.

Cottom, T. M. (2018). *Thick: And other essays*. New York: The New Press.

Crenshaw, K. (1991). Mapping the margins: Intersectionality, identity politics, and violence against women of color. *Stanford Law Review*, 1241–99.

Daigle, M. (2016). Awawanenitakik: The spatial politics of recognition and relational geographies of Indigenous self-determination. *The Canadian Geographer/Le Géographe canadien, 60*(2), 259–69.

——— (2018). Resurging through Kishiichiwan: The spatial politics of Indigenous water relations. *Decolonization: Indigeneity, Education & Society, 7*(1), 159–72.

Daigle, M. & Ramírez, M.M. (2019). Decolonial geographies. In N. Theodore, T. Jazeel, A. Kent, & K. McKittrick (Eds), *Keywords in radical geography: Antipode at 50* (pp. 1–13). London: Wiley.

Elwood, S. (2008). Volunteered geographic information: Future research directions motivated by critical, participatory, and feminist GIS. *GeoJournal, 72*(3–4), 173–83.

Fanon, F. (2008). [1952] *Black skin, white masks*. New York: Grove press.

Faria, C., & Mollett, S. (2016). Critical feminist reflexivity and the politics of whiteness in the "field". *Gender, Place & Culture, 23*(1), 79–93.

Howitt, R. & Stevens, S. (2005). Cross-cultural research: Ethics, methods and relationships. In I. Hay (Ed.), *Qualitative research methods in human geography*, 4th edition, (pp. 30–50). Don Mills, ON: Oxford University Press - Canada.

Kohl, E. & McCutcheon, P. (2015). Kitchen table reflexivity: Negotiating positionality through everyday talk. *Gender, Place & Culture, 22*(6), 747–63.

Kusenbach, M. (2003). Street phenomenlogy: The go-along as ethnographic research tool. *Ethnography, 4*, 455–85.

Lopez, P. J., & Gillespie, K. (2016). A love story: For "buddy system" research in the academy. *Gender, Place & Culture, 23*(12), 1689–700.

Lowe, L. (2015). *The intimacies of four continents*. Durham, NC: Duke University Press.

Lugones, M. (2010). Toward a decolonial feminism." *Hypatia, 25*(4), 742–59.

Matthew, P. A. (2016). *Written/unwritten: Diversity and the hidden truths of tenure*. Chapel Hill, NC: UNC Press Books.

McDuie-Ra, D. (2012). *Northeast migrants in Delhi: Race, refuge, and retail*. Amsterdam: Amsterdam University Press.

Mawani, R. (2012). Law's archive. *Annual Review of Law and Social Science*, 8. 337–65.

Mills, A.J., Durepos, G., & Wiebe, E. (2010). *Encyclopedia of case study research*. Vol. 1. Thousand Oaks, CA: SAGE.

Mohammad, R. (2001). "Insiders" and/or "outsiders": Positionality, theory and praxis. In M. Limb, & C. Dwyer (Eds), *Qualitative methodologies for geographers: Issues and debates* (pp. 101–17). New York: Oxford University Press.

Müller, M. (2007). What's in a word? Problematizing translation between languages. *Area, 39*(2), 206–13.

Nagar, R. (2019). Hungry translations: The world through radical vulnerability: The 2017 Antipode RGS-IBG Lecture. *Antipode,* 51(1), 3–24.

Nagar, R., & Geiger, S. (2007). Reflexivity, positionality and identity in feminist fieldwork revisited. In *Politics and practice in economic geography* (pp. 267–78). SAGE.

Nagar, R., & Shirazi, R. (2019). Radical vulnerability. In N. Theodore, T. Jazeel, A. Kent, & K. McKittrick (Eds) *Keywords in radical geography: Antipode at 50* (pp. 236–42). London: Wiley.

Naylor, L., Daigle, M., Zaragocin, S., Ramírez, M., & Gilmartin, M. (2018). Interventions: bringing the decolonial to political geography. *Political Geography*, *66*, 199–209.

Parker, P., Holland, D., Dennison, J., Smith, S.H., & Jackson, M. (2017). Decolonizing the academy: Lessons from the graduate certificate in participatory research at the University of North Carolina at Chapel Hill. *Qualitative Inquiry, 24*(7), 464–77.

Proudfoot, J. (2015). Anxiety and phantasy in the field: The position of the unconscious in ethnographic research. *Environment and Planning D: Society and Space, 33*(6), 1135–52.

Puwar, N. (2004). *Space invaders: Race, gender and bodies out of place.* Oxford: Berg.

Quijada Cerecer, D.A., Cahill, C., & Bradley, M. (2013). Toward a critical youth policy praxis: Critical youth Studies and Participatory Action Research. *Theory Into Practice, 52*(3), 216–23.

Ramírez, M.M. (2019). City as borderland: Gentrification and the policing of Black and Latinx geographies in Oakland. *Environment and Planning D: Society and Space.* https://doi.org/10.1177/0263775819843924

——— (2020). Take the houses back/take the land back: Black and Indigenous urban futures in Oakland. *Urban Geography, 41*(5), 682–693.

Rose, G. (1997). Situating knowledges: Positionality, reflexivities and other tactics. *Progress in Human Geography, 21*(3), 305–20.

Said, E. (1979). *Orientalism.* New York: Vintage.

Shah, P. (2015). Spaces to speak: Photovoice and the reimagination of girls' education in India. *Comparative Education Review, 59*(1), 50–74.

Sharp, J., & Dowler, L. (2011). Framing the field. In V. J. Del Casino Jr, M.E. Thomas, P. Cloke, & R. Panelli (Eds), *A companion to social geography* (pp. 146–60). Malden, MA: Blackwell Publishing Ltd.

Sharpe, C. (2009). *Monstrous intimacies: Making post-slavery subjects.* Durham, NC: Duke University Press.

Simpson, A. (2007). On ethnographic refusal: Indigeneity, "voice" and colonial citizenship. *Junctures: The Journal for Thematic Dialogue,* no. 9.

——— (2014). *Mohawk interruptus: Political life across the borders of settler states.* Durham, NC: Duke University Press.

Smith, F.M. (1996). Problematising language: Limitations and possibilities in "foreign language" research. *Area,* 160–66.

Smith, F. (2016). Working in different cultures. In N. Clifford, M. Cope, T. Gillespie, & S. French (Eds), *Key methods in geography* (pp. 157–72). Thousand Oaks, CA: SAGE.

Smith, L.T. (1999). *Decolonizing methodologies: Research and Indigenous peoples.* Dunedin and London: University of Otago Press and Zed Books.

Smith, S. (2013). "In the heart, there's nothing": Unruly youth, generational vertigo and territory. *Transactions of the Institute of British Geographers, 38*(4), 272–85.

Sparke, M. (1996). Displacing the field in fieldwork: Masculinity, metaphor and space. In N. Duncan (Ed.), *Bodyspace: Destabilizing geographies of gender and sexuality* (pp. 212–33). London: Routledge.

Stoler, A.L. (1997). Carnal knowledge and imperial power: Gender, race, and morality in colonial Asia. In R. Lancaster and M. di Leonardo (Eds), *The gender/sexuality reader: Culture, history, political economy* (pp. 13–36). New York: Routledge.

——— (2006). *Haunted by empire: Geographies of intimacy in North American history.* Durham, NC: Duke University Press.

Sultana, F. (2007). Reflexivity, positionality and participatory ethics: Negotiating fieldwork dilemmas in international research. *ACME: An International E-journal for Critical Geographies, 6*(3), 374–85.

Sundberg, J. (2014). Decolonizing posthumanist geographies. *Cultural Geographies,* 21, 33–47.

TallBear, K. (2014). Standing with and speaking as faith: A feminist-Indigenous approach to inquiry. *Journal of Research Practice, 10*(2), 17.

Tuck, E. & Yang, K.W. (2012) Decolonization is not a metaphor. *Decolonization: Indigeneity, Education & Society 1*(1).

Valentine, G., Butler, R., & Skelton, T. (2001). The ethical and methodological complexities of doing research with "vulnerable" young people. *Ethics, Place & Environment, 4*(2), 119–25.

van Beek, M. (1999). Hill councils, development, and democracy: Assumptions and experiences from Ladakh. *Alternatives*, 24: 435–60.

Vasudevan, P. (2019). An intimate inventory of race and waste. *Antipode*. https://doi.org/10.1111/anti.12501

______ (2021). Brown Scholar, Black Studies: On Suffering, Witness, and Materialist Relationality. In B. Gökarıksel, M. Hawkins, C. Neubert, & S. Smith (Eds), *Feminist geography unbound: Discomfort, bodies, and prefigured futures*. Morgantown, WV: West Virginia University Press.

Vasudevan, P., & W.A. Kearney. (2016). Remembering Kearneytown: Race, place and collective memory in collaborative filmmaking. *Area, 48*(4), 455–62.

Wang, C., & M.A. Burris. (1994). Empowerment through photo novella: Portraits of participation. *Health Education and Behavior, 21*(2), 171–86.

Whitson, R. (2017). Painting pictures of ourselves: Researcher subjectivity in the practice of feminist reflexivity. *The Professional Geographer, 69*(2), 299–306.

Chapter 4

ACME: An International E-Journal for Critical Geographies. (2003). 2 (1: special issue on practices in feminist research).

Ahmed, S. (2004). *The cultural politics of emotion.* Edinburgh: Edinburgh University Press.

American Indian Culture and Research Journal. (2008). 32 (3: special issue on Mainstreaming Indigenous Geography).

Ansell, N. (2014). Challenging empowerment: AIDS-affected southern African children and the need for a multi-level relational approach. *Journal of Health Psychology, 19*(1), 22–33.

Antipode. (1995). 27: (pp. 71–101: special issue on Discussion and Debate: Symposium on Feminist Participatory Research).

Attanapola, C., Brun, C., & Lund, R. (2013). Working gender after crisis: Partnerships and disconnections in Sri Lanka after the Indian Ocean tsunami. *Gender, Place & Culture, 20*(1), 70–86.

Bailey, M.M., & Shabazz, R. (2013). Gender and sexual geographies of Blackness: New Black cartographies of resistance and survival (part 2). *Gender, Place & Culture (21)*4, 449–52.

Bee, B. (2013). Who reaps what is sown? A feminist inquiry into climate change adaptation in two Mexican *ejidos*. *ACME*: An International E-Journal for Critical Geographies, 12, 131–54.

Billo, E. & Hiemstra, N. (2013). Mediating messiness: Expanding ideas of flexibility, reflexivity, and embodiment in fieldwork. *Gender, Place & Culture 20*(3), 313–28.

Bishop, R. (2005). Freeing ourselves from neocolonial domination in research: A Kaupapa Māori approach to creating knowledge. In N. Denzin & Y. Lincoln (Eds), *The SAGE handbook of qualitative research* (3rd edition) (pp. 109–38). Thousand Oaks, CA: SAGE.

Blanch, F. (2013). Encountering the Other: One Indigenous Australian woman's experience of racialisation on a Saturday night. *Gender, Place & Culture,* 20, 253–60.

Bondi, L. (2003). Empathy and identification: Conceptual resources for feminist fieldwork. *ACME: An International E-journal for Critical Geographies,* 2, 64–76.

Bondi, L., Avis, H., Bankey, R., Bingley, A., Davidson, J., Duffy, R., Einagel, V.I., Green, A-M., Johnston, L., Lilley, S., Listerborn, C., Marshy, M., McEwan, S., O'Connor, N., Rose, G., Vivat, B., & Wood, N. (2002). *Subjectivities, knowledges and feminist geographies: The subjects and ethics of social research.* Lanham, MD: Rowman and Littlefield.

Buang, A., & Momsen, J. (Eds). (2013). *Women and empowerment in the global south.* Bangi, Selangor, Malaysia: University of Malaysia Press.

Cameron, E. (2012). New geographies of story and storytelling. *Progress in Human Geography* 36, 573–92.

Canadian Geographer. (1993). Special issue on Feminism as Method. *37*(1) (Spring).

Canadian Geographer. (2012). 56 (2: special issue on Community-Based Participatory Research Involving Indigenous Peoples in Canadian Geography: Progress?)

Chacko, E. (2004). Positionality and praxis: Fieldwork experiences in rural India. *Singapore Journal of Tropical Geography,* 25: 51–63.

Chakrabarty, D. (2000). *Provincializing Europe: Postcolonial thought and historical difference.* Princeton: Princeton University Press.

Chattopadhyay, S. (2013). Getting personal while narrating the "field": A researcher's journey to the villages of the Narmada Valley. *Gender, Place & Culture, 20*(2), 137–59.

Christensen, J. (2012). Telling stories: exploring research storytelling as a meaningful approach to knowledge mobilization with Indigenous research collaborators and diverse audiences in community-based participatory research. *The Canadian Geographer/Le Géographe canadien,* 56: 231–42.

Coddington, K. 2017. Voice under scrutiny: Feminist methods, anticolonial responses, and new methodological tools. *The Professional Geographer*, *69*(2), 314–20.

Coombes, B. 2012. Collaboration: Inter-subjectivity or radical pedagogy?. *The Canadian Geographer / Le Géographe canadien* 56: 290–1.

Cultural Geographies. (2009). Special issue on Indigenous Cartographies. *16*(2) (April).

Daigle, M. (2019). The spectacle of reconciliation: On (the) unsettling responsibilities to Indigenous peoples in the academy. *Environment and Planning D: Society and Space*, 0263775818824342.

Daigle, M., & J. Sundberg. (2017). From where we stand: Unsettling geographical knowledges in the classroom. *Transactions of the Institute of British Geographers*, *42*(3), 338–41.

Declaration on the Rights of Indigenous Peoples. (2007). www.un.org/esa/socdev/unpfii/documents/DRIPS_en.pdf

de Leeuw, S., E. Cameron, & M.L. Greenwood. (2012). Participatory and community-based research, Indigenous geographies, and the spaces of friendship: A critical engagement. *The Canadian Geographer/Le Géographe canadien,*. *56*(2), 180–94.

DeLyser, D., & D. Sui. (2013). Crossing the qualitative-quantitative chasm III: enduring methods, open geography, participatory research and the forth paradigm. *Progress in Human Geography*, DOI: 10.1177/0309132513479291

Desai, V. (2017). Black and Minority Ethnic (BME) student and staff in contemporary British Geography. *Area*, *49*(3), 320–23.

Dominey-Howes, D., Gorman-Murray, A., & McKinnon, S. (2013). Queering disasters: On the need to account for LGBTI experiences in natural disaster contexts. *Gender, Place & Culture* DOI:10.1080/0966369X.802673

England, K. (2006). Producing feminist geographies: Theory, methodologies and research strategies. In S. Aitkin & G. Valentine (Eds), *Approaches to human geography* (pp. 286–97). London: SAGE.

Escobar, A. (1995). *Encountering development: The making and unmaking of the third world.* Princeton, NJ: Princeton University Press.

Eshun, G., & Madge, C. (2012). "Now let me share this with you": Exploring poetry for postcolonial geography research. *Antipode, 44*(4), 1395–428.

______ (2016). Poetic world-writing in a pluriversal world: A provocation to the creative (re)turn in geography. *Social & Cultural Geography, 17*(6), 778–85.

Esson, J. (2019). "The why and the white": Racism and curriculum reform in British geography. *Area*.

Evans, M., Hole, R., Berg, L.D., Hutchinson, P., & Sookraj, D. (2009). Common insights, differing methodologies toward a fusion of Indigenous methodologies, participatory action research, and White studies in an urban Aboriginal research agenda. *Qualitative Inquiry, 15*(5), 893–910.

Faria, C. (2013). Staging a new South Sudan in the USA: Men, masculinities and nationalist performance at a disaporic beauty pageant. *Gender, Place & Culture,* 20, 87–106.

Faria, C., Falola, B., Henderson, J., & Maria Torres, R. (2019). A long way to go: Collective paths to racial justice in geography. *The Professional Geographer*, 1–13.

Faria, C., & Mollett, S. (2016). Critical feminist reflexivity and the politics of whiteness in the "field". *Gender, Place & Culture*, *23*(1), 79–93.

Freire, P. (2000). *Pedagogy of the oppressed*. New York: Continuum.

Garvin, T., & Wilson, K. (1999). The use of storytelling for understanding women's desires to tan: Lessons from the field. *The Professional Geographer,* 51, 297–306.

Gender, Place & Culture. (2002). 9 (3: special issue on Feminist Geography and GIS).

Geografiska Annaler: Series B: Human Geography. (2006). 88 (3: special issue on Encountering Indigeneity: Re-Imagining and Decolonizing Geography).

Geographical Research. (2007). 45 (2: special issue on Creating Anti-Colonial Geographies: Embracing Indigenous Peoples' Knowledges and Rights).

Gibson-Graham, J.K. (2006). *A postcapitalist politics*. Minneapolis: University of Minnesota Press.

Godlewska, A. & Smith, N. (Eds). (1994). *Geography and empire*. Oxford: Blackwell.

Goeman, M. (2013). *Mark my words: Native women mapping our nation*. Minneapolis: University of Minnesota Press.

Harding, S. (2011). Other cultures' sciences. In S. Harding (Ed.), *The postcolonial science and technologies study reader* (pp. 151–58). Durham and London: Duke University Press.

Harris, L., & J. Wasilewski. (2004). Indigeneity, an alternative worldview: Four R's (relationship, responsibility, reciprocity, redistribution) vs. two P's (power and profit). Sharing the journey towards conscious evolution. *Systems Research and Behavioral Science, 21*(5), 489–503.

He, C. (2013). Performance and the politics of gender: Transgender performance in contemporary Chinese films. *Gender, Place & Culture* DOI:10.1080/0966369X.2013.810595.

Howitt, R. (2019). Unsettling the taken(-for-granted). *Progress in Human Geography, 44*(2), 193–215.

Hunt, S. (2014). Ontologies of Indigeneity: The politics of embodying a concept. *Cultural Geographies,* 21, 27–32.

Jazeel, T. (2017). Mainstreaming geography's decolonial imperative. *Transactions of the Institute of British Geographers, 42*(3), 334–7.

Jenkins, S., Jones, V., & Dixon, D. (2003). Thinking/doing the "F" word: On power in feminist methodologies. ACME*: An International E-Journal for Critical Geographies, 2*(1), 57–63.

Johnson, J.T. (2012). Place-based learning and knowing: Critical pedagogies grounded in indigeneity. *GeoJournal,* 77, 829–36.

Jones, J.P. III, Nast, H.J., & Roberts, S.M. (Eds). (1997). *Thresholds in feminist geography: Difference, methodology, representation*. Lanham, MD: Rowman and Littlefield.

Kovach, M. (2009). *Indigenous methodologies: Characteristics, conversations and contexts.* Toronto: University of Toronto Press.

Kuhn, T.S. (1962). *The structure of scientific revolutions.* Chicago: University of Chicago Press.

Lahiri-Dutt, K. (2013). Bodies in/out of place, masculinities and motherhood of *Kamins* in Indian coal mines. *South Asian History and Culture,* 4, 213–29.

Legg, S. (2017). Decolonialism. *Transactions of the Institute of British Geographers, 42*(3), 345–48.

Louis, R.P. (2007). Can you hear us now? Voices from the margin: Using Indigenous methodologies in geographic research. *Geographical Research,* 45, 130–9.

McLean, J., & Maalsen, S. (2013). Destroying the joint and dying of shame? A geography of revitalised feminism in social media and beyond. *Geographical Research,* 51, 243–56.

Moss, P. (2002a). Taking on, thinking about and doing feminist research in geography. In P. Moss (Ed.), *Feminist geography in practice: Research and methods* (pp. 1–17). Oxford: Blackwell.

______ (2005). A bodily notion of research: power, difference and specificity in feminist methodology. In L. Nelson & J. Seager (Eds), *A companion to feminist geography* (pp. 41–59). Oxford: Blackwell.

Moss, P. (Ed.). (2002b). *Feminist geography in practice: Research and methods.* Oxford: Blackwell.

Moss, P. & Falconer Al-Hindi, K. (2010). Rhizomatic encounters and encountering possibilities. *Thirdspace: Journal of Feminist Theory and Culture*, 9, 1 www.thirdspace.ca/journal/article/viewArticle/385/295 (accessed July 2013).

Mullings, B., & Mukherjee, S. (2019). Reflections on mentoring as decolonial, transnational, feminist praxis. *Gender, Place & Culture*, 1–18.

Mulrennan, M.E., Mark, R., & Scott, C.H. (2012). Revamping community-based conservation through participatory research. *The Canadian Geographer/Le Géographe canadien, 56*(2), 243–59.

Nagar, R. (2013). Storytelling and co-authorship in feminist alliance work: Reflections from a journey. *Gender, Place & Culture, 20*(1), 1–18.

Naylor, L., Daigle, M., Zaragocin, S., Ramírez, M., & Gilmartin, M. (2018). Interventions: Bringing the decolonial to political geography. *Political Geography, 66,* 199–209.

Nicholls, R. (2010). Research and Indigenous participation: Critical reflexive methods. In A. Possamai-Inesedy & G. Gwyther (Eds), *New methods in social justice research for the twenty-first century* (pp. 18–27). London: Routledge.

Niezen, R. (2003). *The origins of Indigenism: Human rights and the politics of identity*. University of California Press.

Noxolo, P. (2017a). Introduction: Decolonising geographical knowledge in a colonised and re-colonising postcolonial world. *Area, 49*(3), 317–19.

______ (2017b). Decolonial theory in a time of the re-colonisation of UK research. *Transactions of the Institute of British Geographers, 42*(3), 342–44.

Noxolo, P., Madge, C., & Raghuram, P. (2012). Unsettling responsibility: Postcolonial interventions. *Transactions of the Institute of British Geographers,* 373, 418–29.

Parr, H. (2001). Feeling, reading, and making bodies in space. *Geographical Review,* 9, 158–67.

Professional Geographer. (1994). 46 (1: special issue on Women in the Field).

Professional Geographer. (1995). 47 (4: special issue on Should Women Count?).

Radcliffe, S.A. (2017a). Decolonising geographical knowledges. *Transactions of the Institute of British Geographers, 42*(3), 329–33.

______ (2017b). Geography and Indigeneity I: Indigeneity, coloniality and knowledge. *Progress in Human Geography, 41*(2), 220–29.

Raghuram, P., & Madge, C. (2006). Towards a method for postcolonial development geography: Possibilities and challenges. *Singapore Journal of Tropical Geography, 27*(3), 270–88.

______ (2007). Feminist theorising as practice. In A-H.K. Falconer & P. Moss (Eds), *Feminisms in geography: Space, place and environment* (pp. 221–9). Lanham: Rowman and Littlefield.

Raju, S. (2005). Gender and empowerment: Creating "thus far and no further" supportive structures. A Case from India. In L. Nelson & J. Seager (Eds), *A companion to feminist geography* (pp. 194–208). Oxford: Blackwell.

Rose, D.B. (1999). Indigenous ecologies and an ethic of connection. In N. Low (Ed.), *Global ethics and environment.* London: Routledge.

Rosenberg, R. (2017). When temporal orbits collide: Embodied trans temporalities in US prisons. *Somatechnics, 7*(1), 74–94.

Scheyvens, R. (2009). Empowerment. In R. Kitchin & N. Thrift (Eds), *International encyclopedia of human geography* (pp. 464–70). Elsevier: Oxford.

Selen, E. (2012). The stage: A space for queer subjectification in contemporary Turkey. *Gender, Place & Culture, 19*(6), 730–49.

Sharp, J. (2005). Geography and gender: feminist methodologies in collaboration and in the field. *Progress in Human Geography, 29*(3), 304–09.

Sharp, J. (2009). Geography and gender: What belongs to feminist geography? Emotion, power and change. *Progress in Human Geography, 33*(1), 74–80.

Sharp, J., Browne, K., & Thien, D. (2004). Introduction. In *Women and Geography Study Group of the RGS-IBG, Geography and Gender reconsidered* 1 (4). London: WGSG.

Smith, L.T. (1999). *Decolonizing methodologies: Research and Indigenous peoples.* Dunedin and London: University of Otago Press and Zed Books.

______ (2005). On tricky ground: Researching the native in the age of uncertainty. In N.K. Denzin & Y.S. Lincoln (Eds), *The SAGE handbook of qualitative research* (pp. 85–107). Thousand Oaks: SAGE Publications.

______ (2013). *Decolonizing methodologies: Research and Indigenous peoples* (2nd edition). Dunedin and London: University of Otago Press and Zed Books.

Sullivan, C. (2018). Majesty in the city: Experiences of an Aboriginal transgender sex worker in Sydney, Australia. *Gender, Place & Culture, 25*(12), 1681–702.

Sundberg, J. (2003). Masculinist epistemologies and the politics of fieldwork in Latin Americanist Geography. *Professional Geographer* 55, 181–91.

______ (2014). Decolonizing posthumanist geographies. *Cultural Geographies* 21, 33–47.

Tolia-Kelly, D.P. (2017). A day in the life of a Geographer: "Lone, black, female". *Area, 49*(3), 324–28.

Tuck, E., & McKenzie, M. (2015). *Place in research: Theory, methodology, and methods.* New York: Routledge.

United Nations Permanent Forum on Indigenous Issues. (2000).

Valentine, G., Jackson, L., & Mayblin, L. (2014) Ways of Seeing: Sexism the Forgotten Prejudice? *Gender, Place & Culture, 21*(4), 401–14.

von Humboldt, A. (1811). *Political essay on the Kingdom of New Spain containing researches relative to the military defense of New Spain: with physical sections and maps.* New York: I. Riley.

Wijnendaele, B.V. (2014). The politics of emotion in participatory processes of empowerment and change. *Antipode* 46, 266–82.

Wilson, S. (2008). *Research is ceremony: Indigenous research methods.* Black Point, NS: Fernwood Pub.

Women and Geography Study Group (Eds). (1997). *Feminist geographies: Diversity and difference.* London: Longmans.

Wrigley-Asante, Charlotte. (2012). Out of the dark but not out of the cage: Women's empowerment and gender relations in the Dangme West district of Ghana. *Gender, Place & Culture 19*(3), 344–63.

Zanotti, L. (2013). Resistance and the politics of negotiation: Women, place and space among the Kayapó in Amazonia, Brazil. *Gender, Place & Culture, 20*(3), 346–62.

Chapter 5

Monk J. & Bedford R. (2016). Writing a compelling research proposal. In I. Hay (Ed.), *Qualitative research methods in human geography* (4th edition). Don Mills, ON: Oxford University Press.

Pulido L. (1996). A critical review of the methodology of environmental racism research. *Antipode,* 282, 142–59.

Ramírez, M.M. (2019). City as borderland: Gentrification and the policing of Black and Latinx geographies in Oakland. *Environment and Planning D: Society and Space.* https://doi.org/10.1177/0263775819843924

Rice, J. (2016). The everyday choices we make matter: Urban climate politics and the postpolitics of responsibility and action. In H. Bulkeley, M. Paterson, & J. Stripple (Eds), *Towards a cultural politics of climate change.* Cambridge, UK: Cambridge University Press.

Rosenberg, R. (2017). When temporal orbits collide: Embodied trans temporalities in US prisons. *Somatechnics, 7*(1), 74–94.

Spiegel, S. (2020). Visual storytelling and socioenvironmental change: Images, photographic encounters, and knowledge construction in resource frontiers. *Annals of the American Association of Geographers, 110*(1), 120–44. DOI: 10.1080/24694452.2019.1613953

Chapter 6

Al-Faham, H., Davis, A.M., & Ernst, R. (2019). Intersectionality: From theory to practice. *Annual Review of Law and Social Science,* 15(Oct), 247–65. https://doi.org/10.1146/annurev-lawsocsci-101518-042942

Anfara, V.A., Brown, K.M., & Mangione, T.L. (2002). Qualitative analysis on stage: Making the research process more public. *Educational Researcher, 31*(7), 28–38.

Anguera, M.T., Blanco-Villaseñor, A., Losada, J.L., Sánchez-Algarra, P., & Onwuegbuzie, A.J. (2018). Revisiting the difference between mixed methods and multimethods: Is it all in the name? *Quality & Quantity, 52*(6), 2757–70.

Bailey, C., White, C., & Pain, R. (1999a). Evaluating qualitative research: Dealing with the tension between "science" and "creativity". *Area, 31*(2), 169–83.

——— (1999b). 'Response'. *Area, 31*(2), 183–4.

Barbour, R.S. (2001). Checklists for improving rigour in qualitative research: A case of the tail wagging the dog? *BMJ, 322*(7294), 1115–17.

Bastian, M. (2017). Towards a more-than-human participatory research. In M. Bastian, O. Jones, N. Moore, E. Roe (Eds), *Participatory research in more-than-human worlds* (pp. 19–37). London: Routledge.

Baxter, J., & J. Eyles. (1997). Evaluating qualitative research in social geography: Establishing "rigour" in interview analysis. *Transactions of the Institute of British Geographers, 22*(4), 505–25.

——— (1999a). The utility of in-depth interviews for studying the meaning of environmental risk. *Professional Geographer, 51*(2), 307–20.

——— (1999b). Prescription for research practice? Grounded theory in qualitative evaluation. *Area, 31*(2), 179–81.

Bradshaw, M. (2001). Contracts and member checks in qualitative research in human geography: Reason for caution? *Area, 33*(2), 202–11.

Brunner, C. (2011). Nice-looking obstacles: Parkour as urban practice of deterritorialization. *AI & Society, 26*(2), 143–52.

Butler, R. (1997). Stories and experiments in social inquiry. *Organisation Studies, 18*(6), 927–48.

Callon, M. (1986). Some elements of a sociology of translation: Domestication of the scallops and the fishermen of St Brieuc Bay. In J. Law (Ed.), *Power, action and belief: A new sociology of knowledge?* London: Routledge.

Carr, J. (2010). Legal geographies—skating around the edges of the law: Urban skateboarding and the role of law in determining young peoples' place in the city. *Urban Geography, 31*(7), 988–1003.

Chow, B.D.V. (2010). Parkour and the critique of ideology: Turn-vaulting the fortresses of the city. *Journal of Dance and Somatic Practices, 2*(2), 143–54.

Clifford, N., Cope, M., Gillespie, S., & French, S. (Eds) (2016). *Key methods in human geography*. Thousand Oaks, London, New Delhi: SAGE Publications.

Clough, A.R., et al. (2004). Emerging patterns of cannabis and other substance use in Aboriginal communities in Arnhem Land, Northern Territory: A study of two communities. *Drug and Alcohol Review, 23*(4), 381–90.

Denzin, N.K. (1978). *The research act: A theoretical introduction to sociological methods*. New York: McGraw-Hill.

Dumas, A., & Laforest, S. (2009). Skateparks as a health-resource: Are they as dangerous as they look? *Leisure Studies, 28*(1), 19–34.

Fallon, L. (2006). The Role of State and Non-State Actors in the Management of the Patagonian toothfish (*Dissostichus eleginoides*). Hobart: Unpublished PhD Thesis, School of Geography and Environmental Studies, University of Tasmania.

Fielding, N. (1999). The norm and the text: Denzin and Lincoln's handbooks of qualitative method. *British Journal of Sociology, 50*(3) 525–34.

Filep, C.V., Turner, S., Eidse, N., Thompson-Fawcett, M., & Fitzsimons, S. (2018). Advancing rigour in solicited diary research. *Qualitative Research, 18*(4), 451–70.

Fish, S. (1980). *Is there a text in this class? The authority of interpretive communities*. London: Harvard University Press.

Flick, U. (2018). *Designing qualitative research*. Thousand Oaks, London, New Delhi: SAGE Publications.

Flowerdew, R. (2005). *Methods in human geography: A guide for students doing a research project* (2nd edition). Harlow: Pearson.

Flyvberg, B. (1998). *Rationality and power: Democracy in practice* (S. Sampson, Trans.). Chicago: University of Chicago Press.

——— 2006. Five misunderstandings about case-study research. *Qualitative Inquiry, 12*(2), 219–45.

Forero, R., Nahidi, S., De Costa, J., Mohsin, M., Fitzgerald, G., Gibson, N., McCarthy, S., & Aboagye-Sarfo, P. (2018). Application of four-dimension criteria to assess rigour of qualitative research in emergency medicine. *BMC Health Services Research, 18*(1), 120–31.

Geertz, C. (1973). *The interpretation of culture: Selected essays*. New York: Basic Books.

Golafshani, N. (2003). Understanding reliability and validity in qualitative research. *The Qualitative Report, 8*(4), 597–607.

Gould, P. (1988). Expose yourself to geographic research. In J. Eyles (Ed.), *Research in human geography*. Oxford: Blackwell.

Herod, A. (1993). Gender issues in the use of interviewing as a research method. *Professional Geographer, 45*(3), 305–17.

Hopkins, P. (2017). Social geography I: Intersectionality. *Progress in Human Geography*, 10.1177/0309132517743677.

Jacobs, J.M. (1999). The labour of cultural geography. In E. Stratford (Ed.), *Australian cultural geographies*. Melbourne: Oxford University Press.

Janesick, V.J. (2000). The choreography of qualitative research design: Minuets, improvisations, and crystallization. In N.K. Denzin & Y.S. Lincoln (Eds), *Handbook of qualitative research*, (2nd edition) (pp. 379–99). Thousand Oaks, London, New Delhi: SAGE Publications.

Jenkins, M. (2013). Parkour classes are helping pensioners stay agile and active, *The Guardian*, Wednesday 28 August, www.theguardian .com/society/2013/aug/28/parkour-classes-pensioners-agile-active accessed on 11 October, 2013.

Kidder, J.L. (2012). Parkour: The affective appropriation of urban space, and the real/virtual dialectic. *City & Community, 11*(3), 229–53.

Kirby, S., & Hay, I. (1997). (Hetero)sexing space: Gay men and "straight" space in Adelaide, South Australia. *Professional Geographer, 49*(3), 295–305.

Le Roux, C.S. (2017). Exploring rigour in autoethnographic research. *International Journal of Social Research Methodology, 20*(2), 195-207.

Livingstone, D.N. (2005). Science, text and space: Thoughts on the geography of reading. *Transactions of the Institute of British Geographers, 30*(4), 391–401.

Lockwood, M., et al. (2007). *Strengths and Challenges of Regional NRM Governance: Interviews with Key Players and Insights from the Literature*. Report no. 4 of the Land and Water Australia Project, Pathways to Good Practice in Regional NRM Governance. Hobart: University of Tasmania.

Mason, J. (2004). *Qualitative researching* (2nd edition). London: SAGE.

Morse, J.M. (2015). Critical analysis of strategies for determining rigor in qualitative inquiry. *Qualitative Health Research, 17*(3), 278–88.

Mould, O. (2009). Parkour, the city, the event. *Environment and Planning D: Society and Space, 27*(4), 738–50.

Murdoch, J. (2006). *Post-structuralist geography.* London: SAGE.

Nind, M. (2017). The practical wisdom of inclusive research. *Qualitative Research, 17*(3), 278–88.

Patton, M.Q. (2015). *Qualitative research and evaluation methods: Integrating theory and practice* (4th edition). Thousand Oaks, London, New Delhi: SAGE Publications.

Peck, B., & Mummery, J. (2018). Hermeneutic constructivism: An ontology for qualitative research. *Qualitative Health Research, 28*(3), 389–407.

Platt, J. (1988). What can case studies do? *Studies in Qualitative Methodology,* 1, 1–23.

Rettke, H., Pretto, M., Spichiger, E., Frei, I.A., & Spirig, R. (2018). Using reflexive thinking to establish rigor in qualitative research. *Nursing Research, 67*(6), 490–97.

Robinson, G. (1998). *Methods and techniques in human geography.* Chichester: John Wiley and Sons.

Roller, M.R., & Lavrakas, P.J. (2015). *Applied qualitative research design: A total quality framework approach.* New York and London: Guilford Publications.

Sayer, A. (2010). *Method in social science: A realist approach* (revised 2nd edition). Milton Park and New York: Routledge.

Stake, R. (1995). *The art of case study research.* Thousand Oaks, CA: SAGE.

Stratford, E. (1998). Public spaces, urban youth and local government: The skateboard culture in Hobart's Franklin Square. In R. Freestone (Ed.), *20th century urban planning experience: Proceedings of the 8th international planning history Conference.* Sydney: University of New South Wales.

——— (2002). On the edge: A tale of skaters and urban governance. *Social & Cultural Geography, 3*(2), 193–206.

——— (2008). Islandness and struggles over development: A Tasmanian case study. *Political Geography, 27*(2), 160–75.

——— (2012). Vantage points: Observations on the emotional geographies of heritage. In G. Baldacchino (Ed.), *Extreme heritage management: Policies and practices from Island Territories* (pp. 1–20). New York: Berghahn Books.

——— (2014). *Geographies, mobilities, and rhythms over the life course: Adventures in the interval.* New York and London: Routledge.

Stratford, E., & Harwood, A. (2001). The regulation of skating in Australia: An overview and commentary on the Tasmanian case. *Urban Policy and Research, 19*(2), 61–76.

Williams, C.C., & Round, J. (2007). Re-thinking the nature of the informal economy: Some lessons from Ukraine. *International Journal of Urban and Regional Research, 31*(2), 425–41.

Williams, G., et al. (2003). Enhancing pro-poor governance in eastern India: Participation, politics and action. *Progress in Development Studies, 3*(2), 159–78.

Wood, L., & Williamson, S. (1996). *Consultants' report on franklin square: Users, activities and conflicts.* Hobart: UNITAS Consulting.

Woodyer, T. (2012). Ludic geographies: not merely child's play. *Geography Compass, 6*(6), 313–26.

Wooldridge, S.W. (1955). The status of geography and the role of fieldwork. *Geography,* 40, 73–83.

Chapter 7

Bryman, A. (2006). Integrating quantitative and qualitative research: How is it done? *Qualitative Research, 6*(1), 97–113.

Campbell, D., & Stanley, J. (1966). *Experimental and quasi-experimental designs for research.* Chicago: Rand McNally.

Castree, N. (2005). The epistemology of particulars: Human geography, case studies and "context". *Geoforum, 36*(5), 541–44.

Dunn, K. (2008). Comparative analyses of transnationalism: A geographic contribution to the field. *Australian Geographer, 39*(1), 1–7.

Flyvberg, B. (2006). Five misunderstandings about case-study research. *Qualitative Inquiry, 12*(2), 219–45.

George, A., & Bennett, A. (2005). *Case studies and theory development in the social sciences.* Cambridge, MA: MIT Press.

Gerring, J. (2004). What is a case study and what is it good for? *American Political Science Review, 98*(2), 341–54.

——— (2007). *Case study research: Principles and practices.* Cambridge: Cambridge University Press.

González-Hidalgo, M. & Zografos, C. (2019). Emotions, power, and environmental conflict:

Expanding the "emotional turn" in political ecology. *Progress in Human Geography*, https://doi.org/10.1177/0309132518824644

Hesse-Biber, S., & Leavy, P. (2004). *Approaches to qualitative research: A reader on theory and practice*. Oxford: Oxford University Press.

Jazeel, T. (2016). Between a area and discipline: Progress, knowledge production and the geographies of Geography. *Progress in Human Geography, 40*(5), 649–67.

Kepkiewicz, L., Chrobok, M., Whetung, M., Cahuas, M., Gill, J., Walker, S., & Wakefield, S. (2015). Beyond inclusion: Toward an anti-colonial food justice praxis. *Journal of Agriculture, Food Systems, and Community Development, 5*(4), 99–104. http://dx.doi.org/10.5304/jafscd2015.054.014

Leszczynski, A. (2018). Digital Methods 1: Wicked Tensions. *Progress in Human Geography, 42*(3) 473–481.

Liebow, E. (1967). *Tally's corner: A study of negro streetcorner men*. Boston: Little Brown.

Lincoln, Y., & Guba, E. (2002). Judging the quality of case study reports. In M. Huberman & M. Miles (Eds), *The qualitative researcher's companion*. London: SAGE.

Platt, J. (1992). "Case study" in American methodological thought. *Current Sociology*, 40, 17–48.

Popper, K. (1959). *The logic of scientific discovery*. London: Hutchinson.

Sayer, A. (2000). *Realism and social science*. London & Newbury Park: SAGE.

Taylor, L. (2016). Case study methodology. In N. Clifford, M. Cope, S. Gillespie, & S. French (Eds), *Key methods in human geography* (pp. 581–95). Thousand Oaks, London, New Delhi: SAGE Publications.

Thomas, W., & Znaniecki, F. (1918). *The Polish peasant in Europe and America*. New York: Dover Publications.

Tuan, Y.F. (1977). *Space and place: The perspective of experience*. Minneapolis: University of Minnesota Press.

Wakefield, S. (2007). Reflective action in the academy: Exploring praxis in critical geography using a "food movement" case study. *Antipode, 39*(2), 331–54.

Weis, T. (2000). Beyond peasant deforestation: Environment and development in rural Jamaica. *Global Environmental Change*, 10: 299–305.

Whyte, W.H. (1943). *Street corner society*. Chicago: University of Chicago Press.

Yin, R. (2003). *Case study research: Design and methods*. Los Angeles: SAGE.

Chapter 8

Basnet, S., Johnston, L., & Longhurst, R. (2018). Embodying "accidental ethnography": Staying overnight with former Bhutanese refugees in Aotearoa New Zealand. *Social & Cultural Geography*, DOI: 10.1080/14649365.2018.1480056

Benson, K., & Richa Nagar, R. (2006). Collaboration as resistance? Reconsidering the processes, products, and possibilities of feminist oral history & ethnography. *Gender, Place, & Culture, 13*(5), 581–92.

Billo, E., & Mountz, A. (2016). For institutional ethnography: Geographical approaches to institutions and the everyday. *Progress in Human Geography, 40*(2).

Brosius, P., & Campbell, L. (2010). Collaborative event ethnography: Conservation and development trade-offs at the Fourth World Conservation Congress. *Conservation and Society*, 8, no. 4, 245–55.

Cook, I. (2004). Follow the thing: Papaya. *Antipode, 36*(4), 642–64.

Crenshaw, K. (1991). Mapping the margins: Intersectionality, identity politics, and violence against women of color. *Stanford Law Review*, 1241–99.

Cuomo, D., & Massaro, V. (2013). Boundary-making in feminist research: New methodologies for "intimate insiders". *Gender Place and Culture, 23*(1), 94–106.

de Jong, A. (2015). Using Facebook as a space for storytelling in geographical research. *Geographical Research, 53*(2), 211–23.

Duggan, M. (2017). Questioning "digital ethnography" in an era of ubiquitous computing. *Geography Compass, 11*(5).

England, K.V.L. (1994). Getting personal: Reflexivity, positionality, and feminist research. *Professional Geographer*, 46, 80–9.

Haraway, D. (1988). Situated knowledges: The science question in feminism and the privilege of partial perspective. *Feminist Studies, 14*(3), 575–99.

Henderson, H. (2016). Toward an ethnographic sensibility in urban research. *Australian Planner, 53*(1), 28–36.

Hiemstra, N. (2016). Periscoping as a feminist methodological approach for researching the seemingly hidden. *The Professional Geographer, 69*(2), 329–36.

Indigenous Peoples Specialty Group, American Association of Geographers. (2010). Declaration of Key Questions about Research

Ethics with Indigenous Communities. www.indigenousgeography.net/IPSG/pdf/IPSGResearchEthicsFinal.pdf

Jansson, D. (2010). The head vs. the gut: Emotions, positionality, and the challenges of fieldwork with a Southern nationalist movement. *Geoforum, 41*(1), 19–22.

Katz, Cindi. (2009). Social systems: Thinking about society, identity, power and resistance. In Clifford et al. (Eds), *Key concepts in geography* (p. 236). London: SAGE.

Kearns, R. (2016). Placing observation in the research toolkit. In I. Hay (Ed.), *Qualitative research methods in human geography* (4th edition). Toronto, Canada: Oxford University Press.

Kobayashi, A. (1994). Coloring the field: Gender, "race", and the politics of fieldwork. *Professional Geographer, 46*(1), 73–80.

______ (2014). The dialectic of race and the discipline of geography. *Annals of the Association of American Geographers, 104*(6), 1101–115.

Maguire, H., McCartan, A., Nash, C.J., & Browne, K. (2019). The enduring field: Exploring researcher emotions in covert research with antagonistic organisations. *Area, 51*(2), 299–306.

Malinowski, B. (1922). *Argonauts of the Western Pacific*. London: Routledge and Kegan Paul.

Mountz, A., Miyares, I., Wright, R., & Bailey, A. (2003). Methodologically becoming: Power, knowledge and team research. *Gender, Place & Culture, 10*(1), 29–46.

Mukherjee, S. (2017). Troubling positionality: Politics of "studying up" in transnational contexts. *The Professional Geographer, 69*(2), 291–98.

Pacheco-Vega, R., & Parizeau, K., (2018). Doubly engaged ethnography: Opportunities and challenges when working with vulnerable communities. *International Journal of Qualitative Methods, 17*(1), 1–13.

Pierce, J., & Lawhon, M. (2015). Walking as method: Toward methodological forthrightness and comparability in urban geographical research. *The Professional Geographer, 67*(4).

Pitt, H. (2015). On showing and being shown plants—a guide to methods for a more-than-human geography. *Area, 47*(1), 48–55.

Rose, G. (1997). Situating knowledges: Positionality, reflexivities and other tactics. *Progress in Human Geography,* 21, 305–20.

Smith, S. (2015). Intimacy and angst in the field. *Gender Place and Culture, 23*(1), 134–46.

Sultana, F. (2007). Reflexivity, positionality and participatory ethics: Negotiating fieldwork dilemmas in international research. *ACME: An International E-journal for Critical Geographies, 6*(3), 374–85.

Sundberg, J. (2003). Masculinist epistemologies and the politics of fieldwork in Latin Americanist Geography. *Professional Geographer,* 55, 181–91.

Thiem, C., & Robertson, M. (2010). Behind enemy lines: Reflections on the practice and production of oppositional research. *Geoforum, 41*(1), 5–6.

Vannini, P. (2015). Non-representational ethnography: New ways of animating lifeworlds. *Cultural Geographies, 22*(2), 317–27.

Watson, A. (2012). I crashed the boat and wept: Localizing the "field" in critical geographic practice. *Emotion, Space and Society, 5*(3), 192–200.

______ (2018). Ethnographic Overview and Assessment of Gates of the Arctic National Park and Preserve: Subsistence Land Use in the Kobuk Preserve, National Park Service, Cultural Resource Report NPS/GAAR/CRR-2018/001. Available from the author or at www.researchgate.net/publication/338263340_Ethnographic_Overview_and_Assessment_of_Gates_of_the_Arctic_National_Park_and_Preserve_Subsistence_Land_Use_of_the_Kobuk_Preserve

Watson, A., & Huntington, O.H. (2008). "They're Here, I Can Feel Them": The epistemic spaces of Indigenous and Western knowledges. *Social & Cultural Geography, 9*(3), 257–81.

Whitson, R. (2017). Painting pictures of ourselves: Researcher subjectivity in the practice of feminist reflexivity. *The Professional Geographer, 69*(2), 299–306.

Chapter 9

Adelman, C., (Ed.). (1981). *Uttering, muttering: Collecting, using and reporting talk for social and educational research*. London: Grant McIntyre.

Ash, J., Kitchin, R., & Leszczynski, A. (2019). Introducing digital geographies. In J. Ash, R. Kitchin, & A. Leszczynski (Eds), *Digital geographies* (pp. 1–10). Thousand Oaks: SAGE.

Babbie, E. (1992). *The practice of social research* (6th edition). Belmont, CA: Wadsworth.

Bampton, R., & Cowton, C.J. (2002). The e-interview. *Forum: Qualitative Social Research Sozialforschung, 3*(2), article 9.

Baxter, J., & Eyles, J. (1997). Evaluating qualitative research in social geography: Establishing "rigour" in interview analysis. *Transactions of the Institute of British Geographers, 22*(4), 505–25.

Buchholz, K. (2019). In 2019, Global Emoji Count is Growing to More Than 3,000. *statista*, July 17, 2019, www.statista.com/chart/17275/number-of-emojis-from-1995-bis-2019/

Burgwyn-Bailes, E., Baker-Ward, L., Gordon, B.N., & Ornstein, P.A. (2001). Children's memory for emergency medical treatment after one year: The impact of individual difference variables on recall and suggestibility. *Applied Cognitive Psychology*, 15, S25–S48.

Callegaro, M. (2011). Videophone interviewing. In P.J. Lavrakas (Ed.), *Encyclopedia of survey research methods*. Thousand Oaks: SAGE Publications.

Chen, P.J., & Hinton, S.M. (1999). Realtime interviewing using the world wide web. *Sociological Research Online, 4*(3).

Cope, M., (2020). Transcribing, coding, and analyzing qualitative data. In A. Kobayashi (Ed.), *International encyclopedia of human geography* (2nd edition, vol. 13) (pp. 357–362). Elsevier. https://dx.doi.org/10.1016/B978-0-08-102295-5.10875-3

Deakin, H. & Wakefield, K. (2014). Skype interviewing: Reflections of two PhD researchers. *Qualitative Research, 14*(5), 603–16.

Dowling, R., Lloyd, K., & Suchet-Pearson, S. (2016). Qualitative methods 1: Enriching the interview. *Progress in Human Geography, 40*(5), 679–86.

Dunn, K. (1993). The Vietnamese concentration in Cabramatta: Site of avoidance and deprivation, or island of adjustment and participation? *Australian Geographical Studies, 31*(2), 228–45.

Dunn, K.M., & Mahtani, M. (2001). Media representations of ethnic minorities. *Progress in Planning, 55*(3), 163–72.

Elman, C. & Kapiszewski D. (2017). Annotating qualitative social science. Parameters: Knowledge under Digital Conditions (SSRC). http://parameters.ssrc.org/2017/04/annotating-qualitative-social-science/

Gardner, R., Neville, H., & Snell, J. (1983). Vietnamese settlement in Springvale. Environmental Report no. 14. Melbourne: Monash University Graduate School of Environmental Science.

Hershkowitz, I., & Terner, A. (2007). The effects of repeated interviewing on children's forensic statements of sexual abuse. *Applied Cognitive Psychology*, 21, 1131–143.

hooks, b. (1990). *Yearning: Race, gender, and cultural politics*. Boston: South End Press.

James, N., & Busher, H. (2006). Credibility, authenticity and voice: Dilemmas in online interviewing. *Qualitative Research*, 6(3), 403–20.

Kearns, R. (1991). Talking and listening: Avenues to geographical understanding. *New Zealand Journal of Geography*, 92, 2–3.

Kong, L. (1998). Refocussing on qualitative methods: Problems and prospects for research in a specific Asian context. *Area, 30*(1), 79–82.

Krueger, R.A., & Casey, M.A. (2014). *Focus groups: A practical guide for applied research* (5th edition). Los Angeles: SAGE.

Goss, J.D., & Leinbach, T.R. (1996). Focus groups as alternative research practice: Experience with transmigrants in Indonesia. *Area, 28*(2), 115–23.

Groves, R.M. (1990). Theories and methods of telephone surveys. *Annual Review of Sociology*, 16, 221–40.

Hay, I. (2016). On being ethical in geographical research, in N. Clifford, M. Cope, T. Gillespie & S. French (Eds), Key *methods in geography*, (3rd edition) (pp. 30–43). Thousand Oaks, London, New Delhi, London: SAGE Publications.

Israel, M. & Hay, I. (2006). *Research ethics for social scientists: Between ethical conduct and regulatory compliance*. London: SAGE Publications.

Lane, R. (1997). Oral histories and scientific knowledge in understanding environmental change: A case study in the Tumut region, NSW. *Australian Geographical Studies, 35*(2), 195–205.

Longhurst, R. (2016). Semi-structured interviews and focus groups. In N. Clifford, M. Cope, S. Gillespie, & S. French (Eds), *Key methods in geography* (pp. 143–56). Thousand Oaks, London, New Delhi: SAGE Publications.

Maccoby, E., & Maccoby, N. (1954). The interview: A tool of social science. In G. Lindzey, (Ed.), *Handbook of social psychology*. Cambridge, MA: Addison-Wesley.

McCoyd, J.L.M., & Kerson, T.S. (2006). Conducting intensive interviews using e-mail: A serendipitous comparative opportunity. *Qualitative Social Work, 5*(3), 389–406.

McGuirk, P.M. (2002). Producing the capacity to govern in global Sydney: A multiscaled account. *Journal of Urban Affairs*, 25, 201–23.

McKay, D. (2002). Negotiating positionings: Exchanging life stories in research interviews. In P. Moss (Ed.), *Feminist geography in practice*. Oxford: Blackwell.

Mann, C., & Stewart, F. (2002). Internet interviewing. In J.F. Gubrium & J.A. Holstein (Eds), *Handbook of interview research: Context and method* (pp. 603–27). Thousand Oaks, CA: SAGE.

Meho, L.I. (2006). E-mail interviewing in qualitative research: A methodological discussion. *Journal of the American Society for Information Science and Technology*, *57*(10), 1284–95.

Minichillo, V., Aroni, R., & Hays, T.N. (1995). *In-depth interviewing: Principles, techniques, analysis* (2nd edition). Melbourne: Longman Cheshire.

Mohammad, R. (1999). Marginalisation, Islamism and the production of the "Other's" "Other". *Gender, Place & Culture*, *6*(3), 221–40.

Muller, S. (1999). Myths, media and politics: Implications for koala management decisions in Kangaroo Island, South Australia. Paper presented to the Institute of Australian Geographers conference, 27 September to 1 October, Sydney.

Nelson, S. (2003). "It's I mean like uh disrespectful". *Times Higher Educational Supplement*, 28 March: 16.

Pearson, L.J. (1996). Place re-identification: The "Leisure Coast" as a partial representation of Wollongong. (School of Geography, University of New South Wales, BSc Honours thesis).

Pulvirenti, M. (1997). Unwrapping the parcel: An examination of culture through Italian home ownership. *Australian Geographical Studies*, *35*(1), 32–9.

Robertson, B.M. (1994). *Oral history handbook*. Adelaide: Oral History Association of Australia.

Schoenberger, E. (1991). The corporate interview as a research method in economic geography. *Professional Geographer*, *43*(2), 180–9.

Tremblay, M.A. (1982). The key informant technique: A non-ethnographic application. In R.G. Burgess (Ed.), *Field research: A sourcebook and field manual*. London: Allen and Unwin.

Valentine, G. (1993). (Hetero)sexing space: Lesbian perceptions and experiences of everyday spaces. *Environment and Planning D: Society and Space*, *11*(4), 395–413.

——— (1997). Tell me about . . . : Using interviews as a research methodology. In R. Flowerdew & D. Martin (Eds), *Methods in human geography: A guide for students doing a research project*. Harlow: Longman.

Vincent, K.A. (2013). The advantages of repeat interviews in a study with pregnant schoolgirls and schoolgirl mothers: Piecing together the jigsaw. *International Journal of Research & Method in Education*, *36*(4), 341–54.

Weller, S. (2017). Using internet video calls in qualitative (longitudinal) interviews: Some implications for rapport. *International Journal of Social Research Methodology*, *20*(6), 613–25.

Winchester, H.P.M., Dunn, K.M., & McGuirk, P.M. (1997). Uncovering Carrington. In R.J. Moore & M.J. Ostwald (Eds), *Hidden Newcastle: Urban memories and architectural imaginaries* (pp. 174–81). Ultimo, Australia: Gadfly Media.

Chapter 10

Cope, M. & Kurtz, H. (2016). Qualitative data analysis. In N. Clifford, M. Cope, T. Gillespie, & S. French (Eds), *Key methods in geography* (pp. 647–64). London: SAGE.

Egoz, S. (2000). Clean and green but messy: The contested landscape of New Zealand's organic farms. *Oral History*, *28*(1), 63–74.

Evans, G. (1965). *Ask the fellows who cut the hay*. London: Faber and Faber.

Evered, K.T. (2011). Traditional ecologies of the opium poppy & oral history in rural Turkey. *Geographical Review*, *101*(2), 164–82.

George, K., & Stratford, E. (2016). Oral history and human geography. In I. Hay (Ed.), *Qualitative research methods in human geography* (4th edition) (pp. 189–201). Ontario: Oxford University Press.

Interview R-0171 with Leroy Beavers Sr. by Kieran Taylor, 7 August 2002, in the Southern Oral History Program Collection (#4007), Southern Historical Collection, Wilson Library, University of North Carolina at Chapel Hill.

Interview K-0013 with Bobby Kirk by Karl E Campbell, 28 October 1985, in the Southern Oral History Program Collection (#4007), Southern Historical Collection, Wilson Library, University of North Carolina at Chapel Hill.

Interview X-0028 with Hattie McCauley by Darius Scott, 31 October 2014, in the Southern Oral History Program Collection (#4007),

Southern Historical Collection, Wilson Library, University of North Carolina at Chapel Hill.

Kwan, M-P. (2008). From oral histories to visual narratives: Re-presenting the post-September 11 experiences of the Muslim women in the USA. *Social & Cultural Geography, 9*(6), 653–669.

Lane, R. (1997). Oral histories and scientific knowledge in understanding environmental change: A case study in the Tumut region, NSW. *Australian Geographical Studies, 35*(2), 195–205.

Maharawal, M.M., & McElroy, E. (2018). The anti-eviction mapping project: Counter mapping and oral history toward Bay Area, housing justice. *Annals of the American Association of Geographers, 108*(2), 380–89.

Matless, D., Oldfield, J., & Swain, A. (2007). Encountering Soviet geography: Oral histories of British geographical studies of the USSR and eastern Europe, 1945–1991. *Social & Cultural Geography, 8*(3), 353–72.

Purifoy, D., (2014). Food policy councils: Integrating food justice and environmental justice. *Duke Environmental Law and Policy Forum,* 24, 375–98.

Raleigh Yow, V. (2014). *Recording oral history: A guide for the humanities and social sciences* (3rd edition). Lanham: Rowman & Littlefield.

Riley, M., & Harvey, D. (2007). Editorial: Talking landscapes: On oral history and the practice of geography. *Social & Cultural Geography, 8*(4), 391–415.

Roane, J.T. (2018). Plotting the Black commons. *Souls, 20*(3), 239–266.

Rogaly, B. (2020). *Stories from a migrant city: Living and working together in the shadow of Brexit.* Manchester University Press.

Rosengarten, T. (1974). *All God's dangers: The life of Nate Shaw.* Chicago: University of Chicago Press.

Sommer, B.W., & Quinlan, M.K. (2018). *The oral history manual* (3rd edition). London: Rowman & Littlefield.

Williams, B. (2018). "That we may live": Pesticides, plantations, and environmental racism in the United States South. *Environment and Planning E: Nature and Space, 1*(1–2), 243–67.

Williams, B., & Riley, M. (2020). The challenge of oral history to environmental history. *Environment and History, 26*(2), 207–231.

Woods, C. (2002). Life after death. *The Professional Geographer, 54*(1), 62–66.

Woods, M. (2010). Performing rurality and practising rural geography. *Progress in Human Geography, 34*(6), 835–46.

Chapter 11

Barbour, R.S. (2007). *Doing focus groups.* Los Angeles: SAGE.

Barr, S., Shaw, G., Coles, T., & Prillwitz, J. (2010). "A holiday is a holiday": Practicing sustainability, home and away. *Journal of Transport Geography, 18*(3), 474–81.

Bedford, T., & Burgess, J. (2002). The focus-group experience. In M. Limb & C. Dwyer (Eds), *Qualitative methodologies for geographers: Issues and debates.* London: Edward Arnold.

Berg, B.L. (1989). *Qualitative research methods for the social sciences.* Boston: Allyn and Bacon.

Bertrand, J.T., Brown, J.E., & Ward, V.M. (1992). Techniques for analyzing focus group data. *Evaluation Review, 16*(2), 198–209.

Burgess, J. (1996). Focusing on fear: The use of focus groups in a project for the Community Forest Unit, Countryside Commission. *Area, 2*(2), 130–5.

Byrne, J. (2012). When green is White: The cultural politics of race, nature and social exclusion in a Los Angeles urban national park. *Geoforum, 43*(3), 595–611.

Cameron, J. (1992). Modern-day tales of illegitimacy: Class, gender and ex-nuptial fertility. (Department of Geography, University of Sydney, MA minor thesis).

Cameron, J., & Hendriks, S. (2014). Narratives of social enterprise: Insights from Australian social enterprise practitioners. In H. Douglas & S. Grant (Eds), *Social innovation, social entrepreneurship and social enterprise: Context and theories.* Victoria, Australia: Tilde University Press.

Carey, M.A. (1994). The group effect in focus groups: Planning, implementing and interpreting focus group research. In J.M. Morse (Ed.), *Critical issues in qualitative research methods.* Thousand Oaks, CA: SAGE.

Carolan, M. (2017). *No one eats alone: Food as a social enterprise.* Washington DC: Island Press.

Collier, M.J., & Scott, M. (2010). Focus group research in a mined landscape. *Land Use Policy, 27*(2), 304–12.

Cousins, J.A., Evans, J., & Sadler, J.P. (2009). "I've paid to observe lions, not road maps!": An emotional journey with conservation volunteers in South Africa. *Geoforum, 40*(6), 1069–80.

Eden, S., Bear, C., & Walker, G. (2008). Mucky carrots and other proxies: Problematising the knowledge-fix for sustainable and ethical consumption. *Geoforum,* 39, 1044–57.

Evans, V., & Sternberg, J. (1999). Young people, politics and television current affairs in Australia. *Journal of Australian Studies,* December, 103–9.

Gibson, K., Cameron, J., & Veno, A. (1999). *Negotiating restructuring and sustainability: A study of communities experiencing rapid social change.* Australian Housing and Urban Research Institute (AHURI) Working Paper. Melbourne: AHURI, Melbourne. Also available: www.communityeconomies.org/papers/comecon/comeconp4.pdf.

Goss, J.D., & Leinbach, T.R. (1996). Focus groups as alternative research practice: Experience with transmigrants in Indonesia. *Area,* 28(2), 115–23.

Hares, A., Dickinson, J.E., & Wilkes, K. (2010). Climate change and the air travel decisions of UK tourists. *Journal of Transport Geography, 18*(3), 466–73.

Holbrook, B., & Jackson, P. (1996). Shopping around: Focus group research in North London. *Area, 28*(2), 136–42.

Hoppe, M.J., et al. (1995). Using focus groups to discuss sensitive topics with children. *Evaluation Review, 19*(1), 102–14.

Jackson, P., & Holbrook, B. (1995). Multiple meanings: Shopping and the cultural politics of identity. *Environment and Planning A,* 27, 1913–30.

Johnson, A. (1996). "It's good to talk": The focus group and the sociological imagination. *The Sociological Review, 44*(3), 517–38.

Kitzinger, J. (1994). The methodology of focus groups: The importance of interaction between research participants. *Sociology of Health and Illness, 16*(1), 103–21.

Kneale, J. (2001). Working with groups. In M. Limb & C. Dwyer (Eds), *Qualitative methodologies for geographers* (pp. 136–50). London: Arnold.

Krueger, R. (1998). *Analyzing and reporting focus group results.* Focus Group Kit 6. Thousand Oaks, CA: SAGE.

Merton, R.K. (1987). The focussed interview and focus groups: Continuities and discontinuities. *Public Opinion Quarterly, 51*(4), 550–66.

Moore, T., McKee, K., & McLoughlin, P. (2015). Online focus groups and qualitative research in the social sciences: Their merits and limitations in a study of housing and youth. *People, Place and Policy, 9*(1), 17–28.

Morgan, D.L. (1996). Focus groups. *Annual Review of Sociology,* 22, 129–52.

——— (1997). *Focus groups as qualitative research* (2nd edition). Thousand Oaks, CA: SAGE.

Myers, G. (1998). Displaying opinions: Topics and disagreement in focus groups. *Language in Society,* 27, 85–111.

O'Brien, K. (1993). Improving survey questionnaires through focus groups. In D.L. Morgan (Ed.), *Successful focus groups: Advancing the state of the art.* Newbury Park: SAGE.

Pratt, G. (2002). Studying immigrants in focus groups. In P. Moss (Ed.), *Feminist geography in practice: Research and methods.* Oxford: Blackwell.

Stewart, D., Shamdasani, P., & Rook, D. (2007). *Focus groups: Theory and practice.* (2nd edition). Thousand Oaks, CA: SAGE.

Teye, J.K. (2012). Benefits, challenges, and dynamism of positionalities associated with mixed methods research in developing countries: Evidence from Ghana. *Journal of Mixed Method Research, 6*(4), 379–91.

Ward, V.M., Bertrand, J.T., & Brown, L.F. (1991). The comparability of focus group and survey results: Three case studies. *Evaluation Review, 15*(2), 266–83.

Zeigler, D.J., Brunn, S.D., & Johnson, J.H. (1996). Focusing on Hurricane Andrew through the eyes of the victims. *Area, 28*(2) 124–9.

Chapter 12

Adams-Campbell, M., Falzetti, A.G., & Rivard, C. (2015). Introduction: Indigeneity and the work of settler archives. *Settler Colonial Studies, 5*(2), 109–116

Ashmore, P., R, Craggs, & Neate, H. (2012). Working with: Talking and sorting in personal archives, *Journal of Historical Geography, 38*(1), 81–89.

Bailey, A., Brace, C., & Harvey, D.C. (2009). Three geographers in an archive: Positions, predilections and passing comment on transient lives. *Transactions of the Institute of British Geographers, 43*(2), 254–69.

Baker, A.R.H. (1997). The dead don't answer questionnaires: Researching and writing historical geography. *Journal of Geography in Higher Education, 21*(2), 231–43.

Beel, D., Wallace, C., Webster, G., & Nguyen, H. (2015). The geographies of community digital archives in rural Scotland. *Scottish Geographical Journal, 131*(3–4), 201–11.

Black, I. (2006). Analysing historical and archival sources. In N. Clifford & G. Valentine (Eds), *Key methods in geography* (pp. 477–500). London: SAGE.

Craggs, R. (2008). Situating the imperial archive: The Royal Empire Society Library, 1868–1945. *Journal of Historical Geography,* 34, 48–67.

Duncan, J.S. (1999). Complicity and resistance in the colonial archive: Some issues in the method and theory of historical geography. *Historical Geography,* 27, 119–28.

Dwyer, C. & Davies, G. (2010). Qualitative Methods III animating Archives, artful interventions and online environments. *Progress in Human Geography,* 34, 317–33.

Falzetti, A.G. (2015). Archival absence: The burden of History. *Settler Colonial Studies, 5*(2), 128–44.

Finer, C. (2000). Researching in a contemporary archive. *Social Policy and Administration, 34*(4), 434–47.

Guelke, L. (1982). *Historical understanding in geography: An idealist approach.* Cambridge, Cambridge University Press.

Hackel, S., & Reid, A. (2007). Transforming an eighteenth-century archive into a twenty-first century database: The Early Californian Population Project. *History Compass, 5*(3), 1013–25.

Harley, J.B. (1992). Deconstructing the map. In T. Barnes & J. Duncan (Eds), *Writing worlds: Discourse, text and metaphor in the representation of landscape.* London: Routledge.

Harris, C. (2001). Archival fieldwork. *Geographical Review, 91*(1–2), 328–34.

Harvey, K. (2006). From bags and boxes to searchable digital collections at the Dalhousie University Archives. *Journal of Canadian Studies, 40*(2), 120–38.

Hodder, J. (2017). On absence and abundance: Biography as method in archival research. *Area, 49*(4), 452–59.

Holland, P.G., & Olson, S. (2016). Ledgers and landscapes: Indicators of rural landscape change in southern New Zealand 1878–1919. *New Zealand Geographer,* 73, 45–56.

Holland, P.S. Olson, S., & Garden, C. (2018). Work schedules on lowland farms in southern New Zealand from the late colonial period to the 1930s. *New Zealand Geographer,* 75, 12–20.

Isaacman, A., Lalu, P., & Nygren, T. (2005). Digitization, history, and the making of a postcolonial archive of southern Africa liberation struggles: The Aluka project. *Africa Today, 52*(2), 54–77.

Johnson, J. (2018). The Palmyra Atoll Digital Archive. *Hawaiian Journal of History,* 52, 143–50.

Keighren, I.M. (2012). Fieldwork in the archive. In R. Phillips & J. Johns (Eds), *Fieldwork for human geography* (pp. 138–40). London: SAGE.

Lovell, G. (2001). "Not a city but a World": Seville and the Indies. *Geographical Review, 91*(1/2), 239–51.

Mason, K., & Zanish-Belcher, T. (2007). Raising the archival consciousness: How women's archives challenge traditional approaches to collecting and use—What's in a name? *Library Trends, 56*(2), 344–59.

Mills, S. (2013). Cultural-historical geographies of the archive: Fragments, objects and ghosts. *Geography Compass,* 7(10), 701–13.

Moore, F.P.L. (2010). Tales from the archive: Methodological and ethical issues in historical geography research. *Area,* 42(3), 262–70.

Moran, J. (2017). *Born digital in New Zealand: Report of survey results.* Wellington: Te Puna Mātauranga o Aotearoa National library of New Zealand.

National Council on Public History. (2003). NCPH ethics guidelines. http://ncph.org/ethics.html.

Ogborn, M. (2006). Finding historical data. In N. Clifford & G. Valentine (Eds), *Key methods in geography* (pp. 101–15). London: SAGE.

Poorthuis, A., Zook, M., Shelton, T., Graham, M., & Stephens, M. (2016). Using geotagged digital social data in geographic research. In N. Clifford, M. Cope, S. Gillespie, & S. French (Eds), *Key methods in human geography* (pp. 248–69). London: SAGE.

Quanchi, M. (2006). Photography and history in the Pacific Islands. *The Journal of Pacific History,* 41(2), 165–73.

Rekrut, A. (2011). Connected constructions, constructing connections, materiality of archival records as historical evidence. In K. Gray & C. Verduyn (Eds), *Archival narratives for Canada re-telling stories in a changing landscape* (pp. 135–57). Halifax and Winnipeg: Fernwood.

Sauer, C. 1942. Foreword to historical geography. *Annals of the Association of American Geographers, 31*(1), 1–24.

Schwartz, J., & Ryan, R. (Eds). (2003). *Picturing place, photography and the geographical imagination.* London: I.B. Tauris.

Summerby-Murray, R. (2011). Marshland memories, constructing narrative in an online archival exhibition. In K. Gray & C. Verduyn (Eds), *Archival narratives for Canada re-telling stories in a changing landscape* (pp. 117–133). Halifax and Winnipeg: Fernwood.

Tibbo, H. (2012). Placing the horse before the cart: Conceptual and digital dimensions of digital curation. *Historical Social Science, 37*(3), 187–200.

Chapter 13

Babbie, E. (2016). *The practice of social research* (14th edition). Belmont, CA: Wadsworth Cengage Learning.

Banaji, S., & Buckingham, D. (2010). Young people, the internet, and civic participation: An overview of key findings from the CivicWeb Project. *International Journal of Learning and Media, 2*(1), 15–24.

Beckett, C., & Clegg, S. (2007). Qualitative data from a postal questionnaire: Questioning the presumption of the value of presence. *International Journal of Social Research Methodology,* 10, 307–17.

Bryman, A. (2016). *Social research methods* (5th edition). Oxford: Oxford University Press.

Chen, S., Hall, G., & Johns, M. (2004). Research paparazzi in cyberspace: The voices of the researchers. In M. Johns, S. Chen, & G. Hall (Eds), *Online social research: Methods, issues and ethics* (pp. 157–75). New York: Peter Lang.

Clifford, N., Cope, M., Gillespie, S., & French, S. (Eds). (2016). *Key methods in human geography*. Thousand Oaks, London, New Delhi: SAGE Publications.

Cloke, P., et al. (2004). Talking to people. In *Practising Human Geography*. London: SAGE.

Crang, M. (2005). Qualitative methods: There is nothing outside the text? *Progress in Human Geography,* 29(2), 225–33.

DeLyser, D., Herbert, S., Aitken, S., Crang, M., & McDowell, L. (Eds). (2010). *The SAGE handbook of qualitative geography*. London: SAGE.

de Vaus, D. (2014). *Surveys in social research* (6th edition). Sydney: Allen and Unwin.

Dillman, D. (2007). *Mail and internet surveys: The total design method* (2nd edition). Hoboken, NJ: John Wiley.

Dillman, D., & Bowker, D. (2001). The web questionnaire challenge to survey methodologists. In U. Reips & M. Bosnjak (Eds), *Dimensions of internet science* (pp. 159–78). Lengerich, Germany: Pabst Science Publishers.

Dillman, D.A., Phelps, G., Tortora, R., Swift, K., Kohrell, J., Berck, J., & Messer, B.L. (2009). Response rate and measurement differences in mixed-mode surveys using mail, telephone, interactive voice response (IVR) and the Internet. *Social Science Research*, 38, 1–18.

Dowling, R., Lloyd, K. & Suchet-Pearson, S. (2016). Qualitative methods1: Enriching the Interview. *Progress in Human Geography, 40*(5), 679–686.

Drozdzewski, D., & Klocker, N. (2019). The visible mother in the workplace. *Mothers/Mothering: Space and Place* (pp. 117–34). Demeter Press.

Fan, W., & Yan, Z. (2010). Factors affecting response rate of the web survey: A systematic review. *Computers in Human Behaviour,* 26, 132–39.

Flowerdew, R., & Martin, D.M. (2005 [2013, ebook]). *Methods in human geography* (2nd edition) London: Taylor & Francis Group.

Fowler, F. (2002). *Social survey methods.* Thousand Oaks, CA: SAGE.

Gibson, C. (2003). Digital divides in New South Wales: A research note on social–spatial inequality using 2001 census data on computer and Internet technology. *Australian Geographer,* 34, 239–57.

Gibson, C., Farbotko, C., Gill, N., Head, L., & Waitt, G. (2013). *Household sustain-ability: Challenges and dilemmas in every day life.* Cheltenham, UK: Edward Elgar.

Gillham, B. (2000). *Developing a questionnaire.* London: Continuum.

Gold, J.R. (1980). *An introduction to behavioural geography*. Oxford: Oxford University Press.

Guthrie, G. (2010). *Basic research methods: An entry to social science research*. London: SAGE.

Herbert, S. (2012). A taut rubber band: theory and empirics in qualitative geographic research. In D. DeLyser, S. Herbert, S. Aitken, M. Crang, & L.McDowell (Eds), *The SAGE handbook of qualitative geography*. Los Angeles: SAGE.

Hewson, C., et al. (2003). *Internet research methods*. London: SAGE.

Hoggart, K., Lees, L., & Davies, A. (2002). *Researching human geography*. London: Arnold.

Lee, R. (2000). *Unobtrusive methods in social research*. Buckingham: Open University Press.

Lumsden, J. (2005). Guidelines for the design of online questionnaires. National Research Council of Canada. Published as NRC/ERC-1127. 9 June, NRC48231. http://iit-iti.nrc-cnrc.gc.ca/iit-publications-iti/docs/NRC-48231.pdf.

McGuirk, P., & Dowling, R. (2011). Governing social reproduction in master-planned estates: Urban politics and everyday life in Sydney. *Urban Studies: an international journal for research in urban studies, 48*(12), 2611–28.

McInroy, L.B. (2016). Pitfalls, potentials, and ethics of online survey research: LGBTQ and other marginalized and hard-to-access youths. *Social Work Research, 40*(2), 83–94. https://dx.doi.org/10.1093%2Fswr%2Fsvw005

McLafferty, S. (2016). Conducting questionnaire surveys. In N. Clifford, M. Cope, S. Gillespie, & S. French (Eds), *Key methods in human geography* (pp. 129–42). London: SAGE.

Madge, C. (2007). Developing a geographer's agenda for online research ethics. *Progress in Human Geography,* 31, 654–74.

Matthews, H., Limb, M., & Taylor, M. (1998). The geography of children: Some ethical and methodological considerations for project and dissertation work. *Journal of Geography in Higher Education,* 22, 311–24.

May, T. (2011). *Social research: Issues, methods and process* (4th edition). Maidenhead: Open University.

Mee, K.J. (2007). "I ain't been to heaven yet? Living here, this is heaven to me": Public housing and the making of home in Inner Newcastle. *Housing, Theory and Society,* 24, 207–28.

Parfitt, J. (2005 [2013, ebook]). Questionnaire design and sampling. In R. Flowerdew, & D.M. Martin (Eds), *Methods in human geography* (2nd edition) London: Taylor & Francis Group.

Patton, M.Q. (2015). *Qualitative evaluation and research* (4th edition). Thousand Oaks, CA: SAGE.

Sarantkos, S. (2012). *Social research* (4th edition). New York: Palgrave McMillan.

Sayer, A. (2010). *Method in social science: A realist approach* (revised 2nd edition). Milton Park and New York: Routledge.

Sikkens, E., van San, M., Sieckelinck, S., Boeije, H., & de Winter, M. (2017). Participant recruitment through social media: Lessons learned from a qualitative radicalization study using Facebook. *Field Methods, 29*(2), 130–39. https://doi.org/10.1177/1525822X16663146)

Sue, V., & Ritter, L. (2012). *Conducting online surveys* (2nd edition). London: SAGE.

Van Selm, M., & Jankowski, N. (2006). Conducting online surveys. *Quality and Quantity,* 40, 435–56.

Vehovar, V., & Manfreda, K.L. (2017). Overview: Online surveys. In R. Fielding, M. Lee, & G. Blank (Eds), *The SAGE handbook of online research methods* (2nd edition). London: SAGE.

Waitt, G., & Knoble, H. (2018). Corporeal geographies of liveability and urban parks. *Urban Studies, 55*(14), 3151–67. doi.org/10.1177/0042098017740080.

Whitson, R. (2017). Painting pictures of ourselves: Researcher subjectivity in the practice of feminist reflexivity. *The Professional Geographer, 69*(2), 299–306.

Chapter 14

Bagheri, N. (2014). Mapping women in Tehran's public spaces: A geo-visualization perspective. *Gender, Place & Culture, 21*(10), 1285–301.

Baxter, J., & Eyles, J. (1997). Evaluating qualitative research in social geography: Establishing "rigour" in interview analysis. *Transactions of the Institute of British Geographers, 22*(4), 505–25.

Bell, S.L., Phoenix, C., Lovell, R., & Wheeler, B.W. (2015). Using GPS and geo-narratives: A methodological approach for understanding and situating everyday green space encounters. *Area, 47*(1), 88–96.

Boschmann, E.E. & Cubbon, E. (2013). Sketch maps and qualitative GIS: Using cartographies of individual spatial narratives in Geographic research. *The Professional Geographer, 66*(2), 236–48.

Brennan-Horley, C., & Gibson, C. (2009). Where is creativity in the city? Integrating qualitative and GIS methods. *Environment and Planning A: Economy and Space, 41*(11), 2595–614.

Brown, M., & Knopp, L. (2008). Queering the map: The productive tensions of colliding epistemologies. *Annals of the Association of American Geographers, 98*(1), 40–58.

Caquard, S., & Dimitrovas, S. (2017). Story Maps & Co. Un état de l'art de la cartographie des récits sur Internet/Story Maps & Co. The state of the art of online narrative cartography. *M@ppe Monde 121*. http://mappemonde.mgm.fr/121_as1/.

Cidell, J. (2010). Content clouds as exploratory qualitative data analysis. *Area, 42*(4), 514–23.

Cieri, M. (2003). Between being and looking: queer tourism promotion and lesbian social space in greater Philadelphia. *ACME: An International E-Journal for Critical Geographies, 2*(2), 147–66.

Cope, M., & Kurtz, H. (2016). Qualitative data analysis. In N. Clifford, M. Cope, S. Gillespie, & S. French (Eds), *Key methods in human geography* (pp. 647–64). London: SAGE.

Duffy, M., & Waitt, G. (2011). Sound diaries: a method for listening to place. *Aether,* 8, 119–36.

Dwyer, S., Piquette, N., Buckle, J., & McCaslin, E. (2013). Women gamblers write a voice: Exploring journaling as an effective counseling and research tool. *Journal of Groups in Addiction and Recovery, 8*(1), 36–50.

Edwards, L.V., Irving, S.M., & Hawkins, A.S. (2011). Till death do us part: Lived experiences of HIV-positive married African American women. *The Qualitative Report, 16*(5), 1361–79.

Eidse, N., & Turner, S. (2014). Doing resistance their own way: Counter-narratives of street vending in Hanoi, Vietnam through solicited journaling. *Area, 46*(3), 242–48.

Elwood, S. (2006). Beyond cooptation or resistance: Urban spatial politics, community organizations, and GIS-based spatial narratives. *Annals of the Association of American Geographers, 96*(2), 323–41.

Evans, J., & Jones, P. (2011). The walking interview: Methodology, mobility and place. *Applied Geography,* 31, 849–58.

Fawaz, M., Gharbieh, A., Harb, M., & Salamé, D. (Eds). (2018). *Refugees as city-makers.* Beirut, Lebanon: American Institute of Beirut.

Filep, C., Thompson-Fawcett, M., Fitzsimons, S., & Turner, S. (2015). Reaching revelatory places: The role of solicited diaries in extending research on emotional geographies into the unfamiliar. *Area, 47*(4), 459–65.

Filep, C.V., Turner, S., Eidse, N., Thompson-Fawcett, M., & Fitzsimons, S. (2018). Advancing rigour in solicited diary research. *Qualitative Research, 18*(4), 451–70.

Hansson, Ulf & Roulston, Stephen. (2017). Evaluations of diaries and GPS-enabled trackers to plot young peoples' geographies – asking the participants what they think. *Children's Geographies.* DOI: 10.1080/14733285.2016.1272915

Harris, J. (2002). The correspondence method as a data-gathering technique in qualitative enquiry. *International Journal of Qualitative Methods, 1*(4), 1–9.

Hayman, B., Wilkes, L., & Jackson, D. (2012). Journaling: Identification of challenges and reflection on strategies. *Nurse Researcher, 19*(3), 27–31.

Hohenthal, J., Minoia, P., & Pellikka, P. (2017). Mapping meaning: Critical cartographies for participatory water management in Taita Hills, Kenya. *The Professional Geographer, 69*(3), 383–95.

Humble, Á.M. & Sharp, E. (2012). Shared journaling as peer support in teaching qualitative research methods. *The Qualitative Report, 17,* 1–19.

Jones, A., & Woolley, J. (2015). The email-diary: A promising research tool for the 21st century? *Qualitative Research, 15*(6), 705–21. doi: 10.1177/1468794114561347.

Kenten, C. (2010). Narrating oneself: Reflections on the use of solicited diaries with diary interviews. *Forum Qualitative Sozialforschung / Forum: Qualitative Social Research, 11*(2), Article 16.

Kim, A.M. (2015). *Sidewalk city: Remapping public space in Ho Chi Minh City.* Chicago and London: The University of Chicago Press.

Knigge, L., & Cope, M. (2006). Grounded visualization: Integrating the analysis of qualitative and quantitative data through grounded theory and visualization. *Environment and Planning A: Economy and Space,* 38: 2021–37.

Knowles, A.K., L. Westerveld, & L. Strom (2015). Inductive visualization: A humanistic alternative to GIS. *GeoHumanities, 1*(2), 233–65.

Kwan, M.P. (2002a). Feminist visualization: Re-envisioning GIS as a method in feminist geographic research. *Annals of the Association of American Geographers, 92*(4), 645–61.

______ (2002b). Is GIS for women? Reflections on the critical discourse in the 1990s. *Gender, Place & Culture,* 9, 271–79.

______ (2008). From oral histories to visual narratives: Re-presenting the post-September 11 experiences of the Muslim women in the USA. *Social & Cultural Geography, 9*(6), 653–69.

Kwan, M.-P., & Ding, G. (2008). Geo-narrative: Extending geographic information systems for narrative analysis in qualitative and mixed-method research. *The Professional Geographer, 60*(4), 443–65.

McCray, T., & Brais, N. (2007). Exploring the role of transportation in fostering social exclusion: The use of GIS to support qualitative data. *Networks and Spatial Economics, 7*(4), 397–412.

McGregor, JoAnn. (2005). Crocodile crimes: People versus wildlife and the politics of postcolonial conservation on Lake Kariba, Zimbabwe. *Geoforum, 36*(3), 353–69

Mennis, J., Mason, M.J., & Cao, Y. (2013). Qualitative GIS and the visualization of narrative activity space data. *International Journal of Geographical Information Science, 27*(2), 267–91.

Meth, P. (2003). Entries and omissions: Using solicited diaries in geographical research. *Area, 35*(2), 195–205.

Milligan, C., Bingley, A., & Gatrell, A. (2005). Digging deep: Using diary techniques to explore the place of health and well-being amongst older people. *Social Science and Medicine, 61*(9), 1882–92.

Morrison, C.-A. (2012). Solicited diaries and the everyday geographies of heterosexual love and home: Reflections on methodological process and practice. *Area, 44*(1), 68–75.

O'Connell, T., & Dyment, J. (2011). The case of reflective journals: Is the jury still out?. *Reflective Practice: International and Multidisciplinary Perspectives, 12*(1), 47–59.

Ostapuk, M. (1997). The dragon and the anchor: using a field experience, journaling, and writing to teach the five geographic skills sets. *Journal of Geography, 9*(4), 196–210.

Pearce, M.W. (2008). Framing the days: Place and narrative in cartography. *Cartography and Geographic Information Science, 35*(1), 17–32.

——— (2014). The last piece is you. *The Cartographic Journal, 51*(2), 107–22.

Quintal-Marineau, M. (2016). *Near the floe edge: Inuit women's role in the Nunavut mixed economy*. Unpublished PhD thesis, Department of Geography, McGill University, Canada.

Singer, M., Stopka, T., Siano, C., Springer, K., Barton, G., Khoshnood, K., Gorry de Puga, A., & Heimer, R. (2000). The social geography of AIDS and hepatitis risk: Qualitative approaches for assessing local differences in sterile-syringe access among injection drug users. *American Journal of Public Health, 90*(7), 1049–56.

Spowart, L., & Nairn, K. (2014) (Re)performing emotions in diary-interviews. *Qualitative Research, 14*(3), 327–40.

Thomas, F. (2007). Eliciting emotions in HIV/AIDS research: A diary based approach. *Area, 39*(1), 74–82.

Tuan, Y.F. (1977). *Space and place: The perspective of experience*. Minneapolis: University of Minnesota Press.

Watts, P.R. (2010). Mapping narratives: The 1992 Los Angeles riots as a case study for narrative-based geovisualization. *Journal of Cultural Geography, 27*(2), 203–27.

Wiles, J.L., Rosenberg, M.W., & Kearns, R.A. (2005). Narrative analysis as a strategy for understanding interview talk in geographic research. *Area, 37*(1), 89–99.

Wiseman, V., Conteh, L., & Matovu, F. (2005). Using diaries to collect data in resource-poor settings: questions on design and implementation. *Health Policy and Planning*, 20, 394–404.

Zuberec, C. (2019). *Hanoi's youth itinerant vendors: An investigation of vending practices, strategies, and experiences in a socialist state.* Unpublished Honours thesis, Department of Geography, McGill University, Canada.

Chapter 15

Alderman, D., & Good, D. (1997). Exploring the virtual South: The idea of a distinctive region on "the web". *Southeastern Geographer, 37*(1), 20–45.

Arnado, J. (2010). Performances across time and space: Drama in the global households of Filipina transmigrant workers. *International Migration, 48*(6), 132–54.

Ash, J. (2010). Architectures of affect: Anticipating and manipulating the event in processes of videogame design and testing. *Environment and Planning D: Society and Space,* 28, 653–71.

Ash, J. (2018). Media and Popular Culture, in James Ash, Rob Kitchin, & Agnieszka Leszczynski (Eds), *Digital geographies* (pp. 143–52). Thousand Oaks: SAGE

Ash, J., Kitchin, R., & Leszczynski, A. (2018a). Digital turn, digital geographies? *Progress in Human Geography, 42*(1), 25–43.

Ash, J., Kitchin, R., & Leszczynski, A. (Eds). (2018b). *Digital geographies*. Thousand Oaks: SAGE.

Bell, S. & Reed, M. (2004). Adapting to the machine: Integrating GIS into qualitative research. *Cartographica, 39*(1), 55–66.

Bonner-Thompson, C. (2017). "The Meat Market": Production and regulation of Masculinities on the Grindr grid in Newcastle-upon-Tyne,

UK. *Gender, Place & Culture, 24*(11), 1611–25.

Cole, J. I., Suman, M., Schramm, P., & Zhou, L. (December 2018). *The World Internet Project – International Report* (9th edition). www.digitalcenter.org/wp-content/uploads/2019/01/World-Internet-Project-report-2018.pdf.

Cope, M. (2018). Qualitative methods and geohumanities. In J. Ash, R. Kitchin, & A. Leszczynski (Eds), *Digital geographies* (pp. 95–105). Thousand Oaks: SAGE,

Crampton, J.W., Graham, M., Poorthuis, A., Shelton, T., Stephens, M., Wilson, M.W., & Zook, M. (2013). Beyond the geotag: Situating "big data" and leveraging the potential of the geoweb. *Cartography and Geographic Information Science, 40*(2), 130–39.

Dean, M., & Laidler, K.A.J. (2014). Leveling the playing field through Facebook: How females construct online playspaces. *Journal of Youth Studies, 17*(1), 113–29.

DeLyser, D., Sheehan, R., & Curtis, A. (2004). eBay and research in historical geography. *Journal of Historical Geography,* 30, 764–82.

Dodge, M. (2019). Rural. In J. Ash, R. Kitchin, & A. Leszczynski (Eds), *Digital geographies* (pp. 36–48). Thousand Oaks: SAGE.

Elwood, S., & Leszczynski, A. (2018). Feminist digital geographies. *Gender, Place & Culture, 25*(5), 629–44.

Emerson, R., Fretz, R., & Shaw, L. (1995). *Writing ethnographic fieldnotes.* Chicago: University of Chicago Press.

Froehling, O. (1999). Internauts and guerilleros: The Zapatista rebellion in Chiapas, Mexico and its extension into cyberspace. In M. Crang, P. Crang, & J. May (Eds), *Virtual geographies: Bodies, space, and relations* (pp. 164–77). New York: Routledge.

Geoghegan, H. (2019). Participatory methods and citizen science." In J. Ash, R. Kitchin, & A. Leszczynski (Eds), *Digital geographies* (pp. 106–17). Thousand Oaks: SAGE.

Graham, M. (2013a). Geography/internet: Ethereal alternate dimensions of cyberspace or grounded augmented realities. *The Geographical Journal, 179*(2), 177–82.

——— (2013b). Social media and the academy: New publics or public geographies? *Dialogues in Human Geography, 3*(1), 77–80.

Green, E., & Singleton, C. (2007). Mobile selves: Gender, ethnicity and mobile phones in the everyday lives of young Pakistani-British women and men. *Information, Communication and Society, 10*(4), 506–26.

——— (2009). Mobile connections: An exploration of the place of mobile phones in friendship relations. *Sociological Review, 57*(1), 125–44.

Gruzd, A., Wellman, B., & Takhteyev, Y. (2011). Imagining Twitter as an imagined community. *American Behavioral Scientist, 55*(10), 1294–318.

Haraway, D. (1985). A manifesto for cyborgs: Science, technology, and socialist feminism in the 1980s. *Socialist Review,* 80, 65–108.

Jones, P., & Evans, J. (2012). The spatial transcript: Analysing mobilities through qualitative GIS. *Area, 44*(1), 92–9.

Juris, J. (2012). Reflections on #Occupy everywhere: Social media, public space, and emerging logics of aggregation. *American Ethnologist, 39*(2), 259–79.

Kinsley, S. (2018). Subject/ivities. In J. Ash, R. Kitchin, & A. Leszczynski (Eds), *Digital geographies* (pp. 153–63). Thousand Oaks: SAGE.

Kitchin, R., Leinhan, D., O'Callaghan, C., & Lawton, P. (2013). Public geographies through social media. *Dialogues in Human Geography, 3*(1), 56–72.

Leurs, K. & Ponzanesi, S. (2011). Communicative spaces of their own: Migrant girls performing selves using instant messaging software. *Feminist Review,* 99, 55–78.

Longhurst, R. (2013). Using Skype to mother: Bodies, emotions, visuality, and screens. *Environment and Planning D: Society and Space,* 31, 664–79.

Madge, C. (2007). Developing a geographer's agenda for online research ethics. *Progress in Human Geography,* 31, 654–74.

Mok, D., Wellman, B., & Carrasco, J. (2010). Does distance matter in the age of the internet? *Urban Studies, 47*(13), 2747–83.

Morse, C., & Mudgett, J. (2018). Happy to be home: Place attachment, family ties, and mobility of contented rural stayers. *The Professional Geographer, 70*(2), 261–69.

Nansen, B., Arnold, M., Gibbs, M., & Davis, H. (2011). Dwelling with media stuff: Latencies and logics of materiality in four Australian homes. *Environment and Planning D: Society and Space, 29*(4), 693–715.

Oman-Reagan, M. (2012). Occupying cyberspace: Indonesian cyberactivists and Occupy Wall Street. *Critical Quarterly, 54*(2), 39–45.

Pfaff, J. (2010). A mobile phone: Mobility, materiality and everyday Swahili trading practices. *Cultural Geographies, 17*(3), 341–57.

Rose, G. (1997). Situating knowledges: Positionality, reflexivities and other tactics. *Progress in Human Geography,* 21, 305–20.

Ruppert, E., Law, J., & Savage, M. (2013). Reassembling social science methods: The challenge of digital devices. *Theory, Culture and Society, 30*(4), 22–46.

Shelton, T., Poorthius, A., Graham, M., & Zook, M. (2014). Mapping the data shadows of Hurricane Sandy: Uncovering the sociospatial dimensions of "big data". *Geoforum,* 52, 167–79.

Thatcher, Jim. (2019). Epistemologies. In J. Ash, R. Kitchin, & A. Leszczynski (Eds), *Digital geographies* (pp. 73–82). Thousand Oaks: SAGE.

Valentine, G., & Holloway, S. (2001). On-line dangers?: Geographies of parents' fears for children's safety in cyberspace. *Professional Geographer, 53*(1), 71–83.

Whatmore, S. (1997). Dissecting the autonomous self: Hybrid cartographies for a relational ethics. *Environment and Planning D: Society and Space, 15*(1), 37–53.

Wilding, R. (2006). "Virtual" intimacies? Families communicating across transnational contexts. *Global Networks, 6*(2), 125–42.

Chapter 16

Askins, K. (2018). Feminist geographies and participatory action research: Co-producing narratives with people and place. *Gender, Place & Culture, 25*(9), 1277–294.

Bingley, A. (2002). Research ethics in practice. In L. Bondi, et al. (Eds), *Subjectivities, knowledges and feminist geographies: The subjects and ethics of research.* London: Rowman and Littlefield.

Brockington, D., & Sullivan, S. (2003). Qualitative research. In R. Scheyvens & D. Storey (Eds), *Development fieldwork: A practical guide.* London: SAGE.

Cahill, C. (2004). Defying gravity: Raising consciousness through collective research. *Children's Geographies, 2*(2), 273–86.

Cahill, C., Sultana, F., & Pain, R. (2007). Participatory ethics: Politics, practices and institutions. *ACME: An International E-Journal for Critical Geographies, 6*(3), 304–18.

Cahill, C., Bradley, M., Castaneda, D., Esquivel, L., Mohamed, N., Organista, J., Sandberg, J., Valerio, M., & Winston, K. (2008). "Represent": Reframing risk through participatory video research. In M. Downing & L. Tenney (Eds), *Video vision: Changing the culture of social science research.* Newcastle, UK: Cambridge Scholars Publishing.

Cahill, C., & Torre, M. (2007). Beyond the journal article: Representations, audience, and the presentation of participatory action research. In S. Kindon, R. Pain, & M. Kesby (Eds), *Participatory action research approaches and methods: Connecting people, participation and place* (pp. 196–205). London: Routledge.

Cameron, J., & Gibson, K. (2005). Participatory action research in a poststructuralist vein. *Geoforum, 36*(3), 315–31.

Cammarota, J. & Fine, M. (2008). Youth participatory action research: A pedagogy for transformational resistance. In *Revolutionizing education: Youth participatory action research in motion* (pp. 1–11). Routledge Taylor & Francis Group.

Chambers, R. (1994). The origins and practice of participatory rural appraisal. *World Development, 22*(7), 953–69.

______ (2017) [2006]. Participatory mapping and geographic information systems: Whose map? Who is empowered and who disempowered? Who gains and who loses? *EJISDC, 25*(1),1–11. https://doi.org/10.1002/j.1681-4835.2006.tb00163.x

Chatterton, P., Fuller, D., & Routledge, P. (2007). Relating action to activism: Theoretical and methodological reflections. In S. Kindon, R. Pain, & M. Kesby (Eds), *Participatory action research approaches and methods: Connecting people, participation and place* (pp. 216–22). London: Routledge.

Cooke, B. (2001). The social psychological limits of participation? In B. Cooke & U. Kothari (Eds), *Participation: The new tyranny?* London: Zed Books.

Cooke, B., & Kothari, U. (2001). The case for participation as tyranny. In B. Cooke & U. Kothari (Eds), *Participation: The new tyranny?* London: Zed Books.

Coombes, B., Johnson, J.T., & Howitt, R. (2014). Indigenous geographies III: Methodological innovation and the unsettling of participatory research. *Progress in Human Geography, 38*(6), 845–54.

Coombs, H.C. (1978). *Kulinma: Listening to Aboriginal Australians*. Canberra: Australian National University Press.

Cornwall, A., & Jewkes, R. (1995). What is participatory research? *Social Science and Medicine*, 41, 1667–76.

Crang, M. (2003). Qualitative methods: Touchy, feely, look-see? *Progress in Human Geography, 27*(4), 494–504.

Gibson-Graham, K.J. (2005). Participatory action research in a poststructuralist vein. *Geoforum*, 36, 315–30.

Goodman, A., Snyder, M., & Wilson, K. (2018). Exploring Indigenous youth perspectives of mobility and social relationships: A Photovoice approach. *The Canadian Geographer / Le Géographe canadien, 62*(3), 314–25.

Hall, B. (2005). In from the cold? Reflections on participatory research from 1970–2005. *Convergence, 38*(1), 5–24.

Halvorsen, S. (2019). Participatory action research. In B. Warf (Ed.), *Oxford bibliographies in geography*. Oxford University Press, New York.

Hickey, S., & Mohan, G. (Eds). (2004). *Participation: From tyranny to transformation*. London: Zed Books.

Hiebert, W., & Swan, D. (1999). Positively fit: A case study in community development and the role of participatory research. *Community Development Journal*, 34, 356–64.

Hume-Cook, G., et al. (2007). Uniting people with place using participatory video in Aotearoa/New Zealand. In S. Kindon, R. Pain, & M. Kesby (Eds), *Participatory action research approaches and methods: Connecting people, participation and place* (pp. 160–9). London: Routledge.

Kale, A. (2017). The Art of Social Connection: Exploring Former Refugee and Host Society Integration via a Collaborative, Participatory Painting Project in Wellington City, unpublished MA Geography thesis, Victoria University of Wellington, Wellington.

Kesby, M. (2000). Participatory diagramming: Deploying qualitative methods through an action research epistemology. *Area, 32*(4), 423–535.

——— (2005). Retheorizing empowerment-through-participation as a performance in space: Beyond tyranny to transformation. *Signs: Journal of Women in Culture and Society*, 30, 2037–65.

——— (2007). Spatialising participatory approaches: The contribution of geography to a mature debate. *Environment and Planning A, 39*(12), 2813–31.

Kesby, M., Kindon, S., & Pain, R. (2005). "Participatory" diagramming and approaches. In R. Flowerdew & D. Martin (Eds), *Methods in human geography* (2nd edition) (pp. 144–66). London: Pearson.

——— (2007). Participation as a form of power: Retheorising empowerment and spatialising participatory action research. In S. Kindon, R. Pain, & M. Kesby (Eds), *Participatory action research approaches and methods: Connecting people, participation and place* (pp. 19–25). London: Routledge.

Kindon, S. (2003). Participatory video in geographic research: A feminist practice of looking? *Area, 35*(2), 142–53.

——— (2012). Thinking-through-complicity with Te Iwi o Ngaati Hauiti: Towards a critical use of participatory video for research. Unpublished PhD thesis. Hamilton: University of Waikato.

Kindon, S. (2010). Participation. In S. Smith, R. Pain, S. Marston, & J.P. Jones III (Eds), *The handbook of social geographies*. London: SAGE.

Kindon, S., & Cupples, J. (2016). "Nothing to declare": Leaving the field. In R. Scheyvens (Ed.), *Development fieldwork: A practical guide*, London: SAGE.

Kindon, S., & Latham, A. (2002). From mitigation to negotiation: Ethics and the geographical imagination in Aotearoa/New Zealand. *New Zealand Geographer, 58*(1), 14–22.

Kindon, S., Pain, R., & Kesby, M. (Eds). (2007a). *Participatory action research approaches and methods: Connecting people, participation and place*. London: Routledge.

——— (2007b). Participatory action research: Origins, approaches and methods. In S. Kindon, R. Pain, & M. Kesby (Eds), *Participatory action research approaches and methods: Connecting people, participation and place* (pp. 9–18). London: Routledge.

Klocker, N. (2012). Doing participatory action research and doing a PhD: Words of encouragement for prospective students. *Journal of Geography in Higher Education, 36*(1), 149–63.

Kothari, U. (2001). Power, knowledge and social control in participatory development. In

B. Cooke & U. Kothari (Eds), *Participation: The new tyranny?*. London: Zed Books.

Leurs, R. (1997). Critical reflections on rapid and participatory rural appraisal. *Development in Practice, 7*(3), 290–93.

Lewin, K. (1946). Action research and minority problems. *Journal of Social Issues,* 1–2, 34–6.

Luchs, M., & Miller, E. (2016). Not so far away: a collaborative model of engaging refugee youth in the dissemination of their digital stories, *Area, 48*(4), 442–48.

McClean, R., Berg, L.D., & Roche, M.M. (1997). Responsible geographies: Co-creating knowledges in Aotearoa. *New Zealand Geographer, 53*(2), 9–15.

Maguire, P. (1987). *Doing participatory research: A feminist approach.* Amherst: Centre for International Education, University of Massachusetts.

Manzo, L., & Brightbill, N. (2007). Towards a participatory ethics. In S. Kindon, R. Pain, & M. Kesby (Eds), *Participatory action research approaches and methods: Connecting people, participation and place* (pp. 33–40). London: Routledge.

Mason K. (2015). Participatory action research: Co-production, governance and care. *Geography Compass, 9*(9), 497–507.

Mitchell, Katharyne (Ed.). (2008). *Practising public scholarship: Experiences and possibilities beyond the academy,* Malden, MA: Wiley-Blackwell.

Monk, J., Manning, P., & Denman, C. (2003). Working together: Feminist perspectives on collaborative research and action. *ACME: An International E-Journal for Critical Geographies, 2*(1), 91–106.

Mosse, D. (1994). Authority, gender and knowledge: Theoretical reflections on the practice of participatory rural appraisal. *Development and Change,* 25, 497–526.

mrs kinpaisby. (2008). Taking stock of participatory geographies: Envisioning the communiversity. *Transactions of the Institute of British Geographers,* 33, 292–99.

mrs c kinpaisby-hill. (2008). Publishing from participatory research. In A. Blunt (Ed.), *Publishing in geography: A guide for new researchers* (pp. 45–7). London: Wiley-Blackwell.

______ (2013). Participatory approaches to authorship in the academy. In A. Blunt (Ed.), *Publishing and getting read: A guide for new researchers in geography* (Section 4.2, p. 24). London: Wiley-Blackwell.

______ (2011). Participatory praxis and social justice: Towards more fully social geographies. In V. DelCasino, M. Thomas, P. Cloke, & R. Panelli (Eds), *A companion to social geography* (pp. 214–34). London: Blackwell.

Pain, R. (2003). Social geography: On action-oriented research. *Progress in Human Geography, 27*(5), 677–85.

Pain, R., & Askins, K. (2011). Contact zones: Participation, materiality and the messiness of interaction. *Environment and Planning D: Society and Space*, 29(5), 803–21.

Pain, R., Finn, M., Bouveng, R., & Ngobe, G. (2013). Productive tensions: Engaging geography students in participatory action research with communities. *Journal of Geography in Higher Education, 37*(1), 28–43.

Pain, R., Kesby, M., & Askins, K. (2011). Geographies of impact: Power, participation and potential. *Area, 43*(2), 183–88.

Pain, R., Finn, M., Bouveng, R., & Ngobe, G. (2013). Productive tensions: Engaging geography students in participatory action research with communities. *Journal of Geography in Higher Education, 37*(1), 28–43.

Pain R., et al. (2016). Mapping alternative impact: Alternative approaches to impact from co-produced research, Centre for Social Justice and Community Action, Durham University. www.dur.ac.uk/resources/beacon/MappingAlternativeImpactFinalReport.pdf

Pain, R., & Francis, P. (2003). Reflections on participatory research. *Area, 35*(1), 46–54.

Pain, R. & Kindon, S. (2007). Participatory geographies. *Environment and Planning A*, 39, 2807–12.

Pain, R., Whitman, G., Milledge, D. & Lune Rivers Trust. (2017). Participatory Action Research Toolkit: An Introduction to Using PAR as an Approach to Learning, Research and Action, Department of Geography, University of Durham. http://communitylearningpartnership.org/wp-content/uploads/2017/01/PARtoolkit.pdf

Parkes, M., & Panelli, R. (2001). Integrating catchment ecosystems and community health: The value of participatory action research. *Ecosystem Health, 7*(2), 85–106.

Parnwell, M. (2003). Consulting the poor in Thailand: Enlightenment or delusion? *Progress in Development Studies, 3*(2), 99–112.

Pratt, G. (2000). Participatory action research. In R. Johnston, D. Gregory, G. Pratt, & M. Watts (Eds), *Dictionary of human geography* (4th edition). Oxford: Blackwell.

Pratt, G., in collaboration with the Philippine Women Centre. (1999). Is this Canada? Domestic workers' experiences in Vancouver, B.C. In J. Momson, (Ed.), *Gender, migration and domestic service*. London: Routledge.

Pretty, J., et al. (1995). *Participatory learning and action: A trainer's guide*. London: International Institute for Environment and Development.

Reason, P. (1988) Three approaches to participatory research. In N.K. Denzin & Y.S. Lincoln (Eds), *Strategies of qualitative inquiry* (pp. 261–91). Thousand Oaks, CA: SAGE Publications.

Ritterbusch, A. (2019). Empathy at knifepoint: The dangers of research and lite pedagogies for social justice movements. *Antipode, 51*(4), 1296–317.

Robinson, J. (2010). Syracuse community geography: evaluating a new approach to public participation geographic information systems, Doctor of Philosophy, University of North Carolina at Chapel Hill. https://doi.org/10.17615/h2ee-pj75

Robinson, J., Block, D., & Rees, A. (2017). Community geography: Addressing barriers in public participation GIS. *The Cartographic Journal, 54*(1), 5–13.

Ruddick, S. (2004). Activist geographies: Building possible worlds. In P. Cloke, P. Crang, & M. Goodwin (Eds), *Envisioning human geographies*. London: Arnold.

Sanderson, E., & Kindon, S. (2004). Progress in participatory development: Opening up the possibility of knowledge through progressive participation. *Progress in Development Studies, 4*(2), 114–26.

Starodub, A. (2019). Horizontal participatory action research: Refugee solidarity in the border zone. *Area, 51*(1), 166–73.

Sultana, F. (2017). Reflexivity. In D. Richardson, N. Castree, M. Goodchild, A. Kobayashi, W. Liu, & R. Marston (Eds), *The international encyclopedia of geography*. New York: John Wiley and Sons Ltd.

Thomas-Slayter, B. (1995). A brief history of participatory methodologies. In R. Slocum, L. Wichart, D. Rochealeau, & B. Thomas-Slayter (Eds), *Power, process and participation: Tools for change*. London: Intermediate Technology Publications

Wadsworth, Y. (1998). What is participatory action research? Action Research International Paper 2. www.scu.edu.au/schools/gecm/ar/ari/p-ywadsworth98.html.

Williams, G., et al. (2003). Enhancing pro-poor governance in eastern India: Participation, politics and action. *Progress in Development Studies, 3*(2), 159–78.

Wynne-Jones, S., North, P., & Routledge, P. (Eds). (2015). Practising participatory geographies: potentials, problems and politics. *Area*, 47.

Chapter 17

Anson, A. (2014). "The world is my backyard": Romanticization, Thoreauvian rhetoric, and constructive confrontation in the tiny house movement. In W.G. Holt (Ed.), *From sustainable to resilient cities: Global concerns and urban efforts* (pp. 289–313). Emerald Group Publishing Limited.

Barrett, M. (1991). *The politics of truth: From Marx to Foucault*. Cambridge: Polity Press.

Binnie, J., & Valentine, G. (1999). Geographies of sexuality—A review of progress. *Progress in Human Geography*, 23, 176–87.

Brown, M. (2000). National closets: Governmentality, sexuality, and the census (with Paul Boyle). In M. Brown, *Closet Space: Geographies of Metaphor from the Body to the Globe* (pp. 88–115). London: Routledge.

Budge, K. (2019). Making in the City: Disjunctures between public discourses and urban policy. *Australian Geographer, 50*(2) 185–89.

Burman, E., & Parker, I. (1993). Against discursive imperialism, empiricism and constructionism: Thirty-two problems with discourse analysis. In E. Burman & I. Parker (Eds), *Discourse analytic research: Repertoires and readings of texts in action*. London: Routledge.

Castree, N. (2013). *Making sense of nature*. London: Routledge.

Crang, M. (1997). Picturing practices: Research through the tourist gaze. *Progress in Human Geography*, 21, 359–73.

Dinhopl, A., & Gretzel, U. (2016). Selfie-taking as touristic looking. *Annals of Tourism Research*, 57, 126–39.

Duncan, J.S. (1987). Review of urban imagery: Urban semiotics. *Urban Geography, 8*(5), 473–83.

Edwards, E. (2003). Negotiating spaces. In J.M. Schartz & J.R. Ryan (Eds), *Picturing place photography and the geographical imagination*. London and New York: I.B. Tauris.

Fairclough, N. (2003). *Analysing discourse: Textual analysis for social research*. New York and London: Routledge.

Foucault, M. (1972). *The archaeology of knowledge* (A. Sheridan Smith, Trans). New York: Pantheon.

______ (1980). *Power/knowledge*. Brighton: Harvester.

Gill, R. (1996). Discourse analysis: Practical implementations. In J.T.E. Richardson (Ed.), *Handbook of qualitative methods for Psychology and the Social Sciences*, 141–56. Leicester: British Psychological Society.

Gorman-Murray, A. (2007). Contesting domestic ideas: queering the Australian home. *Australian Geographer, 38*(3), 195–213.

Goss, J.D. (1993). Placing the market and marketing place: Tourist advertising of the Hawaiian Islands, 1972–1992. *Environment and Planning D: Society and Space*, 11, 663–88.

Grimwood, B.S.R., Yudina, O., Muldhoon, M., & Qiu, J. (2015). Responsibility in tourism: A discursive analysis. *Annals of Tourism Research*, 50, 22–38.

Head, L., & Muir, P. (2006). Suburban life and the boundaries of nature: Resilience and rupture in Australian backyards. *Transactions of the Institute of British Geographers, 31*(4), 505–24

Holloway, S.L, Jayne, M., & Valentine, G. (2008). "Sainsbury's is my local": English alcohol policy, domestic drinking practices and the meaning of home. *Transactions of the Institute of British Geographers*, 33, 532–47.

Johnston, L. (2005). Transformative tans? Gendered and raced bodies on beaches. *New Zealand Geographer*, 61, 110–16.

Larsen, J., & Urry, J. (2011). Gazing and performing. *Environment and Planning D: Society and Space, 29*(6), 1110–25.

Law, L. (2000). *Sex work in Southeast Asia: The place of desire in a time of AIDS*. London and New York: Routledge.

Lawrence, R., & Adams, M. (2005). First Nations and the politics of indigeneity: Australian perspectives on indigenous peoples, resource management and global rights. *Australian Geographer, 36*(2), 265–72.

Longhurst, R. (2005). (Ad)dressing pregnant bodies in New Zealand: Clothing, fashion, subjectivities and spatialities. *Gender, Place & Culture, 12*(4), 433–46.

McGregor, G. (1994). *EcCentric visions: Reconstructing Australia*. Waterloo, ON: Wilfrid Laurier University Press.

McFarlane, T., & Hay, I. (2003). The battle for Seattle: Protest and popular geopolitics in *The Australian* newspaper. *Political Geography*, 22, 211–32.

Markwell, K. (1997). Dimensions of a nature-based tour. *Annals of Tourism Research, 24*(1), 131–55.

Markwell K., & Waitt, G. (2009). Festivals, space and sexuality: Gay pride in Australia. *Tourism Geographies*, 11(2), 143–68.

Penfold, H., Waitt, G., & McGuirk, P. (2018). Portrayals of the tiny house in electronic media: Challenging or reproducing the Australian dream home. *Australian Planner*, 55, 164–73.

Phillips, N., & Hardy, C. (2002). *Discourse analysis: Investigating processes of social construction*. Thousand Oaks, CA: SAGE.

Phillips, L., & Jørgensen, M.W. (2002). *Discourse analysis as theory and method*. London: SAGE.

Potter, J. (1996). Discourse analysis and constructionist approaches: Theoretical background. In J.T.E. Richardson (Ed.), *Handbook of qualitative methods for psychology and the social sciences*. Leicester: British Psychological Society.

Riley, M., & Harvey, D. (2007). Editorial: Talking landscapes: On oral history and the practice of geography. *Social & Cultural Geography, 8*(4), 391–415.

Rose, D.B. (1996a). Histories and rituals: Land claims in the Territory. In B. Attwood (Ed.), *In the age of Mabo: History, Aborigines and Australia* (pp. 35–52). Sydney: Allen and Unwin.

Rose, G. (1997). Situating knowledges: Positionality, reflexivities and other tactics. *Progress in Human Geography*, 21: 305–20.

______ (2001). *Visual methodologies. An introduction to the interpretation of visual materials*. London: SAGE.

Schaffer, K. (1988). *Women and the bush: Forces of desire in the Australian cultural tradition*. Cambridge: Cambridge University Press.

Turner, F.J. (1920). *The frontier in American history*. New York: Henry Holt.

Urry, J. (2002). *The tourist gaze* (2nd edition). London: SAGE.

Waitt, G. (2014) Bodies that sweat: the affective responses of young women in Wollongong,

New South Wales, Australia. *Gender, Place & Culture, 21*(6), 666–82.

Waitt, G., & Head, L. (2002). Postcards and frontier mythologies: Sustaining views of the Kimberley as timeless. *Environment and Planning D: Society and Space,* 20, 319–44.

Waitt, G., & Warren, A. (2008). "Talking shit over a brew after a good session with your mates": Surfing, space and masculinity. *Australian Geographer,* 39(3), 353–65.

Ward, R. (1958). *The Australian legend.* Melbourne: Oxford University Press.

Whittaker, J., & Mercer, D. (2004). The Victorian bushfires of 2002–03 and the politics of blame: A discourse analysis. *Australian Geographer, 35*(3), pp. 259–87.

Wood, L.A., & Kroger, R.O. (2000). *Doing discourse analysis: Methods for studying action in talk and text.* Thousand Oaks, CA: SAGE.

Chapter 18

Ash, J., Kitchin, R., & Leszczynski, A. (Eds). (2019). *Digital geographies.* Thousand Oaks, CA: SAGE.

Baxter, J., & Eyles, J. (1997). Evaluating qualitative research in social geography: Establishing "rigour" in interview analysis. *Transactions of the Institute of British Geographers, 22*(4), 505–25.

Blumen, O. (2002). Criss-crossing boundaries: Ultraorthodox Jewish women go to work. *Gender, Place & Culture, 9*(2), 133–51.

Clifford, N., Cope, M., Gillespie, S., & French, S. (Eds) (2016). *Key methods in human geography.* Thousand Oaks, London, New Delhi: SAGE Publications.

Cope, M. (2008). Patchwork neighborhood: Children's real and imagined geographies in Buffalo, NY. *Environment and Planning, A* 40, 2845–63.

Cope, M., & Elwood, S. (Eds). (2009). *Qualitative GIS: A mixed-methods approach.* London: SAGE.

Cope, M. & Kurtz, H. (2016). Qualitative data analysis. In N. Clifford, M. Cope, S. Gillespie, & S. French (Eds), *Key methods in human geography* (pp. 647–64). London: SAGE.

Cope, M. & Lee, Brian H.Y. (2016). Mobility, communication, and place: Navigating the landscapes of suburban U.S. teens. *Annals of the Association of American Geographers, 106*(2), 311–20.

Crang, M. (2015). The promises and perils of a digital geohumanities. *Cultural Geographies, 22*(2), 351–60.

de Certeau, M. (1984). *The practice of everyday life.* Berkeley: University of California Press.

DeLyser, D., Herbert, S., Aitken, S., Crang, M., & McDowell, L. (Eds). (2010). *The SAGE handbook of qualitative geography.* London: SAGE.

Denzin, N.K., & Lincoln, Y.S. (Eds). (2018). *The SAGE handbook of qualitative research* (5th edition). London: SAGE.

Dey, I. (1993). *Qualitative data analysis: A user-friendly guide for social scientists.* London: Routledge.

Dixon, D., (2010). Analysing meaning. In B. Gomez & J.P. Jones III (Eds), *Research methods in geography* (pp. 392–407). Chichester, UK: Wiley-Blackwell.

Doel, M. (2010). Analysing cultural texts. In D. DeLyser, S. Herbert, S. Aitken, M. Crang, & L. McDowell (Eds), *The SAGE handbook of qualitative geography* (pp. 485–96). London: SAGE.

Fielding, N., & Cisneros-Puebla, C.A. (2009). CAQDAS-GIS convergence: Toward a new integrated mixed method research practice? *Journal of Mixed Methods Research, 3*(4), 349–70.

Housel, J. (2009). Geographies of whiteness: The active construction of racialized privilege in Buffalo, New York. *Social & Cultural Geography, 10*(2), 131–51.

Jung, J.-K. (2009). Computer-aided qualitative GIS: A software-level integration of qualitative research and GIS. In M. Cope & S. Elwood (Eds), *Qualitative GIS: A mixed-methods approach* (pp. 115–35). London: SAGE.

Kane, M., & Trochim, W. (2006). *Concept mapping for planning and evaluation.* London: SAGE Gomez, B., & Jones III, J.P. (Eds). (2010). *Research methods in geography.* Chichester, UK: Wiley–Blackwell.

Goodman, A., Snyder, M., & Wilson, K. (2018). Exploring Indigenous youth perspectives of mobility and social relationships: A Photovoice approach. *The Canadian Geographer/Le Géographe canadien, 62*(3), 314–25.

Lefebvre, H. (1991). *The production of space.* Oxford: Blackwell.

Miles, M.B., & Huberman, A.M. (2013). *Qualitative data analysis: An expanded sourcebook* (3rd edition). Thousand Oaks, CA: SAGE.

Miles, M.B., Huberman, A.M., & Saldaña, J. (2020). *Qualitative data analysis: A methods sourcebook and the coding manual for qualitative researchers.* Thousand Oaks, CA: SAGE.

Pacheco-Vega, R. (2019). Blog: *Suggestions on Qualitative Research Methods Articles for Graduate Students*. www.raulpacheco.org/2019/09/suggestions-on-qualitative-research-methods-articles-and-books-for-graduate-students/

Saldaña, J. (2012). *The coding manual for qualitative researchers* London: SAGE.

Silver, C., & Lewins, A. (2014). *Using software in qualitative research: A step-by-step guide* (2nd edition). Londond: SAGE Publications. http://methods.sagepub.com/book/using-software-in-qualitative-research-2e

Silverman, D. (1991). *Interpreting qualitative data: Methods for analysing talk, text and interaction*. Thousand Oaks, CA: SAGE.

Silverman, D. (2001). *Interpreting qualitative data: Methods for analysing talk, text and interaction* (2nd edition). Thousand Oaks, CA: SAGE.

Strauss, A., & Corbin, J. (1990). *Basics of qualitative research: Grounded theory, procedures and techniques*. Newbury Park: SAGE.

Tuck, E., & McKenzie, M. (2015). *Place in research: Theory, methodology, and methods*. New York: Routledge.

van Hoven, B. (2010). Computer assisted qualitative data analysis. In N. Clifford, S. French, & G. Valentine (Eds), *Key methods in geography* (2nd edition) (pp. 453–65). London: SAGE.

Watson, A., & Till, K. (2010). Ethnography and participant observation. In D. DeLyser, S. Herbert, S. Aitken, M. Crang, & L. McDowell (Eds), *The SAGE handbook of qualitative geography* (pp. 121–37). London: SAGE.

Chapter 19

Alvermann, D.E., O'Brien, D.G., & Dillon, D.R. (1996). On writing qualitative research. *Reading Research Quarterly*, 31, 114–20.

Anzaldúa, G. (1987). *Borderlands/La frontera: The new mestiza*. San Francisco: Spinsters/Aunt Lute Press.

Barnes, T.J. (1989). Place, space and theories of economic value: Contextualism and essentialism. *Transactions of the Institute of British Geographers* NS, 14, 299–316.

______ (1993). Whatever happened to the philosophy of science? *Environment and Planning A*, 25, 301–4.

Barnes, T.J., & Duncan, J. (1992). Introduction: Writing worlds. In T.J. Barnes & J. Duncan (Eds), *Writing worlds: Discourse, text and metaphor in the representation of landscape*. London: Routledge.

Barnes, T.J., & Gregory, D. (1997). Worlding geography: Geography as situated knowledge. In T.J. Barnes & D. Gregory (Eds), *Reading human geography: The poetics and politics of inquiry*. London: Arnold.

Barry, K. (2018). *Everyday practices of tourism mobilities: Packing a bag*. Routledge, London and New York

Baxter, J., & Eyles, J. (1997). Evaluating qualitative research in social geography: Establishing "rigour" in interview analysis. *Transactions of the Institute of British Geographers, 22*(4), 505–25.

Berg, L.D. (1993). Between modernism and postmodernism. *Progress in Human Geography*, 17, 490–507.

Berg, L.D. (1994). Masculinity, place, and a binary discourse of theory and empirical investigation in the human geography of Aotearoa/New Zealand. *Gender, Place & Culture, 1*(2), 245–60.

______ (1997). *Banal geographies*. Paper presented to the Inaugural International Conference of Critical Geographers, Vancouver, 10–13 August.

______ (2004). Scaling knowledge: Towards a critical geography of critical geography. *Geoforum, 35*(5), 553–58.

______ (2012). Geography – (neo)liberalism – white supremacy. *Progress in Human Geography, 36*(4), 508–17.

Berg, L.D., Evans, M., Fuller, D., & the Okanagan Urban Aboriginal Health Research Collective. (2007). Ethics, hegemonic whiteness, and the contested imagination of "Aboriginal community" in social science research in Canada. *ACME: An International E-Journal for Critical Geographies*, (special issue on participatory research ethics) *6*(3), 395–410.

Bissell, D. (2018). *Transit life: How commuting is transforming our cities*. MIT Press.

Bondi, L. (1997). In whose words? On gender identities, knowledge and writing practices. *Transactions of the Institute of British Geographers*, 22, 245–58.

______ (2014). Understanding feelings: Engaging with unconscious communication and embodied knowledge. *Emotion, Space and Society*, 10, 44–54.

Bordo, S. (1986). The Cartesian masculinization of thought. *Signs*, 11, 439–56.

Bradford, M. (2003). Writing essays, reports and dissertations. SAGE N.J. Clifford & G. Valentine (Eds), *Key methods in geography* (pp. 515–32). London: SAGE.

Bryman, A. (2012). *Social research methods* (4th edition). Oxford: Oxford University Press.

Butz, D. (2010). Autoethnography as sensibility. In D. DeLyser, S. Herbert, S. Aitken, M. Crang, & L. McDowell (Eds), *The SAGE handbook of qualitative geography* (pp. 138–56). London: SAGE Publications. doi: http://dx.doi.org/10.4135/9780857021090.n10

Butz, D., & Besio, K. (2004). The value of autoethnography for field research in transcultural settings. *Professional Geographer, 56*(3), 350–60.

Cahill, C. (2010). "Why do they hate us?" Reframing immigration through participatory action research. *Area, 42*(2), 152–61.

Clough, N., & Blumberg, R. (2012). Toward anarchist and autonomous Marxist geographies. *ACME: An International E-Journal for Critical Geographies, 11*(3), 335–51.

Cook, I., et al. (2007). "It's more than just what it is": Defetishising commodities, expanding fields, mobilising change . . .. *Geoforum,* 38, 1113–26.

Creme, P. (2003). Why can't we allow students to be more creative? *Teaching in Higher Education, 8*(2), 273–77.

Dear, M. (1988). The postmodern challenge: Reconstructing human geography. *Transactions of the Institute of British Geographers,* NS 13, 262–74.

de Leeuw, S., Cameron, E., & Greenwood, M.L. (2012). Participatory and community-based research, Indigenous geographies, and the spaces of friendship: A critical engagement. *The Canadian Geographer/Le Géographe canadien, 56*(2), 180–94.

DeLyser, D. & Hawkins, H. (2014). Introduction: writing creatively–process, practice, and product. *Cultural Geographies,* 21, 131–34.

Denzin, N.K. (2003). Reading and writing performance. *Qualitative Research, 3*(2), 243–68.

——— (2008). Emancipatory discourses and the ethics and politics of interpretation. In N.K. Denzin & Y.S. Lincoln (Eds), *Collecting and interpreting qualitative materials research* (3rd edition) (pp. 435–71). Los Angeles: SAGE.

Denzin, N.K., & Lincoln, Y.S. (Eds). (2008). *Collecting and interpreting qualitative materials* (3rd edition). Los Angeles: SAGE.

Derrida, J. (1981). *Dissemination* (B. Johnson, Trans.). Chicago: University of Chicago Press.

de Vaus, D. (2002). *Surveys in social research* (5th edition). Sydney: Allen and Unwin.

Dixon, D.P., & Jones III, J.P. (1996). For a supercalifragilisticexpialidocious scientific geography. *Annals of the Association of American Geographers,* 86, 767–79.

Ellis, C. (2004). *The ethnographic I: A methodological novel about autoethnography.* Walnut Creek, CA: AltaMira Press.

Ely, M. (2007). In-forming re-presentations. In D.J. Clandinin (Ed.), *Handbook of narrative inquiry, mapping a methodology* (pp. 567–98). Thousand Oaks, CA, and London: SAGE.

England, K.V.L. (1994). Getting personal: Reflexivity, positionality, and feminist research. *Professional Geographer,* 46, 80–9.

Evans, M., Hole, R., Berg, L.D., Hutchinson, P., & Sookraj, D. (2009). Common insights, differing methodologies toward a fusion of Indigenous methodologies, participatory action research, and White studies in an urban Aboriginal research agenda. *Qualitative Inquiry, 15*(5), 893–910.

Fairclough, N. 1992. *Discourse and Social Change.* Cambridge: Polity Press.

Fowler, R. (1991). *Language in the news. Discourse and ideology in the press.* London: Routledge.

Flowerdew, R., & Martin, D. (Eds). (2005). *Methods in human geography: A guide for students doing a research project* (2nd edition). Harlow: Pearson.

Frankenberg, R., & Mani, L. (1993). Crosscurrents, crosstalk: Race, "postcoloniality" and the politics of location. *Cultural Studies,* 7, 292–310.

Freeman, M. (2007). Auto-biographical understanding and narrative inquiry. In D.J. Clandinin (Ed.), *Handbook of narrative inquiry, mapping a methodology* (pp. 120–45). London: SAGE.

Foucault, M. (1977). *Language, counter-memory, practice: Selected essays and interviews.* Ithaca: Cornell University Press.

Fowler, R. (1991). *Language in the news: Discourse and ideology in the press.* London: Routledge.

Geertz, C. (1973). *The interpretation of culture: Selected essays.* New York: Basic Books.

Gregory, D. (1978). *Ideology, science and human geography.* London: Hutchinson.

Guelke, L. (1978). Geography and logical positivism. In D. Herbert & R.J. Johnston (Eds), *Geography and the urban environment,* v. 1. New York: Wiley.

Haraway, D. (1991). Situated knowledges: The science question in feminism and the privilege of partial perspective. In D. Haraway (Ed.), *Simians, cyborgs and women: The reinvention of nature.* London: Routledge.

Harrison, R.T., & Livingstone, D.N. (1980). Philosophy and problems in human geography: A pre-suppositional approach. *Area,*12, 25–30.

Hay, I. (2012). *Communicating in geography and the environmental sciences* (4th edition). Melbourne: Oxford University Press.

Hay, I., Allington, P., & Bochner, D. (2021). *Making the grade* (5th edition). Melbourne: Oxford University Press.

Holliday, A. (2007). *Doing and writing qualitative research* (2nd edition). Thousand Oaks, CA, and London: SAGE.

Jay, N. (1981). Gender and dichotomy. *Feminist Studies,* 7, 38–56.

Jones, A. (1992). Writing feminist educational research: Am "I" in the text?. In S. Middleton & A. Jones (Eds), *Women and education in Aotearoa.* Wellington: Bridget Williams Books.

Kitchin, R., & Tate, N. (2000). *Conducting research in human geography: Theory, methodology and practice.* London: Longman.

Kneale, P. (1999). *Study skills for geography students.* London: Arnold.

Kolakowski, L. (1972). *Positivist philosophy: From Hume to the Vienna Circle.* London: Penguin.

Le Doeff, M. (1987). Women and philosophy. In T. Moi (Ed.), *French feminist thought: A reader.* Oxford: Basil Blackwell.

Lloyd, G. (1984). *The man of reason: "Male" and "female" in western philosophy.* London: Methuen.

Madge, C. (2014). On the creative (re)turn to geography: poetry, politics and passion. *Area, 46*(2), 178–85.

Marcus, G.E., & Fisher, M.M.J. (1986). *Anthropology as cultural critique: An experimental moment in the human sciences.* Chicago: University of Chicago Press.

Massey, D. (1993). Power-geometry and a progressive sense of place. In J. Bird, B. Curtis, G. Robertson, & L. Tickner (Eds), *Mapping the futures.* London: Routledge.

Mohanty, C.T. (1991). Cartographies of struggle: Third world women and the politics of feminism. In C. Mohanty, A. Russo, & L. Torres (Eds), *Third world women and the politics of feminism.* Bloomington: University of Indiana Press.

Moss, P. (2016). Writing illness in feminist autobiographical research. In N. Fenton & J. Baxter (Eds), *Practicing qualitative methods in health geographies* (pp. 75–91). London: Routledge.

Nietzsche, F.W. (1969). *On the genealogy of morals* (W. Kaufmann & R.J. Hollingdale, Trans). New York: Vintage.

Pacheco-Vega, R., & Parizeau, K. (2018). Doubly engaged ethnography: Opportunities and challenges when working with vulnerable communities. *International Journal of Qualitative Methods, 17*(1), 1–13.

Pain, R. (2014). Impact: Striking a blow or walking together? *ACME: An International E-Journal for Critical Geographies, 13*(1), 19–23.

Opie, A. (1992). Qualitative research, appropriation of the other and empowerment. *Feminist Review,* 40, 52–69.

Popper, K. (1959). *The logic of scientific discovery.* London: Hutchinson.

Proudfoot, J. (2019). The stakes of situated knowledges. *Dialogues in Human Geography,* 9(2), 158–61. Available: DOI 10.1177/2043820619850270

Radice, H. (2013). How we got here: UK higher education under neoliberalism. *ACME: An International E-Journal for Critical Geographies, 12*(3), 407–18.

Richardson, L., & St. Pierre, E.A. (2008). *Writing: A method of inquiry.* In N.K. Denzin & Y.S. Lincoln (Eds), *The SAGE handbook of qualitative research* (3rd edition) (pp. 473–99). SAGE Publications Ltd.

Rose, G. (1997). Situating knowledges: Positionality, reflexivities and other tactics. *Progress in Human Geography,* 21, 305–20.

Simandan, D. (2019). Revisiting positionality and the thesis of situated knowledge. *Dialogues in Human Geography,* 9(2), 129–49. Available: DOI 10.1177/2043820619850013

Stanton, N. (1996). *Mastering Communication* (3rd edition). London: MacMillan.

Sotoudehnia, Maral. (2016). Making sense of field encounters through intimate encounters. In P. Moss & C. Donovan (Eds), *Writing intimacy into feminist geography* (pp. 33–40). London: Routledge.

Valentine, G. (2003). Geography and ethics: In pursuit of social justice—Ethics and emotions in geographies of health and disability research. *Progress in Human Geography, 27*(3), 375–80.

Valentine, Gill, Butler, Ruth, & Skelton, Tracey. (2001). The ethical and methodological complexities of doing research with "vulnerable" young people. *Ethics, Place & Environment, 4*(2), 119–25.

Whitson, R. (2017). Painting pictures of ourselves: Researcher subjectivity in the practice of feminist reflexivity. *The Professional Geographer, 69*(2), 299–306.

Wright, M.W. (2010). Geography and gender: Feminism and a feeling of justice. *Progress in Human Geography, 34*(6), 818–27. Available: DOI 10.1177/0309132510362931

Chapter 20

Atkinson, C. (2018). *Beyond bullet points. Using PowerPoint to tell a compelling story that gets results* (4th edition). New York: Pearson Education, Inc.

Becker, H.S. (2007). *Writing for social scientists: How to start and finish your thesis, book, or article* (2nd edition). Chicago: University of Chicago Press.

Braun, V., & Clarke, V. (2013). *Successful qualitative research: A practical guide for beginners*. Los Angeles: SAGE.

Cloke, P., & Conradson, D. (2018). Transitional organisations, affective atmospheres and new forms of being-in-common: Post-disaster recovery in Christchurch, New Zealand. *Transactions of the Institute of British Geographers*, *43*(3), 360–76.

Corson, K. (2015). Political agency and citizen journalism: Twitter as a tool of evaluation. *The Professional Geographer*, *67*(3), 364–73.

Cupples, J., & Kindon, S. (2014). Returning to university and writing the field. In R. Scheyvens (Ed.), *Development fieldwork: A practical guide*, 2nd edition (pp. 236–52). Los Angeles: SAGE Publications.

DeLyser, D. (2010). Writing qualitative geography. In D. DeLyser et al. (Eds), *Handbook of qualitative geography* (pp. 341–58). London: SAGE.

Ellison-Collins, T., Ncube, N., Nguyen, K., & Quaid, L. (2017). Memories of the residential red zone, unpublished pamphlet, GEOG 402 Masters class, University of Canterbury.

Forbat, L., & Henderson, J. (2005). Theoretical and practical reflections on sharing transcripts with participants. *Qualitative Health Research*, 15(8), 1114–28.

James, A.M. & Rashed, T. (2006). In their own words: Utilizing weblogs in quick response research. In *Natural Hazards Center, Learning from catastrophe: Quick response research in the wake of Hurricane Katrina* (pp. 71–96). Boulder: Institute of Behavioural Science.

Karnieli-Miller, O., Strier, R., & Pessach, L. (2009). Power relations in qualitative research. *Qualitative Health Research*, *19*(2), 279–89.

Mackinnon, D. & Derickson, K.D. (2013). From resilience to resourcefulness: A critique of resilience policy and activism. *Progress in Human Geography*, *37*(2), 253–270.

Marsh. (2014). Comparing claims from catastrophic earthquakes, February 2014, Marsh & McLennan Companies, http://deutschland.marsh.com/Portals/32/Documents/Marsh%20Risk%20Management%20Research_Earthquakes.pdf [accessed March 22, 2014]

Marshall, C., & Rossman, G.B. (2015). *Designing qualitative research* (6th edition). Thousand Oaks: SAGE.

Mitchell, K. (Ed.) (2008). *Practising public scholarship: Experiences and possibilities beyond the academy*. Malden, MA: Wiley-Blackwell.

Medd, W., Deeming, H., Walker, G., et al. (2015). The flood recovery gap: A real-time study of local recovery following the floods of June 2007 in Hull, North East England. *Journal of Flood Risk Management*, *8*(4), 315–28.

Ovink, H. & Boeijenga, J. (2018). *Too big. Rebuild by design: A transformative approach to climate change*. Nai010 Publishers: Rotterdam.

Pawson, E. (2016). Classrooms without borders. New spaces and places of learning. *Journal of Geography in Higher Education*, *40*(1), 14–30.

Phillips, R. (2015). Playful and multi-sensory fieldwork: Seeing, hearing and touching New York. *Journal of Geography in Higher Education*, *39*(4), 617–29.

Reynolds, G. (2019). *presentationzen: Simple Ideas on Presentation, Design and Delivery* (3rd edition). Berkeley, CA: New Riders.

Reynolds, J., et al. (2011). Quality assurance of qualitative research: A review of the discourse. *Health Research Policy and Systems*, *9*(43).

Robinson, J. & Hawthorne, T. (2018). Making space for community-engaged scholarship in geography. *The Professional Geographer*, 70(2): 277–83.

Richardson, L., & St. Pierre, E.A. (2008). Writing: A method of inquiry. In N.K. Denzin & Y.S. Lincoln (Eds), *The SAGE handbook of qualitative research* (pp. 473–99). SAGE Publications Ltd.

Smith, L.T. (2012). *Decolonizing methodologies: Research and indigenous peoples* (2nd edition). London: Zed Books.

Stewart, E.J., & Draper, D. (2009). Reporting back research findings: A case study of community-based tourism research in northern Canada. *Journal of Ecotourism*, *8*(2), 128–43.

Tengco, S. (2013). An everyday hero: Menchu de Luna Sanchez's view from the First Lady's box. https://obamawhitehouse.archives.gov/blog/2013/02/14/everyday-hero-menchu-de-luna-sanchez-s-view-first-lady-s-box (accessed March 26, 2019)

Willyard, C., Scudellari, M., & Nordling, L. (2018). Partners in science. *Nature*, 562, 4 October, 24–8.

Wolcott, H. (2008). *Writing up qualitative research* (3rd edition). Thousand Oaks: SAGE.

Glossary

Ash, J., Kitchin, R., & Leszczynksi, A. (2016). Digital turn, digital geographies? *Progress in Human Geography, 42*(1), 25–43.

Baxter, J., & Eyles, J. (1997). "Evaluating qualitative research in social geography: Establishing 'rigour' in interview analysis." *Transactions of the Institute of British Geographers, 22*(4): 505–25.

Beauvoir, S., & Parshley, H.M. (1949). *The second sex*. Harmondsworth, Middlesex, U.K: Penguin Books.

Crenshaw, Kimberle (1989) "Demarginalizing the Intersection of Race and Sex: A Black Feminist Critique of Antidiscrimination Doctrine, Feminist Theory and Antiracist Politics," *University of Chicago Legal Forum*: Vol. 1989, Article 8.

England, K.V.L. (1994). "Getting personal: Reflexivity, positionality, and feminist research." *Professional Geographer,* 46: 80–9.

Foucault, M. (1980). *Power/Knowledge*. Brighton: Harvester.

Glaser, B.G., & Strauss, A.L. (1967). *The Discovery of Grounded Theory: Strategies for Qualitative Research*. Chicago: Aldine.

Haraway, D. (1988). Situated knowledges: The science question in feminism and the privilege of partial perspective. *Feminist Studies, 14*(3), 575–99.

Harper, D. (2002). "Talking about pictures: A case for photo elicitation." *Visual Studies,* 17: 13–26.

Harvey, D. (1984). "On the history and present condition of geography: An historical materialist manifesto." *Professional Geographer,* 36: 1–11.

Knigge L., & Cope, M. (2009). Grounded Visualization and scale: A recursive analysis of community spaces. In Cope M. and Elwood S. (eds) *Qualitative GIS: A Mixed Methods Approach*. Thousand Oaks, CA, Sage: 95–114.

Knigge, L., & Cope, M. (2006). 'Grounded visualization: Integrating the analysis of qualitative and quantitative data through grounded theory and visualization'. *Environment and Planning A,* 38: 2021–37.

Nagar, Richa. (2019). Hungry translations: The world through radical vulnerability: The 2017 Antipode RGS-IBG Lecture. *Antipode, 51*(1), 3–24.

Pile, S. (2010). "Emotions and effect in recent human geography." *Transactions of the Institute of British Geographers, 35*(1): 5–20

Rittel, H.W.J., and M.M. Webber. (1973). "Dilemmas in a general theory of planning." *Policy Sciences, 4*(2): 155–69.

Said, E. (1978). *Orientalism*. New York: Vintage Books.

Silver, C., & Lewins, A. (2014). *Using software in qualitative research: A step-by-step guide* (2nd edition). London: SAGE Publications. http://methods.sagepub.com/book/using-software-in-qualitative-research-2e

Index